# FOREWORD

Although the number of new reactors under construction or commissioning worldwide are very few, the field of advanced reactor and fuel design is thriving. There are many research groups actively contributing to the field, ranging from universities and industry to government-sponsored research organisations. The motivations are considerably diverse, including improved economy, improved safety, enhanced proliferation resistance, reduced environmental impact, reduced waste arisings, long-term sustainability and the closing of the fuel cycle by burning long-lived minor actinides. There are indications that the enduring stagnation in new plant orders may soon reverse, and the challenge to the researchers will be to deliver systems that meet the short-term demands of improved economics and safety, and that will in the longer term satisfy the requirement for truly sustainable systems.

In every country the detailed requirements for new systems vary. The specifications are complex and change with time. In each country there are different scenarios where different components of the reactor park can interact in an intricate fashion. The design of new reactor systems can take many years and the development of innovative fuels is an extremely extended process because of the need to carry out extensive irradiation trials to demonstrate good performance. Set against this background, the field of advanced reactor and fuel development is one in which international collaboration is the only viable way forward; this workshop was intended to help promote such collaboration.

In particular, the meeting served to complement regular conferences on advanced reactors and fuels which tend to be broader in scope and have a much larger participation. It offered a less formal forum amongst specialists in reactor and fuel design. As with the first ARWIF workshop (held at the Paul Scherrer Institute in 1998), the response to the call for papers and the attendance were both very good.

A panel discussion session held at the end of the workshop attempted to address four difficult questions concerning the direction of research activities. Some points of agreement were:

1.  A gap is perceived between current research activities in advanced reactors and fuels and the immediate requirements of utilities, evidenced by the low representation of utilities at these two workshops. There is a need to make utilities more aware of the relevance of long-term research and development and its relevance to the long-term sustainability of nuclear generation.

2.  There is a need to develop metrics that would allow the potential benefits of advanced reactors and fuel cycles to be quantified in a meaningful way. It is often difficult for utilities to see the relevance of innovative concepts to their future businesses in the context of, for example, economic savings, reduced environmental impact, etc.

3.  Within a partitioning and transmutation fuel cycle, the optimum strategy for treating curium is not clear. A systematic survey of realistic scenarios involving recycling of curium, interim storage of curium and eventual disposal is needed to help clarify the best strategies.

4. The role of high-temperature gas-cooled reactors (HTGRs) in a long-term sustainable nuclear programme is presently not clear. Although some HTGR fuel cycles are envisaged to involve recycling, most current development efforts are based on open fuel cycles. This might be regarded as an interim step towards a fully sustainable fuel cycle or as one component of a nuclear programme with a mix of reactor types and fuel cycles. The near-term development of HTGR technology may give synergistic benefits for later, long-term sustainable fuel cycles.

Overall, the workshop was very successful and there was much useful exchange of information, views and discussion. I would like to extend my personal thanks to all who participated (including those who were unable to be present) and to all those who contributed to the organisation. In particular, my thanks to Andy Worrall, without whose organisational skills the workshop would not have been a success, and to Enrico Sartori, whose extensive understanding of the field and the researchers helped shape the programme.

*Kevin Hesketh*
Scientific Chairman ARWIF-2001
November 2001

Nuclear Science

# Advanced Reactors with Innovative Fuels

**Second Workshop Proceedings**
**Chester, United Kingdom**
**22-24 October 2001**

**Hosted by**
**British Nuclear Fuels Limited (BNFL)**

NUCLEAR ENERGY AGENCY
ORGANISATION FOR ECONOMIC CO-OPERATION AND DEVELOPMENT

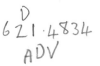

# ORGANISATION FOR ECONOMIC CO-OPERATION AND DEVELOPMENT

Pursuant to Article 1 of the Convention signed in Paris on 14th December 1960, and which came into force on 30th September 1961, the Organisation for Economic Co-operation and Development (OECD) shall promote policies designed:

- to achieve the highest sustainable economic growth and employment and a rising standard of living in Member countries, while maintaining financial stability, and thus to contribute to the development of the world economy;
- to contribute to sound economic expansion in Member as well as non-member countries in the process of economic development; and
- to contribute to the expansion of world trade on a multilateral, non-discriminatory basis in accordance with international obligations.

The original Member countries of the OECD are Austria, Belgium, Canada, Denmark, France, Germany, Greece, Iceland, Ireland, Italy, Luxembourg, the Netherlands, Norway, Portugal, Spain, Sweden, Switzerland, Turkey, the United Kingdom and the United States. The following countries became Members subsequently through accession at the dates indicated hereafter: Japan (28th April 1964), Finland (28th January 1969), Australia (7th June 1971), New Zealand (29th May 1973), Mexico (18th May 1994), the Czech Republic (21st December 1995), Hungary (7th May 1996), Poland (22nd November 1996), Korea (12th December 1996) and the Slovak Republic (14 December 2000). The Commission of the European Communities takes part in the work of the OECD (Article 13 of the OECD Convention).

# NUCLEAR ENERGY AGENCY

The OECD Nuclear Energy Agency (NEA) was established on 1st February 1958 under the name of the OEEC European Nuclear Energy Agency. It received its present designation on 20th April 1972, when Japan became its first non-European full Member. NEA membership today consists of 28 OECD Member countries: Australia, Austria, Belgium, Canada, Czech Republic, Denmark, Finland, France, Germany, Greece, Hungary, Iceland, Ireland, Italy, Japan, Luxembourg, Mexico, the Netherlands, Norway, Portugal, Republic of Korea, Slovak Republic, Spain, Sweden, Switzerland, Turkey, the United Kingdom and the United States. The Commission of the European Communities also takes part in the work of the Agency.

The mission of the NEA is:

- to assist its Member countries in maintaining and further developing, through international co-operation, the scientific, technological and legal bases required for a safe, environmentally friendly and economical use of nuclear energy for peaceful purposes, as well as
- to provide authoritative assessments and to forge common understandings on key issues, as input to government decisions on nuclear energy policy and to broader OECD policy analyses in areas such as energy and sustainable development.

Specific areas of competence of the NEA include safety and regulation of nuclear activities, radioactive waste management, radiological protection, nuclear science, economic and technical analyses of the nuclear fuel cycle, nuclear law and liability, and public information. The NEA Data Bank provides nuclear data and computer program services for participating countries.

In these and related tasks, the NEA works in close collaboration with the International Atomic Energy Agency in Vienna, with which it has a Co-operation Agreement, as well as with other international organisations in the nuclear field.

# TABLE OF CONTENTS

5

# EXECUTIVE SUMMARY

## Motivation, scope and goals

This workshop was organised following recommendations from the OECD/NEA Nuclear Science Committee (NSC) that a follow-up to the first ARWIF workshop, held at PSI in October 1998, would be valuable. The workshop falls under the auspices of the Working Group on the Physics of Plutonium Fuels and Innovative Systems (WPPR) that reports to the NSC.

A new generation of reactor designs is being developed that is intended to meet the requirements of the 21st century. In the short term, the most important requirement is to overcome the relative non-competitiveness of current reactor designs in the deregulated market. For this purpose, evolutionary light water reactor (LWR) designs have been maturing and are being actively promoted. These are specifically designed to be less expensive to build and operate than the previous generation of LWRs, genuinely competitive with alternative forms of generation and at the same time establish higher levels of safety. A new generation of modular, small-to-medium (100-300 MWe/module), integral design water-cooled reactors are under development. These are designed to be competitive with nuclear and non-nuclear power plants, to have significantly enhanced safety, to be proliferation-resistant and to reduce the amount of radioactive waste produced. A different approach to improve competitiveness is the re-emergence of high-temperature reactors (HTRs) using gas turbine technology to obtain higher thermal efficiencies, low construction and operating costs, inherent safety characteristics and low proliferation risk.

In the longer term, assuming that the current stagnation in the market is successfully overcome, other requirements related to long-term sustainability will emerge. Important amongst these will be the need to minimise the environmental burden passed on to future generations (or at least to ensure that the cost to future generations is in balance with the benefits to the current generation), the need to establish sustainability of fuel and the need to minimise stocks of separated plutonium at the minimum possible working level and to minimise accessibility to plutonium.

In this context, topics of interest to the workshop were: reactors consuming excess plutonium, advanced LWRs, HTRs, fast spectrum reactors, subcritical systems, minor-actinide systems and radical innovative systems.

The scope of the workshop comprised reactor physics, fuel performance and fuel material technology, thermal-hydraulics, core behaviour and fuel cycle of advanced reactors with different types of fuels or fuel lattices. Reactor types considered were water-cooled, high-temperature gas-cooled and fast spectrum reactors as well as hybrid reactors with fast and thermal neutron spectra. The emphasis was on innovative concepts and issues related to the reactor and fuel.

The workshop concluded with a wide-ranging panel discussion which considered some difficult questions from which it is hoped that some recommendations for future priorities can be derived. A record of the discussion is included at the end of this summary.

## Workshop organisation

| General chairman | |
|---|---|
| K. Hesketh | |
| **Local organising committee** | |
| A. Worrall, D. Every, S. Crossley | |
| **Scientific advisory committee** | |
| C. Brown (BNFL) | R. Chawla (PSI) |
| M. Mignanelli (AEA-T) | P. D'hondt (SCK•CEN) |
| M. Carelli (Westinghouse) | J. Rouault (CEA) |
| W. Krebs (Framatome-ANP) | P. Alekseev (Kurchatov Institute) |
| L. Walters (ANL-W) | J. Kuijper (NRG-NL) |
| P. Finck (ANL-E) | J. Kim (Univ. Hanyang) |
| J. Gehin (ORNL) | A. Stanculescu (IAEA) |
| R. Konings (EC-ITU) | E. Sartori (OECD/NEA) |

## Participation

The workshop was attended by 64 participants from 13 countries and 3 international organisations. Research laboratories and universities made up approximately 70% of the participants, industry 14%, utilities 5% with the remainder representing international organisations (see Annex 1 for the complete list of participants).

## Technical programme

The workshop was organised into seven plenary sessions (including two parallel sessions on the second day) in which 35 papers were presented. In addition, eight papers were presented in a poster session. Unfortunately, some of the authors were unable to travel following the events of September 11. In two such cases the authors agreed to forward their presentation slides and their papers were presented by the local organisers. In two other cases the papers were not presented, but they have nevertheless been retained within the workshop proceedings and will be published (see Annex 2 for the workshop's technical programme).

There was a good response to the call for papers and more abstracts were submitted than the programme would allow. It was therefore necessary for the Scientific Advisory Committee to reject a number of abstracts. The majority of the rejections were abstracts for papers which emphasised codes and methods rather than the advanced systems, and it was felt that the workshop should emphasise the latter.

The final plenary discussion was devoted to a panel discussion which considered five questions that had been compiled previously. Five panel discussion leaders were appointed in advance and asked to lead the discussion on these questions. Following these prepared presentations the discussion was opened to the floor.

# Session summaries and panel discussion

## Opening session – Chair: K Hesketh

W. Wilkinson opened the proceedings with an invited paper entitled *Barriers and Incentives to Introducing New Reactors in the Deregulated Electricity Market*. The presentation highlighted that from the perspective of a utility operating in a deregulated market, there is only one incentive for new or replacement nuclear build and that is to obtain a commercial rate of return on the investment. However, while new reactor designs currently available or under development promise to achieve significant improvements in total generating cost such that they can be competitive in deregulated markets, there are nevertheless some difficult obstacles to overcome. The presentation highlighted in particular the need for a stable regulatory environment where potential investors can be certain that the regulatory process will not change during the course of construction and also the need for the regulatory processes to be consistent in different countries. A further point was the need for the environmental discharge requirements to be driven by rational cost/benefit approaches and not by demanding near-zero discharges without justification. For sustainable advanced reactors and fuel cycle systems which are intended to stabilise the accumulation of plutonium and/or the minor actinides, getting the economics right will be a considerable barrier. It seems likely that any strategy which is primarily designed to achieve significant benefits in terms of waste reduction, reduced radiotoxicity per GWye, etc. will be economically disadvantaged compared with minimum cost generation strategies. Deregulated markets are not presently set up to deal with anything other than simple cost minimisation as a driver and therefore a major barrier in deregulated markets will be the need to establish mechanisms whereby the non-tangible benefits of advanced fuel cycles can be fully recognised.

A paper by L. Walters and J. Graham, *The Need to Preserve Nuclear Fuels and Materials Knowledge*, considered the precarious state of knowledge preservation in the field of fast reactor fuel design. The authors focused on this area because it reflects their particular expertise, but the situation is similar in many other areas. The authors suggested holding seminars in specific areas such fast reactor fuel design to try and capture past knowledge. The approach would be to invite young scientists and engineers to conduct smart interviews with experienced and retired experts. The former would try to capture the knowledge in writing and the experts would be invited to correct and add to these records. Such exchanges would complement the existing activities of international organisations, such as the NEA, to build up and maintain knowledge preservation databases.

## High-temperature gas reactors – Chair: H. Beaumont, W. Zwermann

Two sessions considered high-temperature gas reactors (HTGRs). The papers presented included one which reviews the European 5th Framework activities in the HTGR field. There are separate programmes covering fuel technology, neutron physics and fuel cycle technology and materials technology. These programmes are broad ranging and include experimental and theoretical studies. There were also two papers describing theoretical fuel cycle studies for the pebble bed modular reactor (PBMR). The first of these develops a design of $B_4C$ burnable poison particles that might be used in a batch refuelling scheme for PBMR and which explores the effects of self-shielding as a function of particle diameter and $^{10}B$ enrichment. The second paper demonstrates the flexibility of the PBMR for utilising different fuels, including U-Th, U-Pu and Th-Pu. Finally, a poster presentation described a fluidised bed reactor concept for which coupled neutronics/multi-phase fluid dynamics calculations have been performed.

### *Design and performance of innovative fuels  – Chairs: R. Thetford, Y. Lee, K. Bakker*

Ten papers were presented under design and performance of innovative fuels, including two posters. Inert matrix fuels featured in several of the papers, specifically zirconia-plutonia, zirconium nitride, cerium-plutonium oxide and rock-like fuels. One paper considered the design of (U-Pu)N fuel for the RBEC lead-bismuth fast reactor. A paper was presented which describes test irradiations of fast reactor uranium-plutonium oxide fuels manufactured using the Sphere-Pac and Vipac processes. Two poster presentations related to the design, fabrication and physical and chemical properties of minor-actinide target fuels.

### *Evolutionary and modular water reactors – Chairs: D. Porsch, P. D'hondt, T. Downar*

Thirteen papers were included in the proceedings under this heading, including four poster presentations. New reactor concepts include small modular PWRs, an upgraded VVER-440, a simplified BWR, a simplified PWR, a PWR with supercritical coolant state and reduced moderation LWRs designed to increase the plutonium breeding ratio. Novel fuel concepts include a PWR using HTGR particle fuel, a PWR partially loaded with inert matrix fuel and four papers on thorium fuel utilisation in LWRs. The research emphasis is on enhanced safety and improved utilisation of plutonium.

### *Fast spectrum reactors – Chairs: P. Alekseev, H. Sekimoto*

Five papers were presented under this heading, which was intended to cover critical fast reactors. The papers included one on a simplified sodium-cooled fast reactor which eliminates the intermediate circuit through the use of novel high integrity steam generators. Other papers described a Pb-Bi-cooled fast reactor and a gas-cooled fast reactor, while one considered the performance of minor-actinide target fuels. A novel concept compatible with a very long life core was presented. This is the candle strategy where only a small axial section of the core undergoes fissions and the fissioning region automatically propagates axially at a rate of a few cm per year. The same concept could apply to a thermal system as well. The emphasis is on fast reactors that are economically competitive with the current generation of LWRs and the evolutionary LWRs derived from them.

### *Molten-salt reactors  – Chair: W. Zwermann*

Three papers on molten-salt reactors were presented, two of which are designed for both having an attractive fissile fuel utilisation and incinerating minor actinides. A third paper, presented as a poster, investigates a thorium-fuelled molten-salt subcritical system intended for primary energy generation with low radiotoxic burdens.

### *Accelerator-driven systems – Chair: C. DeRaedt*

Five papers were presented on subcritical accelerator driven systems (including one poster). One paper described some of the first experimental results for a subcritical system, while another paper described plans for the planned MYRRHA experimental facility. A cascade molten-salt subcritical system was described which uses a super-critical central core surrounded by a subcritical region where the bulk of the minor-actinide transmutation takes place. The super-critical central zone acts to amplify the source neutrons and reduces the current requirements of the accelerator beam. Another paper described a subcritical molten-salt system for minor-actinide transmutation.

*Miscellaneous themes – Chairs: W. Zwermann, K. Hesketh*

Two papers were presented that do not exactly fit with the main session headings. One paper reviews experimental critical mock-up facilities for various reactor systems. Finally, an interesting concept was presented that features a core which is only just subcritical and which uses an accelerator beam to simulate the effect of delayed neutrons. This system has the advantage of being able to load a large fraction of minor actinides, using a coupling to the accelerator current to mimic the effect of an extra delayed neutron group. In this way, many of the difficulties of more "conventional" subcritical systems (such as demanding beam requirements and rapid spatial variations of flux) are avoided.

*Panel discussion – Chair: K Hesketh*

A panel discussion was held at the end of the workshop. Six participants agreed to sit on the panel and lead the discussion on five questions that were established in advance. The panel members were:

| |
|---|
| Pierre D'hondt (CEN•SCK) |
| Richard Sunderland (NNC) |
| Hiroshi Sekimoto (Tokyo Institute of Technology) |
| Joseph Somers (ITU) |
| Henri Mouney (EDF) |

The four questions discussed were:

1) Is there a gap between vendors' and utilities' fuel research programmes designed to support operation and the advanced concept research such as that presented at this conference? If so, what can the research community do to narrow the gap? In other words, does the field need to be made more relevant to the utilities?

2) The benefits of advanced concepts are usually in areas such as safety, proliferation resistance, environmental impact/radiotoxic burden, strategic and so on. A major weakness is that these are "soft" issues for which there is no agreed measure of the benefit. Are there any actions the research community could take to promote agreed metrics in these areas?

3) What should be our strategy for partitioning and transmutation given the intractability of processing and destroying curium? Should there be a policy of encapsulating curium for eventual disposal?

4) In the context of the objectives of initiatives such as Gen IV (particularly sustainability), how would once-though fuel cycles such as those HTGRs fit it? What role would once-through fuel cycles play?

***Question 1 – Is there a gap between vendors' and utilities' fuel research programmes designed to support operation and the advanced concept research such as that presented at this conference? If so, what can the research community do to narrow the gap? In other words, does the field need to be made more relevant to the utilities?***

Pierre D'hondt led the discussion for this question. He suggested that there are two perceptions or models to describe the development process. The first is driven by industry and is associated with "evolutionary" developments on a relatively short time scale. The diagram below illustrates the concept. The development is derived from an actual demand from industry.

The second is driven by the research community and is associated with "revolutionary" developments on a relatively long time scale. The diagram below illustrates this model. Here, a perceived requirement or the study of innovative concepts drives developments. Research over long time scales with or without a specific focus is normally funded through public agencies; it can be very difficult to interest industry in sponsoring research.

In summary, the link from industry to the research community, which is applicable to short-term problem solving and short-term improvements with a direct economic benefit, works well. The link in the opposite direction, concerning long-term development from the research community to industry, does not work as well as it should. Government funding is available to sponsor such research, but it would require the full involvement of industry to bring to fruition and at the moment industrial organisations, particularly the utilities, show very limited interest. (This is illustrated by the fact that the only utility represented at this workshop was EDF.) Other utilities find it difficult to be concerned with anything other than short-term issues (maximum time horizon 5 to 10 years). Therefore the conclusion is that there is indeed a gap between industry and the researchers and that measures need to be taken to narrow the gap for the long-term developments to be implemented industrially.

Points made from the floor during the discussion included:

- The key word is MONEY – the research community should highlight the economic gains of new systems.

- It would be useful to have representatives from the utilities at conferences such as this.

- The utilities have several options including P&T and underground disposal. Those working on underground disposal promote that option and there is a danger that if the experts say that this method is viable then the P&T studies will be sidelined/neglected.

- P&T may reduce the cost of underground disposal.

- A question that needs to be addressed is how an economic figure can be put on savings due to P&T.

- In Japan, issues that might imply that currently operating LWRs may not be safe are at risk of being suppressed.

- In terms of safety and economics, the vendor and the utilities have a conjunction of interest.

- In terms of P&T, the customer is the public/government; they should thus be funding it or at least making a contribution.

*Question 2 – The benefits of advanced concepts are usually in areas such as safety, proliferation resistance, environmental impact/radiotoxic burden, strategic and so on. A major weakness is that these are "soft" issues for which there is no agreed measure of the benefit. Are there any actions the research community could take to promote agreed metrics in these areas?*

Richard Sunderland led the discussion by suggesting a few considerations. On safety he suggested the need for:

- Enhanced safety.

- Improved reliability.

- Inherent safety (e.g. passive systems).

- Advanced control and monitoring systems.

- Plant simplification.

- Improved ISI and maintenance.

- Reduced worker dose.

- Reduced consequences of severe accidents.

On proliferation resistance he noted the need for:

- Proliferation-resistant fuel cycles.

- Fissile material accounting.

- Assured inspection processes (easy accountability and inspection).

- IAEA requirements.

On environmental impact:

- Improved fuel utilisation.

- Reduction in wastes.

- No increase in natural environmental burden.

- Decommissioning.

He saw strategic concerns as being:

- Optimum use of resources.

- Energy costs: capital, financial and fuel cycle.

- Construction time scale.

- Siting.

- Security.

- Profitability.

- National strategies.

In summary:

- What shall we do to develop metrics[1]?

- In what areas should metrics be developed?

- How can metrics be combined to assess reactor concepts on an equivalent basis?

- Who should be involved in metric development: researchers, manufactures, utilities, fuel manufacturers and reprocessors, the public?

- Lead institutions: IAEA, OECD, governments?

As this question is very wide ranging, and many factors need to be considered, it was thus also addressed by Hiroshi Sekimoto:

- The measure of the benefit depends on the consideration of weight of value.

- The consideration of weight of value changes for different persons.

- Even for one person it changes with time for different circumstances and environments.

- It is almost impossible to set a measure of the benefit, which can be applied to everyone.

- Even if the measure is set by the public, specialists can offer the *options* which will be evaluated by such a measure.

Points made from the floor during the discussion included:

- A precedent has been set by the US DOE to answer these questions in terms of the best route for disposal of weapons-grade Pu (e.g. geological disposal or burning). This illustrates the possibility of applying metrics to this type of issue.

- It is difficult to quantify the values of different systems, and this may require an international working group to agree on values. The USA is trying to do this for the Generation IV reactors.

- In terms of risk categorisation it is relatively easy to converge on a value.

---

[1] Metrics – what measures do we adopt to ensure that we meet utility requirements. For example, there is no easy measure to assess whether one system is more proliferation resistant than another. There is a need to develop scales to measure these issues.

- If agreements on metrics could be reached it might be possible to then present these to the public and to shareholders.

- Public opinion is not likely to be favourable if the experts are in disagreement.

The discussion illustrates the difficulty of addressing this question. There is a clear agreement of the need to establish meaningful and useful metrics, but recognition that in most of the areas they are very difficult to define. The economics of different systems is the only area where clear quantitative metrics exist and utilities' decisions are understandably dominated by this aspect. In spite of the difficulties of making progress in this area, it is important enough that it should not be neglected. There may be benefits in attempting to generate suitable metrics even if they are not perfect, because the process may give rise to important new questions and generate new perspectives. There may be benefits to be gained from other fields, such as environmental protection, where there is a need for analogous metrics. Ultimately, it may be the general stakeholders (public, government, shareholders, regulatory bodies, etc.) who decide which metrics will apply, and then it would be the research community's responsibility to provide the specialist inputs needed to apply the chosen metrics.

### Question 3 – What should be our strategy for partitioning and transmutation given the intractability of processing and destroying curium? Should there be a policy of encapsulating curium for eventual disposal?

Joe Somers led the discussion of this question. He began by expanding the question further to ask:

- In the transmutation of Cm, is the radiotoxicity reduction sufficiently high?

- Are Am/Cm separation processes feasible?

- What should the design of the sub-assemblies be?

- How to manufacture Cm targets and sub-assemblies?

- What are the appropriate logistics (co-location of facilities)?

These in turn lead to further questions:

- Encapsulation of Cm – final or interim (the famous 100 years)?

- Simple encapsulation or immobilisation matrices?

- What type of immobilisation matrix [e.g. Pyrochlore (e.g.Gd1.8Cm0.2ZrO7)]?

- Can the interim storage host become the target [e.g. $(ZrYCm)O_2$]?

- Management of He and heat damage to the matrix?

Hiroshi Sekimoto pointed out that $^{244}Cm$ is the curium isotope which initially has the highest concentration and which is a powerful neutron source, but which decays relatively quickly (18.1 year half-life). There would be clear advantages, therefore, in allowing $^{244}Cm$ to decay for a few half-lives before attempting to irradiate curium in targets.

The following additional questions and comments were made from the floor:

- What type of partitioning processes could be applied?

- What is the necessary investment for these partitioning processes?

- It is complicated, chemically, to separate Cm and Am and Am/Cm and the lanthanides? Some promising methods are under development but a lot of work remains before these processes could be applied on an industrial scale.

- In the future it may be necessary to reach a compromise between physics and chemistry.

- It may be more reliable to develop systems that utilise incompletely separated fuels, i.e. an Am/Cm mixture needs to be considered.

- There is a network in place with an OECD/NEA working party to look at these P&T issues.

- P&T may not be so important now but it is important for the far future (2050+); an integrated reactor and fuel cycle may be needed (e.g. molten salt) to avoid the transport of highly active waste.

- Geological disposal should not be ruled out. The USA has an operating geological disposal facility (WIP) for military waste; this demonstrates that geological disposal is viable/possible.

This discussion highlighted the present lack of knowledge in relation to curium and the best strategy for dealing with it. It is clear that the fabrication of Cm transmutation fuels/targets is technically difficult but not impossible. An interim storage period would alleviate these difficulties, but would still require Cm fabrication in a suitable form ensuring chemical durability and management of the considerable heat and He produced. Such an interim storage strategy would necessitate an active nuclear programme remaining long after this storage time. It is important to ensure that all possible scenarios are covered, including a scenario in which nuclear energy is no longer deployed at the end of such an interim storage period. Therefore, it will be necessary to develop strategies which assume final disposal of curium after a possible interim storage period alongside long-term sustainable strategies where curium is brought into balance in a transmuter directly after its separation (or possible interim storage).

***Question 4 – In the context of the objectives of initiatives such as Gen IV (particularly sustainability), how would once-though fuel cycles such as those HTGRs fit it? What role would once-through fuel cycles play?***

Henri Mouney led the discussion of this question, beginning by providing a reminder of the definition of sustainability: "Development that meets the needs of the present generation without compromising the ability of future generations to meet their needs" (WCED, 1987). He then listed the issues to be addressed, which are:

- Sustainability of nuclear energy with the following main requirements:

  - Uranium resources need to be saved. Known resources: 4.3 Mt (NEA, 1997) combined with consumption of 70 kt/year equates to 60 years of supply.

  - Enhanced safety.

- Nuclear waste production should be minimised with an adequate management of spent nuclear fuel. But until now no country has yet implemented a permanent solution such as partitioning and transmutation and/or geological disposal.

- Enhanced resistance to proliferation risks.

- Economic competitiveness which needs to be reinforced.

- Assets and limits of LWR for sustainable development:

  - A mature technology with an irreplaceable experience.

  - Convincing results on economy, safety and reliability.

  - For immediate future – a new generation of PWR (EPR):

    $\Rightarrow$ High burn-up level (65 GWd/t).

    $\Rightarrow$ Plutonium control.

    $\Rightarrow$ Waste volume reduction.

    BUT

    $\Rightarrow$ A non-optimum use of resources – only 1 % of initial U is used (even with reprocessing) in LWR.

- How can HTGRs fit the objectives of sustainability?

  - The French approach (CEA-FRAMATOME/ANP): the choice of coolant appears to be a major element of future nuclear systems:

    $\Rightarrow$ Water is unsuitable to fast neutron systems.

    $\Rightarrow$ Liquid metal leads to a complex fuel handling, and a difficult structure inspection.

    $\Rightarrow$ Gas coolants – their potential must be confirmed particularly in GCR.

- Gas-cooled reactors (GCR); their technological range potentials are the following:

  - Economics:

    $\Rightarrow$ Simplicity of circuits: a single, direct-cycle circuit.

    $\Rightarrow$ High energy performance: gas goes directly to turbo-alternator.

    $\Rightarrow$ Modularity: small modules, standard, assembled in manufacture.

    $\Rightarrow$ Fast construction, less capital outlay.

- Safety:

    ⇒ Robust fuel in the case of accidental transients (passive safety).

    ⇒ Little interaction between fuel and coolant.

    ⇒ Fuel characteristics that are likely to resist to the risk of proliferation.

- Environment protection.

- Optimum use of resources and minimisation of waste (fast spectrum).

- GCR: an evolutionary technological range for sustainability:

    - For the short term: first configuration is aimed at a direct cycle HTR that modern turbines enable (GT MHR or PBMR).

    - For the medium term, specialised GCR allowing:

        ⇒ Very high temperatures and high efficiency.

        ⇒ Optimised configurations for waste transmutation.

    - For the long term, long-lasting energetic development needs:

        ⇒ Fast spectrum for breeding.

        ⇒ Complete uranium consumption.

        ⇒ Integrated cycle transmuting all the actinides.

- Some features of GCRs:

    - Fuel cycle (short term HTR):

        ⇒ Very high burn-up: 120 GWd/t (TRISO particles).

        ⇒ 700 GWd/t (incineration – equivalent to several Pu recyclings in PWR).

    - Once-through fuel cycle attractive from this objective (transmutation of Pu and minor actinides).

BUT

    - U consumption is 13 to 25% higher than for a PWR.

*GCR: The challenge for sustainability is to develop a fast spectrum reactor and an integrated fuel cycle system for effective utilisation of resources and waste production minimisation.*

- Key technology fields for a gas-cooled fast neutron reactor fitting the objectives of sustainability:

  - Fuels must:

    ⇒ Be confining and refractory.

    ⇒ Be able to obtain fast spectra and very high combustion levels.

    ⇒ Authorise different options in reprocessing matter.

  - Reprocessing of the spent fuel as integrated as possible with improvements of existing technologies:

    ⇒ Implementation of dry processing, pyroprocessing.

    ⇒ A good resistance to proliferation.

  - Materials resisting to high temperatures and to fast neutrons allowing passive safety.

  - Technology of high-temperature helium circuits to be developed and improved.

Hiroshi Sekimoto made some additional points on this question:

- Sustainability is an important item for the future of human being, such as the future equilibrium society. However, it will come after a certain period of transition.

- Once-through fuel cycles cannot be accepted for the future equilibrium system, but should be acceptable in the interim. This depends on the burn-up strategy and reprocessing R&D.

- For higher burn-ups such as Pu burner, the once-through option may be acceptable.

- For higher fissile content in the spent fuel, the reprocessing option may be better.

- Temporarily the interim storage option may be attractive.

- The questions of safety and proliferation resistance are the most urgent.

- The question of reducing the radiotoxic burden is less urgent.

- Urgent items should be solved soon and promoted with well-supported R&D.

- However, other items should also be addressed, and supported with enough R&D.

In summary, the research programmes are not yet at the stage where a definite answer can be made to this question. The research programmes are currently at the stage where feasibility of the various technical options are being assessed. The role of HTGRs may be as an interim step towards a fully sustainable fuel cycle or they may find a role as one component of a fuel cycle with a number of different component reactors. There may be synergistic benefits of HTGRs such as establishing gas-cooled technologies that might later find application in long-term sustainable fuel cycles.

*Specific actions*

It is recommended that a summary of ARWIF-2001 should be presented to the next Nuclear Science Committee meeting in June 2002. This presentation should specifically include a discussion of the four questions considered in the panel discussion and the responses to those questions. It is recommended that the Nuclear Science Committee also debate whether it considers that a third ARWIF workshop, to be held in 2004, would be useful.

# OPENING SESSION

*Chair: K. Hesketh, A. Worrall*

# OPENING ADDRESS

**Dr. Sue Ion**
Director of Technology and Operations
Research & Technology
British Nuclear Fuels plc
Springfields, Salwick
Preston PR4 OXJ, UK

Welcome to participants.

On behalf of BNFL, I would like to take this opportunity to welcome you to the ARWIF-2001 workshop.

We at BNFL are very pleased to be hosting this seminar and we hope that you enjoy your stay in the United Kingdom and the historic city of Chester.

I am sure that the next few days will provide an excellent opportunity for experts from around the world to meet, exchange information and update each other on the progress of their work in this challenging field.

BNFL is committed to driving forward the research and development of products and strategies which will ensure a long-term future for nuclear power. In order to achieve this goal, we must continue to develop solutions to the technical and political challenges which face the nuclear industry. We believe that the nuclear option will continue to contribute as a clean source of energy. In the immediate future evolutionary light waters reactors will dominate the commercial market.

However, we must recognise that there are stakeholders with genuine concerns in the areas of safety, non-proliferation, environmental impact and sustainability which need to be addressed further if nuclear's long-term future is to be secured. The innovative reactor and fuel concepts to be presented at this workshop will help to widen the options available. It is increasingly important that organisations involved in the research and development of advanced systems should engage in collaborative exercises such as this workshop. In this way, we can ensure that technology can continue to be developed for future applications.

I wish you an interesting and enjoyable meeting, a pleasant stay here and a safe journey home.

# BARRIERS AND INCENTIVES TO INTRODUCING NEW REACTORS IN THE DEREGULATED ELECTRICITY MARKET

**W.L. Wilkinson**
Dept. of Chemical Engineering and Chemical Technology
Imperial College
Prince Consort Road
London SW7 2BP

## Abstract

After a long period of stagnation, there is now a realistic prospect that a revival of nuclear power building is imminent in the USA and possibly also in the UK. New reactors will need to compete in electricity markets which are being progressively deregulated. This is a very different situation from the one which prevailed when the existing reactors were built. There are strong incentives to pursue new reactors, but there are also difficult barriers to overcome. This paper considers these barriers and incentives in relation to both the evolutionary plants that are most likely to be built in the immediate future and also to more advanced plants that will hopefully follow. While those evolutionary designs currently positioned for new or replacement build have been developed specifically to address the needs of utilities, for many of the advanced concepts that will be presented at this workshop there is a large gulf separating the aspirations of the researchers and the needs of utilities. This gulf will have to be bridged if any such designs are eventually to be deployed commercially.

## The current climate

Nuclear power currently supplies 17% of world electricity production and makes a significant contribution to limiting $CO_2$ emissions. Until recently, the prospects for new or replacement reactor build in Europe and the USA have looked very poor, due in large measure to the perception that nuclear is uneconomic in these increasingly deregulated markets. To be precise, the perception was that *new* nuclear build was uneconomic, but that existing plants were economic to operate given that the initial investment costs in many cases amortised already. Unlike the previous generation of reactors, new or replacement nuclear build will have to be demonstrated to be competitive with the full cost of the initial investment, the associated financing costs and the costs associated with the perceived financial risks. Developments in the US market and elsewhere, combined with concerted efforts from the reactor vendors and the availability of new designs such as AP-600/1000, BWR-90 and NG-Candu, are now starting to transform this perception.

Within the context of world energy requirements, there is a strong case for at least maintaining nuclear's share of the market. Nuclear must be considered as part of a balanced energy portfolio, along with fossil fuels, renewables and energy conservation measures. Suggestions that renewables combined with conservation measures can meet world energy needs are not tenable against the background of increasing world population and energy demand (forecast to double to 2050 and to increase by a factor of 3 to 5 by 2100). It is not claimed that nuclear on its own will provide the complete solution, but faced with such an overwhelming challenge, it should certainly make a significant contribution. The finite fossil reserves and perhaps more importantly, the effect of carbon dioxide emissions on our environment further strengthen the argument not to abandon the nuclear option.

One of the main difficulties is that while the contributions nuclear can make to these global issues are recognised by governments and utilities alike, the newly deregulated markets often lack any specific provisions for accounting for such benefits as reducing global emissions, maintaining diversity and security of energy supply. In the UK a recent climate change levy has been introduced that penalises nuclear even though it has virtually zero $CO_2$ emissions. While the nuclear industry should continue to press governments to develop mechanisms that effectively address the global issues, it is clear that the most convincing strategy in the short term is to develop new plants which are truly competitive in their own right. The challenge is to meet the increasing demand in a sustainable way that does not damage the environment now and that does not limit the options available to future generations.

## New or replacement build – barriers & incentives

From the perspective of a utility operating in a deregulated market, there is only one incentive for new or replacement nuclear build and that is to obtain a commercial rate of return on the investment. In the previous state-controlled monopoly market, utilities had a duty to provide a secure supply, which was sufficient incentive to ensure that a utility would establish diversity of supply. In the deregulated market as it applies now in the UK, the duty to supply no longer applies and market forces are expected to ensure that sufficient supply capacity is provided. For new or replacement build, the main difficulty to address is the large investment cost and the fact that the income stream does not materialise for several years after the initial capital cash flows.

Nuclear plants have high substantive costs, meaning the costs of materials, construction, plant procurement and installation, etc. But the biggest single cost item is actually the financing of these substantive costs over the extended construction period of six or seven years. The situation is further worsened if there is a long process of governmental approvals (especially if the process is not fully clear beforehand). Another major contribution comes from the perceived financial risk associated with the investment. This manifests itself in the form of a high-test discount rate against which the project must show a net return. The high-test discount rate is partly determined by the rate of return investors might expect in the commercial market and partly by the risk element. If the initial investment costs can be reduced sufficiently, then new or replacement build will look attractive almost irrespective of the ongoing costs such as operating and maintenance, fuel procurement, spent fuel management and decommissioning provisions.

Part of the reason that new or replacement build is being seriously considered in the US is that the reactor vendors have been striving very hard to minimise the substantive investment costs and reduce construction times. Existing evolutionary designs, such as AP-600, AP-1000, System 80+, ABWR, NG-Candu and others, are now capable of achieving a specific capital cost as low as $1 000/kWe. Simplifications to the design have helped reduce the substantive costs. At the same time simpler designs equate to designs which have reduced project risk. Ultimately, this should translate into lower investment risk and therefore further lower the overall investment cost.

Every potential market for new or replacement build has its own unique economic boundary conditions. Evolutionary designs are already demonstrably the most competitive in some markets [1], while in others they are marginally competitive. In the UK the electricity selling price is determined by combined cycle gas turbine (CCGT) plants, which presently set the lowest total generating cost. Compared with current CCGT costs, recent assessments have shown that evolutionary plants are marginally competitive [2,3], but only if the full benefits of building a series of plants can be realised.

A major barrier to deploying the evolutionary designs will be meeting the first-of-a-kind costs. These comprise the costs of establishing the regulatory basis and any costs associated with the first time build. It is important that a realistic regulatory process should be established where a new system is approved once and not challenged again for following plants. The AP-600 has made substantial progress towards a more rational regulatory basis, by having gained licensing certification in the US; this licences the design in a generic fashion, leaving only site-specific issues to be considered wherever in the US it is eventually built. Ideally, such generic certification would have general applicability in other countries, where the regulatory process would be defined beforehand (and no possibility of shifting procedural and regulatory requirements).

Governments therefore need to be convinced of the need for realistic regulatory systems. By this means, governments can help create the conditions where new or replacement build can become commercially viable without necessarily having to provide direct subsidies or set up alternative levelising mechanisms such as an emissions levy or emissions trading. Regulatory stability is an absolute prerequisite to attract commercial investment; without this the risk of a new-build project failing to deliver its target investment return will be an immense barrier to external investors. Establishing the appropriate investment environment is an area where there needs to be a partnership between the industry, the regulator and government.

It is worth noting that although the evolutionary designs currently available all feature safety improvements over the current generation of plants (achieved largely because of adopting passive safety systems wherever suitable), in the deregulated market this is only considered a marginal direct benefit and the main benefit is actually the capital and operating cost reduction that the evolutionary approach offers. The utilities, perhaps rightly, consider that the current generation of plants are already sufficiently safe when operated competently. It should not be forgotten however, that maintaining a safe operational record is an important prerequisite for future build and is necessary for continued political support and public acceptability.

After economic viability, the next most important barrier to new or replacement build is the lack, in some countries, of a coherent approach to spent fuel and radioactive waste management. There is an increasing realisation of the need to establish nuclear as a fully sustainable system. A prerequisite will be to establish a clear waste management policy, which should be realistic and driven by rational requirements (i.e. not arbitrarily reducing discharges to zero irrespective of cost/benefit considerations). Only governments are in a position to do this and here is another means by which governments can provide clarity to help to ensure the right conditions for new build.

## The future climate

The future climate will determine whether or not the advanced systems that will be discussed at this workshop will become commercially viable. Although there may be important developments that help to consolidate nuclear as a valid option, any advanced reactor system must satisfy commercial constraints that are unlikely to differ substantially from those that apply currently. It would be unwise to rely on favourable developments in the market in developing and designing advanced systems; a much safer approach would be to assume the current economic constraints will apply.

The future climate depends to a large extent on getting the next generation of nuclear plants built and operating reliably. If the next generation evolutionary designs are not deployed because of failure to establish the correct economic conditions, then the prospects for any advanced reactor concepts will be bleak, as there will effectively be a gradual phase out as existing plants reach the end of their economic lifetimes. Assuming that the next generation plants are built, then the future of advanced reactors will depend on whether nuclear capacity is static, experiencing moderate growth or rapid growth.

In the static scenario, it is quite possible that there will be no great pressure to improve on current designs with respect to fuel utilisation, spent fuel arisings and sustainability issues. The argument would be that the accumulation of spent fuel and waste arisings from the operation of next-generation plants would only perhaps double or triple those which have accumulated already from current plants and could be accommodated within the existing or projected repositories. In this scenario the main pressures for advanced reactor development would be to develop systems with better economics and/or with inherent safety characteristics, such as high temperature reactors (HTRs). The fact that HTRs can offer modest improvements in waste arisings per GWye might be considered sufficient a contribution to sustainability in such a scenario. Combined with the virtual elimination of core degradation scenarios, proliferation resistance benefits and substantially improved economics over evolutionary reactor designs, this might well be considered a perfectly satisfactory position.

The scenarios in which nuclear is growing are those where the opportunities for deploying advanced systems may arise. With world nuclear capacity growing it would be difficult to argue that once-through thermal fuel cycles are sustainable. While there are probably sufficient uranium reserves to sustain a moderate or even a rapid growth scenario for a considerable time, the acceleration in the rate of accumulation of spent fuel and other long-lived wastes will ultimately come to be regarded as unsustainable. A solution could be to deploy new reactor systems and associated fuel cycles which use recycle strategies to limit the future accumulation of long half-life species such as plutonium, neptunium, americium and curium. Fully implementing such sustainable fuel cycles is acknowledged to be a long-term goal that will not be achieved until at least 2050.

Long-term reactor, fuel and fuel cycle research is presently very much focused on such sustainable systems. The emphasis is on fast neutron spectrum systems, either conventional critical reactors or accelerator-driven sub-critical systems. The fast spectrum facilitates the fissioning of plutonium and long-lived minor actinides (MAs) and possibly also the transmutation of the long-lived fission products (LLFPs) into nuclides with shorter half-lives. The favoured approach for MA and possibly LLFP destruction is to concentrate them into special "target assemblies". A consensus is emerging that inert-matrix fuels are advantageous for target fuels because of the reduced generation of fresh plutonium and MAs from neutron captures.

An important factor influencing the evolution of reactors and fuel cycles will be the need to further reduce proliferation risk. Pressure to develop proliferation-resistant systems has been present for some time and indeed the Generation IV Initiative from the US Department of Energy has low proliferation risk as one of its main goals. Following the recent tragic events in the US, it is very likely that the designers will come under renewed pressure to reduce proliferation risk (and also to strengthen plants against missile attack).

**Advanced systems – barriers & incentives**

The best incentive to persuade a utility to adopt advanced reactors will be if the system is demonstrated to have more favourable economics than more conventional nuclear plants and is economically competitive in the utility's market. If the new system also offers inherent safety, reduced waste arisings or reduced proliferation risk, then this will add to the system's attractiveness. But it must be understood that such considerations on their own will not sell new systems in deregulated free markets; the approach must be more economic first with other advantages as add-ons. A prerequisite will also be that the system is demonstrated to be robust to the risk of cost overruns during construction and with demonstrated reliable operational performance. This is a major barrier, since no single utility would wish to be the first to build and operate a new system unless there were some mechanism for insuring against the risk of the project being unsuccessful.

For sustainable advanced reactors and fuel cycle systems which are intended to stabilise the accumulation of plutonium and/or the minor actinides, getting the economics right will be a considerable barrier. The predominance of fast spectrum systems for this purpose, which tend to be more expensive to build and operate than thermal systems, makes it difficult to see how such systems could be economically competitive. The only exception would be in the event of very high uranium prices. Although on the time scale of post-2050 this is a possibility, it would be better not to have to rely on such a scenario as a justification. It seems likely that any strategy which is primarily designed to achieve significant benefits in terms of waste reduction, reduced radiotoxicity per GWye, etc., will be economically disadvantaged compared with minimum cost generation strategies. Researchers have already recognised this and are starting to develop strategies which would minimise the cost penalty by incorporating MA burning reactors as a relatively small component of the reactor mix, with the bulk of the capacity being provided by reactors burning conventional uranium or plutonium fuels. Stabilising plutonium and/or minor actinides will therefore come at a cost and the deregulated markets are not presently set up to deal with anything other than simple cost-minimisation as a driver.

Therefore a major barrier in deregulated markets will be the need to establish mechanisms whereby the non-tangible benefits of advanced fuel cycles can be fully recognised. The continued lack of an agreed mechanism for accounting equitably for $CO_2$ emissions from current power plants illustrates the difficulties that need to be overcome.

Another major consideration is that systems which stabilise plutonium and/or minor actinides are invariably effective only over very extended time scales. The very shortest time scale where effective reduction in plutonium and minor actinide masses is over a reactor lifetime of between 40 and 60 years. In many cases the full benefits will not be seen except over even more extended periods. This will require a high degree of commitment to implement even where there is governmental control over energy generation. The contrast between these extended time scales and the very short time horizon in deregulated markets (often five years or less) cannot be overemphasised.

It is clear that the introduction of sustainable systems will present some very formidable barriers in the current environment where, within the context of deregulated markets, developments are driven almost exclusively by commercial considerations. However, in the final analysis it may turn out that conventional reactors and fuel cycles which cannot be claimed to be sustainable will just not remain an acceptable option. Societal pressure may develop such that a move towards some sort of optimum solution is effectively mandated. It is very important, however, that such a development should be determined by a rational cost/benefit analysis and not driven by misperceptions which would result in a waste of resources. It is equally important that the same approach be applied to alternative energy sources, gas, coal, renewables, etc.

## Concluding remarks

As will be apparent from the papers presented at this workshop, research in the field of advanced reactors, fuels and fuel cycles is polarised into two distinct areas. One area aims primarily to develop systems that meet the immediate requirements of utilities, notably by improving economics and reducing waste arisings. The other aims primarily to meet the longer-term goal of establishing sustainable fuel cycles. The latter goal is a long way from being realised and there is a tendency for researchers to concentrate on the narrow technical issues. At the present stage of development, where the aim is to establish the feasible options, this is probably entirely appropriate. But it is unwise to neglect the ultimate end-users, the utilities, who will eventually need to deploy such systems in a commercially competitive environment. The developers should recognise the practical constraints which will affect the systems at an early stage and demonstrate that these constraints feature in their analyses. This should help to convince utilities of the importance and relevance of long-term research and to gain their support.

## REFERENCES

[1]    R. Tarjanne and S. Rissanen, "Nuclear Power: Least-cost Option for Baseload Electricity in Finland", *Nuclear Energy*, Vol. 40, No. 2, April 2001.

[2]    "BNFL Submission to the Performance and Innovation Units Review of UK Energy Policy", http://www.cabinet-office.gov.uk/innovation/2001/energy/submissions/BNFL.pdf

[3]    British Energy submission to the Government's review of energy policy "Replace Nuclear with Nuclear", http://www.cabinet-office.gov.uk/innovation/2001/energy/submissions/BritishEnergy.pdf

# THE NEED TO PRESERVE NUCLEAR FUELS AND MATERIALS KNOWLEDGE

**Leon C. Walters**
Argonne National Laboratory

**John Graham**
ETCetera Assessments LLP

## Abstract

The demand for nuclear power will likely substantially increase in this century. Developing countries are already including new nuclear plants as an important part of their mix of energy generators. The energy shortage in the United States coupled with the recent improvements in the economic competitiveness of nuclear power is causing a re-evaluation of the nuclear power enterprise. Even more important is the growing concern over $CO_2$ emissions from fossil fuel combustion, the curbing of which could increase the price of the coal option still further. Nuclear energy reduces the $CO_2$ burden directly by displacing fossil energy generation of electricity. In the future, the contribution of nuclear energy to the climate change problem may be even greater if nuclear energy is used for hydrogen generation in the transportation sector.

# Summary

With a reasonable projected growth of nuclear power, the world's supply of $^{235}$U, which can be practically recovered, will be exhausted by mid-century. Therefore, the deployment of the fast breeder reactor to convert the enormous supplies of uranium and thorium to fissile material is inevitable. The question is whether the fast breeder reactor and the associated reprocessing will be ready for deployment when needed. Presently, only Japan and Russia have active programmes, all others being already closed or placed in the process of closure. A review and current status of fast breeder reactor development will be presented in an attempt to address this question.

Most believe that fast breeder reactors and their supporting development and confirmation programmes will be necessary within a few decades. Thus, the issue of having the right information at that time to avoid reinventing the wheel becomes an issue of preserving that information we now possess. In turn this includes gathering pertinent information that might exist only within the minds of ageing and retiring experts as well as accumulating reports, data and samples. Then the information must be stored in an easily accessible and searchable form, and maintained over a long period of time during which management, hardware, software and priorities are likely to change.

Some of this work is being done in other technical areas but in the fast reactor field, the preservation programmes are limited to benchmarked data and published reports. There are no programmes to gather tacit information, material samples or technical failures that provide the basis for development decisions. We summarise the existing state of affairs and make some suggestions for ensuring the success of fast reactor development at a time when they are needed to obviate diminishing fuel supplies in the future.

# Introduction

The incredibly high energy density of nuclear reactor fuels gave the metallurgist (or the more recent designation of the profession, materials scientist) great challenges from the first discovery of nuclear fission. High temperatures and new combinations of fuel-to-cladding and cladding-to-coolant interfaces created a number of compatibility concerns. Soon radiation damage effects appeared that compounded the challenges. Many metal fuel alloys were studied, which were quickly followed by a large number of ceramic combinations of fuel. Along the way combinations of metals and ceramics and combinations of ceramics were investigated as dispersion fuels. A variety of bonding media was used, including mechanical bonds to the cladding, liquid metal and gas bonding. The coolants likewise varied from water, molten salt, liquid metal, to gas.

Fuel and coolant choices were coupled with the selection of absorber, reflector, moderator, component and structural materials. All fuels and materials selections were made to meet the objectives of the reactor. Most objectives for a reactor concept included the highest possible coolant-outlet temperature for the best thermal steam efficiency and the highest possible fuel burn-up for fuel economy. Almost always a compromise had to be made between the two. In addition, fast reactor fuels were also designed to accomplish certain breeding objectives.

For half a century material scientists have been constantly studying and improving nuclear reactor fuels. Coolant-cladding compatibility problems never emerged as significant problems, even though a great deal of effort was spent investigating these potential concerns. This may not be the case with lead and lead alloy coolants. However, coolant-fuel compatibility upon cladding breach was a more significant problem. Metallic fuel and water were not compatible, and oxide fuel and liquid sodium coolant reacted to form a lower density product that aggravated the initial cladding breach. In general, not many fuel systems lent themselves to long-term operation after cladding breach.

Fuel-cladding compatibility was a significant concern in virtually every fuel system. Proper material choices based upon literature data and out-of-reactor tests, of course, assured compatibility between un-irradiated fuel and cladding. However, once the fission reaction started and all the new fission product elements appeared, the compatibility concerns multiplied. Caesium and iodine were the most troublesome as the cause of early cladding failure in ceramic fuel systems, while the accumulation of lanthanide fission products caused concern in metallic fuel systems with the appearance of lower melting phases.

The accumulation of noble fission-product gases created a number of unanticipated problems. For almost all fuel systems fission-product gases had to be accommodated by an increased free plenum volume in the pins. In metallic fuel systems the fission gas bubbles caused the fuel to swell because the metal matrix flows as the pressure in the fission gas bubbles increases. Without allowance for fission gas accumulation early failure could be expected in all fuel systems. For some of the ceramic fuel systems stress due to fission gas accumulation caused the fuel to fracture. When the resulting fuel shards became wedged between the fuel pellet and cladding, early cladding failure also resulted.

In fast neutron spectrum reactors, with high energy and high neutron fluxes, atomic displacement damage created an array of difficult problems. The mechanical properties of all the cladding and structural materials changed dramatically. For most metals there was a loss in ductility, the hardness increased, ductile-to-brittle transition temperatures increased, creep rates were greatly increased and for many metals irradiation swelling appeared to the extent that intolerable dimension changes occurred. The understanding of all these phenomena and the alloy development programmes created to solve them consumed the time of a large fraction of materials scientists around the world.

Many of the problems described and the solutions discovered were sensitive to fabrication techniques. In addition, solutions to some of the problems carried with them safety and operational implications. Thus, the nuclear industry gave birth to extremely restrictive fabrication specifications and a high level of quality assurance unprecedented in any other industry.

## Status

Nuclear research and development activities in the fuels and materials area, as well as other areas, are greatly diminished today from what they were a few decades ago. The facilities in which all the experiments on fuels and materials were conducted have fallen into a state of degradation or have vanished. For example, there is no fast reactor irradiation facility in the United States. More importantly, the researchers who worked and lived through this period of discovery and intense investigation are disappearing from the workforce.

In most industries that have survived there has been continuity from initial discovery to large-scale market penetration with a continuous flow of information and expertise from any generation of workers to the next. Thermal reactor deployment too has enjoyed a semblance of continuity on a world-wide scale. Reactor orders for new thermal reactors are again beginning to accelerate. Thus, thermal spectrum fuels and materials research has progressed with knowledge passing from one generation to the next. However, in the fast reactor area the situation is far different and without precedent.

Interest in the deployment of fast breeder reactors has come to a halt in most countries other than in Japan and Russia. Partly, this is due to proliferation concerns about the fuel cycle but mostly it is due to the lack of near term economic necessity for additional fissile material. Yet most studies indicate an exhaustion of reasonably priced uranium by mid-century. Thus, interest in the fast breeder reactor will most assuredly reappear. The questions addressed in this paper are: What will happen to the enormous amount of fuels, materials, design and operational information that was generated through 50 years of intense and expensive research and development effort? Is it sufficient to believe that it has been documented well enough in the literature that the best of it will survive and will not have to be recreated? Neither the people nor the facilities will be available to recreate the lost information even a few years from now.

In an ongoing programme such as Japan, in which the JOYO programme is going well and MONJU is about to be restarted, there is little incentive to preserve information as must be done in the US and France, for example. This is a pity because the issue of preserving information is easier while it is being developed. Gathering past data, deciding on its relevance and creating new databases of information in a close-out programme is much more difficult. In Russia, even with an ongoing programme, the issue is recognised and would be addressed but for the lack of sufficient funds.

Of what data and information are we speaking? A brief review of a typical fuels irradiation experiment will illustrate the information associated with just such an experiment.

The fuel and cladding had to be fabricated according to some specifications. It may be that both the fuel and cladding were new and thus in the process of fabrication new experience and knowledge were gained. Perhaps several attempts were required to correctly produce the fuel and cladding. The failed fabrication attempts as well as the successes are all valuable knowledge. The irradiation conditions are always important. The neutron flux and temperature are either measured or calculated. The computer codes or measurement techniques are important to be able to assess the validity of the data. After irradiation, the experiment is subjected to both non-destructive and destructive examination. A great deal of data is generated, which is subsequently examined to sort out the good data. These data may be further reduced prior to open literature publication. It may be that no open literature paper exists because the experiment failed or the programme was terminated from lack of funding. All of the information associated with this experiment, as well as the facilities for irradiation and their specifications, could already have been discarded. However, some of it may still exist in the combination of raw data in boxes and filing cabinets, reduced data in computer databases, internal company reports, open literature publications or the in the minds of the scientists and engineers associated with the experiment.

Much of the information is vulnerable to loss, if it has not been lost already. Obviously, an individual's personal experiences are lost when he leaves the workplace. This experience extends to knowing where to look for information as well as what information is valuable and what is not. Information that is stored in boxes and filing cabinets has and will be discarded in a specified time. Even if it has not yet been discarded, it is just as damaging not knowing that the information exists or not having a road map for items of interest. Information on a computer database is a step better, but is still vulnerable to loss. Unless the information on a database is stored under a credible quality assurance programme the data are always suspect. Further, the database must have a reasonable manual so the database can be queried properly. Finally, computer technology is changing so rapidly that the hardware and software to use the database may not exist after a decade or two.

In addition, it is necessary to review the entire fast reactor technology beyond the fuels area to ensure that the whole of the essential fast reactor development is captured and the fuel data is placed in context. For example, fuel information may only be of partial value if decisions made on core assembly design and its seismic behaviour during operational configurations were not also available, since the safety case is made on the fuel performance in off-normal physics spectra during seismic conditions. This US information is presently only encompassed by facility design descriptions while the actual seismic response data may already have been lost.

There may be some doubt in the minds of young scientists that a problem of knowledge loss exists, and this doubt may extend to those in governmental funding positions with little experience with past programmes. However, in the authors' experience hardly a week passes where a search is not initiated to find fast reactor fuels and materials information. This is often of use to a related technology such as the work in progress on the accelerated transmutation of fission products and actinides.

Similarly, for example, information on sodium technology is still required and has a present commercial value in large solar power systems.

Some fast reactor information is being preserved under programmes run by the International Atomic Energy and Nuclear Energy Agency (IAEA/NEA) and by the Department of Energy's Office of Scientific and Technical Information (OSTI.) These programmes encompass the gathering of explicit documents and actual data from successful experiments which have been benchmarked. Very little tacit information (that contained within the minds of retiring experts), or design information or material samples is currently being captured.

Is it worth spending a fraction of what it cost to generate the original information on preserving the most important of it? We feel it is, because to restart a programme of development of such a technology from scratch would be even more expensive in the future.

The United States nuclear weapons establishment believes it is necessary to preserve critical information since they are in danger of losing information for much the same reasons that fast reactor technology is losing information. The cessation of nuclear weapons testing came coincidentally with the ageing of the workforce. It was recognised that as the scientists and engineers retired not enough was being done to capture their years of experience. As is true in the fast reactor technology, there was no perceived need to pass their knowledge on to a new generation. This problem was identified by a Congressional study as a national security concern. Therefore, the United States national laboratories involved in weapons programs were directed to rectify this situation. A number of techniques were put in place to recover and preserve the information. Retired and retiring staff were video interviewed often two or more at a time to stimulate one another in the extraction and gathering of valuable experience. A pre-planned format was used in all interviews. Also, key staff was asked to document the areas they perceived to be critical. Some of this information was included in special courses taught to young and promising engineers and scientists. Their work is far from over, yet there is much to be learned that could equally be applied to the preservation of fast reactor fuels and materials knowledge.

Other nations have similar programmes: BNFL plc, for example, also employs "smart" interviewing to gather tacit, rather than explicit, information from retiring employees where a commercial reason exists. A national laboratory, however, has to act in the nation's interests within its best judgement of what will be needed and when.

**Possible solutions**

In November 2001 the International Atomic Energy Agency will hold a consultancy to consider the problem of knowledge loss over all areas of fast reactor technology, with fuels and materials issues included. The consultancy is in response to a general understanding that action must be taken soon. Experts from all the countries that have been involved in fast reactor research and development will begin the work of determining what information should be preserved, how it should be preserved, what funds are needed and from where the funds to do the work might come. The funds are not small because, besides gathering varied information, there is also the technical issue of maintaining it for 50 years in the face of software and hardware that become obsolete on a much shorter time scale.

Building upon the "smart" interviewing techniques used by the Defence Department and by commercial organisations such as BNFL plc, a process which extends the extraction of expertise of retiring scientists and experts is proposed. It includes a direct transfer of knowledge to the younger generation at the same time as the information is gathered. The following scheme could yield significant results if conducted within the next few years:

- Assemble at the same time about double that number of young and intelligent engineers and scientists aspiring to be nuclear fuel experts.

- An agenda should be derived that contained the main subjects to be covered in the ensuing discussion but not in much detail. For example, mixed oxide fuel and fabrication might be an agenda item. The oxide fuel expert(s) would lead the discussion and would have the responsibility of focusing on real information and expertise rather than on reminiscences.

- Once the meeting progresses there would be a great deal of cross stimulation and perhaps some good debate and argument.

- The entire week would be video taped.

- Pairs of young engineers and scientists would be assigned to take copious notes in particular areas of discussion. Questions would be limited to points of clarification.

- After the week of discussion with experts the young scientists and engineers would have the responsibility of producing a document that captures discussions. Their notes and the video would be used as aids in the preparation of the document(s).

- The summary document(s) would then be cycled back to the experts for editing.

Such a meeting would be expected to yield information that has not been published or is difficult to find. Discussions on the rationale of why some paths were rejected and others pursued would be encouraged. Subtle fabrication techniques and design decisions could be explored. Failures that were never published could be captured. Furthermore, in the process of producing such documents, which could be used for teaching and future reactor development, a number of young engineers and scientists would receive a good start on their own education.

**Conclusion**

We have identified a concern that unless something is done now, information needed to use the fast reactor option in the future will not be available. The serious consequences of huge additional costs for regenerating information coupled with years of delay are recognised by experts in a number of countries and international agencies. We have proposed one solution to the gathering of tacit information in the minds of ageing experts and coupled that with a direct transfer to young engineers and scientists. Similar proposals need to be developed on an international basis and funds found to carry out the work in those countries that have so valuably contributed in past decades to existing fast reactor, and particularly fast reactor fuels' information. Meetings at Reno, Nevada, this year and a subsequent IAEA workshop will start the work.

# HIGH-TEMPERATURE GAS REACTORS

*Chairs: H. Beaumont, W. Zwermann*

# EUROPEAN COLLABORATION ON RESEARCH INTO HIGH-TEMPERATURE REACTOR TECHNOLOGY*

**T.J. Abram** (BNFL), **D. Hittner** (FRAMATOME ANP),
**W. von Lensa** (Fz-Jülich), **A. Languille** (CEA), **D. Buckthorpe** (NNC),
**J. Guidez** (EC-JRC), **J. Martín-Bermejo** (EC DG-Research)

## Abstract

Europe has a long history of innovation with regard to high-temperature reactor (HTR) systems. The HTR was first suggested here in the mid-1950s, and the first experimental units of both the prismatic and pebble-bed core designs were operated in the UK in 1966 and in Germany in 1967, respectively. Recent years have seen considerable renewed world-wide interest in HTR systems, with new experimental reactors recently attaining criticality in Japan (HTTR) and China (HTR-10), and international programmes established for the design of modular HTRs, such as the PBMR and the GT-MHR. Within Europe, a number of collaborative R&D programmes have been established under the auspices of the European Commission's Framework-5 Programme, co-ordinated by a Technology Network (HTR-TN). This network is currently comprised of 18 organisations from eight European countries, together with the EC's Joint Research Centre (JRC). The purpose of these programmes is to secure the wealth of historical European expertise in HTRs, and building on this knowledge, to further pursue the development of HTR technology in order to underpin its eventual commercial application within Europe. Research programmes have been instigated in the following areas:

- *Fuel technology (HTR-F)*. The original programme of R&D on HTR fuel technology included activities concerning the compilation of historical fuel irradiation behaviour, the development of an analytical model for in-reactor fuel performance, the development of fuel manufacturing technology, and an irradiation experiment in which both German fuel spheres and American fuel compacts will be irradiated in the High Flux Reactor (HFR) at Petten to a target burn-up of 200 GWd/tU. The EC has now also agreed to support a follow-up programme, aimed at performing post-irradiation examination of the fuel, including fault simulation studies in the Cold-finger Apparatus (KÜFA) at the JRC's Transuranium Institute.

- *Core neutronics and fuel cycle technology (HTR-N)*. This programme is divided into studies of HTR core physics and methods validation, including an assessment of the need for new nuclear data, analysis of alternative core designs and fuel cycles aimed for example at the burning of Pu and/or minor actinides and assessments of HTR waste treatment and disposal, including long-term characteristics of coated particle fuels. The EC has also agreed to support a follow-up programme, allowing more detailed studies in this area.

---

* The full paper was unavailable at the time of publication.

- *Materials (HTR-M).* Materials technology will be key to the success of the proposed modular HTR systems. In this programme, activities have been divided into four work packages: design and materials for the reactor pressure vessel, materials for high-temperature applications (e.g. reactor internals, core support structure, turbo-machinery), properties of irradiated graphite and graphite oxidation.

This paper will present an overview of the status of the 5th Framework HTR technology programmes, and will describe the principal results obtained to date.

# DESIGN OF $B_4C$ BURNABLE PARTICLES MIXED IN LEU FUEL FOR HTRs

**V. Berthou, J.L. Kloosterman, H. Van Dam, T.H.J.J. Van der Hagen**
Delft University of Technology
Interfaculty Reactor Institute
Mekelweg 15, 2629 JB Delft, The Netherlands
Email: v.berthou@iri.tudelft.nl

## Abstract

The purpose of this study is to design burnable particles in the fuel elements of high temperature reactors (HTRs) in order to control reactivity as a function of burn-up. We focus on heterogeneous poisoning in which burnable particles (particles containing only burnable poison) are mixed with fuel particles in a graphite matrix. There are many degrees of freedom in the design of the burnable particle and the emphasis is put here on the optimisation of its geometry (spherical or cylindrical), its size (different radius) and its composition ($B_4C$ with either natural boron or 100% enriched in $^{10}B$). As a result, we have designed burnable particles that considerably reduce the reactivity swing during irradiation (up to 2.5%).

# Introduction

During the operation of a nuclear reactor, the reactivity effect of fuel burn-up must be compensated by some means of long-term reactivity control. One means for such control is the use of burnable poison in the fuel elements. In this way, it is possible to balance the reactivity loss due to fuel burn-up and fission-product poisoning by the reactivity gain due to the disappearance of the burnable poison.

With homogeneous poisoning it seems impossible to obtain a flat reactivity curve as a function of burn-up [1]. More promising is heterogeneous poisoning, in which burnable particles (particles containing only burnable poison) are being mixed with the fuel particles in the graphite matrix. Because of the many degrees of freedom, correct dimensioning of the burnable particles and the ratio of burnable particles and fuel particles is not straightforward. In this study, the emphasis is put on optimisation of the geometry, size and composition of the burnable particle as well as on the optimal ratio of burnable particles and fuel particles, in order to minimise the reactivity swing as a function of irradiation time.

The first part of this paper presents the description of the fuel and the burnable particles as well as the computational model. The second part shows the calculations for two different geometries: spherical and cylindrical.

## Computational model and reactor type

### Computational model

The reactor physics codes used are from the SCALE system [2]. The BONAMI code, applying the Bondarenko method, is used for the resonance treatment in the unresolved energy region, and the NITAWL code, applying the Nordheim integral method, is used for the resonance treatment in the resolved energy region. XSDRNPM, a 1-D discrete-ordinates transport code, is used for the cell-weighting calculations, while the actual burn-up calculations are performed by the ORIGEN-S fuel depletion code. All data used are based on the JEF-2.2 nuclear data library [3].

In all calculations, a macro cell is modelled containing one burnable particle surrounded with several (typically two or three) fuel layers. The total thickness of all the fuel layers determines the effective number of fuel particles per burnable particle. In the macro-cell calculation, the burnable particle is subdivided into burnable layers, each layer having its own characteristic nuclide composition.

Because the SCALE system cannot treat the double heterogeneity of the fuel explicitly, the cell-weighting procedure is split into two parts. First, the homogenised (resonance-shielded and cell-weighted) cross-sections of each fuel layer are calculated in a micro-cell calculation. The micro unit cell consists of a three-zone coated fuel particle with a nuclide composition (particle and graphite) characteristic for the actual burn-up time step. Secondly, if necessary, the resonance shielding calculations for each layer of the burnable particle are performed. Finally, a macro-cell calculation is done to calculate the neutron flux density and the neutron spectrum in each burnable layer, and the fission power density in each fuel layer. In these calculations, the average core power density is 3 MW.m$^{-3}$, and the temperature is 900°C.

Besides the resonance-shielded cell-weighted cross-sections to be used in the macro-cell calculation, the resonance-shielded zone-averaged nuclide cross-sections are also calculated for each fuel layer, and passed to the burn-up data library. Subsequently, burn-up calculations are performed for each fuel layer and for each layer of the burnable particle using the nuclide cross-sections updated at each burn-up time step. For the burnable layers, the depletion calculation is done using the constant flux approximation, while for the fuel layers the constant power approximation has been applied. As mentioned before, both the neutron flux density and spectrum, and the fission power density are calculated in the macro-cell calculation.

The whole sequence of micro-cell calculations, macro-cell calculation and burn-up calculations for both the fuel layers and the burnable layers are repeated many times (typically 20) to calculate the composition of each layer and the k-infinite of the system as function of burn-up. Furthermore, at each burn-up time step, the temperature coefficient of reactivity is calculated with the VAREX code [4].

### *Reactor type*

For the fuel considered, the ESKOM PMBR [5,6] has been chosen with an average core power density of 3 MW.m$^{-3}$. The fuel particles contain 8% enriched $UO_2$ with a uranium mass of 9 grams per pebble. One pebble contains more than 13 000 fuel particles, each with a radius of 0.046 cm. For the poison in the burnable particle, we have chosen $B_4C$ with either natural boron or 100% enriched $^{10}B$.

The chosen poison has a high thermal absorption cross-section, and the absorption-to-scattering ratio is more than 1 000. This implies that the burnable particle can be assumed to be purely absorbing. If the radius of the particles is large in units of thermal neutron absorption mean free path, then we can consider the particles as "black" particles, which means that every neutron hitting its surface will be absorbed. The effective absorption cross-section for a small "black" particle is related to its geometrical cross-section, i.e. for a sphere or a cylinder (with a radius R), it is calculated as [1]:

$$A_{sphere} = \pi R^2 \qquad A_{cylinder} = \frac{\pi}{2} R \qquad \text{(per unit of length)}$$

The target is to reach a k$_{infinitive}$ of 1.05 constant during the whole irradiation time.

### Spherical burnable particles

### *Evolution of reactivity as a function of burn-up*

The burnable particles considered here are similar to the fuel particles: spherical geometry with a radius of 0.046 cm. The burnable poison is $B_4C$ with either natural boron or enriched boron in 100% of $^{10}B$.

### *$B_4C$ with natural boron*

The evolution of the reactivity during the irradiation of the fuel has been studied (Figure 1) for different volume ratios between fuel and burnable particles. The case without any burnable particle in the fuel is taken as a reference.

It can be seen that after 1 300 equivalent full power days (EFPD), the different curves do not converge: burnable particles are not fully burnt yet. The drop of the reactivity at the beginning of the irradiation is due to the short-lived fission products (Xe and Sm effect).

**Figure 1. Reactivity as a function of burn-up for different values of the volume ratio between fuel and burnable particles, for spherical burnable particles (B₄C with natural boron)**

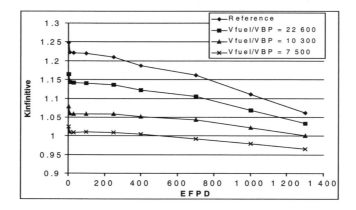

The case with a ratio of 10 300 between the fuel volume and the poison volume starts with a correct reactivity level of 1.05, but the reactivity swing is still too large during the irradiation: 7% up to 1 300 EFPD.

The different cases studied correspond to an insertion of only one or two burnable particles in the pebble; few burnable particles already have a significant impact on the reactivity.

*B₄C with enriched boron (100% of $^{10}$B)*

The same calculations have been done with burnable particles made of B₄C with enriched boron (100% of $^{10}$B).

Because the neutron mean free path for this poison is smaller (0.0064 cm), the burnable particle has been divided into more zones for the calculations (14) so as to have two layers per mean free path.

Figure 2 shows the reactivity as a function of burn-up for a volume ratio of 22 600 between the fuel and the poison in order to compare the burnable particles made of natural and enriched boron.

**Figure 2. Reactivity for spherical burnable particles as a function of burn-up for a volume ratio between the fuel and the burnable poison of 22 600 (B₄C with natural and enriched boron)**

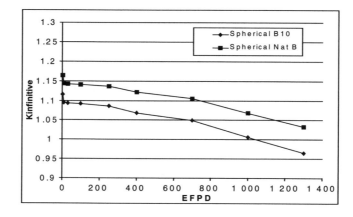

The two curves do not start at the same reactivity level, as it would be the case for a completely black particle. At first sight, this contradicts a previous study [1] in which gadolinium was used as a poison. In that study, the burnable particle (Gd) had a neutron mean free path of $8 \cdot 10^{-3}$ cm, which means that the ratio between the burnable particle diameter and the neutron mean free path is about $\frac{D}{\lambda} = 120$. Consequently the particle is black.

In our study, the neutron mean free path in natural boron is $3.24 \cdot 10^{-2}$ cm, and in enriched boron, $6.4 \cdot 10^{-3}$ cm. That leads to ratios of $\frac{D}{\lambda} = 3$ and 14, respectively.

It is clear that a burnable particle with natural boron is not black, which is why we notice a gap between the two curves at the beginning of life (BOL).

This feature can also be observed on the spectrum of the burnable particle inner zone (Figure 3). For a $B_4C$ poison with enriched boron, there is no thermal neutron in the inner zone of the burnable particle, but for a $B_4C$ poison with natural boron, there are still some thermal neutrons that can reach the inner zone of the burnable particle. This means that this particle cannot be considered as a black particle.

**Figure 3. Spectrum of the burnable particle inner zone at the beginning of life (BOL) for $B_4C$ with natural boron and enriched boron**

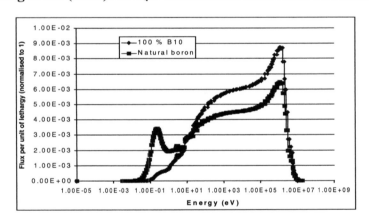

## Evolution of $^{10}B$ concentration as a function of the burnable particle radius

Figures 4 and 5 show the concentration of $^{10}B$ as a function of the distance from the centre of the burnable particle for each burn-up step (given in EFPD), for respectively the case with natural boron and enriched boron. The volume ratio between the fuel and the poison is 10 300.

In the case of $B_4C$ with natural boron, we notice that the burning of the particle is quite homogeneous whatever the distance from the centre of the burnable particle, which confirms the fact that the burnable particle is not black for thermal neutrons.

In the case of $B_4C$ with enriched boron, the external zone burns much faster than the internal zone; the thermal neutrons are all absorbed when they hit the surface of the burnable particle. We can also notice that the burnable particle is not fully burnt at the end of life (EOL) as previously pointed out.

**Figure 4. $^{10}$B concentration as a function of distance from the centre of the burnable particle, with the burn-up as a parameter, for a B$_4$C spherical burnable particle made of natural boron**

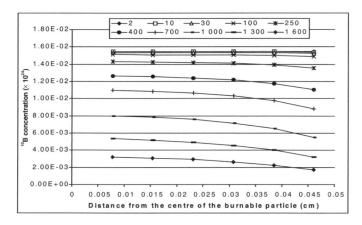

**Figure 5. $^{10}$B concentration as a function of distance from the centre of the burnable particle, with the burn-up as a parameter, for a B$_4$C spherical burnable particle made of enriched boron**

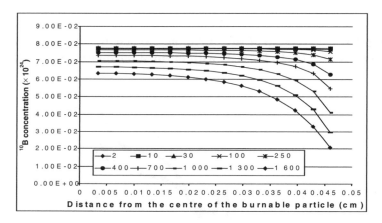

### *Variation of the burnable particle diameter*

We have seen that the effective macroscopic absorption cross-section of a black particle is related to its geometrical cross-section, which means that particles with different radii would have different behaviour.

In spherical geometry, different radii of the burnable particle are considered with a constant volume ratio of the fuel and the burnable particles. This means that the number of burnable particles per pebble increases when the radius of each particle decreases. The smaller the burnable particle, the more homogeneous the poison is distributed, and the faster the particles will burn.

In Figure 6, the reactivity is shown as a function of time for the reference case without poison and for two radii of the burnable particles, 0.046 cm and 0.03 cm. The initial reactivity is lower in the case of smaller particles because of more burnable particles present in the pebble that are distributed more homogeneously. Also, the effective absorption cross-sections are respectively $A = \pi R^2 = 0.0066$ cm$^2$ for a radius of 0.046 cm and $A = \pi R^2 = 0.0028$ cm$^2$ for a radius of 0.03 cm.

**Figure 6. K$_{infinitve}$ as a function of irradiation time for the reference case without poison and for spherical burnable particles with different radii**

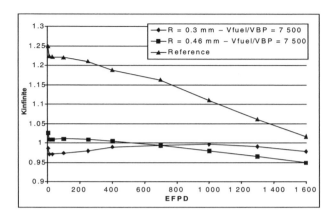

Consequently, the reactivity swing is smaller for the burnable particle with a radius of 0.03 cm (2.5% compared to 6% with a radius of 0.046 cm).

Different volume ratios between the fuel and the poison have been studied, as well as other burnable particle radii (0.046, 0.03, 0.02 cm). The case of burnable particles with a small radius (0.02 cm) and a volume ratio of 10 300 has a reactivity swing of 3.6% up to 1 600 EFPD. The burnable particles are almost fully burnt at 1 600 EFPD, but the level of reactivity is too low (between 1 and 1.03).

## Uniform temperature coefficient (UTC)

Since the kernel of each fuel particle has a high surface-to-volume ratio, an effective heat transfer from the fuel to the moderator is realised. This means that there is only a small temperature difference between the kernel and the surrounding graphite, specifically 1 K [7]. Since the temperature of the kernel and the graphite is almost equal, this allows us, in view of reactivity feedback effects, to consider a uniform temperature coefficient.

Figure 7 shows the UTC as a function of the irradiation time for the reference case without burnable particles and for two different values of volume ratio between fuel and poison (the case of spherical burnable particles with a 0.046 cm radius, and containing B$_4$C with enriched boron).

**Figure 7. Uniform temperature coefficient (UTC) for the reference case and for two different volume ratio between the fuel and the spherical BP (containing B$_4$C with enriched boron)**

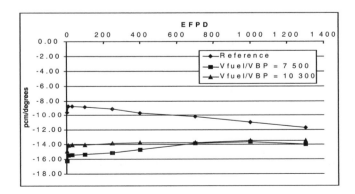

The UTC is relatively constant during the irradiation, around -10 pcm/K. In the presence of burnable particles, the UTC is more constant, and a little bit more negative due to a slight shift of the spectrum towards higher energies. At the end of the irradiation the spectra are the same, and then the UTC are equal.

## Cylindrical burnable particles

### *Evolution of reactivity as a function of burn-up*

The burnable particles considered here have a cylindrical geometry with a diameter of 0.092 cm. The burnable poison is $B_4C$ with either natural boron or boron 100% enriched in $^{10}B$.

#### *$B_4C$ with natural boron*

The evolution of the reactivity during the irradiation of the fuel has been studied (Figure 8) for different values of the volume ratio between fuel and burnable particles. The case without any poison in the fuel is taken as a reference.

**Figure 8. Reactivity as a function of burn-up for different values of
the volume ratio between the fuel and the burnable poison and for the
reference case, for cylindrical burnable particles ($B_4C$ with natural boron)**

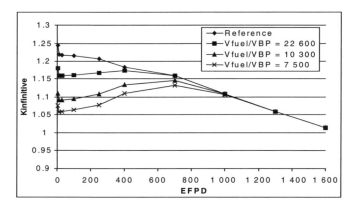

It can be seen that, after already 1 000 EFPD, the different curves converge: the burnable particles are fully burnt.

The case with a volume ratio of 22 600 between the fuel and the poison has a lower reactivity swing: 6% up to 1 000 EFPD.

If we now consider two different shapes of burnable particles (spherical and cylindrical) but with the same effective absorption cross-section (meaning a smaller radius for the cylindrical particle) and the same ratio of fuel volume and poison volume, the reactivity swing is smallest for the cylindrical particle (much more cylindrical burnable particles homogeneously distributed in a pebble).

#### *$B_4C$ with enriched boron (100% of $^{10}B$)*

The burnable particle containing $B_4C$ with natural boron is fully burnt after 1 000 EFPD. With enriched boron, a flatter curve can be obtained, with a reactivity swing of 2.2% (for a volume ratio between fuel and poison of 10 300) up to 1 600 EFPD. Then the burnable particle is fully burnt at 1 600 EFPD (Figure 9). The only drawback is the level of reactivity, around 1.03, which is a little bit too low.

**Figure 9. Reactivity as a function of burn-up for different values of
the volume ratio between the fuel and the burnable poison and for the
reference case, for cylindrical burnable particles (B₄C with enriched boron)**

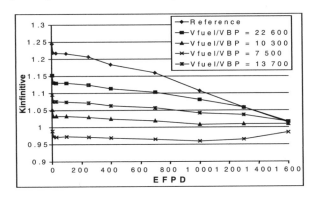

For the same burnable particle radius, and for a black particle, the effective absorption cross-section is larger for a cylindrical shape than for a spherical one:

$$A_{cylinder} = \frac{\pi}{2}R = 0.072 \text{ cm}^2 \text{ and } A_{sphere} = \pi R^2 = 0.0066 \text{ cm}^2$$

As a result, the cylindrical particle is fully burnt at the end of the irradiation while the spherical particle still contains significant amounts of $^{10}$B after 1 600 EFPD.

*Evolution of $^{10}$B concentration as a function of the burnable particle radius*

Figure 10 shows the concentration of $^{10}$B as a function of the distance from the centre of the burnable particle for each burn-up step (given in EFPD) for the case of enriched boron. The volume ratio between the fuel and the poison is 10 300. Also here we can notice that the cylindrical burnable particle is fully burnt faster than the spherical one: after 1 000 EFPD the $^{10}$B concentration has decreased both in the outer and inner zone of the burnable particle.

**Figure 10. $^{10}$B concentration as a function of the distance from the centre of the burnable particle, with the burn-up as a parameter, for a B₄C cylindrical burnable particle with enriched boron**

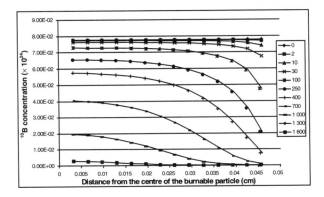

**Conclusion**

The purpose of this study is the use of burnable particles in the fuel elements of HTR in order to control the reactivity effect of the fuel burn-up. Heterogeneous poisoning in which burnable particles (particles containing only burnable poison) are mixed with the fuel particles in a graphite matrix is considered. In order to obtain the flattest reactivity-to-time curve numerous parametrical studies have been carried out to tailor the burnable particle.

The general trends are the following:

- The use of $B_4C$ with enriched boron (100% of $^{10}B$) is required in order to have a "black" particle.

- The smaller the radius for the burnable particle, the more homogeneous the poison is distributed, and the faster the particle will be burnt.

- For a spherical shape, a small reactivity swing is obtained (2.5%) with a burnable particle radius of 0.03 cm.

- The cylindrical shape leads to the lowest reactivity swing as a function of irradiation time: 2.2% up to 1 600 EFPD for burnable particles containing $B_4C$ 100% enriched in $^{10}B$ with a radius of 0.046 cm, and for a volume ratio between fuel and poison of 10 300).

Heterogeneous poisoning of HTR fuel seems quite promising. We have designed burnable particles containing $B_4C$ that considerably reduce the reactivity swing. Further studies will focus on burnable particles that contain graphite or fuel surrounded with a thin layer of poison, on different poison materials (e.g. Er or Gd) and on different fuel compositions. The ultimate aim is to abandon active long-term reactivity control in HTRs.

*Acknowledgements*

The authors would like to acknowledge the European Commission for co-funding this research that is performed within the project "High Temperature Reactor Physics and Fuel Cycle Studies (HTR-N)" [8] in collaboration with other European organisations.

# REFERENCES

[1]    H. Van Dam, "Long-term Control of Excess Reactivity by Burnable Particles", *Annals of Nuclear Energy*, 27, 733-743 (2000).

[2]    SCALE-4.2, "Modular Code System for Performing Standardized Computer Analyses for Licensing Evaluations", Oak Ridge National Laboratory, Tennessee (1994).

[3]    JEF-2.2, "The JEF-2.2 Nuclear Data Library", Report JEFF17, Nuclear Energy Agency, Paris (2000).

[4]    J.L. Kloosterman, J.C. Kuijper, "VAREX, A Code for Variational Analysis of Reactivity Effects: Description and Examples", M&C 2001, Salt Lake City, USA, September 2001.

[5]    J.H. Gittus, "The ESKOM Pebble-bed Modular Reactor", *Nuclear Energy*, 38, No. 4, 215-221 (1999).

[6]    D.R. Nicholls, "Status of the Pebble-bed Modular Reactor", *Nuclear Energy*, 39, No. 4, 231-236 (2000).

[7]    E.E. Bende, "Plutonium Burning in a Pebble-bed Type High Temperature Nuclear Reactor", Ph.D thesis, Delft University of Technology, ISBN 90-9013168 (2000).

[8]    "High Temperature Reactor Physics and Fuel Cycle Studies (HTR-N)", contract no. FIKI-CT-2000-00020, http://www.cordis.lu/fp5/projects.htm.

# THORIUM AND PLUTONIUM UTILISATION IN PEBBLE-BED MODULAR REACTOR

**U.E. Sıkık, H. Dikmen, Y. Çeçen, Ü. Çolak, O.K. Kadiroğlu\***
Hacettepe University Nuclear Engineering Department
06532 Beytepe Ankara, Turkey
\*okk@alum.mit.edu

## Abstract

Thorium and plutonium utilisation in a high temperature gas-cooled pebble-bed reactor is investigated with the aim to predict the economic value of vast thorium reserves in Turkey. A pebble-bed reactor of the type designed by PBMR Pty. of South Africa is taken as the investigated system. The equilibrium core of a PBMR is considered and neutronics analyses of such a core are performed through the use of the SCALE-4.4 computer code system KENOV.a module. Various cross-section libraries are used to calculate the criticality of the core. Burn-up calculations of the core are performed by coupling the KENOV.a module with the ORIGEN-S module. Calculations are carried out for various U-Th, U-Pu-Th and U-Pu combinations. The results are preliminary in nature and the work is currently proceeding as planned.

## Introduction

With the advent of a new generation nuclear reactors and especially HTGRs, nuclear power is becoming more popular with first-time users of nuclear technology. The promising new design of the pebble-bed modular reactor of PBMR Pty. is an interesting new concept based on proven technology and having superb safety characteristics, smaller power output, short construction time and low generation cost. With the new liberalised energy markets in many countries, and especially in Turkey, large nuclear power plants with very large capital investment requirements are rather difficult to realise. Private utilities that will be active in the market will find PBMR an investment worthy of consideration in the near future.

Turkey has one of the largest thorium reserves in the world and the nuclear technology that will be transferred to the country should be able to utilise this resource. Decision-makers of the utilities and government organisations need to have some idea about the possible applications of this technology, and thus it is of importance to study the possible uses of PBMR with different fuel materials.

A detailed study of the effects of different fuel elements on the economics of the PBMR fuel cycle must be performed using sophisticated Monte Carlo techniques. Such a study is very time consuming, however, if the boundary values of the fuel properties are not determined in advance. In the introductory phase of the study the limiting values of the fuel properties are sought and preliminary burn-up calculations are performed to determine these limits. Based on the findings of these preliminary calculations an acceptable range of fuel enrichment, amount of Pu and Th added to the fuel pebbles are determined. In the following phases of the research more refined calculations, such as multi-group 3-D full core Monte Carlo calculations will be performed within these obtained limits.

## Modelling of the PBMR core

The reactor core that is investigated in this study is a standard PBMR [1]. The thermal power of the reactor is 265 MW and the average operating temperature 600°C. The core height and radius are 8.44 m and 3.5 m, respectively. Nine (9) grams of 8% enriched uranium is used in the form of TRISO particles that make up the 6 cm diameter standard fuel pebbles. The central part of the core is made of graphite pebbles, creating an annular cylindrical core with conical upper and lower sections.

The criticality conditions of the above-mentioned core are calculated by using the SCALE-4.4 computer code system KENOV.a Monte Carlo module [2]. The 27-group cross-sections for HTGR applications that are present in the SCALE-4.4 system are used. The full core is modelled and the neutron flux distribution and the power distribution of PBMR are obtained from KENOV.a runs. For this critical configuration a representative unit cell is created so that the infinite medium criticality constant is taken as the reference for future unit cell calculations. For each different fissile material used in the core, a new unit cell infinite medium criticality search is performed.

Based on the calculated power distribution an average power density is obtained and used as the input for the ORIGEN [3] burn-up calculations. In the burn-up calculations a homogenised zero dimensional core is considered and non-linear reactivity model is used.

The linear reactivity model is a simplified model for describing nuclear reactor behaviour; it is a collection of algorithms and methods collectively designated the linear reactivity model [4]. The model is restricted to situations in which reactivity is a linear function of burn-up. An extended model exists for on-line refuelling cases. However, the non-linear reactivity model deals with high order polynomial function fittings of burn-up. This model is useful for CANDU and pebble-bed reactors. The model, in our case, simply treats pebbles as batches and as the number of batches goes to infinity the discharge burn-up is related to the integral of the reactivity with respect to the burn-up.

The model simplifies burn-up into an equal power sharing level of complexity for the steady-state reload case. The most commonly used fitting functions are third degree polynomials, however there is no restriction on using higher orders, and any non-pathological reactivity curve could be fitted and its implied cycle burn-up could be computed.

## Results

Burn-up calculations are performed for uranium-thorium systems in which the enrichment and the amount of the heavy metal in the pebbles varied. The burn-up values for different thorium concentrations are calculated. Similar calculations are performed for plutonium; weapons grade as well as LWR discharged fuel, combined with uranium and thorium.

Since the aim of this preliminary study is to define the possible boundary values, attention is only paid to the burn-up and other features such as the value of the negative temperature coefficient; thermal limits, fuel behaviour, economics, etc. are not taken into account.

### *Uranium-thorium mixtures*

The enrichment of uranium in the fuel varied between 8% to 19.9% and it is found that the burn-up increases with the amount of thorium added to the pebbles. The addition of thorium, a fertile material, to the pebbles requires increased enrichment as seen in Figure 1.

**Figure 1. Effects of the amount of thorium and uranium enrichment on burn-up**

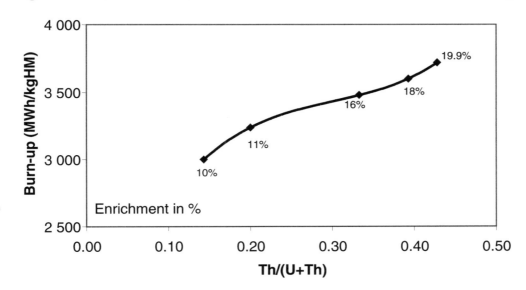

The burn-up can be increased by approximately 25% if the enrichment is doubled. On the other hand, if the amount of the heavy metal in the pebble is increased from 9 grams to 30 grams while the amount of thorium in the heavy metal is kept almost constant, the burn-up increases as presented in Figure 2 for two limiting uranium enrichments.

**Figure 2. Effects of the amount of heavy metal in the pebble on the burn-up**

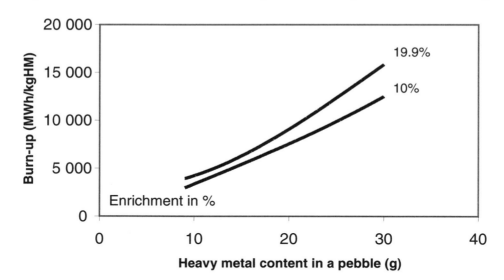

*Plutonium and plutonium mixtures*

Plutonium can be used as the only fissile material in the pebble, but this option is rather unrealistic, so instead of enriched uranium a mixture of 8 grams of natural uranium and 1 gram of weapons grade plutonium is considered. In this case, the burn-up for such a fuel is calculated to be 4 056 MWh/kgHM, which is larger than the standard PBMR burn-up values found in the literature [1].

If similar calculations are performed for a mixture of $^{239}$Pu and $^{232}$Th, the value of the burn-up is 3 816 MWh/kgHM.

We have considered plutonium discharged from a typical 1 000 MW light water reactor (LWR). Nevertheless, it is found that if mixed with thorium, criticality of the system could not be achieved. Due to the relatively high absorption of $^{240}$Pu and $^{241}$Pu, a system fuelled with LWR discharged plutonium can only be critical if the ratio of $^{239}$Pu is around 95% of the plutonium introduced into the pebble.

On the other hand, if a small amount of LWR discharged plutonium is mixed with enriched uranium, criticality can be achieved. For various enrichments of uranium, 10% to 19.9%, the burn-up value increases with the amount of plutonium inserted into the pebble as presented in Figure 3.

**Summary**

It is shown that addition of thorium to the pebbles increases the burn-up, provided that the enrichment of uranium is also increased. Approximately 25% increase in burn-up can be achieved by doubling the enrichment of uranium.

The use of LWR discharged plutonium as a way to reduce the nuclear waste can also be done at much lower burn-up values than uranium-thorium systems. Again an increase in the uranium enrichment is necessary and doubling the enrichment doubles the burn-up values.

**Figure 3. Effects of the amount of plutonium in the pebble on the burn-up**

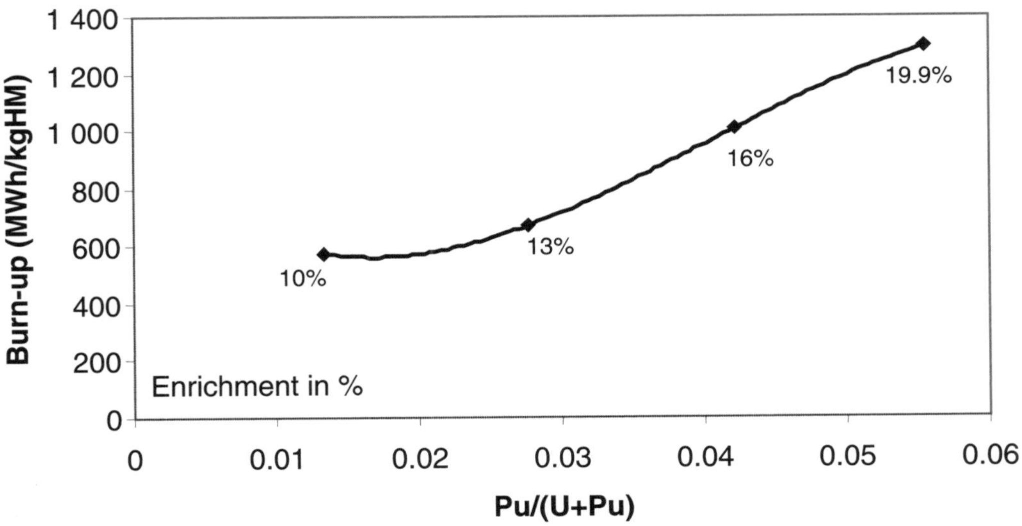

REFERENCES

[1] "Review of the Pebble-bed Modular Reactor (PBMR) Plant" Current Status and Future Development of Modular High Temperature Gas-cooled Reactor Technology, p. 21, IAEA-TECDOC-1198, 2000.

[2] L.M. Petrie, N.F. Landers, "KENO V.a: An Improved Monte Carlo Criticality Program With Supergrouping", NUREG/CR-0200, Revision 6, Vol. 2, Section F11, ORNL/NUREG/CSD-2/R6, September 1998.

[3] O.W. Hermann, R.M. Westfall, "ORIGEN-S: Scale System Module to Calculate Fuel Depletion, Actinide Transmutation, Fission-product Build-up and Decay, and Associated Radiation Source Terms", NUREG/CR-0200, Revision 6, Vol. 2, Section F7, ORNL/NUREG/CSD-2/V2/R6, September 1998.

[4] M.J. Driscoll, T.J. Downar, E.E. Pilat, "The Linear Reactivity Model for Nuclear Fuel Management", American Nuclear Society, 1990.

# A CONCEPTUAL FLUIDISED PARTICLE-BED REACTOR APPLICATION OF SPACE-DEPENDENT KINETICS

**C.C. Pain, J.L.M.A. Gomes, M.D. Eaton, C.R.E. de Oliveira, A.P. Umpleby, A.J.H. Goddard**
Computational Physics and Geophysics Group
Department of Earth Science and Engineering
Imperial College of Science, Technology and Medicine
Prince Consort Rd, London SW7 2BP, UK

**H. van Dam, T.H.J.J. van der Hagen and D. Lathouwers**
Interfaculty Reactor Institute (IRI)
Delft University of Technology
Mekelweg 15, NL 2629 JB Delft, Netherlands

## Abstract

In this paper we describe research conducted into the dynamics of a conceptual helium-cooled fluidised-bed thermal nuclear reactor. The reactor consists of an inner reactor cavity in which the uranium particles are fluidised. This cavity is surrounded by moderating graphite walls. The physics behind this reactor is extremely complex, involving the fluidisation of particles and the associated voidage fluctuations feeding back into the criticality of the system. Detailed axi-symmetric and three-dimensional numerical simulations suggest that the reactor has passive control of criticality features, and that it can have a steady thermal power output.

## Introduction

In this paper we study a conceptual helium-cooled fluidised-bed thermal reactor which has been shown in earlier work to have potentially attractive passive safety characteristics [1]. The reactor consists of an axi-symmetric bed surrounded by graphite moderator. TRISO-coated fuel particles 1 mm in diameter are fluidised with helium. Improvements in computer performance and in numerical methods now allow numerical simulation of the complex non-linear criticality feedback mechanisms. A coupled neutronics /multi-phase fluids dynamics code has been developed, FETCH (finite element transient criticality), which is based on finite element modelling for both fluids and radiation transport. The option of this code for fluidised particles has been used. We present further evidence that this system should be dynamically stable, under normal operating conditions through modelling extreme transients, where the helium coolant flow is started impulsively at full flow rate with an initially settled bed. We also demonstrate that, although this may be subject to significant short-term fluctuations, the temperature distribution inside the reactor remains homogeneous without substantial time variation. In addition, it is shown that, due to the mixing features of a fluidised bed, all particles yield similar fission-heat sources over time scales of the order of 10 seconds. This would result in uniformity in burn-up of particles.

The numerical model FETCH used to study this reactor has been extensively validated for criticality transients involving liquids with careful comparison with experimental results [2,3,4]. The gas-particle fluidised bed modelling is performed using a two-fluid granular temperature approach which is the most computationally feasible of the detailed spatial particle modelling approaches. This model and the resulting code have been validated for fluidised beds in [5,6]. Thus the authors believed that the numerical model is a good representation of the physics of such a reactor system.

## Modelling method

This section outlines the approach to solving the radiation transport and multi-phase fluid equations adopted here. The transient space-dependent kinetic simulations were performed using geometry conforming finite elements for both the neutron radiation transport (modelled in full phase space) and the multi-phase (gas-particle) models embodied in the FETCH computer code used in this work. The three-fluid conservation equation sets cover mass, momentum and thermal energy. The mass equation represents the void fraction, densities and velocities of liquids in given phases. In addition to momentum transfer terms between phases, the momentum equation has terms for interfacial drag forces realised through the Ergun equation [6].

Scalar field equations include those for delayed neutron precursors which are generated and advected within the particle phase. Delayed neutron precursors are assumed to exist only in the particle phase and are in six delayed neutron precursor group form. The continuity equation for the particles depends on an equation of state embodying the response of particle density to pressure and temperature (via the speed of sound and the expansion coefficient respectively). This, together with the use of the ideal gas equation, for helium, allows the pressure to be calculated. Boundary conditions are those associated with specified shear stress [6] and no normal flow at the walls of the container and for both the particle and gas phases. At the topmost boundary of the domain we specify the temperature, density and pressure (60 bars) for the inflowing gas. The boundary conditions for gas outflow are specified normal and zero shear stress conditions. The finite element discretisation and solution of the multi-phase flow equations are described in [6].

The Boltzmann neutron transport equations are solved using the even-parity principle, see [7]. To discretise these equations finite elements in space, $P_N$ approximations in angle, finite difference in energy and a two-level discretisation in time were used. In the examples shown six energy groups are used. The associated material cross-sections are obtained by interpolating group condensed (by WIMS, see [8]) cross-section sets in temperature, see [4]. Also, the $P_1$ approximation in neutron angle direction is used here.

## Reactor description

The bed consists of 1 mm in diameter TRISO-coated uranium particles [1] which are fluidised with helium at 60 bars pressure and an inlet gas temperature of 220°C. The inner cylindrical fluidised bed core (about 1 m in diameter) is surrounded by neutron-moderating graphite reflectors. Figure 1 shows the reactor with a quarter of its cylindrical geometry removed to reveal the inner core. As the particles are fluidised and they move up into the upper parts of the reactor the criticality of the system increases due to an increase in moderation and reflection furnished by the graphite walls. The criticality $K_{eff}$ versus uniform core expansion is shown in Figure 2(a). This is such that at the collapsed bed state the system is subcritical and in the expanded bed state the system is supercritical and thus starts to deposit heat energy into the system. However, the overall temperature coefficient of the reactor is negative [as shown in the $K_{eff}$ versus uniform temperature graph, Figure 2(b)] and thus as the reactor heats up the criticality of the system decreases until in a time-averaged sense the heat produced by the system matches the heat removed from the particles by the fluidising helium gas.

## Reactor dynamics explored

In this section we explain the physics behind the initial part of the transient and that associated with the quasi-steady state after a few minutes of reactor operation.

### Physics of the initial fission pulse

As the power initially rises it has a doubling time which is dictated largely by $\beta_{eff}$, and at this stage the dynamics have no non-linear feedback mechanism. However when the solution temperature rises the negative temperature coefficient terminates the rise. Since the initial $K_{eff}$ is relatively large due to the particles being initially a cool 220°C, the largest peak in fission power [see Figure 3(a)] is seen just after reactor start-up. The reactors' robustness is severely tested here as it is started in relatively extreme conditions by impulsively initialising the fluidising gas velocity so that the reactivity ramp is large. As usual it is prudent to use a large neutron source at reactor start-up.

The heat energy deposited just after reactor start-up is deposited mainly next to the bottom and side graphite walls. This heat is rapidly dissipated through the reactor with the rapid mixing of the particles and the heat transfer equalisation influence of the fluidising gas. This makes the maximum temperature versus time curve initially highly peaked, see Figure 3(b). The rapid decrease of maximum temperature is due to this dissipation, which results in the reactor bed temperature being remarkably uniform [see Figure 4(b)] at all other times.

Since the particles just after the initial fission spike have a relatively large temperature, the criticality of the reactor decreases, resulting in a large decrease in the fission rate. This then allows the fluidising gases to gradually remove the particle heat which is the reason why there is a gentle slope in the maximum temperature versus time curve [see Figure 3(b)] just after the maximum temperature has been reached. Eventually, this temperature becomes low enough for the criticality of the system to increase sufficiently to initiate the depositing of significant quantities of energy.

*Dynamics after the initial fission pulse and towards quasi-steady state*

The reactor then moves into a regime where there can be a relatively gentle oscillation in time-averaged (over tens of seconds) power output. This is due to the non-linear feedback of heat removal and the role of delayed neutrons in providing a delay in the response to an increase in criticality. This slow oscillation in power is very similar to a more conventional reactor following a reactivity change. However, superimposed onto this fission power are the short power fluctuations which have a frequency of the order of 1 Hz.

Finally, the reactor moves into the quasi-steady state regime where the fission power, in a time-averaged sense, is matched by heat removed from the reactor by the fluidising gases. The power output, in a time-averaged sense, of the simulation whose results are shown in Figure 3(b) is about 10 MW thermal. This equals the increase in heat flux over the reactor, that is $\delta T C_p \rho A U$ in which $\delta T$ is the increase in gas temperature over the bed, $C_p$ is the heat capacity of the gas at the inlet (bottom of fluidised bed), $\rho$ is the density of helium at bed inlet, $A$ is the cross-section area of the fluidised bed and $U$ is the inlet gas velocity (= 120cm/s). In addition, the mass of 1 mm diameter fuel particles is $1.6 \times 10^6$ g and they have a collapsed bed height of 130 cm.

The rapid mixing of the particles in the bed means that the particles are uniformly burnt in the reactor. This can be seen in the similarity of the longest-lived delayed neutron precursor concentration (which is a measure of the time-averaged heat deposited to the particles) to the particle concentration [see Figures 4(a) and 4(c)]. The power output which has a very similar profile to the shortest-lived delayed neutron precursor concentration is seen in Figure 4(d).

At this time level it is seen that the particles are highly concentrated near the central axis [left hand boundary of Figure 4(a)]. This is also observed in the time-averaged particle volume fractions. However, the validity of this result is questionable. Experiments of fluidising similar particles in similar geometries suggest that the voidage in the centre is usually relatively low. We believe this discrepancy between the simulated results and expected results is due to imposing axi-symmetry on the flow dynamics. To test this supposition we have conducted similar experiments in full 3-D geometry. A snapshot of the particle volume fraction at 22 seconds into such a simulation is shown in Figure 5. At this time level no large concentration of particles in the centre of the reactor is observed. This is consistent with the time-averaged particle volume fractions (not shown here).

## Conclusions

The fluidised bed nuclear reactor investigated here has been shown through detailed spatial and temporal modelling to have a power output that varies by about an order of magnitude. However, the thermal power output of the reactor is fairly steady. In addition, the temperature throughout the reactor was found to be remarkably homogeneous, due to its superb particle mixing characteristics. In addition, this mixing is shown to lead to uniform burn-up of the uranium fuel particles. Future work will look at reactor safety in extreme scenarios such as the removal of decay heat when helium is no longer flowing through the reactor. Also, work is required to reliably predict via simple parameterisations the thermal power output and temperature of the reactor for a given fuel loading and fluidising gas velocity.

# REFERENCES

[1] V.V. Golovko, J.L. Kloosterman, H. van Dam and T.H.J.J. van der Hagen, "Analysis of Transients in a Fluidized Bed Nuclear Reactor", PHYSOR 2000, Pittsburg, Pennsylvania, USA (2000).

[2] C.C. Pain, C.R.E. de Oliveira, A.J.H. Goddard, A.P. Umpleby, "Transient Criticality in Fissile Solutions – Compressibility Effects", *Nucl. Sci. Eng.*, 138, 78-95 (2001).

[3] C.C. Pain, C.R.E. de Oliveira, A.J.H. Goddard, A.P. Umpleby, "Criticality Behaviour of Dilute Plutonium Solutions", *Nucl. Tech.*, 135, 194-215 (2001).

[4] C.C. Pain, C.R.E. de Oliveira, A.J.H. Goddard, "Non-linear Space-dependent Kinetics for Criticality Assessment of Fissile Solutions", *Prog. Nucl. Energy*, 39, 53-114 (2001).

[5] C.C. Pain, S. Mansoorzadeh, C.R.E. de Oliveira, "A Study of Bubbling and Slugging Fluidized Beds using the Two-fluid Granular Temperature Model", *Int. J. Multiphase Flow*, 27, 527-551 (2001).

[6] C.C. Pain, S. Mansoorzadeh, C.R.E de Oliveira, A.J.H. Goddard, "Numerical Modelling of Gas-solid Fluidised Beds using the Two-fluid Approach", *Int. J. Numer. Methods Fluids*, 36, 91-124 (2000).

[7] C.R.E. de Oliveira, "An Arbitrary Geometry Finite Element Method for Multi-group Neutron Transport with Anisotropic Scattering", *Prog. Nucl. Energy*, 18, 227 (1986).

[8] WIMS7 User Guide, AEA Technology Report ANSWERS/WIMS(95)4, June (1996).

**Figure 1. The nuclear fluidised bed reactor with quarter of the domain removed to reveal the inner fluidised bed core. The core is surrounded by dense nuclear graphite.**

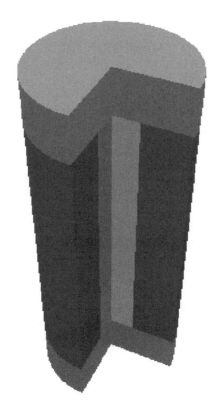

**Figure 2. Sensitivity of $K_{eff}$ for different expanded bed heights and different temperatures**

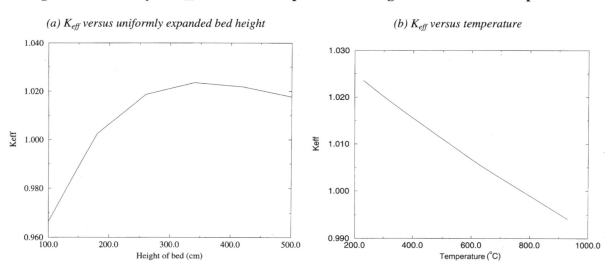

(a) $K_{eff}$ versus uniformly expanded bed height

(b) $K_{eff}$ versus temperature

# Figure 3. Main dynamic parameters

*(a) Fission rate and accumulative fission versus time*          *(b) Maximum temperature versus time*

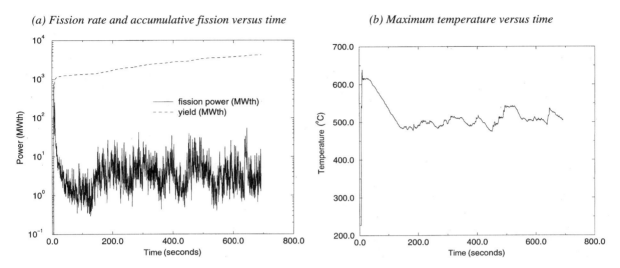

# Figure 4. Various fields at 80 seconds. The diagrams show a cross-section of the inner core of the reactor modelled in rz-geometry.

(a) Volume fraction          (b) Liquid temperature          (c) $1^{st}$ delayed group          (d) $6^{th}$ delayed group

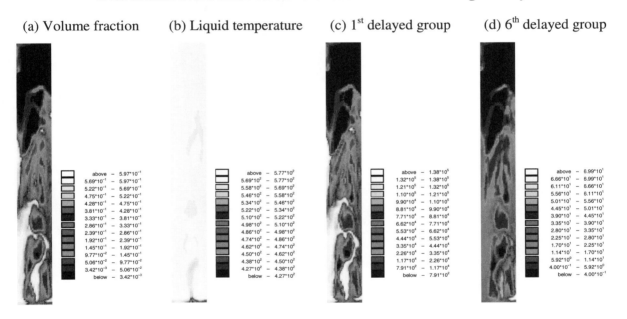

**Figure 5. Simulated 3-D nuclear fluidised bed reactor at 22 seconds into the simulation. The particle volume fraction is shown in the inner reactor cavity on five plains through the reactor.**

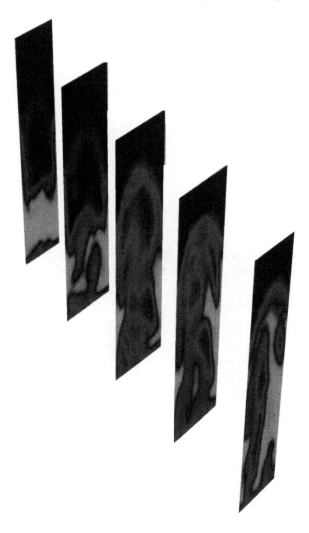

# DESIGN AND PERFORMANCE OF INNOVATIVE FUELS

*Chairs: R. Thetford, Y-W. Lee, K. Bakker*

# EXTENSIVE CHARACTERISATION OF A MATERIAL FOR UNDERSTANDING ITS BEHAVIOUR AS A NUCLEAR FUEL: THE CASE OF A ZIRCONIA-PLUTONIA INERT-MATRIX FUEL

**C. Degueldre, F. Ingold, C. Hellwig**
LWV, Paul Scherrer Institute
CH-5232 Villigen-PSI, Switzerland

**S. Conradson**
MST, Los Alamos National Laboratory
NM 87545, USA

**M. Döbeli**
PSI-ETH-Hönggerberg
CH-8093 Zurich, Switzerland

**Y.W. Lee**
KAERI
P.O. Box 105, Yuseong, Taejon, 305-600 Korea

## Abstract

Recent developments of an inert-matrix fuel (IMF) have led to the selection of yttria-stabilised zirconia (YSZ) doped with erbia and plutonia at PSI, Switzerland [1,2]. This IMF is foreseen to utilise plutonium in LWRs and to destroy it in a more effective manner than MOX. After utilisation in LWR, the plutonium isotopic vector in the spent IMF foreseen for direct disposal will be devaluated far beyond the standard spent fuel. The physical properties of the material depend on the choice of stabiliser as well as on other dopants, e.g. burnable poison or fissile material. As a result of an iterative study, a $(Er,Y,Pu,Zr)O_2$ solid solution with a defined fraction of fissile and burnable poison was selected. Among others this material has been fabricated for irradiation experiments in the Petten High Flux Reactor [3], and in the Halden Reactor [4]. For the material qualification, relevant fuel properties were considered. Among them, the material lattice parameter, fuel and component densities, micro/nano structural studies, thermal conductivity, stability under irradiation, efficient retention of fission products and solubility are key properties for the fuel in reactor as well as for the geologic disposal of the spent fuel.

## Fabrication of the fuel material

Dopants currently used in the industry are for example $Y_2O_3$ or $Ln_2O_3$. In addition to yttria and erbia, plutonium dioxide was added to the zirconium dioxide matrix. Two fabrication routes were followed for the preparation of the IMF material. The first applied a wet route including co-precipitation of the oxi-hydroxide phase by internal gelation [5] starting with the nitrate solutions. The second applied a dry route using an attrition-milling unit adapted to the zirconia material [6] starting with the powder of the constituent oxides. The materials were then compacted and sintered under controlled atmosphere.

## Lattice parameter modelling and determination by XRD

The densities of the components, the sample and of the material are key parameters for fuel qualification. The latter is calculated from the crystallographic structure. The lattice parameter of cubic zirconia strongly depends on the choice of stabiliser and other dopants, i.e. burnable poison. However, the amount of added fissile material is crucial. Kim's [7] empirical equation [8] and Vegard law are used to calculate the lattice parameters of the IMF material and compared with the experimental data obtained by XRD analysis, e.g. Degueldre and Conradson [9]. The agreement between theoretical and experimental data is good (see Figure 1). The data may be used to calculate the theoretical density of the material or of constituents as well as relevant parameters such as relative density or porosity. This work was recently completed for porosity and porous feature determination by applying complementary techniques such as ceramography and X-ray tomography [10].

**Figure 1. Testing experimental data and formal relationships for quaternary zirconias**
**$Er_{0.05}Y_{0.100}(Ce/Pu)_xZr_{0.85-x}O_{1.925}$ for the lattice parameter (a) and for the theoretical density (TD)**

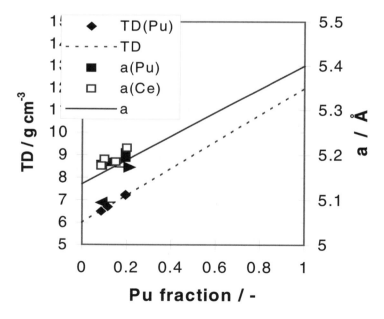

## Atomic environment analysis by XAS and reconstruction

Conventional crystallographic analysis of YSZ may show oxygen displacement along lattice vectors and cation disorder. However, because diffraction is only sensitive to the average long-range order and biased towards periodic and symmetric distortions, it is essential to supplement it with probes of short-range order that may reveal additional, element-specific details about local ordering. This information is gained by X-ray absorption spectroscopy (XAS) on zirconia samples utilising synchrotron radiation [11]. Figure 2 shows the impact of U or Ce dopants on the cation environment in quaternary $Er_{0.05}Y_{0.10}(Ce/U)_{0.10}Zr_{0.75}O_{1.925}$ solid solution. It is striking to see that the Ce/U substitution also has an impact on the radial function of the other cations.

**Figure 2. Radial distribution functions (rdf) around indicated elements in the quaternary $Er_{0.05}Y_{0.10}(Ce/U)_{0.10}Zr_{0.75}O_{1.925}$ CSZ compounds, showing element specific changes in local environments resulting from substituting Ce by U**

*The larger size of U(IV) relative to Ce(IV) [cf. nearest neighbour (nn) O positions in Ce/U plot] results in an expansion of the second and third nn for Zr and Er, whose rdfs are otherwise very similar in both the Ce and U compounds. The substantial differences in the Y environment, including splitting of the O shell at 2.3 Å and increased separation of the more distant nn O set, and in Ce/U environment including the splitting of the second shell cations, imply strong interaction between the Y and U sites that results in preferential Y-U second nn aggregation.*

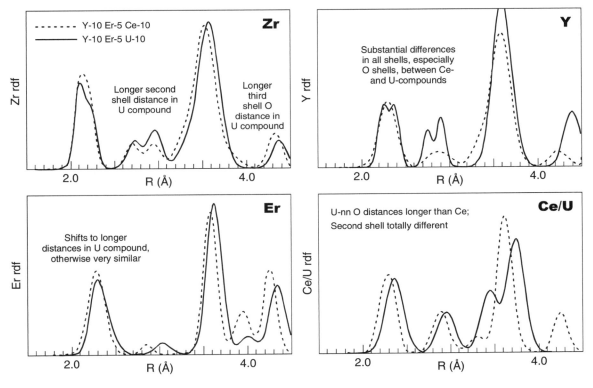

A special role for Y in co-operative behaviour is also implied by the radical changes in its radial distribution function prompted by the replacement of Ce with U. Atomic scale collective rather than isolated behaviour and resulting nanoscale structures do appear to be an important parameter in cubic stabilised zirconia.

In summary, in the zirconia-based solid solutions, some metal ions are not randomly, isomorphically substituted into the lattice but instead have tendencies to form nanometer scale clusters or networks and to yield distortion as depicted in Figure 3. In the zirconia IMF with Ce, U or calculated for Pu, distortions are described around the Y ions for example. This type of nanometer-scale heterogeneity is typically coupled to the phase stability and to other properties such as thermal conductivity.

**Figure 3. Suggested unit cell for doped cubic zirconia: $Ce_{0.10}Y_{0.10}Er_{0.05}Zr_{0.75}O_{1.925}$**

*Note: The displacement pattern of oxygens vs. a zirconium is presented, the vacancy is depicted as an empty sphere, and the considered zirconium cation is surrounded by seven oxygens and moved toward six oxygens creating a local distortion. Similar distortions are recorded for the U-substituted material and are expected for the Pu-doped IMF.*

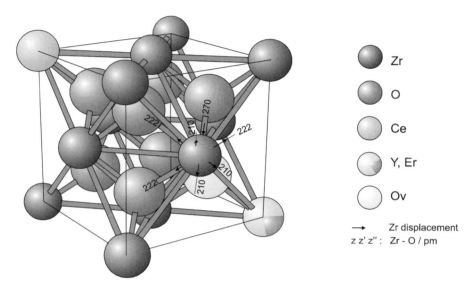

## Thermal conductivity

The thermal conductivity is simply the product of the density, the specific heat capacity (both intrinsic thermodynamic properties) and of the thermal diffusivity. The latter was measured as a function of the temperature. The inverse of the diffusivity of Pu-IMF follows a linear trend with temperature in the range 500 to 2 000 K. The thermal conductivity of the (Pu,Ce)-IMF samples lies between 2.2 and 1.5 W $K^{-1}$ $m^{-1}$ (see Figure 4). These values are much smaller than for pure dense zirconia, which is comparable to that of urania. The addition of yttria, definitively required for the stabilisation of zirconia, has a tremendous effect on the thermal conductivity of zirconia at low temperature. The effect is comparable to that observed by the addition of gadolinia to urania [12]. Since the thermal conductivity of a zirconia-based inert-matrix fuel is rather low, other fuel designs such as cercer or cermet are suggested for optimum utilisation of the fissile material in LWR. The solid solution IMF may also be utilised as hollow pellet for safety reasons.

## Fission-product retention

To study the retention of caesium, xenon and iodine in IMF their diffusion properties in inert-matrix are investigated. To quantify their diffusion properties, Xe, Cs and I were implanted into yttria-stabilised zirconia respectively to depths of around 100 and 200 nm from the surface. These implantations in the solid solution did not amorphise the material as described earlier, e.g. Degueldre, *et al.* [1]. After successive heat treatments to a maximum temperature of 1 973 K, quantitative depth profiles were determined by Rutherford backscattering. No profile modification by diffusion was observed for Xe [14]. For Cs and I the diffusion coefficient follows an Arrhenius trend with the temperature (Figure 5). The behaviour of Xe was investigated at the sub-nanoscopic level and compared with results obtained with zirconia samples implanted with Cs or I, as well as with Xe, Cs and I in $UO_2$ (see Figure 5).

# Figure 4. Thermal conductivity of IMF samples compared to MOX and pure dioxide material

*The relative density of the Ce IMF $Er_{0.07}Y_{0.10}Ce_{0.15}Zr_{0.68}O_{1.913}COP$ and the relative density of the Pu IMFs $Er_{0.04}Y_{0.14}Pu_{0.09}Zr_{0.73}O_{1.91}ATT$ and $Er_{0.04}Y_{0.14}Pu_{0.08}Zr_{0.68}O_{1.91}COP$ are respectively 90.0, 94.4 and 86.3 %. Note that the conductivity decreases with the dopant concentration and with the porosity, also the peculiar behaviour of the Ce-IMF with conductivity increase due to photonic conductivity. $ZrO_2$ data from Raghavan, et al. [13].*

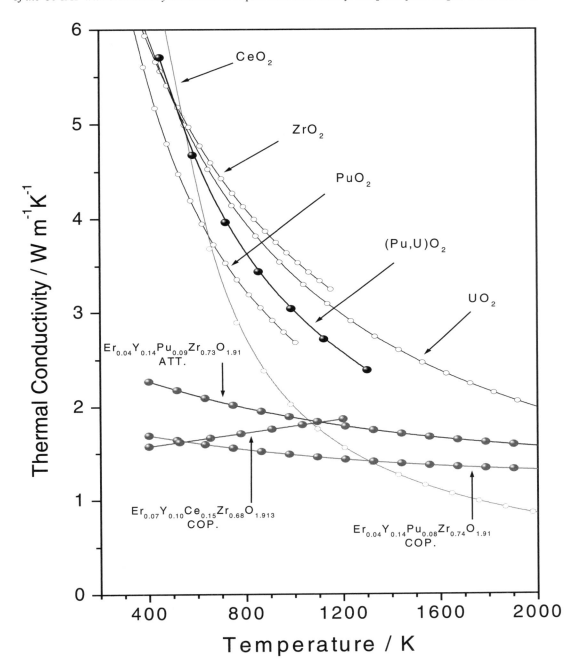

**Figure 5. Comparison of diffusion coefficient (D) for various elements (relevant fission products) with the intrinsic ion diffusions in zirconia**

*Note: For (Y,Zr)O₂, Cs and I follow Arrhenius law, the diffusion coefficient of Xe is below the detection limit. Data for UO₂ from Busker, et al. [15] and Prussin, et al. [16].*

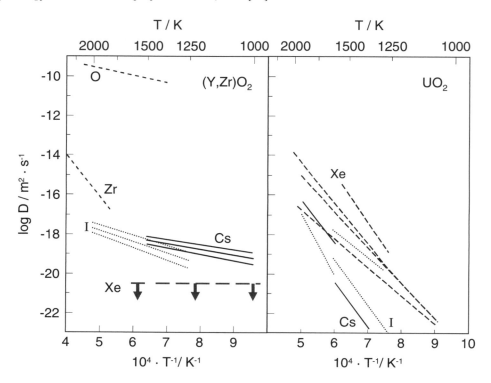

## Zirconia-based spent fuel as waste material for geological disposal

The zirconia inert-matrix offers from its chemical properties a double advantage: first its solubility is extremely low and secondly it is not redox sensitive. As it can be seen in Figure 6, its solubility is slightly enhanced by the presence of carbonate complexes in neutral condition [17]; however, tetravalent uranium behaves similarly. Zirconia is redox insensitive, which is advantageous in comparison to uranium that in oxidising conditions may be oxidised to the hexavalent state, increasing the solubility by several orders of magnitude.

## Concluding remarks

As a consequence of the extensive experimental studies performed above, a coherent picture is emerging on the relevant properties of the zirconia-based IMF material. The suggested zirconia-based IMF has a very strong thermodynamic stability which it maintains even under irradiation conditions. The material density (or porosity) and the density of its constituents may be calculated and derived from the single components concentration utilising the experimental lattice parameter gained by XRD. The local distortions identified by XAS yield decreases of the thermal conductivity, which was measured for the Pu-IMF and was found rather low compared to pure zirconia or to classical MOX. However, the retention of safety-relevant fission products and the low solubility of the redox-insensitive zirconia IMF make this fuel material very attractive for plutonium utilisation in LWR applying a "once-through then out" strategy.

**Figure 6. Solubility of monoclinic zirconia and oxy-hydroxide precipitate as a function of total carbonate concentration: comparison of experimental data with computed solubility curves**

*[]: concentration in mole per litre (M); beta0(1,I) stands for the stability constant of the complex $Zr(CO_3)_i^{(4-2i)}$. Note that the pore water around the waste package is expected to reach a $\log [CO_3]_{tot}$ between -3 and -2 corresponding to a solubility ranging from $10^{-9}$ to $10^{-8}$ M.*

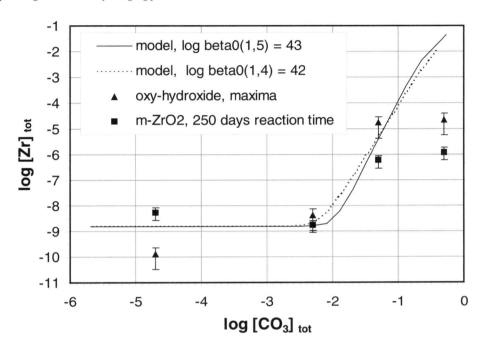

**REFERENCES**

[1]    C. Degueldre, U. Kasemeyer, F. Botta, G. Ledergerber, "Plutonium Incineration in LWR by a Once-through Cycle with a Rock-like Fuel", *Mater. Res. Soc. Proc.*, 412, 15 (1996).

[2]    C. Degueldre, J.M. Paratte, "Basic Properties of a Zirconia-based Fuel Material for Light Water Reactors", *Nucl. Technol.*, 123, 21 (1998).

[3]    R.P. Schram, K. Bakker, H. Hein, J.G. Boshoven, R. van der Laan, C. Sciolla, T. Yamashita, C. Hellwig, F. Ingold, R. Conrad, S. Casalta, "Design and Fabrication Aspects of a Plutonium Incineration Experiment using Inert Matrices in a "Once-through-then-out" mode", Proc. 6th IMF, *Progr. Nucl. Energy*, 38, 259 (2001).

[4]    U. Kasemeyer, C. Hellwig, Y.W. Lee, D.S. Song, G.A. Gates, W. Wiesenack, "The Irradiation Test of Inert-matrix Fuel in Comparison to Uranium Plutonium Mixed Oxide Fuel at the Halden Reactor", Proc. 6th IMF, *Progr. Nucl. Energy*, 38, 309 (2001).

[5]   G. Ledegerber, C. Degueldre, P. Heimgartner, M. Pouchon, U. Kasemeyer, "Inert-matrix Fuel for the Utilisation of Plutonium", Proc. 6th IMF, *Progr. Nucl. Energy*, 38, 301 (2001).

[6]   Y.W. Lee, H.S. Kim, S.H. Kim, C.Y. Joung, S.H. Na, G. Ledergerber, P. Heimgartner, M. Pouchon, M. Burghartz, "Preparation of Simulated Inert-matrix Fuel with Different Powders by Dry Milling Method", *J. Nucl. Mater.*, 274, 7 (1999).

[7]   D.J. Kim, "Lattice Parameters, Ionic Conductivities and Solubility Limits in Fluorite-structure $MO_2$ Oxide (M = Hf, Zr, Ce, Th, U) Solid Solutions", *J. Am. Ceram. Soc.*, 72, 1415 (1989).

[8]   M. Burghartz, G. Ledergerber, F. Ingold, P. Heimgartner, C. Degueldre, "X-ray Diffraction Analysis and Data Interpretation of Stabilised Inert-matrix, Impact on the Material Qualification", Proc. 6th IMF, *Progr. Nucl. Energy*, 38, 247 (2001).

[9]   C. Degueldre, S. Conradson, "Characterisation of Ternary and Quaternary Zirconias by XRD and EXAFS, Result Comparison and Data Modelling", *Appl. Phys. A* (in press) (2001).

[10]  C. Degueldre, M. Pouchon, M. Streit, O. Zaharko, M. DiMichel, "Analysis of Porous Features in Zirconia-based Inert-matrix, Impact on the Material Qualification", Proc. 6th IMF, *Progr. Nucl. Energy*, 38, 241 (2001).

[11]  S. Conradson, C. Degueldre, F.J. Espinosa, S. Foltyn, K. Sickafus, J. Valdez, P. Villela, "Complex Behavior in Quaternary Zirconias for Inert-matrix Fuel: What does this Material Look Like at the Nanometer Scale?", Proc. 6th IMF, *Progr. Nucl. Energy*, 38, 221 (2001).

[12]  M. Hirai, S. Ishimoto, "Thermal Diffusivities and Thermal Conductivities of $UO_2$-$Gd_2O_3$," *J. Nucl. Mater.*, 28, 995 (1991).

[13]  S. Raghavan, H. Wang, R. Dinwiddie, W. Porter, M. Mayo, "The Effect of Grain Size, Porosity and Yttria Content on the Thermal Conductivity of Nanocrystalline Zirconia", *Script. Materialia*, 39, 1119 (1998).

[14]  C. Degueldre, M. Pouchon, M. Döbeli, K. Sickafus, K. Hojou, G. Ledergerber, S. Abolhassani-Dadras, "Behaviour of Implanted Xenon in Yttria-stabilised Zirconia as Inert-matrix of a Nuclear Fuel", Proc. 102nd Am. Ceram. Soc., *J. Nucl. Mater.*, 289, 115 (2001).

[15]  G. Busker, R.W. Grimes, M.R. Bradford, "The Diffusion of Iodine and Caesium in $UO_{2\pm x}$ Lattice", *J. Nucl. Mater.*, 279, 46 (2000).

[16]  S. Prussin, D. Olander, W. Lau, L. Hansson, "Release of Fission Products (Xe, I, Te, Cs, Mo and Tc) from Polycrytalline $UO_2$," *J. Nucl. Mater.*, 154, 25 (1988).

[17]  M. Pouchon, E. Curti, C. Degueldre, L. Tobler, "The Influence of Carbonate on the Solubility of Zirconia: New Experimental Data", Proc. 6th IMF, *Progr. Nucl. Energy*, 38, 443 (2001).

# ANNULAR PLUTONIUM-ZIRCONIUM NITRIDE FUEL PELLETS

**Marco Streit\*, Franz Ingold**
Paul Scherrer Institute, Laboratory for Materials Behaviour
CH-5232 Villigen PSI, Switzerland

**Ludwig J. Gauckler**
Swiss Federal Institute of Technology, Department of Materials
CH-8092 Zurich, Switzerland

**Jean-Pierre Ottaviani**
Commissariat à l'Énergie Atomique, CE Cadarache
F-13108 Saint-Paul-lez-Durance Cedex, France

## Abstract

With direct coagulation casting (DCC), it may be possible to fabricate specially shaped fuel pellets to reach higher burn-ups. This work aims at using the DCC process for mixed plutonium-zirconium nitride annular pellets. Zirconium nitride has been proposed as an inert matrix material to burn plutonium or to transmute long-lived actinides in accelerator-driven subcritical systems or fast reactors. Plutonium-zirconium nitride microspheres were fabricated using a combination of the sol-gel method and carbothermic reduction. Zirconium nitride annular pellets were shaped using DCC with commercially available zirconium nitride.

---

\* This work is performed as the Ph.D. thesis of Marco Streit.

## Introduction

To burn plutonium or to transmute long-lived actinides in accelerator-driven subcritical systems (ADS) or fast reactors (FR), zirconium nitride has been proposed as inert matrix material by several authors [1-3]. The nitrides were synthesised by carbothermic reduction of the oxides. Therefore different non-plutonium containing solid solution nitrides have been produced to investigate their chemical and physical behaviour before producing the plutonium containing fuel.

One of the most interesting questions in the production of nuclear fuel is to have the possibility to fabricate specially shaped fuel pellets (e.g. barrel with a hole) to reach higher burn-ups. To obtain an annular pellet (as a more simple shape) a new shaping method, called direct coagulation casting (DCC), used by Gauckler, *et al.* [4-10], was introduced.

In addition, thermodynamic calculations for the system Pu-Zr-N were carried out to predict the behaviour of the material.

## Theoretical background

### Production routes

To yield mixed nitride powder three different routes (Figure 1) were investigated. The first route is using the PSI sol-gel method [1,11,12], which ends up with plutonium-zirconium oxide microspheres containing carbon black. The spheres were converted into the nitride by carbothermic reduction under nitrogen atmosphere. The second route uses directly a mixture of plutonium oxide powder, zirconium oxide powder and carbon black for the carbothermic reduction. In the third route commercially available zirconium nitride powder was mixed with plutonium nitride powder obtained from the reduction of the oxide with carbon. All the resulting nitride powders were used to cast annular pellets via the DCC. As a back-up shaping method the conventional method is envisaged.

**Figure 1. Flow sheet of the production route of annular (Pu,Zr)N barrels**

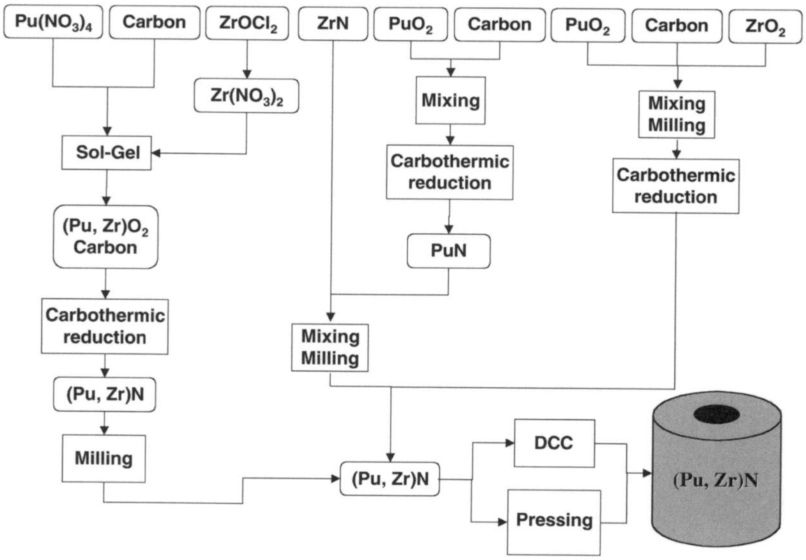

## Sol-gel reaction

Ledergerber [1,13,14] has given a good overview of the so-called "sol-gel" process. It is based on the internal gelation of pure Newtonian solutions (feed solutions) containing urea, hexamethylen-eteramine (HMTA), nitric acid and a metaloxonitrate. In the following the process is explained using uranium.

In a first step urea is added to an uranylnitrate solution to complex the uranyl ions:

$$UO_2(NO_3)_2 \quad + \quad H_2N-\overset{\overset{\displaystyle O}{\|}}{C}-NH_2 \quad \rightleftharpoons \quad [UO_2(NH_2CONH_2)_2](NO_3)_2$$

This prevents the hydrolysis of the uranyl ions as well as the complexation by the HMTA, which would end up in a precipitation in the feed solution. Droplets of the feed solution are heated up in hot silicon oil which accelerates the protonation and decomposition of HMTA:

$$(CH_2)_6N_4 \quad \underset{}{\overset{+ HNO_3}{\rightleftharpoons}} \quad (CH_2)_6N_4 \cdot HNO_3 \quad \underset{- HNO_3}{\overset{+ 10H_2O}{\longrightarrow}} \quad 4NH_4OH + 6CH_2O$$

Ammonium is produced during this decomposition reaction, which initialises the gelation process. The oxide precipitates and the droplets become solid:

$$3[UO_2(NH_2CONH_2)_2](NO_3)_2 + 7NH_4OH + H_2O$$

$$\downarrow$$

$$(UO_3)_3 \cdot NH_3 \cdot 5H_2O + 6NH_4NO_3 + 3CO(NH_2)_2$$

$$2[UO_2(NH_2CONH_2)_2](NO_3)_2 + 5NH_4OH$$

$$\downarrow$$

$$(UO_3)_2 \cdot 2H_2O + 4NH_4NO_3 + 2CO(NH_2)_2$$

The obtained microspheres are washed to eliminate the silicon oil and other reaction substances. Ammonia and water are eliminated during calcination of the yield material.

## Carbothermic reduction

Different authors [1,2,13,15-18] have summarised the assumed reactions during the carbothermic reduction of uranium or plutonium dioxide in a nitrogen/hydrogen atmosphere. Converted to a general metal oxide the following equation was obtained.

$$\Delta_r G° = m1\Delta_f G°\left(^1MN\right) + m2\Delta_f G°\left(^2MN\right) - \Delta_f G°\left(\left(^1M_{m1}\right)O_{\frac{m1(o+u)}{m}}\right) - \Delta_f G°\left(\left(^2M_{m2}\right)O_{\frac{m2(o+u)}{m}}\right)$$

$$+ \frac{o}{m}\Delta_f G°(CO) + \frac{u}{m}\Delta_f G°(H_2O) + \frac{o}{m}RT\ln\frac{p(CO)}{p°} + \frac{u}{m}RT\ln\frac{p(H_2O)}{p°} - \frac{1}{m}RT\ln\frac{(m2*m1*m2)}{(m1*m2)^m}$$

$$- \frac{u}{m}RT\ln\frac{p(H_2)}{p°} - \frac{1}{2}RT\ln\frac{p(N_2)}{p}$$

With this equation, using the values given in Table 1, it is possible to calculate the Gibbs free energy of reaction for the carbothermic reduction. Figure 2 shows TGA-GC measurements of the inactive materials. The reaction starts at about 1 200°C where the weight loss (upper diagram) and the gas release (lower diagram) starts; this fits very well with the calculations, using the equation, shown in Figure 3. The reaction maximum is at about 1 400°C.

**Table 1. Symbols and variable in the used equation**

| Symbol/variable | Meaning/stands for |
|---|---|
| M | $^1M_{m1}\,^2M_{m2}$ or $^2M$ |
| $^1M$ | Zr |
| $^2M$ | Zr, Ce, Nd, U, Pu |
| m | 2 for $^2M$ = Nd otherwise 1 |
| m1, m2 | m1 + m2 = 1, here: m1 = 0.8, m2 = 0.2 |
| o | 3 for $^2M$ = Nd otherwise 2 |
| u, x, y | >0; <1 |

**Figure 2. TGA-GC measurements of mixed zirconium oxides containing 20 at.% Zr, Ce, Nd or U**

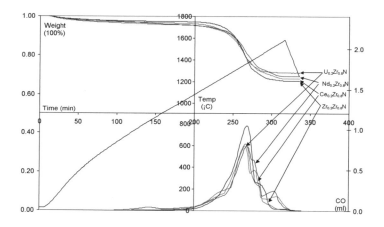

**Figure 3. Gibbs energy of reaction calculation for carbothermic reduction of mixed zirconium oxides containing 20 at.% Zr, Ce, U or Pu**

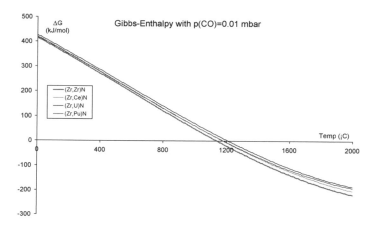

*Direct coagulation casting*

The direct coagulation process is based on the destabilisation of electrostatic stabilised ceramic suspension by time-delayed *in situ* reactions [4]. Enzyme catalysed reactions (as urea/urease hydrolysis) can be used to create salt to increase ionic strength and/or to shift the pH of the suspension to the isoelectric point (IEP) of the powder. In both cases the viscosity of the ceramic suspension increases and a wet green body is obtained. The mechanical strength of these bodies depends on the solid loading and the particle size of the used suspension. It can be increased by using small additions of alkaline swellable thickeners (AST). These macromolecules change their conformation during a pH shift to the alkaline regime during the enzymatic reaction. The reaction kinetics depends on the temperature and the enzyme concentration used. In this work the hydrolysis of urea by the enzyme urease is used. At temperatures between 0 and 60°C in the pH range 4 to 9 it is according to the following reaction: $NH_2CONH_2 + H_2O \xrightarrow{\text{urease}} HCO_3^- + 2NH_4^+$. Thus the enzyme produces 3 mol ions from 1 mol substrate. This reaction increases the ion concentration in the suspension and/or shifts the pH from acid regime to 9.

With this method it is possible to cast complex shapes with ceramic powder suspensions. The ceramic processing routes are well developed and reported for alumina powders and the urea/urease hydrolysis. Other enzyme systems are possible but not as successful when used for the DCC method [4,19-24].

In this work the DCC method is adapted to potential nuclear fuel using zirconium nitride powders as inert matrix. A major problem in the nuclear field is the influence of irradiation on the enzyme. It is known that "…all enzymes can be inactivated when irradiated in solution, although the radiation doses necessary to inactivate different enzymes vary greatly…" [25]. Before working with plutonium containing samples the influence of α-irradiation and of heavy metals on urease will be tested.

## Experimental results and discussion

*Sol-gel*

In a first step HMTA and urea were dissolved in water, and carbon black and a dispersant called "Lyokol" were added; after dispersing the "HMTA-C-solution" was ready to use. Metallic solutions were prepared by dissolving the metallic nitrate salt in water. For gelation the different metallic solutions were mixed in a fixed ratio to obtain the "M-solution" with the calculated ratio of 80 at.% zirconium and 20 at.% of a second metal (Zr, Ce, Nd, U, Pu) in the desired nitride spheres.

The HMTA-C-solution was cooled down to about 1°C in the feed reservoir of the gelation system. While stirring the M-solution was slowly added. This feed solution with the ratios shown in Table 2 was stirred and cooled during the whole gelation.

**Table 2. Used ratios for feed solution**

|  | Ratio |
|---|---|
| HMTA/Metal | 1.20 mol% |
| Urea/Metal | 1.05 mol% |
| Carbon/Metal | 2.50 mol% |

The solution was passed through a vibrated nozzle with a diameter of 550 or 600 microns. The droplets obtained fell in silicon oil with a temperature of 110°C. The gelation reaction takes place in the warm oil and microspheres were thus obtained. The spheres were washed with actrel, ammonia and methanol to remove silicon oil, water and organics. The green microspheres were dried in a rotary kiln at 110°C. The dried material was calcined at 600°C for four hours under an argon-hydrogen (8%) atmosphere.

## Carbothermic reduction

The calcined microspheres were heated up with 300°C/h to a reaction level at 1 400°C. After eight hours the furnace was heated up to 1 600°C for an additional eight hours. This whole procedure was done in a nitrogen-hydrogen (7%) atmosphere. The cooling down was done in an argon-hydrogen (8%) atmosphere. Figure 4 shows the measured CO release during the carbothermic reduction of the inactive test material.

**Figure 4. CO release during carbothermic reduction of inactive metal zirconium oxide**

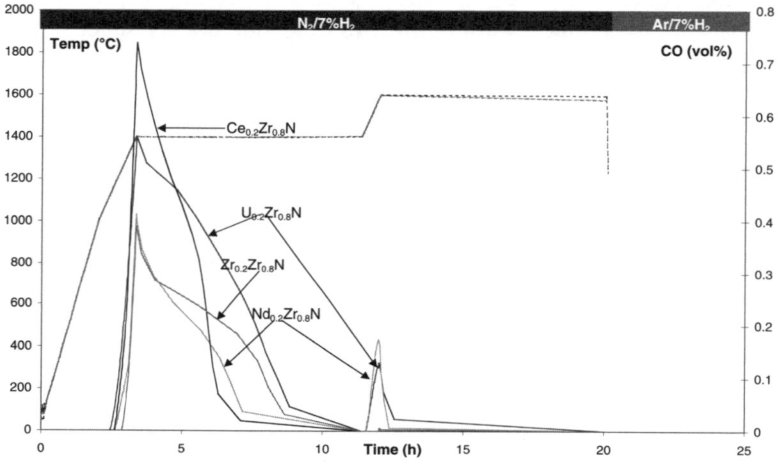

The golden microspheres of $ZrN$, $Ce_{0.2}Zr_{0.8}N$, $Nd_{0.2}Zr_{0.8}N$, $U_{0.2}Zr_{0.8}N$, $Pu_{0.2}Zr_{0.8}N$ have densities between 91 and 103% TD. The microstructures of the spheres (Figure 5) show rather big pores due to the high gas release during reduction and a second (black) phase. This phase consist most likely of unreacted material, e.g. of metal oxide and carbon black. This is emphasised by the results of the chemical analysis of two inactive samples shown in Table 3. The neodymium-containing material was oxidised during two months of the storage time. This material seems to be highly reactive with oxygen from the air. It is predicted that the material could not be used with the DCC method employing water.

## Direct coagulation casting

### Experiments to prepare DCC

The IEP of commercially available zirconium nitride powder (AlfaAesar, 99.5%, $Hf < 3\%$, Lot 63560) was found, measuring the electric sonic amplitude (ESA), at pH 5.6. This is too low to use the urea/urease hydrolysis when shifting the pH. However, as we also can create salt at the buffer pH=9 of the urea decomposition reaction the destabilisation of the suspension by the ionic strength was used.

**Figure 5. Ceramography of U$_{0.2}$Zr$_{0.8}$N microspheres**

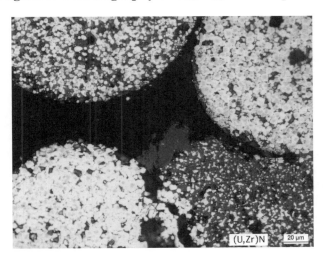

**Table 3. C, O, N analysis of ZrN and Ce$_{0.2}$Zr$_{0.8}$N microspheres**

| Material | O | C | N | Calc. N |
|---|---|---|---|---|
| ZrN | 1.22 wt.% | 2.84 wt.% | 9.76 wt.% | 13.31 wt.% |
| Ce$_{0.2}$Zr$_{0.8}$N | 4.79 wt.% | 2.67 wt.% | 6.97 wt.% | 12.18 wt.% |

A foaming effect was found in early experiments with solid loadings between 60 and 64 vol.%. The gas release during the urea hydrolysis was not possible because of the high surface tension of the suspension. In the following casting experiments solid loadings between 58 and 60 vol.% were used.

Figure 6 shows the effect of increasing the ionic strength by salt addition on the viscosity of zirconium nitride suspension. There is a slight decrease of viscosity at the beginning of the addition. This effect could be used to shorten the solidification time during DCC by adding a quantity of salt (e.g. ammonium chloride) before using the urea hydrolysis.

**Figure 6. Viscosity versus ion concentration of a
64 vol.% zirconium nitride suspension during salt addition**

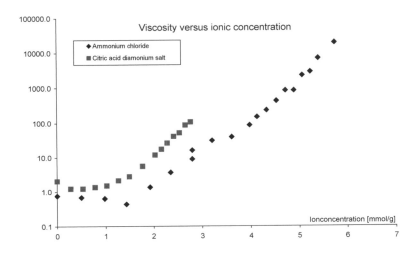

*DCC with rising the ionic concentration*

For casting experiments commercial (AlfaAesar) zirconium nitride powder was added to a solution made from urea (8.75 mol%), citric acid diammonium salt (1 mol%) and water. The slurry was dispersed with an ultrasonic horn or a POLYTRON-disperser to yield a 60 vol.% zirconium nitride suspension with pH between 7.5 and 8.5. 1 mmol ammonium chloride per gram suspension was added to increase the ionic concentration.

After adding urease (10-20 units per gram powder) the suspension was shortly stirred and left for about 30 minutes before filling the greased shaping mould with the material. The mould was protected against drying for 24 hours. After that time the pellets were dried in the mould for another 24 hours. Then the pins were removed and the annular pellets were de-moulded. The green pellets were dried in air for three days and afterwards dried at 100°C for another 24 hours. Figure 7 shows the first green annular zirconium nitride pellets obtained with this method. Their microstructure still contained macropores due to a too-high concentration of urease and therefore a too-fast urea decomposition reaction producing too much $CO_2$ locally.

**Figure 7. ZrN annular green pellets shaped with DCC
(outer diameter is 15 mm, inner diameter is 3 mm)**

*DCC with using AST*

Commercial zirconium nitride was added to an emulsion made from urea (8.75 mol%), citric acid diammonium salt (1 mol%), Aculyn (2.5 wt.% of polymere in emulsion) and water. After dispersing the suspension four droplets of octanol were added to reduce the surface tension. After adding urease the suspension was shaken for 30 minutes before filling the moulds.

**Figure 8. First ZrN green pellets shaped with DCC
using AST (outer diameter 15 mm, inner diameter 3 mm)**

The moulds were protected against drying out for 19 hours. After drying for four days the pins were removed and after an additional day the pellets were de-moulded. After drying two days in the exsiccator the pellets were dried for 24 hours at 110°C.

## Conclusions and outlook

It has been shown that the calculated Gibbs energy for carbothermic reduction fits well with the experimental results. The main parts of the planed production routes, e.g. the sol-gel process, the carbothermic reduction and the ceramic shaping method DCC, have been tested. Direct coagulation casting with zirconium nitride works in principle, but has to be modified for the nuclear application. The main problem in the future will be the stability of the urease enzyme in a suspension containing an α-emitter. The next steps in this work will concentrate on the optimisation of the DCC method for radioactive material and the production of annular pellets for irradiation experiments.

## REFERENCE

[1]  G. Ledergerber, "Internal Gelation for Oxide and Nitride Particles", Japan Atomic Energy Research Institute, Tokai-mura, JAERI-Review 96-009, 27, pp. 1-21 (1996).

[2]  S. Daumas, "Étude et réalisation de support-matrices inertes par le procédé sol-gel pour l'incinération des actinides mineures", in Univertité Aix-Marseille I, Marseille, p. 130 (1997).

[3]  M. Burghartz, et al., "Some Aspects of the use of ZrN as an Inert Matrix for Actinide Fuels", Journal of Nuclear Materials, 288, pp. 233-236 (2001).

[4]  L.J. Gauckler, T.J. Graule and F.H. Baader, "Ceramic Forming Using Enzyme Catalysed Reactions", Materials Chemistry and Physics, 61, pp. 78-102 (1999).

[5]  T.J. Graule, et al., "Direct Coagulation Casting (DCC) – Fundamentals of a New Forming Process for Ceramics", Ceramics Transactions, 51, pp. 457-461 (1995).

[6]  T.J. Graule, F.H. Baader and L.J. Gauckler, "Enzyme Catalysis of Ceramic Forming", Journal of Material Education, 16, pp. 243-267 (1994).

[7]  T.J. Graule, F.H. Baader and L.J. Gauckler, "Shaping of Ceramic Green Compacts Direct from Suspensions by Enzyme Catalyzed Reactions", Cfi/Ber, DKG, 71(6), pp. 317-323 (1994).

[8]  T.J. Graule, F.H. Baader and L.J. Gauckler, "Casting Uniform Ceramics with Direct Coagulation", Chemtech, pp. 31-37 (1995).

[9]  T.J. Graule and L.J. Gauckler, "Process for Producing Ceramic Green Compacts by Double Layer Compression", in US Patent Office, USA, USP 5,667,548 (1997).

[10]  T.J. Graule, L.J. Gauckler and F.H. Baader, "Verfahren zur Herstellung keramischer Grünkörper durch Doppelschicht-Kompression", in Schweizer Patentamt, Schweiz, CH 685 493 A5 (1995).

[11]   G. Ledergerber, "Improvements of the Internal Gelation Process", American Nuclear Society, Trans ANS, pp. 55-56 (1982).

[12]   G. Ledergerber, "Internal Gelation using Microwaves", International Atomic Energy Agency, Wien, IAEA-TECDOC-352, pp. 165-174 (1985).

[13]   G. Ledergerber, et al., "Preparation of Transuranium Fuel and Target Materials for the Transmutation of Actinides by Gel Coconversion", Nuclear Technology, 114, pp. 194-204 (1996).

[14]   G. Ledergerber, et al., "Experience in Preparing Nuclear Fuel by the Gelation Method", in ENC-4, Transactions Vol 4, Genf, ENS/ANS, pp. 225-232 (1986).

[15]   G. Pautasso, K. Richter and C. Sari, "Investigation of the Reaction $UO_{(2+x)}$ +PuO$_2$ + C + N$_2$ by Thermogravimetry", Journal of Nuclear Materials, 158, pp. 12-18 (1988).

[16]   P. Bardelle, "Synthèse du mononitrure mixte d'uranium-plutonium par réduction des oxydes", in Université Aix-Marseille III, Marseille, p. 114 (1988).

[17]   P. Bardelle and H. Bernard, "Process for the Preparation of Uranium and/or Plutonium Nitride for use as a Nuclear Reactor Fuel", in Europäisches Patentamt, Europa, EP880402257 19880908 (1989).

[18]   P. Bardelle and D. Warin, "Mechanism and Kinetics of Uranium-plutonium Mononitride Synthesis", Journal of Nuclear Materials, 188, pp. 36-42 (1992).

[19]   B. Balzer, M.K.M. Hruschka and L.J. Gauckler, "Coagulation Kinetics and Mechanical Behavior of Wet Alumina Green Bodies Produced via DCC", Journal of Colloid and Interface Science, 216, pp. 379-386 (1999).

[20]   L.J. Gauckler, et al., "Enzyme Catalysis of Alumina Forming", Key Engineering Materials, Trans Tech Publications, Switzerland, 159-160, pp. 135-150 (1999).

[21]   L.J. Gauckler, et al., "Enzyme Catalysis of Ceramic Forming: Alumina and Silicon Carbide", in 9th CIMTEC – World Ceramics Congress, Ceramics: Getting into the 2000s, P. Vincenzini, Part C, Techna Srl., 15-40 (1999)

[22]   M.K.M. Hruschka, et al., "Processing of b-silicon Nitride from Water-based a-silicon Nitride, Alumina and Yttria Powder Suspensions", Journal of the American Ceramic Society, 82(8), pp. 2039-2043 (1999).

[23]   W. Si, et al., "Direct Coagulation Casting of Silicon Carbide Components", Journal of the American Ceramic Society, 82(5), pp. 1129-1136 (1999).

[24]   D. Hesselbarth, E. Tervoort and L.J. Gauckler, "Destabilisation of High Solids Loading Al$_2$O$_3$ Particle Suspension with Alkali Swellable Polymers via Enzyme Assisted Reactions", in Proceedings of the Advanced Research Workshop on Engineering Ceramics, Workshop: Engineering Ceramics 99, "Multifunctional Properties – New Perspectives", Smolenice, Slovakia, P. Sajgalik and L.Z., Key Engineering Materials 175-176, Trans Tech Publications: 31-43 (1999)

[25]   A.P. Casarett, "Radiation Biology", C. University, ed., New Jersey: Prentice-Hall (1968).

# ROCK-LIKE OXIDE FUELS FOR BURNING EXCESS PLUTONIUM IN LWRs

**Toshiyuki Yamashita, Ken-ichi Kuramoto, Hiroshi Akie, Yoshihiro Nakano,
Noriko Nitani, Takehiko Nakamura, Kazuyuki Kusagaya and Toshihiko Ohmichi***
Department of Nuclear Energy Systems, Japan Atomic Energy Research Institute (JAERI)
Tokai-mura, Ibaraki 319-1195, Japan
*Research Organisation for Information Science and Technology
Tokai-mura, Ibaraki 319-1195, Japan

## Abstract

JAERI has performed research on the plutonium rock-like oxide (ROX) fuels and their once-through burning in light water reactors (the ROX-LWR system) in order to establish an option for utilising or effectively disposing of the excess plutonium. Features of the ROX-LWR system are almost complete burning of plutonium and the direct disposal of spent ROX fuels without reprocessing. The ROX fuel is a sort of inert matrix fuel, and consists of mineral-like compounds such as yttria-stabilised zirconia (YSZ), spinel ($MgAl_2O_4$) and corundum ($Al_2O_3$). Several difficulties must be overcome to demonstrate the feasibility of the ROX-LWR system from the reactor physics and the materials science points of view. Described are activities concerning development of particle dispersed fuels, the in-pile irradiation behaviour of these fuels, improvement of the ROX fuel core characteristics and fuel behaviour under reactivity-initiated accident condition.

# Introduction

The intentions of an advanced reactor system such as a fast reactor and a reduced-moderation water reactor are to breed [239]Pu and to effectively utilise limited nuclear resources. On the other hand, plutonium stock in commercial reactor spent fuels amounted to about 1 300 tonnes at the end of 1999 [1], and about 150 tonnes of plutonium of that were recovered by reprocessing. Because of the moratorium on fast reactor programmes, the amount of plutonium stock is becoming significantly excess throughout the world. Moreover, about 100 tonnes of pure weapons-grade plutonium are to be recovered from dismantled warheads in the United State and Russia. Research activities on the inert matrix fuels (IMF) were initiated in an effort to reduce the excess plutonium [2-5].

A possible way to use the excess plutonium seems to be MOX fuels for light water reactors (LWRs). However, this option is not very effective from the plutonium annihilation rate point of view, because a full MOX core will transmute only about 30% of the initially loaded plutonium and a partial loaded MOX core (1/3 MOX + 2/3 $UO_2$) will produce a small amount of plutonium [6]. On the other hand, the IMF can effectively annihilate plutonium; more than 80% of [239]Pu or more than 70% of plutonium in a once-through scenario [2].

The Japan Atomic Energy Research Institute (JAERI) has performed a R&D study on plutonium rock-like oxide (ROX) fuels and their once-through burning in light water reactors (the ROX-LWR system). Features of the ROX-LWR system are almost complete burning of plutonium and the direct disposal of spent ROX fuels without reprocessing. The ROX fuel is a sort of inert matrix fuel and consists of mineral-like compounds such as yttria-stabilised zirconia (YSZ), spinel ($MgAl_2O_4$) and corundum ($Al_2O_3$), which are very stable chemically and physically. Plutonium is incorporated in the YSZ phase making a solid solution. This treatment is of importance from the non-proliferation point of view. Once such a solid solution of plutonium and YSZ is formed, it is practically impossible to separate plutonium from the solid solution. Moreover, plutonium in spent ROX fuels will be degraded and will be useless for weapons. In the present paper an outline of the research concerning ROX-LWR systems is described with emphasis on the recent in-pile irradiation results of ROX fuels.

# ROX fuel development

YSZ is known to be very stable against both neutron and fission fragment irradiations and has a high melting temperature (~2 960 K), and therefore is the most hopeful candidate for the inert matrix. No amorphisation of YSZ was observed in the 60 keV-Xe irradiation test and the YSZ swelling rate was only 0.72% at an Xe dose of $1.8 \times 10^{16}$ cm$^{-2}$ Xe at 925 K [7]. A recent 72 MeV-I irradiation test showed that amorphisation of YSZ was not observed, while the crystal size of YSZ became much smaller [8]. However, its thermal conductivity is extremely low (~2 W/m·K); less than a half of that of $UO_2$. Although the melting temperature of YSZ is higher than the expected fuel temperature (~2 500 K) under normal operating condition with a peaking factor of 2.1 [6], it is desirable to reduce the fuel temperature from the view point of fuel performance such as fission gas release, fuel restructuring and so on. Improvement in the thermal conductivity of YSZ can be attainable by mixing YSZ with spinel or corundum, both of which have higher thermal conductivity. Figure 1 shows thermal conductivities for such materials as YSZ, spinel, corundum and a simulated ROX fuel composite (YSZ:$UO_2$:spinel =54.5:15.2:30.3 in mol%) [9]. The thermal conductivity of $UO_2$ is also depicted for comparison. The measured thermal conductivity of the YSZ/spinel composite is very much improved and is similar to or slightly higher than that of $UO_2$.

## Figure1. Comparison of thermal conductivity of various materials

*Solid line indicates data of UO₂ [10], and dotted line data of corundum [11]*

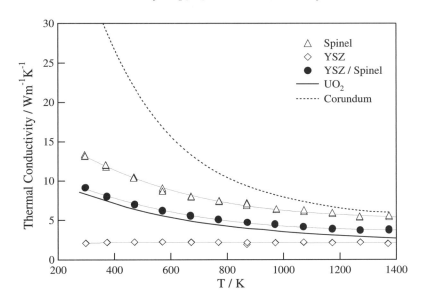

Corundum has the advantage of a high thermal conductivity, as seen in Figure 1. Its high thermal conductivity, however, will be degraded rapidly even by neutron irradiation. Clinard and Parkin reported that the thermal conductivity of corundum decreased to 1/2-1/3 of the original level at a neutron dose of $10^{21}$ cm$^{-2}$ [12]. Neutron irradiation also caused a large swelling of 6-10% [13]. Fission fragments (FFs) caused a great amount of severe damage in corundum and rendered it amorphous. It should be noted that this amorphisation was observed to occur at a relatively low fission density of $2 \times 10^{16}$ cm$^{-3}$ [14]. The eutectic temperature between pure $ZrO_2$ and corundum is 2 150 K, but that of YSZ and corundum has not yet been reported. A low melting temperature of the nuclear fuel leads to difficulties in reactor design. Some disadvantages which might be caused by corundum were also recognised in our previous irradiation study on ROX fuel disks which consisted of YSZ, spinel and a large amount of corundum [15]; about 30% of swelling, a new hibonite formation with plutonium ($PuMgAl_{11}O_{19}$) and the melting temperature decrease by making a ternary eutectic melt of YSZ, spinel and corundum. From these findings, it is concluded that corundum is not suitable for an inert matrix unless a special engineering design is developed.

Spinel reacts well in combination with YSZ. The YSZ/spinel composite has a eutectic temperature of 2 200 K [16], a higher thermal conductivity than UO₂ and high stability against neutron irradiation [12,17]. A fairy high swelling of spinel was observed in recent 72 MeV-Xe irradiation tests; 22% at room temperature and 15% at 773 K [18]. An amorphisation of spinel was also observed in 60 keV-Xe irradiation and crystal recovery needed to anneal it at 1 197 K [19]. Furuno, *et al.* also performed a Xe gas release experiment during annealing the Xe-irradiated spinel by thermal adsorption spectroscopy and observed no Xe gas release up to 1 473 K, suggesting that the implanted Xe was completely trapped in the spinel matrix.

### *Particle dispersed fuel concept and its preparation*

Although spinel shows good stability against neutron irradiation, it seems not to be very stable against high energy FFs. It is preferable to localise FF damage to a small region to avoid swelling and degradation of thermal conductivity. A new particle dispersed ROX fuel concept has been developed, in which small YSZ particles containing fissile materials are homogeneously dispersed in a spinel matrix [20,21]. In this configuration the FF damaged area will be limited to a thin layer around the YSZ particle surface. A simple calculation shows that the swelling rate is 3.2% for the 200 μm YSZ particle dispersed ROX fuel, assuming that only the damaged regions contribute to the swelling, the swelling rate of these regions is 25% and the FF track length is 8 μm. The damaged volume and therefore the swelling rate decrease with increasing particle diameter. However, a larger size particle will produce a large temperature difference in the particle due to its low thermal conductivity resulting in unfavourable irradiation behaviour such as hair crack generation, high fission gas release and so on. Vialland, *et al.* made an attempt to optimise a particle size and a gap between the particle and spinel matrix [22]. The optimum particle size seems to be between 200 and 300 μm and will be determined by considering damaged volume, local heating, thermal stress and so on.

Particle dispersed ROX fuel pellets have been successfully fabricated for irradiation tests [20,23]. About 250 μm YSZ particles containing plutonium or 20% enriched uranium as fissile were dispersed in the spinel matrix. Well-defined YSZ particles can be prepared using the sol-gel technique which was employed in the kernel fabrication for the fuel of the high temperature gas-cooled reactor. External appearance and microstructure of a particle dispersed fuel pellet are shown in Figure 2. No cracks or harmful defects are visible in the external appearance (left-hand figure). In the right-hand figure, narrow gaps are seen between YSZ particles (white) and spinel matrix (grey). These gaps may reduce the thermal stress and act as a sink for fission gases. However, wide gaps may cause local heating and a low apparent pellet density. Optimisation of gap width is one of keys in the fuel pellet fabrication.

**Figure 2. Appearance and microstructure of the particle dispersed ROX fuel pellet**

 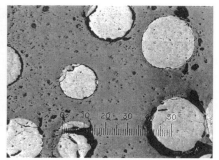

Mechanical blended ROX fuel pellets, prepared by mixing YSZ and spinel powders, and YSZ single phase fuel pellets were also fabricated for the irradiation tests. The apparent pellet densities of these pellets were more than 90% of theoretical density.

*In-pile irradiation test*

Irradiation tests of these ROX fuel pellets, for which 20% enriched uranium was used as fissile, were carried out in Japan Research Reactor 3 (JRR-3) in JAERI. The test fuel matrix is listed in Table 1. A total of five pins were irradiated, of which two pins contain spinel, two pins corundum and one pin YSZ as an inert matrix. These five pins were irradiated at a nominal reactor power of 20 MW for 100 days and the maximum neutron fluence was about $7 \times 10^{24}$ m$^{-2}$.

## Table 1. Test fuel matrix for the JRR-3 irradiation

| Pin | Composition (mol%) | | | | YSZ inclusion size | $^{235}$U density $(10^{20}/cm^3)$ |
| | YSZ[a] | $UO_2$ | $MgAl_2O_4$ | $Al_2O_3$ | | |
|---|---|---|---|---|---|---|
| SD | 20.0 | 37.1 | 42.9 | – | 250 µm | 13.00 |
| SH | 20.0 | 37.1 | 42.9 | – | 10-50 µm | 13.10 |
| Z | 80.0 | 20.0 | – | – | Solid solution | 8.62 |
| CD | 16.5 | 30.6 | – | 52.9 | 250 µm | 13.54 |
| CH | 16.5 | 30.6 | – | 52.9 | 10-50 µm | 13.17 |

[a] YSZ = 79.9 mol% $ZrO_2$ + 20.1 mol% $YO_{1.5}$

The estimated irradiation conditions are summarised in Table 2. The linear power was calculated from fluence monitor analysis. Pellet surface and centre temperatures were estimated from the linear power values obtained. As can be seen from the table, all pins except for the Z pin were irradiated at remarkably high temperatures. Burn-up analysis was performed using the SRAC95 code system [24] and calculated burn-ups are indicated in the bottom row. The value of 100 GWd·m$^{-3}$ corresponds to 10 GWd·t$^{-1}$ of the conventional $UO_2$ fuel.

## Table 2. Estimated irradiation conditions

| Pin | SD | SH | Z | CD | CH |
|---|---|---|---|---|---|
| Linear power/kW·m$^{-1}$ | 23.0 | 23.4 | 13.9 | 24.9 | 20.7 |
| Pellet surface temperature/K | 1 250 | 1 440 | 990 | 1 300 | 1 290 |
| Pellet centre temperature/K | 1 740 | 1 940 | 1 490 | 1 820 | 1 730 |
| Burn-up/GWd·m$^{-3}$ | 100 | 103 | 59 | 105 | 88 |

Pellet swelling due to irradiation was evaluated from changes in fuel stack lengths and diameters. The relative length of each fuel stack against total pin length was determined from X-ray photos taken before and after irradiation. No stack elongation was observed for all pins within the experimental error of 0.1 mm. The profilometry of each pin showed that the diameter expansion was very small, less than 10 µm, except for the SH pin, for which a diameter increase of 60 µm was recorded at the maximum value. The diameter change was evaluated from the initial pellet diameter, a gap between the pellet and cladding and profilometry value. Radial and volumetric swellings are summarised in the third and fourth columns of Table 3. The YSZ single phase fuel (Z pin) shows the lowest swelling and the particle dispersed fuels (SD and CD pins) swell less than the powder mixture fuels (SH and CH pins). It should be noted that swelling of the corundum-based fuels (CD and CH pins) was very small compared with the literature value of about 30% [14]. In that case, the fuel specimen was irradiated at a low temperature (about 560 K) and a large swelling due to amorphisation of the corundum phase was found. In the present study, on the other hand, fuel temperature was estimated to be above about 1 300 K. X-ray diffraction analysis of the irradiated fuels clearly showed the existence of a crystalline corundum phase although some peak broadening was observed. The low swelling of corundum-based fuels may be attributed to rapid recovery from amorphous state due to high-irradiation temperature.

Gases in the fuel pin were collected and analysed through a puncture test. The fractional gas release (FGR) of Xe is the ratio of the measured Xe volume and the calculated one which was obtained through burn-up analysis using the SRAC95 code, and the values are listed in the last column of Table 3. The burn-up calculation showed that the Xe/Kr gas ratio was 7.2, which agreed well with the measured ratio of 7.0 to 7.2. The YSZ single phase fuel shows the lowest FGR of 2.2%. Contrary to the swelling behaviour, the FGR of particle-dispersed fuels (SD and CD pins) is larger than that of

**Table 3. Dimensional variation and fractional gas release of ROX fuel pins**

| Pin | Max. temperature /K | $\Delta\Phi/\Phi$ (%) | $\Delta V/V$ (%) | FGR of Xe (%) |
|-----|---------------------|------------------------|-------------------|----------------|
| SD | 1 850 | 2.7 | 5.5 | 38 |
| SH | 2 080 | 5.0 | 10.2 | 22 |
| Z | 1 580 | 2.0 | <4.0 | 2.2 |
| CD | 1 930 | 2.1 | 4.3 | 22 |
| CH | 1 830 | 2.8 | 5.7 | 7.8 |

the powder mixture fuels (SH and CH pins). The high FGR values indicate that the FP gas may release directly through open paths of hair cracks and not by diffusion through matrices. Such hair cracks may occur due to the stress caused from a thermal expansion difference between the large YSZ fuel particles and matrices. A similar FGR behaviour was reported recently by Bakker, *et al.* [25] and Neeft, *et al.* [26]. They observed higher FGR for the "macro" dispersed fuels, where large fissile inclusions of diameter about 150 µm were dispersed in spinel matrix, than the "micro" dispersed ones of which typical fissile inclusion size was less than 1 µm.

Ceramographic photos of the cross-sections of irradiated Z, SH and CH fuels are shown in Figure 3. Several radial cracks are observed in Z and CH fuels as observed in the conventionally-irradiated $UO_2$ fuel. No apparent microstructure change was recognised in these fuels. On the other hand, a significant change of microstructure is observed mainly in the central region of the SH fuels. The microstructure change in spinel matrix fuel was observed for the first time in the present irradiation experiment. A precise SEM image of the SH fuel is shown in Figure 4 together with existing phases analysed by

**Figure 3. Cross-sectional views of irradiated Z, SH and CH fuels**

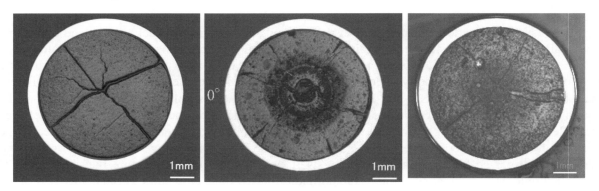

**Figure 4. SEM image of the irradiated SH fuel**

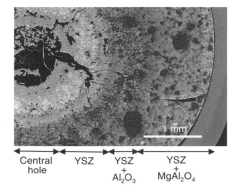

EPMA. A hole of about 1 mm in diameter was formed in the centre of the pellet. A bright region (I) around the central hole was found to be a YSZ phase containing uranium. An outer ring region (II) of about 0.5 mm thick around the YSZ region consisted of YSZ and corundum phases. The outermost region (III) was a mixture of YSZ and spinel and its microstructure was nearly the same as that of a fresh fuel. The temperature between regions (II) and (III) was calculated to be about 1 700 K. In the region blow 1 700 K, neither spinel decomposition nor restructuring was observed.

Spinel may decompose to $Al_2O_3$ and MgO by heavy fission fragment irradiation. Above 1 700 K and under high thermal gradient (2 000 $K \cdot cm^{-1}$), MgO vaporised incongruently from the spinel matrix and transported to a cooler region because the vapour pressure of MgO was higher by about 5 orders of magnitude than that of $Al_2O_3$. Vaporisation of MgO and subsequent restructuring may be a driving force of the central hole formation by pushing closed pores in the matrix to the centre. This restructuring seems to cause the high FGR of spinel matrix fuels compared with that of the corundum matrix fuels where appreciable microstructure change was not observed.

## Reactor physics study

The ROX fuel core has several inherent disadvantages which emerge from the fact that it contains only plutonium as fissile and no fertile materials. They have a small Doppler reactivity and a large burn-up reactivity swing compared to the $UO_2$ fuel core. The small Doppler reactivity may cause a large energy deposition to the fuel under a reactivity-initiated accident (RIA) condition and extremely severe fuel damage. The large burn-up reactivity swing will cause a large power peak which should be avoided, especially in the event of a loss-of-coolant accident (LOCA).

### *Improvement of ROX core characteristics*

Two approaches were undertaken in order to improve the Doppler reactivity of the ROX fuel core to a level comparable to that of the $UO_2$ fuel core: the addition of resonant nuclides such as $^{232}Th$, $^{238}U$ and natural Er to the ROX fuel, and composing a heterogeneous core of 1/3ROX + 2/3$UO_2$.

The analytical results are listed in Table 4. By adding 15 at.%U or 24 at.%Th to weapons-grade plutonium, or 8 at.%U or 15 at.%Th to reactor-grade plutonium, the Doppler reactivity and the peaking factor could be successfully improved. In the case of uranium addition, a small amount of natural erbium (about 0.6 at.%) was also added to the ROX fuel. The addition of uranium in combination with erbium was found to be very effective in flattening the burn-up reactivity swing and reducing the power peaking [6]. As for the partial ROX core, some improvement in the Doppler reactivity is attained but not in the peaking factor. Further improvement will be necessary for this core configuration by adjusting fuel composition (fissile and additives densities) near the interface of ROX and $UO_2$ subassembly.

### *Accident analyses*

The RIA analysis was performed with the EUREKA-2 code [27] for improved ROX cores and summarised in Table 5 together with those for the $UO_2$ core. Fuel enthalpy (H) and fuel temperature ($T_F$) of the original ROX fuel core well exceed the limiting value of 960 $kJ \cdot kg^{-1}$ for the $UO_2$ core and the ROX fuel melting temperature of 2 200 K. The improved ROX fuel cores clear the limitations. Among them, the U(Er) added ROX fuel core shows nearly the same values as the $UO_2$ core. It should be noted here that the fuel enthalpy given in unit volume energy is more practical than that in unit

### Table 4. ROX fuel core characteristics

| Core | Doppler reactivity (900 K → 1 200 K) | Peaking factor |
|---|---|---|
| Weapons-grade Pu | | |
|    Original ROX | -0.098 | 2.7 |
|    ROX + 15 at.%U(Er) | -0.61 | 2.1 |
|    ROX + 24 at.%Th | -0.56 | 2.4 |
|    1/3ROX + 2/3UO$_2$ | -0.48 | 2.8 |
| Reactor-grade Pu | | |
|    ROX + 8 at.%U(Er) | -0.59 | 2.2 |
|    ROX + 15 at.%Th | -0.64 | – |
| UO$_2$ | -0.75 | 2.0 |

mass energy when the value is compared with that of UO$_2$, because of the large difference in density between the ROX and UO$_2$ fuels. Here again, both H and T$_F$ of the partial ROX fuel core are still high and further improvement will be required. The detailed description of the RIA analysis has been given elsewhere [28].

Peak cladding temperatures of various ROX fuel cores during a LOCA event, analysed using the RETRAN-2 code [29], are given in the last column of Table 5. The analysed system and event were a four-loop type 1 100 MWe class PWR and a cold-leg large break LOCA event at BOC and EOC, respectively [6]. All improved ROX cores clear the limiting value of 1 473 K. Because the peaking factor of the ROX + U(Er) cores could be reduced to a low level, T$_{PC}$ of these cores are as low as that of the UO$_2$ core.

### Table 5. Maximum fuel enthalpy (H), fuel temperature (T$_F$) and peak cladding temperature (T$_{PC}$) in RIA and LOCA events

| Core | H/kJ·kg$^{-1}$ | H/10$^6$ kJ·m$^{-3}$ | T$_F$/K | T$_{PC}$/K |
|---|---|---|---|---|
| Weapons-grade Pu | | | | |
|    Original ROX | >>960 | – | >>2 200 | >1 470 |
|    ROX + 15 at.%U | 810 | 4.5 | 1 700 | 1 090 |
|    ROX + 24 at.%Th | 940 | 5.2 | 1 950 | 1 240 |
|    1/3ROX + 2/3UO$_2$ | 1 020 | 5.7 | 2 100 | 1 240 |
| Reactor-grade Pu | | | | |
|    ROX-8 at.%U | 800 | 4.5 | 1 700 | 1 090 |
| UO$_2$ | 390 | 4.3 | 2 080 | 1 080 |

### *Pulse irradiation tests under RIA conditions*

Along with these calculation analyses, pulse irradiation experiments were performed on the ROX fuels in the Nuclear Safety Research Reactor (NSRR) in JAERI to investigate the fuel behaviour under RIA conditions [30,31]. Three ROX fuels, a mechanically blended YSZ/spinel (SH) fuel, a YSZ single phase (Z) fuel and a particle dispersed YSZ/spinel (SD) fuel, were fabricated into short test rods with a $17 \times 17$ PWR type specification and were irradiated with a short pulse of half widths ranging from 4.4 ms to about 10 ms. Inserted energy depositions were in the range between 0.5 and 2.2 MJ·kg$^{-1}$.

Peak fuel enthalpy and rod failure are summarised in Figure 5. The horizontal dotted line of 9.6 GJ·m$^{-3}$ indicates the current rod failure threshold for the UO$_2$ fuel rod. Open marks show that the fuel rods were intact, while filled marks show that the fuel rods were failed. The rod failure threshold for the ROX fuel seems to be above 10 GJ·m$^{-3}$ and slightly higher that that of the UO$_2$ fuel.

**Figure 5. ROX fuel pin failure threshold enthalpy in terms of GJ·m$^{-3}$**

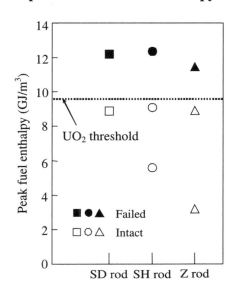

Fuel failure behaviour for the ROX fuel is quite different from that of the UO$_2$ fuel. The failure of the UO$_2$ fuel occurs during the quenching. A thinner part of the once molten and embrittled cladding was broken mainly due to thermal stress and the fuel itself remains solid. A large mechanical energy is released by the reaction between hot fuel and coolant water. On the other hand, the failure of the ROX fuel occurred due to cladding burst when the cladding temperature peaked. A significant fraction, of about 50 to 70%, of partly molten fuel was ejected through the burst opening into the surrounding water. However, generation of pressure pulse or water hammer due to molten fuel/water interaction was not detected. Relatively mild fuel ejection during high temperature burst might have prevented the mechanical energy generation.

**Summary and conclusions**

Research on the plutonium rock-like oxide (ROX) fuels and their once-through burning in light water reactors has been performed in order to establish an option for utilising or effectively disposing of excess plutonium. The following are conclusions so far obtained.

Yttria-stabilised zirconia is the most hopeful candidate for an inert matrix when its low thermal conductivity is improved. The combined use of YSZ with spinel, especially YSZ particles dispersed in the spinel matrix, is a possible means of improvement.

From the irradiation test of the ROX fuels, the particle dispersed fuels showed lower swelling and higher FGR than those of powder mixture fuels. Because a mismatch in thermal expansion between a YSZ particle and spinel matrix led to hair cracks and the resultant high FGR, further optimisation in fuel design must be carried out to improve FGR behaviour.

Core characteristics of the ROX fuel were sufficiently improved by adding resonant nuclei such as $^{232}$Th, $^{238}$U and natural Er. The improved ROX fuel core will have almost the same behaviour as the conventional $UO_2$ fuel core under LOCA or RIA conditions. Pulse irradiation experiments simulating RIA conditions showed that the failure threshold enthalpy of ROX fuel rods was similar to that of conventional $UO_2$ fresh fuel rods. These results in reactor physics studies clearly show that the ROX-LWR system is a feasible option for effective burning of excess plutonium.

# REFERENCES

[1]     Nuclear Technology Review, IAEA General Conf. GC (44)/9, Vienna, Aug. 2000.

[2]     H. Akie, T. Muromura, H. Takano, S. Matsuura, *Nucl. Technol.*, 107, 182 (1994).

[3]     J.M. Paratte, R. Chawla, *Ann. Nucl. Energy*, 22 (7), 471 (1995).

[4]     C. Lombardi, *et al.*, Proc. Int. Conf. on Evaluation of Emerging Nuclear Fuel Cycle Systems (GLOBAL-95), Versailles, France, 11-14 September 1995, Vol. 2, p. 1374 (1995).

[5]     C. Prunier, *et al.*, Proc. Int. Conf. on Evaluation of Emerging Nuclear Fuel Cycle Systems (GLOBAL-95), Versailles, 11-14 September 1995, Vol. 2, p. 1409 (1995).

[6]     H. Akie, H. Takano, Y. Anoda, *J. Nucl. Mater.*, 274, 139 (1999).

[7]     G. Ledergerber, *et al.*, Proc. Int. Conf. on Future Nuclear Systems (GLOBAL-97), Yokohama, Japan, 5-10 October 1997, Vol. 2, p. 1068 (1997).

[8]     K.E. Sickafus, *et al.*, *J. Nucl. Mater.*, 274, 66 (1999).

[9]     T. Yamashita, *et al.*, Proc. Int. Conf. on Future Nuclear Systems (GLOBAL-99), Jackson Hole, USA, 29 August-3 September (1999).

[10]    D.L. Hargman, MATPRO-Ver. 11, Idaho National Engineering Laboratory (1981).

[11]    Y.S. Touloukian, R.W. Powell, C.Y. Cho, P.G. Klemens, "Thermophysical Properties of Matter", The TCRP Data Series, Vol. 2, p. 119, IFI/Plenum, New York (1970).

[12]    F.W. Clinard, Jr., D.W. Parkin, USDOC Report, CONF-811237 (1981).

[13]    G.W. Keilholtz, R.E. Moor, H.E. Robertson, *Nucl. Technol.*, 17, 234 (1973).

[14]    R.M. Berman, M.I. Bleiberg, W. Yeniscavich, *J. Nucl. Mater.*, 2, 129 (1960).

[15]    T. Yamashita, *et al.*, *J. Nucl. Mater.*, 274, 98 (1999).

[16]    N. Nitani, *et al.*, *J. Nucl. Mater.*, 274, 15 (1999).

[17]   F.W. Clinard, Jr., G.E. Hurley, L.W. Hobbs, *J. Nucl. Mater.*, 108/109, 655 (1982).

[18]   Hj. Matzke, V.V. Rondinella, T. Wiss, *J. Nucl. Mater.*, 274, 47 (1999).

[19]   S. Furuno, *et al.*, *Nucl. Instr. and Meth.*, B 127/128, 181 (1997).

[20]   T. Shiratori, *et al.*, *J. Nucl. Mater.*, 274, 40 (1999).

[21]   N. Chauvin, R.J.M. Konings, Hj. Matzke, *J. Nucl. Mater.*, 274, 105 (1999).

[22]   I. Vialland, J.M. Bonnerot, M. Bauer, Proc. Int. Conf. on Future Nuclear Systems (GLOBAL-97), Yokohama, Japan, 5-10 October 1997, p. 1131 (1997).

[23]   R.P.C. Schram, *et al.*, *Prog. Nucl. Energy*, 38, 259 (2001).

[24]   K. Okumura, K. Kaneko, K. Tsuchihashi, Japan Atomic Energy Research Institute Report JAERI-Data/Code 96-015 (1996) (in Japanese).

[25]   K. Bakker, *et al.*, *Prog. Nucl. Energy*, 38, 313 (2001).

[26]   E.A.C. Neeft, *et al.*, *Prog. Nucl. Energy*, 38, 427 (2001).

[27]   N. Ohnishi, T. Harami, H. Hirose, M. Uemura, Japan Atomic Energy Research Institute Report JAERI-M-84-074 (1984) (in Japanese).

[28]   H. Akie and T. Nakamura, *Prog. Nucl. Energy*, 38, 363 (2001).

[29]   C.E. Peterson, J.H. McFadden, M.P. Paulson, G.C. Gose, Electric Power Research Institute Report EPRI NP-1850 CCM, 1 (1981).

[30]   K. Okonogi, *et al.*, *J. Nucl. Mater.*, 274, 167 (1999).

[31]   T. Nakamura, *et al.*, *Prog. Nucl. Energy*, 38, 379 (2001).

# APPLICATION OF CERAMIC NUCLEAR FUEL MATERIALS FOR INNOVATIVE FUELS AND FUEL CYCLES

**Young-Woo Lee**
Korea Atomic Energy Research Institute
P.O. Box 105 Yuseong, Taejon, Korea 305-600

## Abstract

Fuels for nuclear power reactors currently in operation for generating electricity are mostly the ceramic $UO_2$ and the so-called mixed oxide (MOX) with uranium-plutonium oxide. The relevant technologies developed up to now are industrially matured, and still continue to be developed from the aspects of economy and safety, including high burn-up. Recently, fuel cycles with innovative fuels have been conceived and preliminary results from basic studies already exist with regard to the ever-increasing interest in the incineration of excess Pu materials from warheads and surplus reactor Pu stocks as well as minor actinide (MA) transmutation. These concepts employ mostly ceramic nuclear fuel materials such as $ZrO_2$ and spinel-based oxides and non-oxides with ceramic matrices (nitrides and carbides). Problems that the relevant technologies face for fuels currently used in power reactors are first briefly treated in connection with their high burn-up behaviour. Issues concerning materials for the innovative fuels and fuel cycles mentioned above are subsequently discussed in a more detailed manner in view of their developmental status, material properties and irradiation behaviours studied up to the present time.

## Introduction

Currently, 438 fission nuclear reactors are in commercial operation, supplying some 17% of the world's electricity generation [1]. These power reactors are mostly water-cooled reactors and are fuelled almost exclusively with ceramic uranium dioxide and some with uranium-plutonium mixed oxide (MOX), the advantage of which lies in their phase stability up to their high melting points, good compatibility with cladding and coolants as well as stability against radiation. The relevant technologies of these fuels are at a commercially mature stage, and their irradiation behaviours in the reactors are extensively understood in view of their in-reactor performance.

Examination of the in-reactor behaviour of light water reactor (LWR) fuels at high burn-up (e.g. > 40 GWd/tHM), in particular, has been an objective since the 1980s, and a particularity recently observed with the fuel high burn-up is the so-called "polygonisation" of the pellet peripheral region, or rim, typically with a thickness of less than about 0.5 mm, a grain-subdivided structure called the rim structure. The material problems associated with this phenomenon and their solutions are not systematically and quantitatively assessed as a whole, though attempts to characterise the structure were recently summarised [2,3,4].

Plutonium management is currently a crucial issue world-wide in any discussion on the future of nuclear energy, since it is an inevitable by-product of nuclear energy generation with the currently established fuel technology. A long-term reprocessing strategy as well as recent decisions on the dismantling of nuclear weapons will give rise to an excess of separated Pu stocks. Utilising and/or burning Pu produced in power reactors is one of the options to solve the problems caused by this surplus Pu. In addition, the spent fuels discharged from the reactors contain several minor actinides (MAs, i.e. Am, Cm and Np) and long-lived fission products (LLFP, i.e. Tc, I and Cs). Due to their radiotoxicity, these minor actinides from high level waste (HLW) containing fission products have to be separated (partitioned) for long-term HLW disposition, and then further broken down into non-radiotoxic and stable elements (transmutation).

Recently, innovative fuels have been conceived, responding to these needs and changes in fuel cycle options. R&D works have been initiated among the nuclear research organisations in an increasing number of countries in order to solve the problems regarding the surplus plutonium from both warheads and world-wide reprocessing campaigns, as well as for the transmutation of MAs and LLFPs. Several fuel concepts – including the so-called inert matrix fuel (IMF) – for these purposes are currently being developed for their material and in-pile performance abilities and fabrication technology for the various candidate materials in connection with the remaining stages in the nuclear fuel cycle.

A number of ceramic oxide fuels and oxide inert matrices including stabilised $ZrO_2$ [5] and $MgAl_2O_4$ [6], and non-oxide fuels and inert matrices including AlN and ZrN have been investigated up to the present time. In this paper, problems that the relevant technologies face for fuels currently used in power reactors are first briefly treated in connection with their high burn-up behaviour. Issues concerning potential candidate materials for these specific purposes in the innovative fuels and fuel cycles mentioned above are subsequently discussed in view of their developmental status, material properties and irradiation behaviours as have been studied up to the present time.

## Fuel technologies for current ceramic fuels and their improvement

### Problems to be solved in current fuel technologies

Nowadays, nuclear electric power is produced based upon the current fuel technologies, mostly of water-cooled reactors by the fission of $^{235}U$ and $^{239}Pu$ in the form of ceramic oxides, $UO_2$ and

(U-Pu)$O_2$, with fissile contents of 3-8 wt.%. (U-Gd)$O_2$ with Gd as a burnable poison of the neutron is also used in oxide fuel with higher enrichment. During their burn-up, a number of fission products are formed in large amounts, as dissolved oxides, metallic precipitates, oxide inclusions and volatile element-forming bubbles, the effect of which is the swelling of the fuels [7]. The amount of these fission products will increase when the burn-up increases. A complete understanding of the behaviour of these fission products during irradiation from the viewpoint of fuel performance still remains to be attained. In addition, the formation of these fission products lowers the thermal conductivity of the oxide fuel, which is of prime importance in their thermal performance. Thermal conductivity of the fuel during irradiation decreases as the burn-up increases, mainly due to increasing amounts of fission products, and hence the centre temperature of the fuel, inducing a larger release of gaseous fission products, especially at EOL. This is also true for MOX fuel at higher burn-up (> 40 GWd/tHM) in particular, as the decrease in reactivity with burn-up is smaller than that for $UO_2$, and thus operates in higher temperatures, leading to a larger fission gas release, which should be taken into account for MOX fuel rod design. Recent results of the irradiation tests compare the fission gas releases of commercial $UO_2$ and MOX fuel [8]. Overall in-reactor performance of current commercial MOX is satisfactory, and currently 33 LWRs are loaded with MOX [9], a number which is expected to increase, as is shown Table 1. However, a specific feature of MOX fuel also requires further improvement, due to the presence of Pu-rich particle agglomerates resulting from non-homogeneity in the microstructure of pellets during the commercial fabrication campaign, for its in-reactor and thermal performance.

### *Fuels for increased burn-up and the rim structure*

Examination of the in-reactor behaviour of light water reactor (LWR) fuels at high burn-up (e.g. > 40GWd/tHM), in particular, has been an objective since the 1980s for fuel research organisations and power utilities. This is because, by increasing burn-ups, fuels can be economically utilised, reducing the amount of spent fuel. One of the associated phenomena recently observed with the fuel high burn-up is the so-called "polygonisation" of the pellet peripheral region, or rim, typically with a thickness of less than about 0.5 mm, a grain-subdivided structure called the rim structure. The material problems associated with this phenomenon and their solutions have not yet been systematically and quantitatively assessed as a whole.

Early in the 1990s, an international joint collaboration work was launched, called the High Burn-up Rim Project (HBRP). By 2000, a series of research results had been published (e.g. [2]), elucidating the cause as well as attempting physical models for basic understanding of the mechanisms. Such works are still ongoing in a number of related international and national research organisations. Analysing the results obtained up to the present, this effect is thought to be due to the resonance absorption of epithermal neutrons by [238]U and is generally agreed to be caused by a combination of the high burn-ups and low operating temperatures. The characteristics of the rim structure have been relatively well described [10] and can be summarised as:

- Xe depletion in the matrix.

- Development of pores with a typical diameter of 1-2 μm.

- Development of subdivided grain structure with a typical grain size of 0.2-0.3 μm, which occurs successively at different local burn-ups.

However, more extensive work is obviously needed, experimentally as well as theoretically, to attain a complete knowledge of this high burn-up effect.

An important safety issue when increasing burn-up is to mitigate the larger fission gas release, which is expected due to degradation of fuel thermal conductivity and formation of the rim structure mentioned above. In this respect, a number of studies have been devoted, for the past several decades, to the fabrication of large-grained pellets with controlled porosity and their irradiation tests, recently giving some positive preliminary results [e.g. 11,12]. The technologies regarding large-grained pellets with controlled porosity involve various sintering techniques such as an increase of oxygen potential in the sintering atmospheres, including the use of sintering additives, or dopants, and pore formers with appropriate levels. A number of dopants have been tested, consequently, for both $UO_2$ and MOX pellets, aiming at producing pellets for higher burn-ups and less fission gas release, examples being MgO, $TiO_2$ and $Nb_2O_5$ [13]. It has been observed, however, that some of these dopants contribute to the enlargement of grains and, at the same time, to the enhancement of the diffusion of fission gas bubbles in the fuel matrix. Therefore, very careful attention should be paid to the selection of the dopants for such a purpose.

## Materials in recent development of innovative fuels for advanced fuel cycles

Nuclear power steadily produces a mass of spent fuels which contain a significant amount of minor actinides and fission products with very long half-lives (long-lived fission products, LLFPs) with high radiotoxicity. The conventional recycling option involves only major actinides, i.e. U and Pu, while these MAs and LLFPs remain with other fission products which are vitrified before being buried in deep repositories. Partitioning of these MAs and some selected LLFPs is a method which would reduce the long-term radiotoxicity of the residual waste components. The recovered MA nuclides would be recycled in the fuel cycle activities and returned to the reactor inventory of fissile and fertile material for transmutation to stable and/or shorter-lived isotopes. In addition, apart from the surplus Pu separated from the spent commercial reactor fuels, there is a need for the disposition of the Pu from dismantled warheads, recently decided between the US and Russia, aiming at reducing the total amount of world-wide Pu. This Pu must enter into current fuel cycle activities, along with burning (incineration) the surplus Pu existing all over the world.

### Fuel and target materials for Pu dispositioning and transmutaion of MAs and LLFPs

At present, several kinds of surplus Pu exist in the world, which can be categorised into [14]:

- Pu from reprocessing spent LWR fuels (commercial or reactor Pu, R-Pu).

- Pu from the dismantling of nuclear warheads (ex-weapons or military Pu, W-Pu).

- Pu contained in spent reactor fuel from electricity-producing nuclear power stations (spent fuel Pu, SF-Pu).

Materials aspects for this surplus Pu dispositioning can be related to the technologies available or newly developed to diversify for efficiency or to accommodate the difficulties arising from the different materials problems, which could include:

- Pu burning using conventional MOX technology.

- Pu burning in inert matrices.

- Pu burning in a thoria matrix.

- Pu burial by vitrified waste disposal [14].

The materials aspect for the last two means is not discussed in detail here, as Pu burial by vitrified waste disposal would be out of the scope of this paper and Pu burning in thoria fuel is still in the preliminary experimental stage, though the concept has long been discussed and thoria itself and thoria mixed with $PuO_2$ and $UO_2$ as fission fuels have been extensively tested.

Conventional MOX technology involves the mixing of U and Pu in ceramic oxide form, achieving a rather complete solid solution, with an appropriate content of Pu depending on the reactor type used to burn the Pu. The relevant technology, currently commercially available, gives no difficulties, even with W-Pu. However, this option is not to eliminate Pu but rather to gain energy from R-Pu or to transform W-Pu into spent fuel by burning in the reactor, thereby meeting the spent fuel standard of proliferation resistance. One of the specific features of MOX technology for W-Pu is gallium removal in the stream of process, the technology of which still needs to be developed to appropriately reduce the Ga content in the final MOX pellet [15,16].

One of the fuel materials of the newly developed concept for Pu burning, as well as for MA transmutation, is ceramic inert matrix fuel, and the results obtained are being published and presented in a number of international scientific journals and conferences [e.g. 17]. This fuel concept is based on "U-free" (i.e. for excluding continuous U-Pu conversion) Pu or MA compounds embedded in a material matrix (sometimes also called a "support" or a "diluent") by means of solid solution formation or by forming a two-phase microstructure. The stable matrix will then, after burning in a power reactor, be directly disposed of for final fuel disposal [18]. Potential candidate materials for the matrix have recently been reviewed and their material aspects compared [19].

Materials for the ceramic inert matrix require particular properties with regard to nuclear fuel material [20,21], including: high melting points, low vapour pressure, good thermal properties (thermal conductivity, heat capacity), good mechanical properties, good compatibility with cladding and coolant, low neutron capture cross-sections as well as availability and fabricability of the materials. Furthermore, materials should be stable, on a long-term basis, against radiation, and interaction with fuel materials during burn-up and after irradiation should be minimised in view of the in-reactor fuel performance and final disposal of spent fuel. An important point is also that thermodynamic entities such as oxygen potential, solubility and phase relationship in different chemical states must be known to reliably predict the in-reactor behaviour under normal and off-normal conditions.

*Physico-chemical properties*

A number of inert matrix materials have been pre-selected [19,20], mostly among the ceramic oxides and non-oxide materials. Some of the important property data for the selected materials available up to the present time, affecting the selection of materials for this innovative fuel concept, are summarised in Table 2, with regard to some physico-chemical properties mostly related to their use in nuclear reactor with fuel materials. In this table, some of the nitride ceramics are included for their unique and potential use in fast reactors with PuN as a fuel material and consequently, examination for their compatibility with Pu and MA nitrides is desirable, whereas most oxide ceramics can be commonly used for both thermal reactors and fast reactors. Certain nitrides such as ZrN can also be considered as matrix materials for oxide fuel materials since oxygen can possibly be dissolved in these nitride ceramics to form a certain oxynitride phase. The formation of such phases is still unknown, and thus a study on the relevant phases should be carried out if the material is to be applied.

*Irradiation behaviour and interaction with fission products*

Apart from the material properties investigated on the topics described above, studies on in-reactor behaviour have been initiated using various techniques such as fission-product implantation [22,23], irradiation with Xe and I ions [23,24] through the use of accelerator and/or ion implanter and neutron irradiation [5,26,27], and some significant results came to be published and show indications that:

- Spinel ($MgAl_2O_4$) and $CeO_2$ are relatively stable against radiation (though spinel swells at the fission energy level and is polygonised [19]).

- The volume change is smaller than 1% for spinel ($MgAl_2O_4$) and excessive (about 4.2%) for $\alpha$-$Al_2O_3$ at a neutron fluence of $1.7 \times 1\,026\ m^{-2}$ [28].

- For $ZrO_2$, the monoclinic phase in partially-stabilised zirconia (PSZ) was observed to disappear by neutron irradiation while the cubic phase remained intact [24].

- Zircon ($ZrSiO_4$) has high dissociation at a high temperature and monazite has poor radiation stability and thermal conductivity.

These results are likely to lead to a preliminary selection of spinel ($MgAl_2O_4$) and stabilised zirconia as inert matrix materials among ceramic oxides for further studies.

### *Development of fabrication technology*

Even though the inert matrix fuel concept still remains in the development stage, i.e. far from being commercialised, it is worthwhile to consider the technology for fuel fabrication. Nearly all the specimens were prepared for small-scale laboratory evaluation, either by taking the conventional pellet fabrication route, i.e. powder preparation, pressing and sintering or by modifying it to the least extent [25]. However, as mentioned briefly above, there are two different fuel classes in this fuel concept – homogeneous and heterogeneous – for different purposes. Homogeneous fuel is that with the fuel material which forms a single solid solution with the matrix material, and hence has a single phase achievable by selecting an inert matrix with a large solubility for fuel material. An example is stabilised zirconia with Pu in solution. Meanwhile, heterogeneous fuel has a two-phase structure either by the dispersion of a fissile ceramic in an inert ceramic matrix (CERCER), an example being $AmO_2$ particles in spinel $MgAl_2O_4$, or by the dispersion of fissile ceramic in a metal(CERMET) such as $PuO_2$ dispersed in Mo or W. For CERCER, if the inert matrix material shows no solubility of fuel material the added fuel material is present as a second phase. The fuel material can either be added in the form of fine particles or in the form of macro-inclusions. For Pu incineration, homogeneous inert matrix fuel is preferred, as it can easily be adapted to the conventional pellet fabrication route already commercialised in the nuclear industry. However, for two-phase heterogeneous fuel, other techniques than simple pelletising technology should be established in view of the feasibility for a larger scale with mass production. For heterogeneous fuel, ceramic particles to be dispersed in the matrix can be successfully prepared in general following the sol-gel route. Konings, *et al.* [29] discussed the "ideal" characteristics of the dispersion-type fuel in terms of size, distribution and its "host" matrix.

## Concluding remarks

In this paper, problems that the relevant technologies face for fuels currently used in power reactors are first briefly treated in connection with their high burn-up behaviour. Potential candidate materials for Pu dispositioning and MA transmutation were then reviewed on the current issues being discussed and compared in terms of the material properties and the behaviour against irradiation with the currently available relevant information.

The trend in improvement for current oxide fuels is mainly going for higher burn-ups, attempting to solve the associated materials problems, i.e. high burn-up structure formation, degradation in thermal conductivity and increased fission gas release. For the incineration of plutonium and transmutation of MA, there are two potential candidate materials for oxide fuel materials: stabilised zirconia and magnesium aluminium spinel, which were focused on at the 6th IMF Workshop [30]. For nitride fuels, which are supposed to be burnt more favourably in fast reactors, AlN and ZrN can be considered.

*Acknowledgement*

This research has been performed in part under the auspices of the Nuclear R&D Programme of the Ministry of Science and Technology in Korea.

## REFERENCES

[1]    International Atomic Energy Agency, Power Reactor Information System home page, address: www.iaea.org/programmes/a2/index.html, accessed Sept. 2001.

[2]    M. Kinoshita, *et al.*, "High Burn-up Rim Project (II) Irradiation and Examination to Investigate Rim-structured Fuel", ANS Topical Meeting on LWR Fuel Performance, Park City, Utah, 10-13 April 2000.

[3]    J. Spino and D. Papaioannou, "Lattice Parameter Changed Associated with the Rim-structure Formation in High Burn-up $UO_2$ Fuels by Micro X-ray Diffraction", J. *Nucl. Mater.*, 281, 146 (2000).

[4]    K. Une, *et al.*, "Rim Structure Formation of Isothermally Irradiated $UO_2$ Fuel Discs", *J. Nucl. Mater.*, 288, 20 (2001).

[5]    G. Ledergerber, *et al.*, "Inert Matrix Fuel for the Utilization of Plutonium", *Prog. Nucl. Energy*, 38, 301 (2000).

[6]    K. Bakker, *et al.*, "Fission Gas Release in a Spinel-based Fuel used for Actinide Transmutation", *Prog. Nucl. Energy*, 38, 313 (2000).

[7]    H. Kleykamp, "The Chemical State of the Fission Products in Oxide Fuels", *J. Nucl. Mater.*, 131, 221 (1985).

[8]    R.J. White, *et al.*, "Measurement and Analysis of Fission Gas Release from BNFL's SBR MOX Fuel", *J. Nucl. Mater.*, 288, 43 (2001).

[9]    K. Fukuda, *et al.*, "Global View on Nuclear Fuel Cycle – Challenges for the 21st Century", GLOBAL'99, Jackson Hole, Wyoming, 29 Aug.-3 Sept. 1999.

[10]  J. Spino, K. Vennix and M. Coquerelle, "Detailed Characterization of the Rim Microstructure in PWR Fuels in the Burn-up Range 40-67 GWd/tM", 231, 179 (1996).

[11]  T. Fujino, *et al.*, "Post-irradiation Examinations of High Burn-up Mg-doped $UO_2$ Fuel", ANS Top. Mtg. on LWR Fuel Performance, Park City, Utah, 10-13 April 2000.

[12]  K. Une, *et al.*, "Rim Structure Formation and High Burn-up Fuel Behavior of Large-grained $UO_2$ Fuels", *J. Nucl. Mater.*, 278, 54 (2000).

[13]  T. Fujino, *et al.*, "Post-irradiation Examinations of High Burn-up Mg-doped $UO_2$ in Comparison with Undoped $UO_2$, Mg-Nb-doped $UO_2$ and Ti-doped $UO_2$", *J. Nucl. Mater.*, 297, 176 (2001).

[14]  J. van Geel, Hj. Matzke and J. Magill, "Bury or Burn? Plutonium – The Next Nuclear Challenge", *Nucl. Energy*, 36, 305 (1997).

[15]  David Kolman, *et al.*, "Thermally-induced Gallium Removal from Plutonium Dioxide for MOX Fuel Production", GLOBAL'99, Jackson Hole, Wyoming, 29 Aug.-3 Sept. 1999.

[16]  Y.S. Park, *et al.*, "Thermal Removal of Gallium from Gallia-doped Ceria", *J. Nucl. Mater.*, 280, 285 (2000).

[17]  Hj. Matzke, "Inert Matrix Fuels for Incineration of Plutonium and Transmutation of Americium", Plutonium Futures – The Science, Santa Fe, New Mexico, 10-13 July 2000.

[18]  C. Degueldre, *et al.*, "Plutonium Incineration in LWRs by a Once-through Cycle with a Rock-like Fuel", Mat. Res. Soc. Symp. Proc., 412,15 (1995).

[19]  H. Kleykamp, "Selection of Materials as Diluents for Burning of Plutonium Fuels in Nuclear Reactors", *J. Nucl. Mater.*, 275,1 (1999).

[20]  M. Burkhartz, *et al.*, "Inert Matrices for the Transmutation of Actinides: Fabrication, Thermal Properties and Radiation Stability of Ceramic Materials", *J. Alloy and Compds.*, 271-273, 544 (1998).

[21]  M. Beauvy, *et al.*, "Actinide Transmutation: New Investigation on some Actinide Compounds", *ibid., idem.*, 557.

[22]  M.A. Pouchon, *et al.*, "Behavior of Caesium Implanted in Zirconia Based Inert Matrix Fuel", *J. Nucl. Mater.*, 274, 61 (1999).

[23]  Hj. Matzke, V.V. Rondinella and T. Wiss, "Material Research on Inert Matrix: A Screening Study", *ibid., idem.*, 47.

[24]  K.E. Sickafus, *et al.*, "Radiation Damage Effects in Zirconia", *ibid., idem.*, 66.

[25]  E.A.C. Neeft, *et al.*, "Neutron Irradiation of Polycrystalline Yttrium Aluminate Garnet, Magnesium Aluminate Spinel and α-alumina", *ibid., idem.*, 79.

[26]  N. Chauvin, *et al.*, "In-pile Studies of Inert Matrices with Emphasis on Magnesia and Magnesium Aluminate Spinel", *ibid., idem.*, 91.

[27] T. Yamashita, *et al.*, "In-pile Irradiation of Plutonium Rock-like Fuels with Yttria Stabilized or Thoria, Spinel and Corundum", *ibid., idem.,* 98.

[28] Y-W. Lee, *et al.*, "Preparation of Simulated Inert Matrix Fuel with Different Powders by Dry Milling Method", *ibid., idem.,* 7.

[29] R.J.M. Konings, *et al.*, "Transmutation of Actinides in Inert-matrix Fuels: Fabrication Studies and Modeling of Fuel Behaviors", *ibid., idem.,* 84.

[30] L. Wang, *et al.*, "Effect of Xe Ion Implantation in Spinel and Yttria Stabilzed Cubic Zirconia", *Prog. Nucl. Energy,* 38, 295 (2000).

**Table 1. Status of large-scale MOX fuel utilisation in thermal reactors (as of end 1998) [9]**

| Countries | Number of thermal reactors | | | |
| --- | --- | --- | --- | --- |
| | Operating | Licensed for MOX | Loaded with MOX | Applied for MOX license |
| Belgium | 7 | 2 | 2 | – |
| France | 57 | 20 | 17 | 8 |
| Germany | 19 | 12 | 10 | 4 |
| Japan | 52 | 3 | 1 | 1 |
| Switzerland | 5 | 3 | 3 | – |
| Total | 130 | 40 | 33 | 13 |

**Table 2. Some important properties of selected candidate materials for the inert matrix [19]**

| Properties | $Al_2O_3$ | $MgAl_2O_4$ | $ZrO_2$ | ZrN | AlN |
| --- | --- | --- | --- | --- | --- |
| Melting point (°C) | 2 054 | 2 105 | 2 170 | 2 960 | Incongruent |
| Vapour pressure (bar) | $\sim10^{-6}$ | – | $\sim10^{-6}$ | $\sim10^{-7}$ | $\sim10^{-1}$ |
| Thermal conductivity(W/K·m) | 8.2 (1 000°C) 5.8 (1 500°C) | 7.7 (1 000°C) 8.0 (1 500°C) | 2.2 (1 000°C) 1.5 (1 500°C) | 23 (1 000°C) 26 (1 500°C) | 36 (1 000°C) 27 (1 500°C) |
| Remarks | No reaction with coolant | No reaction with coolant | No reaction with coolant | – | Compatible |
| | No reaction with cladding | No reaction with cladding | No reaction with cladding | Possible formation of $Zr_4M_2N$ | Compatible |

# INNOVATIVE MOX FUEL FOR FAST REACTOR APPLICATIONS

**K. Bakker\* and H. Thesingh**
Nuclear Research and Consultancy Group (NRG)
P.O. Box 25, 1755 ZG Petten
The Netherlands
\*k.bakker@nrg-nl.com

**T. Ozawa, Y. Shigetome, S. Kono and H. Endo**
Japan Nuclear Cycle Development Institute (JNC)
Japan

**Ch. Hellwig, P. Heimgartner, F. Ingold and H. Wallin**
Paul Scherrer Institut (PSI)
Switzerland

## Abstract

JNC, PSI and NRG have started a project to investigate the properties of MOX Sphere-pac and MOX Vipac fuels under fast reactor conditions and to compare these properties with those of MOX pellet fuels and with the results of computer models. The project consists roughly of the following phases:

- Testing of various production techniques, production and characterisation of the various fuel pins.

- Irradiation at high power of these various types of fuels.

- Post-irradiation examination and validation of computer models.

This paper describes the details of the current stage of the project:

- The fuel segments to be irradiated.

- The layout of the irradiation experiment.

- The irradiation scenarios.

- The results of the neutronics and fuel behaviour computations.

## Introduction

Japan Nuclear Cycle Development Institute (JNC), Paul Scherrer Institut (PSI) and Nuclear Research and Consultancy Group (NRG) have signed an agreement to compare the behaviour under irradiation of Sphere-pac fuel, vibration packed (Vipac) fuel and pellet fuel. The name of this project is FUJI (Fuel irradiations for JNC and PSI). The comparison of the various types of fuel should provide information on their suitability as fast reactor fuel. The results will also serve to validate and optimise fuel behaviour models. The fuel and the fuel pins will be fabricated by PSI. These pins will be transported to NRG at Petten and will be irradiated in the High Flux Reactor (HFR) at Petten. The main part of the PIE will also be performed by NRG. The research is mainly focused on Sphere-pac fuel. Sphere-pac fuel is one of the candidates for simple and remote manufacturing of advanced fuel for minor actinide (MA) and plutonium burning in a fast reactor. The inherent advantages of the sphere fabrication and the packing of the spheres in a fuel pin are the absence of dust and the suitability of the process steps for remote operation [1]. Irradiation tests in fast reactors with carbide Sphere-pac fuel were conducted in the 80s [2,3]. Later, oxide Sphere-pac fuel was also tested in LWRs [4,5]. However, there exists only little technical knowledge concerning the fabrication and irradiation behaviour of Sphere-pac fuels containing minor actinides and high concentrations of plutonium in the form of uranium plutonium oxide (MOX).

The objectives of this programme are as follows:

- Compare the fuel performance of Sphere-pac fuel, pellet fuel and Vipac fuel under the same irradiation conditions.

- Study the restructuring of fresh (un-irradiated) Sphere-pac fuel up to melting temperatures.

- Study the MA and plutonium burning in a reactor irradiation with a radial temperature profile close to that expected in a fast reactor spectrum.

- Study the plutonium and MA migration and the restructuring behaviour at high linear heat ratings in a fuel irradiation experiment with a radial temperature profile close to that expected in a fast neutron flux.

In order to achieve the above-mentioned objectives the fuel will be irradiated during a relatively short period (maximum 132 hours) at relatively high power (maximum 800 W/cm) in the FUJI irradiation rig. This rig is based on the long-standing experience at the HFR with similar fast reactor fuel irradiation rigs [6].

## Fuel segments

Within the FUJI project, a total of eight fuel pins will be irradiated, each pin consisting of two segments. Thus in total 16 segments will be irradiated. Before irradiation the two segments will be connected by a screw and fixed to each other by spot welding, thus forming one fuel pin (Figure 1). Each segment has its own gas plenum, causing the segments to behave independently of each other. Each fuel pin will be mounted in one sample holder. Since the FUJI device can hold two irradiation sample holders simultaneously, in total four successive irradiations will be performed. Fourteen segments contain $(U_{0.75}Pu_{0.25})O_2$ fuel, while two segments contain $U_{0.7}Pu_{0.25}Np_{0.05}O_2$ spheres. Neptunium is inserted in the fuel in order to test the possibility to transmutate neptunium as a part of an overall recycling strategy. Nine segments contain Sphere-pac fuel, two segments contain Vipac fuel and five segments contain pellet fuel. The Sphere-pac fuels will be prepared with two different densities caused

**Figure 1. Schematic drawing of two disconnected fuel segments.**
**The upper segment contains pellet fuel and the lower segment contains**
**Sphere-pac fuel. After connection of the two segments a fuel pin is formed.**

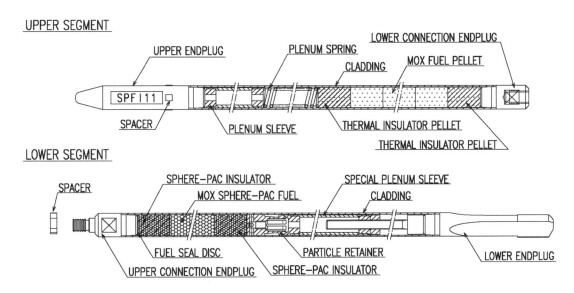

by different filling techniques. The details of the segments to be irradiated are shown in Table 1. The fuel is contained in a stainless steel cladding (PNC 1520) with an inner diameter of 6.7 mm and an outer diameter of 7.5 mm. The lengths of the fuel stacks are 250 mm, except for three of the pellet stacks that have a length of only 50 mm. The main specifications of the three types of fuel to be irradiated are described in the following sections.

**Table 1. The details of the segments to be irradiated**

| | Initial sintering $\leq 550$ W/cm | | Restructuring (1) $\leq 550$ W/cm | | Restructuring (2) $\leq 550$ W/cm | | PTM $< 900$ W/cm | |
|---|---|---|---|---|---|---|---|---|
| | *Upper segments* | | | | | | | |
| Fuel type | MOX | MOX | MOX | MOX | MOX | MOX | MOX | MOX |
| Fuel form | Pellet | Sphere | Pellet | Sphere | Pellet | Sphere | Pellet | Sphere |
| Smear density [%TD] | 86.6 | 78-81 | 86.6 | 78-81 | 86.6 | 78-81 | 86.6 | 78-81 |
| Fuel density | 92%TD | 95%TD | 92%TD | 95%TD | 92%TD | 95%TD | 92%TD | 95%TD |
| Fuel column length [mm] | 50 | 250 | 50 | 250 | 50 | 250 | 250 | 250 |
| | *Lower segments* | | | | | | | |
| Fuel type | MOX | MOX | MOX | Np-MOX | MOX | MOX | MOX | Np-MOX |
| Fuel form | Sphere-pac | Vipac | Sphere-pac | Sphere-pac | Sphere-pac | Vipac | Pellet | Sphere-pac |
| Smear density [%TD] | 78-81 | ? | 78-81 | 78-81 | 72-73 | ? | 86.6 | 78-81 |
| Fuel density | 95%TD | 95%TD | 95%TD | 95%TD | 95%TD | 95%TD | 92%TD | 95%TD |
| Fuel column length [mm] | 250 | 250 | 250 | 250 | 250 | 250 | 250 | 250 |

MOX = $(U_{0.75}Pu_{0.25})O_2$
Np-MOX = $(U_{0.7}Pu_{0.25}Np_{0.05})O_2$

## Sphere-pac fuel

Some initial fabrication tests are performed jointly by JNC and PSI, but the final fabrication of the 16 fuel segments will be done by PSI. The internal gelation technique will be used for the production of the spheres. Three batches of sphere sizes will be made for the $(U_{0.75}Pu_{0.25})O_2$ fuel (being 800 micron, 190 micron and 70 micron). The Sphere-pac fuel will consist of two sphere sizes. Using two size fractions increases the density of the fuel that can be obtained compared to fuel that contains only one size fraction. Two methods will be used to fill the two sphere sizes of the $(U_{0.75}Pu_{0.25})O_2$ fuel into the cladding:

1) Simultaneous filling of the 800 micron and the 190 micron fraction resulting in a smear density of 72-73%.

2) Infiltration filling of the 70-micron fraction between the 800-micron spheres, giving a smear density of 78-81%.

The $(U_{0.7}Pu_{0.25}Np_{0.05})O_2$ fuel will be made by infiltration filling of 800 micron and 70 micron fuel. Therefore in total three types of Sphere-pac fuel will be irradiated: $(U_{0.75}Pu_{0.25})O_2$ fuel with two densities and $(U_{0.7}Pu_{0.25}Np_{0.05})O_2$ fuel with one density. It has been confirmed experimentally that it is feasible to make these approximate densities using these two filling techniques. The smear density is a parameter that has a large impact on the fuel behaviour, due to its large influence on the thermal conductivity of the fuel. For this reason it is interesting from a modelling point of view to vary the density of the fuel in a controlled manner. The ends of the Sphere-pac fuel stacks will be thermally isolated using depleted $UO_2$ spheres. Tests have shown that the filling of fuel and insulation spheres can be done in such a manner that the intermixing of the fuel spheres and the insulation spheres is negligible. Fuel seal discs made of tungsten are mounted between the end plugs and insulator spheres to prevent the finest particles entering into the clearance between the end plugs and the cladding. After filling of the spheres a so-called particle retainer is placed on the spheres to keep them in position.

## Vipac fuel

Vipac fuel consists of more or less randomly-shaped fragments fabricated by sintering granules from granulated pre-compacts. The Vipac fuel will consist of 3-6 size fractions, which will be obtained by sieving. These fragments will be filled and vibrated after filling in a similar manner as Sphere-pac fuel. The two Vipac segments to be made will be identical and their density will be similar to that of the Sphere-pac fuel. The internal geometry of the Vipac fuel segments is generally identical to that of the Sphere-pac fuel segments, except that in the Vipac fuel segment, fuel seal discs are placed between insulator spheres and Vipac particles to prevent intermixing.

## Pellet fuel

Five fuel segments will be filled with pellet fuel. $UO_2$ and $PuO_2$ powder are mixed in an attrition mill and zinc stearate and pore-former are added afterwards. The powder is pre-compacted, granulated, pressed to pellets and sintered. The pellets will be ground to the right outer dimensions. The homogeneity of the thus-obtained Pu/U distribution will be checked using alpha autoradiography. The fuel for the five pellet segments will be identical. The internal components in the pellet segments (spring, depleted $UO_2$ insulation pellets, etc.) are similar to those in standard fuel pins.

## The design of the irradiation experiment

The irradiation will be performed in the High Flux Reactor (HFR) in Petten in the FUJI irradiation rig. This rig is optimised to allow for irradiation of the segments at high power under well-controlled conditions. The power generated in the fuel segments will be continuously monitored. The irradiation of the 16 segments will be done in four separate irradiation runs. In each of these irradiation runs four segments will be irradiated under almost identical conditions. The predicted fuel behaviour (temperature distribution, restructuring, melting, etc.) will also be discussed. The complete irradiation experiment consists schematically of the following components:

- Fuel segments.

- Fuel pins (two fuel segments make up one fuel pin).

- The sample holder. In each sample holder one fuel pin is positioned.

- The irradiation rig. One irradiation rig contains two sample holders.

The layout of the fuel segments and the connection of two segments into one fuel pin have been described in the previous sections. The sample holder which contains the fuel pin and the irradiation rig in which two sample holders are placed are described hereafter.

### The sample holder

The fuel cladding is surrounded by stagnant sodium. This sodium is surrounded by a molybdenum shroud tube, which carries the instrumentation (thermocouples and neutron fluence monitor sets) and which homogenises the temperature of the fuel cladding. Stagnant sodium is also positioned outside of the molybdenum shroud. In order to assure the clearance between a fuel pin and the molybdenum shroud tube, there will be a spacer on the upper end plug of the segments and at the connector between the two segments. All is placed in two stainless steel containments separated by a small gas gap. This gap is filled with a binary mixture, helium/neon or neon/nitrogen, in order to adjust the temperature of the fuel cladding. The width of the gap is varied axially to maintain the cladding outside temperature approximately constant over the length of the cladding. The outside of the outer stainless steel containment is water-cooled. A schematic horizontal cross-section of the sample holder is shown in Figure 2. The sample holder is inserted in the irradiation rig as is discussed in the next section.

**Figure 2. Schematic horizontal cross-section of the sample holder.
The sample holder is surrounded by flowing coolant water.**

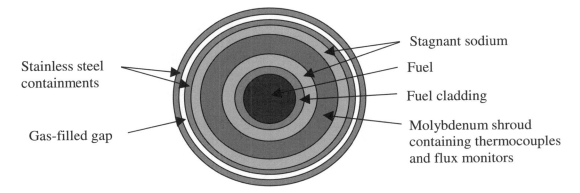

*The irradiation rig*

The irradiation rig (Figure 3) is designed to contain two sample holders. Each sample holder is surrounded by coolant water. The fuel power is determined by measuring the flow and temperature increase of the coolant in each channel. The irradiation rig is positioned on a trolley outside the reactor core. Moving the irradiation rig with respect to the reactor core varies the power in the fuel pins. The irradiation rig is made of aluminium with stainless steel and silver neutron absorbers to homogenise the circumferential power distribution in the fuel. The HFR has a core with fuel elements that have a fuelled height of 60 cm, which causes a significant axial flux buckling. This axial flux buckling is beneficial to the present experiment, since it causes an axial power variation over the height of the fuel stacks, which allows studying the restructuring at various powers under irradiation conditions that are otherwise identical.

**Figure 3. Schematic horizontal cross-section of the irradiation rig. The rig is mainly surrounded by water. Details of the sample holder are shown in Figure 2.**

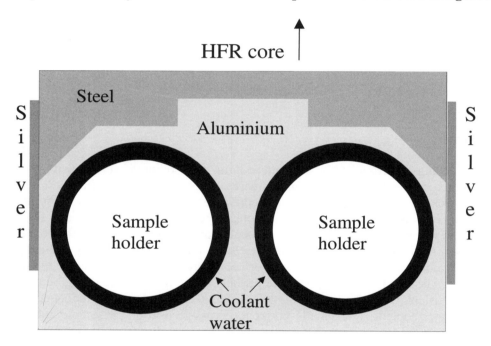

**Irradiation scenario**

In total four different irradiations will be done, designated Initial Sintering, Restructuring (1), Restructuring (2) and Power to Melt (PTM). The duration of all four irradiations is rather short (a few days), since the phenomena to be studied (restructuring, necking, sintering, etc.) occur on this short time scale. The segments that will be irradiated in the various irradiations are shown in Table 1. All irradiations will start with a power increase from 0 W/cm to about 530 W/cm in 36 hours. After this the Initial Sintering irradiation will be stopped. The Restructuring (1) irradiation and the Restructuring (2) irradiation will continue at about 530 W/cm for 48 hours and 96 hours, respectively, and then stopped. The Power to Melt irradiation will be performed under identical conditions as Restructuring (1) before the irradiation will be interrupted. After this a neutron radiogram will be made. After this phase a ramp-type irradiation up to a maximum of 800 W/cm will be performed, which will be interrupted several times in order to make intermediate neutron radiograms. Prior and after all four irradiations a neutron radiogram will be made. The neutron radiograms will be made to study the formation of a central void in the fuel and to study possible fuel melting. By making neutron radiograms during the interruptions of the irradiation, it is feasible to obtain additional information on the behaviour of the fuel during irradiation.

### *The irradiation conditions*

Two types of computations are discussed hereafter.

1) Neutronics computations.

2) Fuel behaviour computations.

It should be kept in mind that the computational results shown hereafter serve as examples of the large set of design computations that have been done up to now.

### *Neutronics computations*

Neutronics computations have been done to determine the radial and circumferential power distribution in the fuel pins under irradiation. These computations have been done using both the Monte Carlo code MCNP and the neutronics code package WIMS7 combined with the diffusion code HFR-TEDDI. These two codes (which are independent of each other) have been used in order to obtain a good knowledge on the accuracy of the results. Very good agreement was observed between the outcome of the two codes. The main result of the computations was that: i) the circumferential power variation in the fuel pins is very small, which is important to obtain a good experiment, and ii) a radial power variation in the fuel pins is present (Figure 4), due to the fact that the HFR is a light-water-cooled reactor. This radial power profile is taken into account in the fuel behaviour computations.

**Figure 4. Radial power profile in Sphere-pac fuel as used for the fuel
behaviour computations. Relative radius equals zero is the centre of the fuel.**

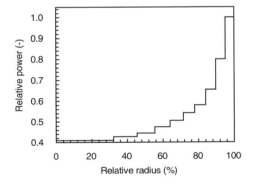

### *Fuel behaviour computations*

The fuel behaviour computations have been done with the JNC code CEPTAR [7]. This code was originally meant to describe the behaviour of pellet fuel. It has been adapted in order to be able to describe Sphere-pac fuel as well, by removing the gap between the pellets and the cladding and by using a low density for the fuel in the code. The thermal conductivity in the CEPTAR code has been calibrated for the use of 77-81% dense Sphere-pac fuel, using Sphere-pac fuel irradiation data from earlier Power-to-Melt tests. After this calibration of CEPTAR, a comparison with the thermal computation results of the PSI code SPHERE [8] was performed, and it was confirmed that CEPTAR provided similar results to SPHERE for this particular type of experiment. Since SPHERE uses a mechanistic method to calculate the change of thermal conductivity due to in-pile sintering, it can be generally applied to Sphere-pac performance calculations. The advantage of using the adapted CEPTAR code without any dedicated Sphere-pac models for this particular investigation is that it can predict both the pellet and the Sphere-pac thermal behaviour. This is of importance to the FUJI experiment in which the behaviour of both types of fuels is compared.

The radial power profile is different in the FUJI case compared to the fast reactor case. This would lead to a lower central temperature in the case of the FUJI experiment compared to the fast reactor case, when the power in both cases would be equal. Since this difference is unacceptable for our objects, the fuel power, the pellet-cladding gap width and the cladding temperature of the FUJI segments are adapted in such a manner that the resulting radial temperature profile in the fuel in both cases is similar. An example of the results of these computations is shown in Figure 5, from which it can be concluded that the fast reactor temperature distribution is approached in the FUJI case within approximately 50°C.

**Figure 5. The radial temperature profiles in Sphere-pac fuel in the case of the FUJI experiment**

*○ – 530 W/cm, cladding temperature 400°C and in a fast reactor ● – 410 W/cm, cladding temperature 510°C.*
*The temperature difference (×) between both profiles is shown on the right axis (SPHERE-results).*

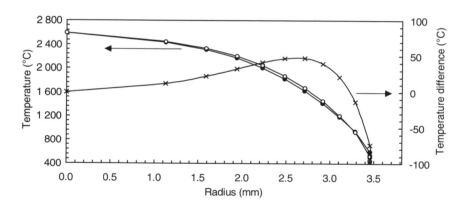

The radial temperature profile is the parameter that determines phenomena such as fuel restructuring and fuel melting. Therefore it is much more important to have a temperature profile that is identical to fast reactor conditions than to have a power profile that is identical to fast reactor conditions. For this reason in the FUJI design the temperature profile has been optimised.

With the determination of the required fuel power, the pellet-cladding gap width and the cladding temperature of the FUJI segments the behaviour of the complete fuel segments have been predicted (Figure 6). The data shown in Figure 6 are for the axial maximum flux position. Due to the axial flux variation the behaviour of each axial location is different, which is shown in Figure 7. This figure shows the prediction of the axial dependence of the various radial zones present in the fuel at the end of the irradiation at a maximum power of 700 W/cm. The widths of the following three regions are predicted: the columnar grain region, the molten fuel region and the central void region. In the neutron radiograms, which are made during and after the irradiation, these regions will be studied in order to

**Figure 6. The time dependence of the predicted fuel central temperature and the relative central void diameter for pellet and Sphere-pac fuel during the Restructuring (2) irradiation**

*For the axial location of the present computations the fuel power is increased from*
*0 W/cm to 530 W/cm in 36 hours and then kept constant (CEPTAR and SPHERE results)*

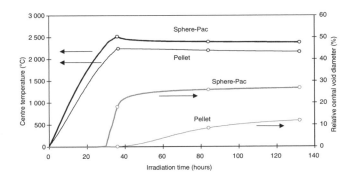

**Figure 7. Axial variation of the various radial zones and the variation of the fuel power (○). The various radial zones are: columnar grain region (■), molten fuel (▲) and central void (+).**

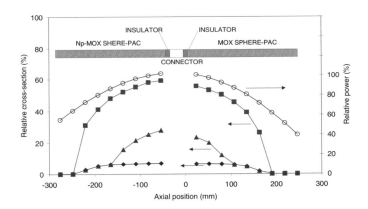

compare the predicted with the actual situation. Included in Figure 7 is the axial variation of the fuel power, which is due to the flux buckling in the HFR core. This axial power variation, inducing an axial temperature variation, causes the axial variation of the widths of the various regions.

In parallel with the above-mentioned CEPTAR computations, SPHERE computations have also been done. The SPHERE code contains various models that are dedicated for Sphere-pac fuel, such as models that describe the sintering and necking behaviour of spheres. The interaction of these microscopic phenomena with macroscopic phenomena such as thermal conductivity is taken into account in detail.

**Post-irradiation examination**

The post-irradiation examination (PIE) that will be performed consists, amongst others, of the following parts:

- Neutron radiography, which will be performed prior, during and after irradiation.

- Profilometry.

- X-ray photography.

- Ceramography.

- Electron probe micro analysis.

The results of the PIE will be used, amongst others, for the optimisation and the validation of the codes CEPTAR and SPHERE.

# REFERENCES

[1]    G. Ledergerber, Ch. Hellwig, F. Ingold, Z. Kopajtic, H. Endo, M. Morihira, Y. Nakajima, S. Nomura, "Simplifying the Process Flow Sheet for Actinide Fuel Fabrication, Preparation for an Irradiation Test in JOYO", Proc. Int. Conf. on Future Nuclear Systems, GLOBAL'99, Jackson Hole, USA, 29 August-3 September (1999).

[2]    A. Delbrassine, L. Smith, "Pellet and Sphere-pac (U,Pu)C Fuel Comparative Irradiation Tests", *Nucl. Techn.*, 49,129-135 (1980).

[3]    R.E. Mason, C.W. Hoth, R.W. Stratton, F. Botta, "Irradiation and Examination Results of the AC-3 Mixed-carbide Test", *Trans. Am. Nucl. Soc.*, 66, 215-217 (1992).

[4]    R.W. Stratton, F. Botta, R. Hofer, G. Ledergerber, F. Ingold, C. Ott, J. Eindl, H.U. Zwicky, R. Bodmer, F. Schlemmer, "A Comparative Irradiation Test of $UO_2$ Sphere-pac and Pellet Fuel in the Goesgen PWR", Int. Topical Meeting on LWR Fuel Performance, Avignon, France, 174-183, 21-24 April (1991).

[5]    L-Å. Nordström, H. Wallin, Ch. Hellwig, "A Comparative Irradiation Test of Pellet and Sphere-pac Fuel (IFA-550.9)", Enlarged Halden Programme Group Meeting on High Burn-up Fuel Performance, Safety and Reliability, Loen, Norway, 24-29 May 1999, OECD Halden Reactor Project, HPR-351/25.

[6]    J. Ahlf and A. Zurita, "High Flux Reactor (HFR) Petten – Characteristics of the Installation and the Irradiation Facilities", European Commission report EUR 15151 EN (1993).

[7]    T. Ozawa, H. Nakazawa and T. Abe, "Development and Verifications of Fast Reactor Design Code "CEPTAR"", ICONE-9, No. 346, Nice, France, 8-12 April 2001.

[8]    H. Wallin, L.Å. Nordström, Ch. Hellwig, "The Sphere-pac Fuel Code SPHERE-3", Proc. IAEA TCM on Nuclear Fuel Behaviour Modelling at High Burn-up, Windermere (UK), June 2000.

# BEHAVIOUR OF ROCK-LIKE OXIDE FUELS UNDER REACTIVITY-INITIATED ACCIDENT CONDITIONS

**Kazuyuki Kusagaya, Takehiko Nakamura, Makio Yoshinaga,
Hiroshi Akie, Toshiyuki Yamashita and Hiroshi Uetsuka**
Japan Atomic Energy Research Institute (JAERI)

## Abstract

Pulse irradiation tests of three types of un-irradiated rock-like oxide (ROX) fuel – yttria-stabilised zirconia (YSZ) single phase, YSZ and spinel ($MgAl_2O_4$) homogeneous mixture and particle-dispersed YSZ/spinel – were conducted in the Nuclear Safety Research Reactor to investigate the fuel behaviour under reactivity-initiated accident conditions. The ROX fuels failed at fuel volumetric enthalpies above 10 $GJ/m^3$, which was comparable to that of un-irradiated $UO_2$ fuel. The failure mode of the ROX fuels, however, was quite different from that of the $UO_2$ fuel. The ROX fuels failed with fuel pellet melting and a part of the molten fuel was released out to the surrounding coolant water. In spite of the release, no significant mechanical energy generation due to fuel/coolant thermal interaction was observed in the tested enthalpy range below ~12 $GJ/m^3$. The YSZ type and homogenous YSZ/spinel type ROX fuels failed by cladding burst when their temperatures peaked, while the particle-dispersed YSZ/spinel type ROX fuel seemed to have failed by cladding local melting.

## Introduction

Rock-like oxide (ROX) fuel, a kind of inert matrix fuel, has been developed at the Japan Atomic Energy Research Institute (JAERI) as an optional method for burning excess plutonium (Pu) in light water reactors (LWRs) [1]. One of the advantages of the ROX-LWR system is almost complete Pu burning because it is free of $^{238}$U transformed into Pu. However, also due to the absence of fertile nuclides such as $^{238}$U, the system will have extremely small Doppler feedback (negative reactivity coefficient), which may cause severe condition for fuel rods in reactivity-initiated accidents (RIAs).

Some remedies for improving the coefficient are thus considered: addition of resonant nuclides in ROX fuel, or heterogeneous core configuration with ROX and UO$_2$ fuels. Akie, *et al.* [2] show that, through a reactor physics study, the coefficient can be improved to the level of conventional UO$_2$ fuel core by adding resonant nuclides such as $^{232}$Th or $^{238}$U to ROX fuel. Since the addition of such nuclides reduces the Pu transmutation rate, the amount of the additives should be optimised. To determine the optimum condition, we must first obtain the limiting RIA condition where the ROX fuel can survive, i.e. the failure threshold in fuel enthalpy.

The ROX fuel has different properties from those of UO$_2$ fuels in many respects: melting temperature, density, thermal conductivity and thermal expansion coefficient, etc. [3]. Therefore, the failure threshold and the failure mechanism of ROX fuel are expected to be different from those of UO$_2$ fuel. These should be examined in order to assess the applicability of the ROX-LWR system. To this end, ROX fuel behaviour has been experimentally investigated under simulated RIA conditions [4,5].

## Experimental

### Test fuel rod

Three kinds of un-irradiated ROX fuel, simulated by replacing Pu with U, were tested:

(1) Type Z: single phase of solid solution of UO$_2$ and yttria-stabilised zirconia (YSZ).

(2) Type SH: nearly homogeneous mixture of UO$_2$-YSZ phase and spinel (MgAl$_2$O$_4$).

(3) Type SD: particle-dispersed type where UO$_2$-YSZ particles with diameter of ~300 μm are dispersed in spinel matrix.

The material composition for each kind is shown in Table 1. The $^{235}$U enrichment is 19.5% for each case. The mechanical and thermal properties of the U-ROX fuels were confirmed to be similar to those of Pu-ROX fuel [3], indicating that results of the present study using the U-ROX fuels are valid for the original Pu-ROX fuel.

### Table 1. Specifications of ROX test fuel rod for NSRR pulse irradiation tests

| Element | Overall/stack length | | 279 mm/135 mm |
|---|---|---|---|
| Pellet | Material composition (mol%) | Z | UO$_2$: YSZ = 20 : 80 |
| | | SH | UO$_2$: YSZ : MgAl$_2$O$_4$ = 25 : 27 : 48 |
| | | SD | UO$_2$: YSZ : MgAl$_2$O$_4$ = 37 : 20 : 43 (YSZ = ZrO$_2$: Y$_2$O$_3$ = 88-81 : 12-19) |
| | $^{235}$U enrichment | | 19.5 wt.% |
| | Diameter/length | | 8.05 mm/9.0 mm (P/C radial gap = 0.085 mm) |
| Cladding | Material | | Zircaloy-4 |
| | Outer/inner diameter | | 9.5 mm/8.22 mm (thickness = 0.64 mm) |

Fourteen ROX fuel pellets were enclosed in a Zircaloy-4 cladding tube. The structure of the test fuel rod is shown in Figure 1. The pellet stack length is 135 mm and the rod whole length is 279 mm. The other dimensional specifications of the cladding and the pellets are the same as those of 17 × 17 PWR type-B fuel. The fuel rod was filled with helium gas at pressures of 0.1-0.4 MPa. The specifications of the test fuel rod are summarised in Table 1.

**Figure 1. Schematic drawing of the ROX fuel rod for the NSRR pulse tests**

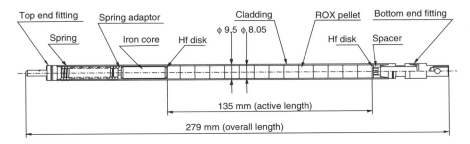

## Pulse irradiation

Reactivity-initiated accident (RIA) simulating experiments have been performed for the ROX fuel rods using the Nuclear Safety Research Reactor (NSRR) in JAERI. The NSRR is a TRIGA-Annular Core Pulse Reactor (ACPR) utilising U-ZrH fuel-moderator elements. The configuration of the NSRR core is illustrated in Figure 2. The reactor has a dry experimental cavity of 220 mm in inner diameter penetrating the core centre. The test fuel rod confined in an experimental capsule (Figure 3) is set in the experimental cavity and exposed to a high pulsed neutron flux simulating RIA conditions in commercial LWRs. The capsule is a sealed container of about 200 mm in outer diameter and 1 230 mm in height, and is filled with coolant water. A maximum reactivity insertion of 3.36%$\Delta k/k$ ($4.6) from zero power is allowed to produce a peak reactor power of 21 GW, a total reactor energy release of 108 MJ, and the full width at half maximum (FWHM) of 4.4 ms.

**Figure 2. Horizontal cross-section of the NSRR core**

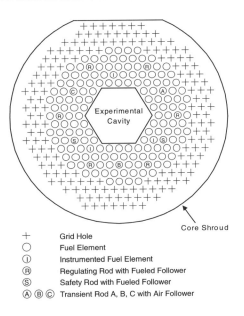

125

**Figure 3. Experimental capsule and instruments for pulse irradiation tests for ROX fuels**

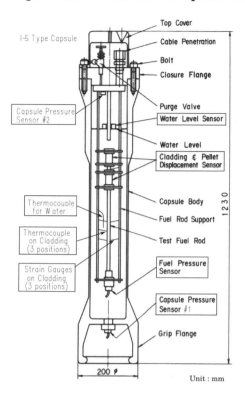

Eight pulse irradiation tests have been performed. The test conditions are listed in Table 2. The principal test parameter is peak fuel enthalpy ($E_p$) which ranged from 3.2 to 12.4 GJ/m$^3$. The FWHM of the pulses are ranged from 5-14 ms (narrower width for stronger pulse). Rod power histories for the particle dispersed (type SD) ROX fuel rods are shown in Figure 4 for examples. The fuel rods were irradiated under the condition of room temperature, atmospheric pressure and stagnant water cooling.

**Table 2. Pulse irradiation condition of the ROX fuel tests**

| ROX fuel type | Test ID | Inserted reactivity ($\%\Delta k/k$) | Peak fuel enthalpy $E_p$ (GJ/m$^3$) | (kJ/g) |
|---|---|---|---|---|
| Z | 945-1 | 1.47 | 3.2 | 0.5 |
| | 945-2 | 2.69 | 9.0 | 1.5 |
| | 945-3 | 3.20 | 11.5 | 1.9 |
| SH | 943-1 | 1.97 | 5.6 | 1.0 |
| | 943-2 | 2.74 | 9.1 | 1.6 |
| | 943-3 | 3.29 | 12.4 | 2.2 |
| SD | 947-1 | 2.35 | 8.9 | 1.6 |
| | 947-2 | 2.83 | 12.2 | 2.2 |

During a pulse irradiation test, transient behaviour of the test fuel rod was monitored with instruments shown in Figure 3. These include: R-type thermocouples for cladding surface temperature, K-type thermocouples for surrounding coolant water temperature, strain gauge type pressure sensors for capsule internal pressure, a strain gauge type pressure sensor connected to the rod bottom end for

**Figure 4. Examples of power history of the test fuel rod**

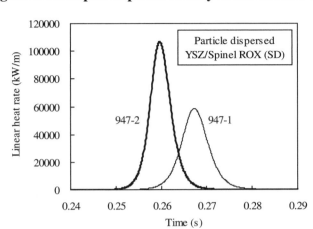

fuel rod internal pressure, linear variable differential transformer (LVDT) sensors for pellet stack and cladding axial elongation, strain gauges for cladding hoop strain, and a float type level sensor for coolant water surface movement which follows fuel rod failure. Depending on the test condition, some of the equipments were selected to use.

After pulse irradiation, the test rod was disassembled from the experimental capsule and subjected to the following examinations: appearance observation, cladding diameter measurement with a laser type profilometer, optical microscopy observations of the rod cross-sections. In the case of fuel rod failure, total weight and size distribution of the released fuel pellet fragments were measured.

## Results and discussion

### *Transient behaviour*

The transient records of cladding surface temperature ($T_{cs}$), rod internal pressure ($P_{rod}$), capsule internal pressure ($P_{cap}$) and water column movement sensor signal ($E_{wcm}$) are shown in Figure 5 for the tests with the largest enthalpy inserted for each kind of ROX fuel. For $T_{cs}$, the data from the thermocouple at the centre of the pellet stack are shown for tests 945-3 and 943-3, while at 20 mm above the centre for test 947-2 due to the centre thermocouple failure. The temperature rose immediately after the pulse insertion at the time of ~0.2 s, and peaked at about 1-2 s. The maximum temperatures were about 1 850, 1 750 and 1 650 K for types Z, SH and SD, respectively. The maximum cladding temperatures are shown in Figure 6 for all the tests and the un-irradiated $UO_2$ data as a function of peak fuel enthalpy. Naturally, the maximum temperatures increase with peak fuel enthalpy. The dependence for type Z is almost identical to that of $UO_2$ fuel, while the temperature of the two kinds of spinel matrix ROX fuels (type SH and SD) were lower by about 200 K than that of type Z and $UO_2$ fuel at the same enthalpy level, because of the lower fuel melting (eutectic) point of the YSZ/spinel system (2 210 K) as compared to those of YSZ (3 000 K) and $UO_2$ (3 110 K). The solid and dotted lines in the figure represent temperatures calculated by a transient analysis code FRAP-T6 [6] to which ROX fuel property tables were added. The code can well estimate the peak temperature for ROX fuels both with and without spinel matrix.

As seen in Figure 5, fluctuations in the water column movement sensor signal ($E_{wcm}$) were observed in the three tests. A half wavelength of the signal means the time interval during which the water level ascends or descends by 3 mm. Since the water level movement should result from the fuel

**Figure 5. Examples of transient records. The mark ▲ indicates the rod failure timing.**

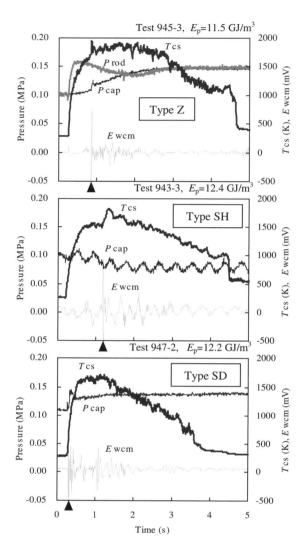

**Figure 6. Relationship between cladding peak temperature and peak fuel enthalpy of ROX and UO₂ fuels**

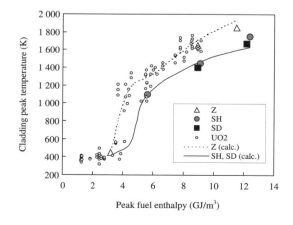

rod failure, the beginning of the fluctuation should be the rod failure timing (this was confirmed by capsule internal pressure spike). The failure timings are determined to be 0.60, 0.97 and 0.04 s after the pulse insertion for types Z, SH and SD ROX fuels, respectively. It is noted that the cladding temperature became as high as ~1 600 K at the failure time for type Z and SH, while it was still low as ~500 K for type SD ROX fuel. The $E_{wcm}$ signals shown in Figure 5 are too small to evaluate the water level escalation quantitatively, meaning only ripples of the water surface. This indicates that mechanical energy generation due to thermal interaction between released hot fuel and surrounding coolant water is not significant.

Transient records of fuel pellet stack elongations are shown in Figure 7 for three un-failed ROX fuel rods with $E_p \sim 9$ GJ/m$^3$. The maximum elongation of the type Z fuel was larger than those with spinel matrix. This can be attributed to the higher temperature reached and larger thermal expansion coefficient of Z type ROX fuel pellet. One can note a strange behaviour for the Z type fuel – fast shrinkage at ~0.7 s. This is presumably due to fuel pellet melting. The melt fraction estimated by the FRAP-T6 code for the rod is 35%, which is larger than those of the others. The results of the transient measurements are summarised in Table 3 with calculation results by the code.

*Fuel rod failure*

The ROX fuel rods failed at the highest enthalpy tested for each kind: 11.5 GJ/m$^3$ for type Z, 12.4 GJ/m$^3$ for type SH and 12.2 GJ/m$^3$ for type SD, as shown in Figure 5. The other test rods with lower fuel enthalpies remained intact. The appearances of the failed fuel rods are shown in Figure 8. A black oxide layer was observed on the cladding surface for all the failed and intact fuel rods excluding one rod with the lowest enthalpy inserted (test 945-1, $E_p = 3.2$ GJ/m$^3$). Wavy deformation of the cladding surface was observed only for the 945-3 test rod (type Z), which probably results from higher cladding peak temperature as described in the previous section. For each failed rod, there was one failure opening. Their axial positions were 65, 30 and 1 mm from the bottom end of the fuel stack for types Z, SH and SD, respectively. Note that the failure position of the type SD fuel rod is the bottom end of the fuel stack. Through the opening, a part of molten fuel pellet was released out to coolant water. The weight ratios of ejected fuel are 50, 70 and 78% for types Z, SH and SD, respectively. The most typical size of dispersed fuel fragments are 2-4 mm (32 wt.%) for type Z, and 0.5-1 mm for types SH (37 wt.%) and SD (39 wt.%). Such fragment size is relatively larger than that observed for an un-irradiated UO$_2$ fuel which failed at $E_p \sim 19$ GJ/m$^3$ and generated large mechanical energy in the NSRR experiment where 40 wt.% of the fragments was smaller than 0.5 mm [7].

**Figure 7. Transient behaviour of fuel pellet stack elongation for un-failed ROX fuel rods**

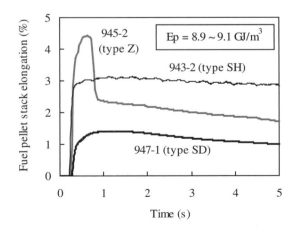

**Table 3. Summary of transient measurement records and FRAP-T6 calculation results**

| Fuel type | YSZ single phase (Z) | | | Homogeneous YSZ/spinel (SH) | | | Particle-dispersed YSZ/spinel (SD) | |
|---|---|---|---|---|---|---|---|---|
| Test ID | 945-1 | 945-2 | 945-3 | 943-1 | 943-2 | 943-3 | 947-1 | 947-2 |
| Peak fuel enthalpy (GJ/m$^3$) | 3.2 | 9.0 | 11.5 | 5.6 | 9.1 | 12.4 | 8.9 | 12.2 |
| Failure time after pulse insertion (s) | – | – | 0.60 | – | – | 0.97 | – | 0.04 |
| Max. cladding outer surface temperature (K) | 450 [460] | 1 650 [1 710] | 1 850 [1 940] | 1 100 [1 100] | 1 450 [1 470] | 1 750 [1 630] | 1 400 [1 510] | 1 650 [1 670] |
| Cladding outer surface temperature at failure (K) | – | – | ~1 600 | – | – | ~1 600 | – | ~500 |
| Max. pellet centre temperature (K) | [1 770] | [2 820] | [2 940] | [1 890] | [2 210] | [2 210] | [2 210] | [2 210] |
| Pellet release fraction (%) [Pellet melt fraction (%)] | – [0] | – [35] | 50 [84] | – [0] | – [14] | 70 [55] | – [8] | 78 [54] |
| Max. pellet stack elongation (%) | 1.2 [1.9] | 4.4 [3.9] | – [3.9] | 2.1 [1.7] | 3.1 [2.1] | – [2.1] | 1.4 [2.1] | – [2.1] |
| Max. cladding elongation (%) | 0.0 [0.7] | 1.0 [2.9] | – [3.1] | – [0.3] | 0.6 [0.9] | – [1.0] | 0.8 [0.9] | – [1.1] |
| Max. cladding hoop strain (transient) (%) | > 0.2 [0.5] | > 0.6 [2.5] | > 0.5 [2.7] | 0.05 [0.5] | – [0.9] | – [1.1] | >0.2 [2.0] | – [4.6] |
| Max. cladding hoop strain (residual) (%) * | ~0 [0.0] | ~1 [1.8] | – [1.8] | ~0 [0.0] | ~1.5 [0.4] | ~4 [0.4] | ~3 [1.4] | – [4.1] |

[ ] – calculated by FRAP-T6 code, * – measured by diameter profilometry

**Figure 8. Appearance of failed ROX fuels after pulse irradiation.
The arrow indicates the position of failure opening.**

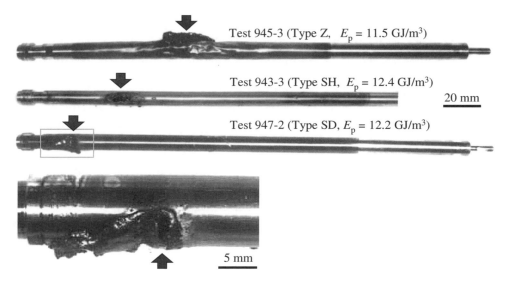

Test 945-3 (Type Z, $E_p$ = 11.5 GJ/m$^3$)

Test 943-3 (Type SH, $E_p$ = 12.4 GJ/m$^3$)

20 mm

Test 947-2 (Type SD, $E_p$ = 12.2 GJ/m$^3$)

5 mm

The failure threshold for the three kinds of ROX fuels are summarised in Figure 9 in terms of peak fuel enthalpy together with that for fresh UO$_2$ fuels obtained from many experiments in the NSRR. When one compares the failure thresholds of ROX fuels with those of UO$_2$ fuels, it is appropriate to use the enthalpy per unit fuel volume rather than per unit mass, because of a large difference in fuel density between the ROX (~6 g/cm$^3$) and UO$_2$ fuels (~10 g/cm$^3$). For the three kinds of ROX fuel, the failure threshold enthalpy is found to be comparable to (or somewhat large than) that of UO$_2$ fuel (~10 GJ/m$^3$). This indicates that the current UO$_2$ failure limit can be applied to ROX fuel designing.

**Figure 9. Failure threshold enthalpy of ROX fuel compared with that of fresh UO₂ fuel**

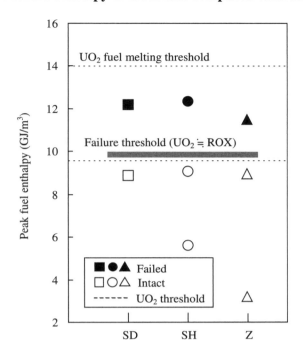

The failure mechanism of ROX fuels, however, is found to be quite different from that of UO₂ fuel. The cross-sections of the failed ROX fuels are shown in Figure 10 together with that of typically failed UO₂ fuel [8]. The cutting positions are at or just above (for 947-2) the failure openings. Above the failure threshold, fresh UO₂ fuel typically fails by cracking of embrittled cladding thinned by melting. The failure occurs at the moment of quenching (when the boiling condition goes back to nucleate boiling from film boiling) which causes thermal stress on the embrittled cladding. Fuel pellets remain solid and are not released from the rod below the UO₂ fuel melting threshold of 14 GJ/m³. If the peak fuel enthalpy goes over this threshold, the UO₂ fuel pellet will be molten and the cladding tube will break into pieces. Then, all the molten fuel will be instantly dispersed into coolant water, generating significant mechanical energy such as water hammer and pressure impulse due to fuel/coolant thermal interaction.

On the other hand, ROX fuel pellets are already molten at the failure threshold due to their lower melting point (in fact, fuel pellet partial melting is observed at a lower enthalpy of ~9 GJ/m³), and the molten fuel pellets are released from a cladding failure opening. In spite of the release, no significant mechanical energy generation due to fuel/coolant interaction was observed in the transient measurements as described above. This is probably because the ejection speed of molten fuel was extremely low due to a limited opening area and a small increase of rod internal pressure (in the successfully measured case test 943-3, it was 0.16 MPa at maximum as shown in Figure 5). A slow ejection of molten fuel will lead to a small rate of steam generation via thermal interaction with coolant, which will result in a small mechanical energy generation. This result on mechanical energy generation for the ROX fuels is encouraging for its application in LWR. However, in the case of irradiated ROX fuels, fission gases released in the fuel rod could cause higher rod internal pressure, which might lead to significant mechanical energy generation.

The failure mechanism of types Z and SH ROX fuels must be "high temperature cladding burst" which results from a decrease in cladding strength due to high temperature and from increased internal pressure by elevating internal temperature. This is judging from the shapes of the failure openings,

**Figure 10. Cross-sections of failed UO₂ and ROX fuel rods. The cutting positions are at the failure cracking or opening for (a)-(c), and just above the opening for (d).**

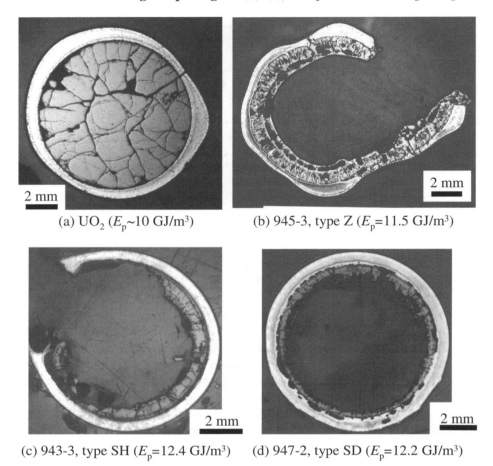

(a) UO$_2$ ($E_p$~10 GJ/m$^3$)  (b) 945-3, type Z ($E_p$=11.5 GJ/m$^3$)

(c) 943-3, type SH ($E_p$=12.4 GJ/m$^3$)  (d) 947-2, type SD ($E_p$=12.2 GJ/m$^3$)

which have significant residual deformations, as well as from the measured cladding temperatures – as high as ~1 500 K – at the moment of the failures. In the case of such burst-like failure, the failure threshold will be influenced by the rod internal pressure. Therefore, the threshold might be decreased for irradiated ROX fuel rods with increase in fission gas release.

For type SD ROX fuel, the failure mechanism seems to be "cladding local melting". This is because the failure opening at the bottom of the fuel stack looks like a melting hole (Figure 8) where the residual strain was very limited, while the measured cladding temperature near mid of the fuel stack was as low as ~500 K at the time of failure. Relatively large axial power peaking at the bottom pellet (13%, estimated from neutron calculation) and the inhomogeneous fuel particles clustering in as-fabricated fuel pellet could have contributed to the cladding local heating.

**Conclusions**

Fuel behaviour under RIA conditions was investigated by pulse irradiation tests in the NSRR for three kinds of ROX fuels – single phase YSZ type (Z), homogenous YSZ/spinel type (SH) and particle dispersed YSZ/spinel type (SD). The failure threshold of the ROX fuels is at least 10 GJ/m$^3$, which is comparable to that of UO$_2$ fuel. There is no apparent difference in the threshold among the three kinds of ROX fuel, though the cladding peak temperature of type Z is higher than those of types SH and SD.

The failure mechanisms of the ROX fuels are quite different from that of $UO_2$ fuel. The types Z and SH ROX fuels failed by high temperature cladding burst, while the type SD ROX fuel seemed to have failed by cladding local melting. These ROX fuels failed together with fuel pellet melting, and a large amount of molten fuel was released into the surrounding coolant water. In spite of the release, no significant mechanical energy generation due to fuel/coolant thermal interaction was observed. The test results – equivalent failure threshold to that of $UO_2$ and no significant mechanical energy generation – indicate that the ROX fuels, at least the fresh ones, can be utilised in LWRs from the viewpoint of fuel behaviour under RIA conditions.

# REFERENCES

[1]    H. Akie, T. Yamashita, N. Nitani, H. Kimura, H. Takano, T. Muromura, A. Yasuda and Y. Matsuno, Workshop Proceedings of Advanced Reactors with Innovative Fuels, Villigen, Switzerland, pp.199-208, 21-23 October (1998).

[2]    H. Akie and T. Nakamura, *Progress in Nuclear Energy*, 38, 363-370 (2001).

[3]    N. Nitani, T. Yamashita, T. Matsuda, S.-i. Kobayashi and T. Ohmichi, *J. Nucl. Mater.*, 274, 15-22 (1999).

[4]    T. Nakamura, H. Akie, K. Okonogi, M. Yoshinaga, K. Ishijima and H. Takano, Workshop Proceedings of Advanced Reactors with Innovative Fuels, Villigen, Switzerland, pp. 299-309, 21-23 October (1998).

[5]    T. Nakamura, K. Kusagaya, M. Yoshinaga, H. Uetsuka and T. Yamashita, *Progress in Nuclear Energy*, 38, 379-382 (2001).

[6]    L.J. Siefken, C.M. Allison, M.P. Bohm and S.O. Peck, NUREG/CR-2148, EGG-2104 (1981).

[7]    T. Tsuruta, S. Saito and M. Ochiai, JAERI-M 84-235 (1985) (in Japanese).

[8]    T. Fuketa, K. Ishijima and T. Fujishiro, *J. Nucl. Sci. Technol.* 33, 43-51 (1996).

# THEORETICAL REQUIREMENTS TO TOLERANCES TO BE IMPOSED ON FUEL ROD DESIGN PARAMETERS FOR RBEC-M LEAD-BISMUTH FAST REACTOR

**A. Vasiliev, P. Alekseev, K. Mikityuk, P. Fomichenko, A. Shestopalov**
Russian Research Centre "Kurchatov Institute", Kurchatov sq., 123182, Moscow, Russia
Phone: +7-095-196-70-16, Fax: +7-095-196-37-08, E-mail: avas@dhtp.kiae.ru

## Abstract

Development of advanced reactors with innovative materials requires comprehensive analysis of fuel rod design parameters as well as tolerances to be imposed on these parameters. Currently, it is considered traditional to estimate uncertainties in core neutronics parameters on the basis of known tolerances imposed on fuel rod design parameters. However, requirements to some core neutronics parameters of advanced reactors can be first formulated and then taken into account, while developing the technologies for innovative fuel rod manufacturing, i.e. an "inverse" problem can be solved. The aim of this problem is to find combinations of fuel rod design tolerances which provide that selected core neutronics parameters remain within specified deviations during base irradiation.

## Introduction

Increased requirements concerning the safety and reliability of advanced nuclear reactors impose rather strict limitations on possible uncertainties with regard to a number of reactor parameters. The reactor technical project should include proof that the important reactor parameters remain within permissible limits during reactor operation, in particular accounting for available fuel rod design tolerances.

The solution to this particular problem first requires a selection of fuel rod design parameters to be optimised. An analysis of uncertainties in reactor parameters and sensitivities of these uncertainties to the selected fuel rod design parameters would then be required, as would be the development of recommendations for upgrading of requirements of design and manufacturing technology for some reactor components.

At the same time, there currently exists neither industrial manufacturing experience or extended test data on deviations of design parameters for innovative fuel rods of advanced reactor. Therefore, tolerances for the manufacture of these fuel rods are not known. However, the tolerances, required from the viewpoint of reactor safety and reliability, can be estimated at the stage of development of new technologies. These estimates can be based on calculations of fractions which certain fuel rod parameters contribute in total uncertainties of major reactor parameters. Such an analysis can be useful in particular for determination of technological stages which should be improved, and for cost optimisation of innovative fuel rod manufacturing stages.

Thus, the following task can be formulated: taking into account tolerances which are well-known for available fuel rod manufacturing industry, find required tolerances for those design parameters of fuel rod, which relate to new fuel type. This search is aimed at provision of specified deviations in core neutronics parameters under base irradiation conditions.

The main stages of the proposed analysis can be the following:

- Choice of important reactor operational and safety parameters to be optimised (reactivity balance, core power peaking factors, etc.); specification of ultimate permissible uncertainties for these parameters.

- Specification of fuel rod design parameters for which required tolerances are searched (influence of the chosen parameters on core neutronics are not well-known because of new fuel type: fuel density, plutonium content in fuel and fuel stoichiometry, etc.). Tolerances for other fuel rod parameters are fixed based on technological experience.

- Calculational estimates of tolerances, which could be imposed on fuel rod design parameters, while manufacturing, to provide specified reactor parameters during base irradiation.

- Verification of the results by a statistical analysis based on multiple direct calculations.

The proposed theoretical analysis of requirements for fuel rod design parameters for the RBEC fast reactor is presented in the paper. An option of RBEC reactor with Pb-Bi coolant and nitride fuel (U-Pu)N, specified as RBEC-M, is considered.

## RBEC reactor

The RBEC reactor [1-4] is an advanced fast modular reactor of medium power with lead-based liquid coolant. The project of RBEC reactor facility with 900 MWt and 340 MWe was developed in Russia by OKB Gidropress, RRC "Kurchatov Institute" and IPPE, with the participation of the Bochvar Institute and RIAR (Dimitrovgrad). The aim of the RBEC project was the creation of a nuclear steam supply system on the basis of Russian experience in design and operation of fast reactors and liquid-metal technology. The RBEC design and thermal-hydraulic parameters are based, as much as possible, on the technical decisions proved in BN-type sodium-cooled fast reactors, light water thermal reactors and transport nuclear power units. These decisions correspond to existing experience in the use of fuel, structural materials and technology of liquid-metal coolant.

In the framework of optimisation studies, a number of RBEC reactor options with different fuel compositions, core and primary circuit layouts were considered. In particular, an option of the RBEC-M reactor with Pb-Bi coolant and mixed nitride fuel (U-Pu)N is currently considered. Nitride with an enrichment of 99.9% by $^{15}$N is used. A fuel cycle was studied with the use of depleted uranium nitride and reactor-grade plutonium as a fuel. The isotopic composition of the reactor-grade plutonium is given in Table 1 [5].

**Table 1. Isotopic composition of reactor-grade plutonium in fresh FAs of RBEC-M reactor**

| Isotope | $^{238}$Pu | $^{239}$Pu | $^{240}$Pu | $^{241}$Pu | $^{242}$Pu | $^{241}$Am |
|---|---|---|---|---|---|---|
| Content, mass % | 1.32 | 60.32 | 24.27 | 8.33 | 4.95 | 0.81 |

This isotopic composition corresponds to plutonium generated in irradiation of enriched uranium fuel in a typical PWR reactor of 900 MWe up to a burn-up of 33 MWd/kgU. Fuel is reprocessed after 10 years of cooling and loaded in the reactor for two years [5].

Three zones with different fuel rod diameters are used in the RBEC-M core to flatten fields of power, coolant temperature and velocity in the core. The forth type of pin is used as a fertile rod in the lateral blanket. A mixed uranium-plutonium nitride fuel with a density of 13.3 g/cm$^3$ and a plutonium content of 13.7% is used in all core zones. Axial and lateral blankets contain pellets of depleted uranium nitride. Fuel rods have claddings of ferritic-martensitic stainless steel EP-823 (12%Cr-Si).

A fuel cycle accepted in the RBEC-M reactor consists of six partial fuel cycles of 300 effective full-power days (EFPD) each. The reactor is shut down for refuelling for 60 days. FAs are not reshuffled for all fuel cycle duration (1 800 EFPD).

## Mixed uranium-plutonium nitride fuel

The basis of interest in the application of nitride fuel is the following important and attractive properties:

- High theoretical density of nitride fuel (14.3 g/cm$^3$) allows for increase of fuel mass fraction in the core and, hence, increase of reactor breeding ratio and decrease of reactivity change with burn-up.

- Nitrides compared to oxides have higher thermal conductivity, which grows with temperature increase. This leads to a decrease in the centreline fuel temperature, reduces core energy deposition, drops radial temperature gradient in fuel and allows to significantly increase the linear heat generation rate of fuel rods.

- Nitride fuel is satisfactorily compatible with steel claddings of fuel rods, and compares well with oxide fuel on this parameter.

A number of negative properties of individual nitrides of uranium and plutonium, as well as mixed nitride, are noted in literature.

The properties of nitrides of individual heavy nuclides and their mixtures, which require special attention when they are used as a reactor fuel, include the following:

- The high capture cross-section of $^{14}N$, of which natural nitrogen contains 99.62%, deteriorates reactor economic indices. Generation of radioactive $^{14}C$ as a result of nuclear reactions of $^{14}N$ can complicate fuel reprocessing and solutions for ecological problems.

- The relatively high swelling rate of mixed uranium-plutonium fuel due to accumulation of fission products restricts ultimate fuel burn-up. Besides, it should be noted that nuclear reactions $^{14}N(n,\alpha)$ and $^{14}N(n,p)$ occur in nitride fuel with participation of $^{14}N$, as a result of which helium and hydrogen are generated. This leads to the formation of a large number of pores. At high temperatures the sizes of these pores grow and the fuel-swelling rate abruptly increases. When nitrogen enriched by $^{15}N$ is used, the negative impact of these reactions correspondingly reduces.

- Fission gas release from nitride fuel significantly depends on temperature. Gas release from mixed nitride fuel grows with temperature increase. The availability of oxygen and carbon admixtures in nitride can lead to a significant increase in gas pressure inside the fuel rod.

The application of a heterogeneous FA design based on fuel rods with $(U-Pu)O_2$ and fertile rods with dense fuel was considered a measure for increasing the average fuel density for the basic RBEC concept. Such an approach was considered as a step toward mastering the technologies for fuel compositions of increased density.

The following features characterise an option of RBEC-M reactor design, proposed in the present study:

- Mixed uranium and plutonium nitride is used as fuel charged in the core in traditional quasi-homogeneous layout. Nitride of depleted uranium is used as fertile material of breeding zones.

- The design value of peak fuel burn-up is limited by the value of 13-14% h.a.

- The nitride fuel is assumed to be manufactured by technology, which provides the level of admixtures of oxygen and carbon in fuel below 0.1% mass each [6], and which allows for manufacturing of fuel pellets with density of 90% of theoretical density and higher.

- Nitrogen is used with enrichment of 99.9% by $^{15}N$, providing both good neutron-physical parameters and acceptable amounts of $^{14}C$, generated in-reactor for the fuel irradiation period.

## Considered tolerances imposed on design parameters of RBEC-M reactor fuel rods

Tolerances imposed on the design parameters of RBEC-M fuel rods and considered at the given stage of reactor development are displayed in Table 2. The tolerances for cladding diameter were assumed to correspond to the tolerances of the unified fuel rod of BN-600 and BN-350 reactors [7]. The content of oxygen and carbon admixtures in mixed mononitride fuel is assumed to be below 0.1 mass%.

The following design parameters directly relate to nitride fuel utilisation and are interesting for estimating the contribution which their deviations introduce in the total uncertainty in the main core neutronics parameters: fuel density, diameter of fuel pellets, plutonium content in fuel, fuel stoichiometry.

The importance of deviations in fuel density and diameter of fuel pellets is determined by the high density of nitride fuel. Even insignificant deviations of these parameters from their nominal values introduce considerable perturbations in core neutronics parameters.

Plutonium content in fuel also significantly impacts core neutronics parameters, primarily because of the considerable difference in reactivity contributions between plutonium isotopes and $^{238}$U.

Fuel stoichiometry has an important value for nitride fuel, based on natural nitrogen or nitrogen with low-enrichment by $^{15}$N, because of the high neutron capture cross-section of $^{14}$N. The conducted calculations showed that the use of nitrogen with enrichment of 99.9% by $^{15}$N in the RBEC-M reactor project renders the sensitivity of reactor neutronics parameters to fuel stoichiometry very insignificant. In this analysis the deviation of the mass content of nitrogen in fuel was conservatively assumed to be 5%.

Thus, the following tolerances were accepted as optimised tolerances for design parameters of RBEC-M fuel rods:

- Fuel density ($\pm x$).

- Plutonium content in fuel ($\pm y$).

- Diameter of fuel pellets ($-z$).

In general, probabilistic distribution law for tolerances should be specified on the basis of results of measurements during fuel rod manufacturing and assembly (empirical distribution of parameters of a specific manufacturing batch of fuel rods). Normal (Gauss) and gamma distribution laws were assumed in the present calculations (see Table 2). A confidence interval of the tolerances was assumed to be 99.7%.

Due to the current absence of data on the statistical distribution of design parameters of RBEC-M fuel rods, the following conservative assumptions were made in this study:

1) Design parameters of fuel rods are constant along the height and are independent of each other.

2) FAs are manufactured by batches for each core zone. The number of FAs in a manufacturing batch is equal to the number of FAs loaded in the corresponding core zone at the beginning of each partial fuel cycle. All fuel rods of all FAs from one manufacturing batch are assumed to have the same design parameters.

## Table 2. Main design parameters and tolerances for the manufacture of RBEC-M reactor fuel rods

| Parameter | Nominal value and tolerance | | | Supposed distribution law |
| --- | --- | --- | --- | --- |
| | Core zone 1 | Core zone 2 | Core zone 3 | |
| Outer cladding diameter, mm | 7.0±0.03 | 7.5±0.03 | 8.6±0.03 | Normal |
| Inner cladding diameter, mm | 6.5±0.03 | 7.0±0.03 | 8.1±0.03 | Normal |
| Total O and C content in fuel, mass % | $0_{+0.1}$ | | | Gamma |
| Atom ratio N/(U+Pu), rel. units | 1±0.05 | | | Normal |
| Fuel density, g/cm$^3$ | 13.3±x | | | Normal |
| Mass fraction of PuN in fuel | 0.1370±y | | | Normal |
| Fuel diameter, mm | $5.7_{-z}$ | $6.2_{-z}$ | $7.2_{-z}$ | Gamma |

## Studied core neutronics parameters of the RBEC-M reactor

Parameters such as core reactivity, breeding ratio, power peaking factors, control rod efficiencies, reactivity effects, etc. can be considered as the main core neutronics parameters, specified uncertainties of which could be used for determining the requirements of tolerances of fuel rod design parameters.

Uncertainties introduced in estimates of core reactivity margins by uncertainties in fuel rod design parameters can be provided by two reasons: direct core reactivity perturbation caused by perturbation of a fuel rod design parameter, and deviation in reactivity swing with burn-up caused by a corresponding change in fuel breeding parameters.

In the present study the core reactivity in equilibrium fuel cycle was chosen for analysis as a main neutronics parameter of the RBEC-M reactor. The uncertainty in fuel burn-up reactivity effect caused by the impact of deviations in fuel rod design parameters on the core breeding ratio (CBR) was found to be insignificant because the RBEC-M CBR is close to 1. It was assumed in the study that the total uncertainty in reactor criticality should not exceed the effective fraction of delayed neutrons. This value was estimated to be 0.0037 for the RBEC-M reactor.

The core power peaking factor is another parameter, which, to a considerable extent, determines the reliability of fuel rods and reactor safety, and rather significantly depends on tolerances in fuel rod design parameters. The criteria for ultimate uncertainties in core power peaking factors can be obtained from estimates of fuel rod failure thresholds based on a statistical thermal-mechanical analysis of fuel rod behaviour under base irradiation. It was assumed in this study that the searched values of tolerances for the design parameters of RBEC-M fuel rods should provide uncertainties in volumetric power peaking factor $K_v$ to be below 5%.

Thus, the following criteria for selection of tolerances of studied design parameters of RBEC-M reactor fuel rods were chosen in the given study:

$$\begin{cases} \Delta\rho \leq \beta_{eff} \\ \Delta K_v / K_v \leq 0.05 \end{cases} \tag{1}$$

where $\rho$ is the core reactivity $(k_{eff} - 1)/k_{eff}$ and $K_v$ is the core volumetric power peaking factor.

## Calculational technique

The calculational technique is based on the assumption that algorithms of Classic and Generalised Perturbation Theory of first order are applicable in the case when fuel rod design parameters deviate in the range of corresponding tolerances. Applicability of these algorithms to such tasks were checked by comparison with results obtained by multiple direct statistical calculations of core neutronics parameters for advanced fast reactors of BREST and RBEC types with lead-based coolant and nitride fuel. The deviations in estimates of uncertainties in core reactivity and power peaking factor obtained by first-order Classic and Generalised Perturbation Theory and the direct statistical calculations lay in the range of 5% and 10%, respectively.

The assumption that deviations of design parameters do not correlate with each other, and between different manufacturing batches of fuel rods, and the hypothesis that core neutronics parameters are distributed according to the normal law allow for the estimation of the resulting uncertainties in the core neutronics parameters caused by possible deviations of fuel rod parameters. These uncertainties can be written in a general form as:

$$D_F^2 = \sum_K \left(d_{FK}^x\right)^2 + \sum_K \left(d_{FK}^y\right)^2 + \sum_K \left(d_{FK}^z\right)^2 + \sum_{K,m} \left(d_{FK}^{m \neq x,y,z}\right)^2 \tag{2}$$

where $D_F$ is the permissible uncertainty in the considered core neutronics parameter $F$, $K$ is the fuel rod manufacturing batch, $m$ is the perturbed design parameter, $d_{FK}^m$ is the uncertainty in $F$ caused by deviations of $m$. The first three terms in Eq. 2 are unknowns ($x$ is the fuel density tolerance, $y$ is the Pu content tolerance and $z$ is the fuel pellet diameter tolerance), the forth term corresponds to the known tolerances (see Table 2). According to this equation all combinations of studied tolerances ($x,y,z$) can be found, which provides that the requirement of Eq. 1 is met.

The perturbation theory algorithms are based on calculations of sensitivity coefficients of the core neutronics parameters to nuclear concentrations of reactor materials [8]. As already noted, fuel rods from one manufacturing batch were assumed to have the same design parameters, which are constant over the fuel rod height. Deviations of the core neutronics parameters caused by deviations in fuel rod design parameters for each manufacturing batch are calculated with the use of sensitivity coefficients under the linear assumption:

$$\begin{cases} \Delta \rho_K^m = \sum_{jeK} \int_{Vj} \sum_i SC_{i,j}^{k_{eff}} \cdot \Delta \gamma_{i,j}^m \\ \left(\dfrac{\Delta K_v}{K_v}\right)_K^m = \sum_{jeK} \int_{Vj} \sum_i SC_{i,j}^{K_v} \cdot \Delta \gamma_{i,j}^m \end{cases} \tag{3}$$

where $\rho$ is the core reactivity $(k_{eff} - 1)/k_{eff}$; $K_v$ is the core volumetric power peaking factor, $m$ is the considered design parameter, $K$ is the considered fuel rod batch, $j$ is the calculational volume, $i$ is the isotope which nuclear concentration changes with the change of parameter $m$, $SC_{i,j}^{k_{eff}}$ and $SC_{i,j}^{K_v}$ are the sensitivity coefficients of k-effective and $K_v$ to concentration of isotope $i$ in volume $j$, $\Delta \gamma_{i,j}$ is the variation of nuclear concentration of isotope $i$ in volume $j$. The calculational volume is an axial section of the fuel assembly.

The sensitivity coefficients of the core neutronics parameters considered were calculated with the use of the JAR-FR code system [8] under multi-group diffusion approximation. Changes in volume fractions of core materials caused by deviations of fuel rod design parameters were calculated on the basis of data given in Table 2.

## Calculational results

If there are no uncertainties in fuel density, Pu content and fuel pellet diameter ($x = 0$, $y = 0$, $z = 0$), the uncertainties provided by non-optimised parameters (the forth term in Eq. 5), are $0.137\%\Delta(1/k_{eff})$ and $0.62\%$ for core reactivity $\rho$ and power peaking factor $K_v$, respectively.

The calculational results, obtained taking tolerances $x$, $y$, $z$ into account, are shown in Figure 1 and Figure 2. The areas limited by the given surfaces correspond to permissible deviations (tolerances) of the studied fuel rod design parameters, providing the conditions of Eq. 1.

The resulting combination of permissible values of fuel rod design tolerances, required for meeting the requirements of Eq. 1, is given in Figure 3.

Thus, theoretical recommendations for permissible deviations of fuel density, Pu content and fuel pellet diameter can be obtained on the basis of Figure 3. The given combinations of tolerances for these fuel rod parameters provide that the requirement of Eq. (1) be met and imposed on the chosen core neutronics parameters. For instance, when deviation of fuel density is $\pm 0.15$ g/cm$^3$ and deviation of plutonium content is $\pm 0.10$ mass%, then permissible tolerance for fuel pellet diameter should be not higher than -0.05 mm.

To select the optimal combination of permissible tolerances of fuel rod parameters, Figure 3 can be superimposed on a similar surface obtained from requirements for thermal-mechanical parameters (permissible fuel temperature, cladding stresses, etc.), and/or on a similar surface based on cost estimates of provision of required tolerances, etc.

**Figure 1. Surface of ultimate tolerances for fuel density,**
**Pu content and fuel pellet diameter, which provide permissible**
**deviation of core reactivity in equilibrium fuel cycle of RBEC-M reactor**

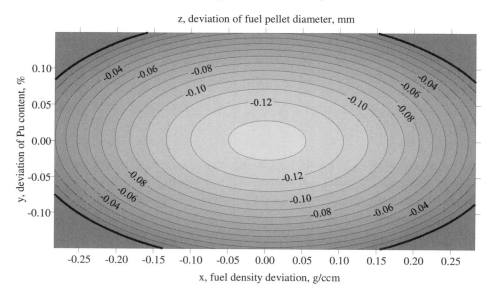

**Figure 2. Surface of ultimate tolerances for fuel density, Pu content and fuel pellet diameter, which provide permissible deviation of power peaking factors in equilibrium fuel cycle of RBEC-M reactor**

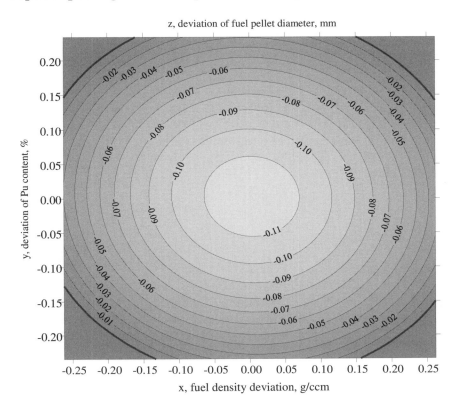

**Figure 3. Surface of ultimate tolerances for fuel density, Pu content and fuel pellet diameter, which provide permissible deviations of core reactivity and power peaking factors in equilibrium fuel cycle of RBEC-M reactor**

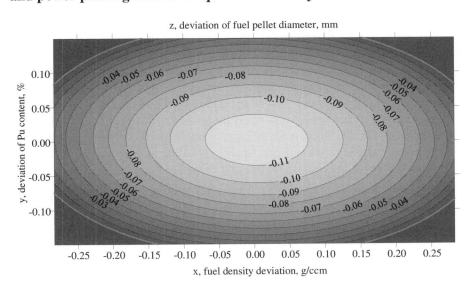

# Conclusions

Currently, it is considered traditional to estimate uncertainties in core neutronics parameters on the basis of known tolerances imposed on fuel rod design parameters. However, requirements for some core neutronics parameters of advanced reactors can be first formulated and then taken into account, while developing the technologies for innovative fuel rod manufacturing, i.e. an "inverse" problem can be solved.

An example of the development of recommendations for the quality of manufacturing of fuel rods for the advanced fast reactor RBEC-M with Pb-Bi coolant and mixed (U-Pu)N nitride fuel is considered in this paper. This analysis is aimed at searching for combinations of tolerances for fuel rod design parameters, which provide specified uncertainties in core criticality (below $\beta$) and volumetric power peaking factor (below 5%). Those fuel rod design parameters, which directly relate to the use of new fuel types, were studied:

- Fuel density.

- Plutonium content in fuel.

- Diameter of fuel pellets.

As a result of the study, a surface of the ultimate tolerances for fuel density, Pu content and fuel pellet diameter was obtained. Thus, all combinations of the tolerances are obtained which provide permissible deviations of core reactivity and power peaking factors in the equilibrium fuel cycle of the RBEC-M reactor.

No doubt the considered criteria of selecting the ultimate permissible tolerances for fuel rod design parameters are not unique. Similar criteria can be developed from the viewpoint of fuel rod thermal mechanics (permissible fuel temperature, cladding stresses, etc.), cost parameters of fuel manufacturing procedures, etc. The theoretical requirements for the quality of fuel rod manufacturing can be superimposed and, thus, a combination of fuel rod design tolerances which are optimal for a number of considered parameters can be found.

A more traditional method of evaluating the uncertainties in neutronics parameters is based on multiple statistical calculations. This method provides values of deviations of core neutronics parameters for certain specified tolerances of fuel rod design parameters, i.e. one point on the surface identical to Figure 3. The value of the proposed technique is that the whole surface of permissible tolerances is built analytically on the basis of algorithms of Classic and Generalised Perturbation Theory and a single calculation of sensitivity coefficients for neutronics parameters. A series of such surfaces can be plotted for more and more strict requirements for uncertainties of core neutronics parameters. This approach could provide a practically complete understanding of uncertainties in core neutronics parameters for any combination of tolerances for design parameters. It seems impossible to conduct such an analysis through direct calculations.

# REFERENCES

[1] P. Alekseev, P. Fomichenko, K. Mikityuk, V. Nevinitsa, T. Shchepetina, S. Subbotin, A. Vasiliev, "RBEC Lead-bismuth Cooled Fast Reactor: Review of Conceptual Decisions", Proc. of the Workshop on Advanced Reactors with Innovative Fuels, ARWIF-2001, Queen Hotel, Chester, UK, 22-24 October 2001.

[2] V. Orlov, N. Ponomarev-Stepnoi, I. Slesarev, P. Alekseev, *et al.*, "Concept of the New Generation High Safety Liquid-metal Reactor (LMFR)", Proc. of Int. Conf. on Safety of New Generation Power Reactors, USA, Seattle, May 1988.

[3] P. Alekseev, S. Subbotin, *et al.*, "Potential Possibilities of a Three-circuit Scheme for the Enhancement of Lead-cooled Reactor Safety", presented at ARS'94 – Int. Topical Meeting on Advanced Reactors Safety, Pittsburgh, PA, USA, April 1994.

[4] V. Orlov, I. Slesarev, P. Alekseev, *et al.*, "Two- and Three-circuit Nuclear Steam Generating Plant With Lead-cooled Fast Reactor (RBEC)", Proc. of the 7th All-Union Seminar on Reactor Physics Problems, "Volga-1991", p. 62, Moscow, USSR, September 1991.

[5] "Le combustible au plutonium. Une évaluation", OCDE, Paris (1989).

[6] R.B. Kotelnikov, S.N. Bashlikov, *et al.*, "High-temperature Nuclear Fuel", Moscow, *Atomizdat*, 1978 (in Russian).

[7] F.G. Reshetnikov, Yu.K. Bibilashvili, I.S. Golovnin, *et al.*, "Development, Fabrication and Operation of Fuel Rods of Power Reactors, Moscow, *Energoatomizdat*, 1995 (in Russian).

[8] A.V. Vasiliev, P.N. Alekseev, P.A. Fomichenko, L.N. Yaroslavtseva, "Development and Evaluation of an Effective Nodal Diffusion Method for Perturbation Theory", Proc. of Annual Meeting on Nuclear Technology-97, Aachen, Germany, 13-15 May 1997.

# ADVANCED PLUTONIUM ASSEMBLY (APA): EVOLUTION OF THE CONCEPT, NEUTRON AND THERMAL-MECHANIC CONSTRAINTS

**Jacques Porta, Bernard Gastaldi, Cécile Krakowiak-Aillaud, Laurence Buffe**
DEN/Cad/DER/SERSI
Bât. 212, CE Cadarache
F-13108 St. Paul-lez-Durance Cedex, France

## Abstract

The APA concept was developed with the aim of increasing the PWR capacity to burn plutonium emerging from the recycling of irradiated fuels in the French park of nuclear power plants.

At first, a concept using annular pins was optimised to allow a good consumption of plutonium while preserving an acceptable neutron control.

To cope with the technological problems and those posed by the manufacture of these annular pins, an alternative concept is presented here. It poses as initial conditions the conservation of both the plutonium balance and the respect of the reactivity control.

## Introduction

The development of the advanced plutonium assembly (APA) concept [1-4] lies within the scope of the plutonium cycle, minor actinides and of reprocessed uranium mastery. This concept brings brief replies to three major problems while allowing:

- Reduction in the radiotoxicity inventory resulting from UOX/PWR cycles.

- Economy of natural resources, by replacing uranium with plutonium.

- Reduction in proliferation risks, by maintaining significant quantities of plutonium in-core.

APA in PWR includes two types of fuel pins. The first is identical to standard $UO_2$ pins; the second contains plutonium on inert matrix. For these fuel pins, the first selected configuration was the annular geometry. It is cladded using an M5 alloy composed of Zr and 1% Nb, and is cooled by the cladding internal and external faces [2].

The core includes only APA assemblies, differing from the current MOX-loaded core in which only 30% of the assemblies of the core contain MOX. This APA assembly makes it possible to master plutonium flows produced by the PWRs and presents the advantage of only concerning approximately 30% of the nuclear park [4,5].

## Strategy

The APA cycle starts with the loading of plutonium from MOX recycling in all the assemblies of a batch. This fuel is discharged after five years of irradiation and reprocessed after five years of cooling. The complete cycle lasts approximately 12 years.

The concept, at the beginning, aims primarily at an increased plutonium burning in the water reactors, by the improvement of *in situ* burning and multi-recycling. The major idea is to separate uranium and plutonium and to place them in different pins, then to replace $^{238}U$ by an inert matrix to remove the source of plutonium per conversion of $^{238}U$ out of $^{239}Pu$ and to increase the thermal component of the neutron spectrum in order to increase plutonium fissions by increasing the moderation ratio near the plutonium pins (more significant volume of water). The needs for neutron control direct towards a local over-moderation in order to increase the efficiency of soluble boron, which also increases the plutonium fissions. This neutron optimisation resulted in designing an annular fuel pin (APA-a), thin because the low thermal conductivity of ceramics induces a reduced thickness to be able to evacuate the power and to limit the heat gradient (Figure 1).

**Figure 1. Original annular APA design**

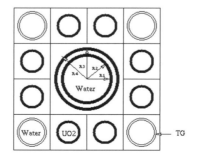

From this concept, the studies carried out show an excellent level of plutonium burning and especially a very great flexibility of the cycle, thanks to many degrees of additional freedom like the content (and the quality, i.e. the isotopic composition) of plutonium but also enrichment in $^{235}U$ of the standard $UO_2$ pins. Neutron control is completely satisfactory (in particular the soluble boron efficiency), and the set of kinetic coefficients is acceptable. From the neutron standpoint, annular configuration led to a good plutonium inventory while limiting the number of APA-loaded reactors in the park.

However, these very positive qualities are somewhat counterbalanced by other less favourable characteristics of the concept, particularly with regard to ceramic (CERCER) fuel pins:

- Manufacturing is delicate: indeed, if on the laboratory level some successes were very quickly obtained in the manufacture of thin rings, the transposition at the industrial level does not seem obvious because of the brittleness of these thin rings which render machining and grinding very difficult. Moreover, the realisation of the pins' higher and lower sealings constitutes a difficult problem for which a solution has not yet been found. Finally, the preliminary thermal-mechanical studies show the need for connecting the internal and the external cladding in order to avoid the buckling of the internal cladding.

- The thermal-hydraulics of ceramic pins with a gap between the pellet and the cladding also poses a problem related to the design of the fuel pin. The hydraulic diameter of the internal channel is much larger than that of the external channel, which leads to an increased internal flow and thus to an overcooling of the internal cladding and consequently to a risk of differential buckling in the case of ceramic fuel pin. Thermal-mechanics show that the ceramic solution, even mean, is not viable because it leads to heat gradients which are too significant. The natural flow redistribution between the inner and outer hydraulic channels because of the significant variations of the hydraulic diameters and of the presence of the grids in the outer zone leads to a reduction (with respect to the standard assemblies) of the margin to the boiling crisis under normal operation, and during the transient of total stop of primary pumps. The solutions to this problem can lie on the one hand in the power reduction in the annular hottest pins during the cycle (in agreement with neutron) and on the other hand in the use of pressure reducers in the inner channel and low resistive grids.

- Thermal-mechanics of the assembly also highlight the need for revisiting the mechanical design of the grids concerning the vibratory behaviour and the seism resistance. Lastly, the design of the assembly lower plate must be completely re-examined in order to homogenise the flow distributions so as to minimise the transverse flows and to correctly cool the plutonium pins.

- The aspects concerning the stiffness of the assembly must be studied with precision insofar as they represent selection criteria to determine and choose some different options. The use of large fuel pins increases the general assembly stiffness whereas the use of elements of a section close to that of the standard pins preserves a stiffness close to that of the current assemblies.

It arises by what precedes that to circumvent the difficulties associated with the annular CERCER concept, the CERMET can be one of the elements of APA concept re-orientation on metal inert matrix, zircaloy for example, insofar as this alloy is well qualified in PWRs.

The good thermal conductivity of CERMET opens possibilities for the choices of the configuration (or configurations) to retain, the cruciform geometry in particular. Nevertheless several criteria must be integrated and in particular that of the zirconium mass which is possible to load in the core, of the volumetric content of fuel in the matrix and the production of minor actinides (MA) in CERMET, and of the in-operation fuel temperature.

In order to be able to carry a total preliminary appreciation on the "small cross" option, a study gathering the evaluations and/or reflections in neutron, thermal-hydraulics, thermal-mechanics and accidents was carried out. The major results are presented here.

**Neutron studies**

*Definition and optimisation of the cross*

One defines a cross by the three following sizes (Figure 2):

- e = cladding thickness.

- E = fuel thickness.

- L = wing length.

**Figure 2. Definition of cross sizes**

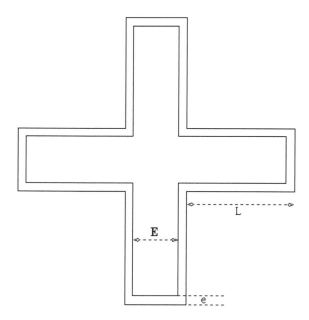

Considering criteria such as: manufacturing properties, Zr mass, total Pu mass in the assembly, then from keeping at best the moderation ratio, the following values are obtained:

- e = 0.25 mm.

- E = 1.50 mm.

- L = 3.50 mm.

A smaller cross (1.5 × 1.5 mm) is also studied; it allows an increase in the local moderation ratio.

Table 1 allows a comparison of various significant parameters of the two cruciform options with a standard pin and an annular APA pin.

The small cross 15 × 35 presents a local moderation ratio very near that of APA-a. A very significant local over-moderation for the smaller crosses 15 × 15 can be noted, as can the fact that the (15 × 15) small crosses and the annular pin have the same average cord overall. In the same neutron spectrum, this implies very similar neutron performances, which is assured here taking into account the very close moderation ratio.

**Table 1. Some major parameters for the four options**

|  | Standard pin | Annular APA pin | Small cross (1.5 × 3.5 mm) | Small cross (1.5 × 1.5 mm) |
|---|---|---|---|---|
| Moderation ratio | 1.6697 | 5.9364 | 5.4520 | 12.6898 |
| Wet perimeter (cm) | 2.9782 | 13.0307 (×36) | 3.6000 (×80) | 2.0000 (×80) |
| Average cord | 0.8200 | 0.2522 | 0.2735 | 0.2500 |

*Calculation methodology*

The exact treatment of the cruciform geometry requires a 2-D exact calculation. The calculation of the evolution of a 17 × 17 assembly with this exact option (including the self-shielding) would represent an important computing time. An "equivalent" geometry was thus required to allow much faster calculations, with adequate precision, in order to provide reliable conclusions.

This equivalence was checked on a simplified lattice including an evolution calculation up to an average 60 GWd/t burn-up rate.

A very significant reduction of the computing time can be noted (Table 2), without a conspicuous degradation of the accuracy of the results.

**Table 2. Comparison of some important parameters
for exact calculation and two approximations**

|  | 2-D exact calculation 2-D exact self-shield | 2-D exact calculation UP0/UP1 self-shield | UP1 calculation UP0/UP1 self-shield |
|---|---|---|---|
| $K_\infty$ BOL | 1.30212 | 1.30235 | 1.30978 |
| $K_\infty$ EOL | 1.04194 | 1.04200 | 1.04432 |
| Pu consumption (kg) | 13.013 | 13.017 | 13.008 |
| Pu production Pu (kg) | 2.741 | 2.734 | 2.731 |
| Pu balance (kg) | -10.272 | -10.283 | -10.276 |
| Calc running time | 16 h | 5 h | 9 mn |

## Results

Four configurations were calculated, with crosses in a $17 \times 17$ assembly. They are presented in Figures 3, 4, 5 and 6.

- *Figure 3* – Configuration 1 = 80 (0.15 × 0.35 cm) crosses on the assembly periphery.

- *Figure 4* – Configuration 2 = 80 (0.15 × 0.35 cm) crosses on the assembly.

- *Figure 5* – Configuration 3 = 76 (0.15 × 0.35 cm) crosses in the centre of the assembly.

- *Figure 6* – Configuration 4 = Configuration 1 with smaller crosses 0.15 × 0.15 cm.

**Figure 3. Configuration 1**

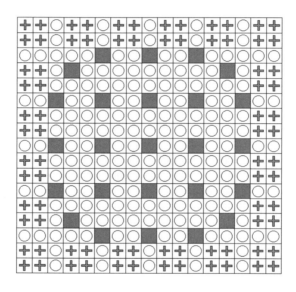

**Figure 4. Configuration 2**

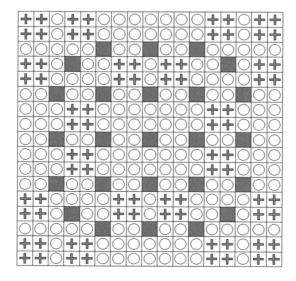

## Figure 5. Configuration 3

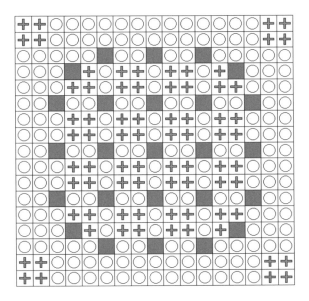

## Figure 6. Configuration 4

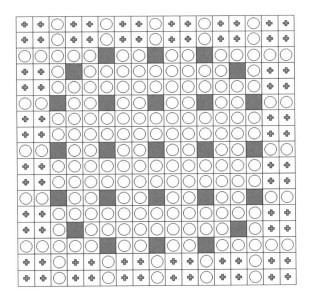

One can notice that the reactivity is not very sensitive to the number of crosses in the assembly, nor to their distribution (Figure 7). On the other hand the size of the cross wing has a considerable effect on the assembly infinite K. Indeed the over-moderation of a smaller cross (0.15 × 0.15 cm) is responsible for a reactivity excess which prolongs the cycle length, representing approximately 3 000 pcm of positive reactivity.

The power distribution is on the other hand very sensitive to the both factors, dimensions and positions of the crosses. Figures 8-11 give the power distributions of the various configurations. One notices high values of the power peaking, about 1.7, quite higher than the usual $F_{xy}$.

## Figure 7. K∞ – all configurations

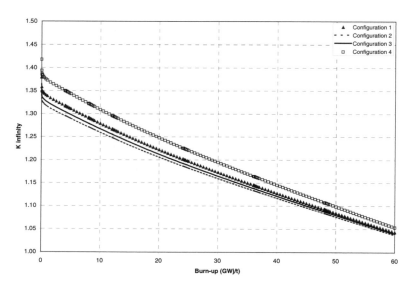

Figure 7. $K_\infty$ – all configurations

Legend:
- ▲ Configuration 1
- - - - Configuration 2
- —— Configuration 3
- □ Configuration 4

## Figure 8. Power distribution of Configuration 1

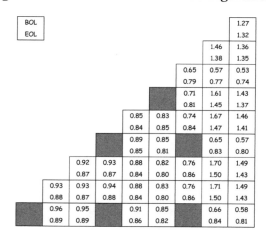

Figure 8. Power distribution of Configuration 1

BOL / EOL values (each cell shows BOL above, EOL below):

|      |      |      |      |      |      |      | 1.27 |
|------|------|------|------|------|------|------|------|
|      |      |      |      |      |      |      | 1.32 |
|      |      |      |      |      |      | 1.46 | 1.36 |
|      |      |      |      |      |      | 1.38 | 1.35 |
|      |      |      |      |      | 0.65 | 0.57 | 0.53 |
|      |      |      |      |      | 0.79 | 0.77 | 0.74 |
|      |      |      |      |      | 0.71 | 1.61 | 1.43 |
|      |      |      |      |      | 0.81 | 1.45 | 1.37 |
|      |      |      | 0.85 | 0.83 | 0.74 | 1.67 | 1.46 |
|      |      |      | 0.84 | 0.85 | 0.84 | 1.47 | 1.41 |
|      |      |      | 0.89 | 0.85 |      | 0.65 | 0.57 |
|      |      |      | 0.85 | 0.81 |      | 0.83 | 0.80 |
|      | 0.92 | 0.93 | 0.88 | 0.82 | 0.76 | 1.70 | 1.49 |
|      | 0.87 | 0.87 | 0.84 | 0.80 | 0.86 | 1.50 | 1.43 |
| 0.93 | 0.93 | 0.94 | 0.88 | 0.83 | 0.76 | 1.71 | 1.49 |
| 0.88 | 0.87 | 0.88 | 0.84 | 0.80 | 0.86 | 1.50 | 1.43 |
|      | 0.96 | 0.95 |      | 0.91 | 0.85 | 0.66 | 0.58 |
|      | 0.89 | 0.89 |      | 0.86 | 0.82 | 0.84 | 0.81 |

## Figure 9. Power distribution of Configuration 2

Figure 9. Power distribution of Configuration 2

BOL / EOL values (each cell shows BOL above, EOL below):

|      |      |      |      |      |      |      | 1.25 |
|------|------|------|------|------|------|------|------|
|      |      |      |      |      |      |      | 1.24 |
|      |      |      |      |      |      | 1.43 | 1.33 |
|      |      |      |      |      |      | 1.27 | 1.20 |
|      |      |      |      |      | 0.64 | 0.57 | 0.53 |
|      |      |      |      |      | 0.74 | 1.05 | 1.00 |
|      |      |      |      |      | 0.70 | 1.61 | 1.44 |
|      |      |      |      |      | 0.80 | 1.38 | 1.31 |
|      |      |      | 0.78 | 0.78 | 0.74 | 1.75 | 1.59 |
|      |      |      | 0.82 | 0.83 | 0.83 | 1.46 | 1.46 |
|      |      |      | 0.77 | 0.76 |      | 0.73 | 0.69 |
|      |      |      | 0.86 | 0.85 |      | 0.81 | 0.80 |
|      | 0.80 | 0.78 | 1.88 | 1.87 | 0.76 | 0.76 | 0.76 |
|      | 0.83 | 0.87 | 1.55 | 1.53 | 0.84 | 0.78 | 0.79 |
| 0.85 | 0.82 | 0.78 | 1.85 | 1.83 | 0.76 | 0.79 | 0.80 |
| 0.86 | 0.84 | 0.87 | 1.54 | 1.52 | 0.85 | 0.80 | 0.81 |
|      | 0.89 | 0.86 |      | 0.72 | 0.71 | 0.82 | 0.81 |
|      | 0.87 | 0.85 |      | 0.83 | 0.82 | 0.81 | 0.82 |

## Figure 10. Power distribution of Configuration 3

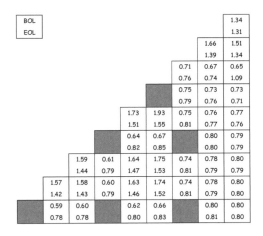

## Figure 11. Power distribution of Configuration 4

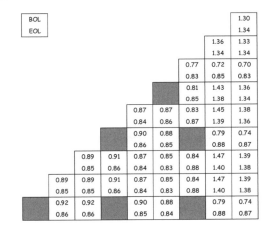

As shown in Table 3, the Pu balance is roughly the same for all the configurations.

### Table 3. Pu balance for the studied configurations

|  | Consumption | Production | Balance |
|---|---|---|---|
| Configuration 1 | 12.124 | 4.407 | -7.717 |
| Configuration 2 | 12.328 | 4.999 | -7.329 |
| Configuration 3 | 12.599 | 4.967 | -7.631 |
| Configuration 4 | 11.310 | 4.068 | -7.242 |

The extreme variations range from 10% on consumption (Configurations 3 and 4), 18% on production (Configurations 2 and 4) to lead to almost 8% on the balance (Configurations 1 and 4).

The first studies show that the $PuO_2$-Zr CERMET fuel in the form of small crosses is an attractive alternative for APA-a, for improved Pu consumption.

The simplified calculation methodology using an annular cylindrisation of the cross shows a good adequacy computer running time and accuracy of the results to obtain the various parameters making it possible to characterise the studied configurations.

## Thermal analysis

At first, the studies focus on the particular thermal behaviour of these fuels, which are innovative considering two standpoints: first because of their composition (material bipolarity: CERamic and METal) but also because of their geometry (which introduces the by-design bipolarity at the assembly level).

The first calculations were performed using the CASTEM 2000 code, in forecast of later mechanical studies.

### *Thermal evaluation APA annular CERMET (20% PuO$_2$/80% Zr)*

Annular CERMET APA is considered, with the following assumptions:

- Composition of CERMET: 20% of PuO$_2$, 80% of zircaloy.

- Annular geometry with following dimensions:

  - Thickness of inner cladding = 0.070 cm.

  - Inner fuel radius = 0.895 cm.

  - External fuel radius = 1.112 cm.

  - Thickness of the external cladding = 0.080 cm.

- Maxwell-Eucken mixture law for the CERMET thermal conductivity:

$$\lambda(CERMET) = \lambda(Zr) * \frac{2\lambda(Zr) + \lambda(PuO_2) - 2\alpha[\lambda(Zr) - \lambda(PuO_2)]}{2\lambda(Zr) + \lambda(PuO_2) + \alpha[\lambda(Zr) - \lambda(PuO_2)]}$$

where $\alpha$ is the volumetric proportion of PuO$_2$ in the CERMET.

- Volumetric power density of $1.2 \times 1.3 \times 741$ W/cm$^3$, corresponding to the APA assembly core hottest pellet (computed value for APA CERCER 280 EFPD)

- Surface temperature of inner and external cladding in contact with the coolant imposed at 347°C (value resulting from thermal-hydraulic calculations with annular CERCER APA).

The interest of this calculation is to ensure coherence between thermal studies already carried out by CEA on annular APA, and the continuation of the investigations in this field.

For this study, the CASTEM 2000 data file was elaborated so as to be able to easily change the composition of the CERMET composite and the thermal properties of materials.

Results: this calculation led to a maximum fuel temperature of 450°C (CERMET centreline temperature). This value is in agreement with the results previously obtained.

## Thermal evaluation for APA small crosses

### CERMET (20% PuO$_2$/80% Zr)

The APA "small crosses" was evaluated from the thermal point of view, with the same assumptions concerning the definition of fuel and its properties (volume fractions: 20% plutonium and 80% zircaloy in the CERMET, Maxwell-Eucken law for the CERMET thermal conductivity).

Concerning the volumetric power density in the fuel, and the temperature imposed on the cladding outer surface, the values of annular CERCER APA are kept (i.e. P = 1.2 × 1.3 × 741 W/cm$^3$ for the hottest pellet, and T = 347°C), while waiting for the results of the first thermal-hydraulic calculations, specific to APA cruciform concept, currently in progress. The CASTEM 2000 data file was parameterised in order to easily modify dimensions of the cross (length and thickness of the wings, cladding thickness, etc.).

For calculation the neutron optimised dimensions are retained: 1.5 × 3.5 mm.

The results are in conformity with what was expected, namely a maximum temperature of fuel lower than that of annular configuration APA: T$_{max}$ = 424°C.

### CERMET (20% PuO$_2$/80% SiAl)

In order to estimate the temperature gain allowed by the choice of a metal having a better thermal conductivity than the zircaloy, exactly the same calculation is performed but with silumin (alloy of 13% silicon and 87% aluminium).

For this calculation, the thermal conductivity of silumin was taken as a constant, equal to 180 W/m/°C (value at 300°C).

The maximum fuel temperature is then of 371°C, whereas the imposed temperature on the external face of the cladding is 347°C!

### CERMET (40% PuO$_2$/60% Zr)

Lastly, the APA "small crosses" is evaluated by considering a different CERMET composition: 40% plutonium and 60% zircaloy (volumetric fraction).

Since the volumetric fraction of plutonium is doubled, the volumetric power density generated in fuel is increased. Thus, P = 1.2 × 1.3 × 1 475 W/cm$^3$ for the hottest pellet.

For the imposed external temperature, T = 347°C is kept.

The maximum fuel temperature is obviously higher than that obtained in the APA "small crosses" case with previous CERMET fuel and reaches T$_{max}$ = 521°C.

## Double bipolarity and neutron/thermal duality

Table 4 gathers the fuel and cladding maximum temperatures obtained for the various studied cases (CERMET 20% Pu vol.).

**Table 4. Fuel and cladding temperatures**

|  | UO$_2$ standard | APA-a CERMET | APA cross Zr | APA cross SiAl |
|---|---|---|---|---|
| **Fuel T$_{max}$** | ~900°C | 450°C | 424°C | 371°C |
| **Clad. T$_{max}$** | ~400°C | 403°C | 373°C | 366°C |

The APA CERMET concept leads to the idea of having UO$_2$ standard fuel pins and CERMET pins in the same assembly (bipolarity in terms of materials at the local scale) with greatly improved thermal conductivity (even in the case of a zircaloy matrix). This results in a much lower T$_{max}$, from which arises the second bipolarity, at the assembly scale.

A new phenomenon appears in the APA CERMET assembly which arises from the double bipolarity of this concept: the power peaking factor is located in the coldest fuel pin because of the difference in geometry and thermal conductivity.

This neutron/thermal duality opens prospects in terms of optimisation for fuel and assembly (Pu pin loading, geometry, number of CERMET pins per assembly, geometric distribution of the pins, etc.) but shows the possibility for creating new problems in terms of behaviour during incidental or accidental situations such as reactivity-insertion accident (RIA) transients, transient of reactivity due to the fast withdrawal of an control rod, cold hazard, etc.

## Safety studies– severe accidents

### Safety

Cold hazard transients appear to be a priority to evaluate, or of insertion of reactivity accidents, in order to identify possible difficult issues. The neutron/thermal duality of the APA concept could indeed make these transients dimensioning (UO$_2$ pins becoming "fuses" in the event of RIA!!).

The APA concept is currently being studied on the basis of operating conditions in regular PWRs.

The optimisation of the concept could result in proposing different operating conditions because of the fuel performances and of its thermal potential.

The safety aspects, and of course severe accident scenarios, should be entirely re-examined.

### Severe accidents

#### Bipolar accident/cold accident

The reflection on the scenarios of the most probable severe accidents initiated for a whole CERMET core [6], is to be included within the APA framework, by taking into account the double bipolarity (material on the level of the CERMET pins, and thermal duality on the level of the assembly) of the concept. This reflection is to be carried out in parallel with a study related to possible materials, by integrating the probable physicochemical interactions.

It is already envisaged to study the CERMET pin degradation using analytical tools (ICARE2). The cruciform geometry is not currently easy to model, though the degradation kinetic should be able to be understood while varying the initial conditions, involved materials, etc., the idea being to exploit the material bipolarity of the concept, and *in fine* to direct the choice of these materials.

*Hydrogen risk*

The principal argument against the APA CERMET ($PuO_2/Zr$) concept is the increase in the Zr mass in the core, which leads to an increase in the potential hydrogen risk (fast formation of hydrogen in the event of severe accident induced by reaction of the vaporised water in contact with Zr at a temperature higher than 1 200°C).

The analysis of this risk is to be entirely re-examined knowing that even if the quantity of Zr in the heart is increased:

- On the one hand this metal will be present in massive geometry thus not easily nor instantaneously oxidisable in its totality (see results of QUENCH tests).

- In addition the temperatures will be much lower in the CERMET pins, which will increase by as much the margin compared to the temperature of departure of the reaction of metal oxidation, moreover the kinetics of heating will be very different (improved thermal conductivity).

## Conclusions

The studies presented here contribute to show that the APA cruciform geometry has a significant neutron and thermal-mechanical potential. This geometry offers significant degrees of freedom to the level of the moderation ratio which can be modulated at will while exploiting simple geometrical characteristics. The plutonium balances are very well preserved compared to the initial solution; in the same way, neutron control is preserved at the same level. Various criteria will be able to intervene for the exact definition of the crosses' distribution in the assembly, such as for example the power distribution and the maximum mass of plutonium in the assembly and also the number of crosses allowed in the assembly. The thermal studies show that the temperatures in the crosses are particularly low and that the behaviour in the event of severe accident is very specific to these composite fuels. Double bipolarity thus defines results in opening a new field of investigation. Physical behaviours are very different from those usually met in PWRs loaded with standard fuels.

In order to define the concept best, thermal-hydraulics studies must now bring the necessary elements to the definition of the assembly behaviour. In the same way studies making it possible to better determine the problems involved in manufacture are in progress. These various studies must lead to the argued choice of a reference APA concept by the end of 2003.

# REFERENCES

[1]     A. Puill, J. Bergeron, "Improved Plutonium Consumption in a Pressurized Water Reactor", Proc. GLOBAL'95, Versailles, France, Sept. 1995.

[2]     A. Puill, J. Bergeron, "Advanced Plutonium Fuel Assembly: An Advanced Concept for using Plutonium in Pressurized Water Reactors", *Nuc. Tech.*, Vol. 119, August 1997.

[3]     J. Porta, A. Puill, M. Bauer, P. Matheron, "APA: U-free Plutonium Pin in a Heterogeneous Assembly to Improve Plutonium Loading in a PWR – Neutron, Thermal-hydraulic and Manufacturing Studies", IAEA TCM, Victoria (Canada), IAEA TECDOC 1122, May 1998.

[4]     A. Bergeron, J. Bergeron, S. Bourreau, P. Matheron, A. Puill, M. Rohart, "Integral Multi-recycling of Plutonium in Pressurized Water Reactors", Top Fuel'99, International Topical Meeting, LWR Nuclear Fuel Lights at the Beginning of the Third Millenium, Avignon, France, September 1999.

[5]     J. Porta, S. Baldi, J. Bergeron, P. Dehaudt, "Composite Fuels in Inert Matrix for a Plutonium Multi-recycling Strategy in PWRs", 2000 ANS Annual Meeting, San Diego, USA, 4-8 June 2000 (invited paper).

[6]     J. Porta, C. Aillaud, S. Baldi, "Composite Fuels: Neutron Criteria for Selection of Matrix, Core Control, Transients and Severe Accidents", ICONE 7, 7110, Tokyo, Japan, April 1999.

# THERMOPHYSICAL AND CHEMICAL PROPERTIES OF MINOR-ACTINIDE FUELS

**Mike A. Mignanelli[1], Roger Thetford[2]**
[1]AEA Technology
[2]Serco Assurance
Harwell, Didcot, OX11 0QJ, United Kingdom

## Abstract

A new generation of fuels is proposed for plutonium management and minor-actinide burning. Fuel performance models are based on the physical and chemical properties of these fuels, and how these properties vary with parameters such as temperature, composition (including oxygen content) and burn-up. However, the effort so far applied to studying the properties of these fuels is far less than was applied in the past to $UO_2$ and $(U,Pu)O_2$. This paper critically surveys the available data for thermophysical properties of minor-actinide and diluent oxides, and makes recommendations for thermal expansion, density and lattice parameter, melting points, enthalpy and specific heat capacity, thermal conductivity and elastic constants.

# Introduction

As the world's nuclear power programme evolved, a great amount of theoretical and experimental effort was applied to studying the basic properties of the fuels: first uranium dioxide and then mixed uranium-plutonium dioxide. Recommendations were produced for all the relevant physical properties of $UO_{2+x}$, as a function of temperature up to and beyond the melting point, and also as functions of burn-up and of the oxygen-to-metal ratio (O/M, often characterised by the deviation $x$ from stoichiometry). Experiments were also carried out to determine many of the properties of the mixed oxide $U_{1-y}Pu_yO_{2+x}$ (where $x$ can be negative as well as positive). Where data on the effect of Pu fraction were unavailable, the uranium dioxide properties were often applied to the mixed oxide.

The properties of advanced fuels proposed for minor-actinide burning are much less well-known. A generalised uranium-free oxide fuel with a diluent or inert matrix might be $(Pu,Np,Am,Cm,Zr)O_{2+x}$ or $(Pu,Np,Am,Cm,Mg)O_{2+x}$. Pins containing 10% or more of the minor actinides are proposed for actinide-burning fast reactors, and up to 40% minor actinides in accelerator-driven systems [1]. Three extra actinides plus possible diluents of $ZrO_2$, $Y_2O_3$ or $CeO_2$, or an MgO inert matrix, give seven more dimensions of parameter space to be explored in addition to Pu fraction, O/M ratio, temperature and burn-up. Moreover, the high gamma activity of Cm demands heavily-shielded experimental facilities. Not surprisingly, many of the properties of these advanced fuels presently have to be estimated.

With such a spectrum of possible fuels, modelling plays a vital part in selecting a candidate for future development. The modelling codes need these physical properties cast in a usable form. This paper critically surveys the available data for thermophysical and chemical properties of minor-actinide and diluent oxides, and makes recommendations for thermal expansion, density and lattice parameter, melting points, enthalpy and specific heat capacity, thermal conductivity and elastic constants. In several cases, the available data seem to be poor and judgements are required. Some data are missing; the most important missing items are highlighted. Often, the best approximation for missing data is merely to extrapolate the $UO_2$ values.

# Data recommendations

## Thermal expansion

To avoid the problems of different definitions of "linear expansion coefficient" we specify that thermal expansion is defined by $L(T) = L(293)f(T)$. Many of the references define an instantaneous linear expansion coefficient $c(T) = (1/L)dL/dT = f'(T)/f(T) = d(\ln f(T))/dT$.

### Actinides

No data are available for the thermal expansion of the minor-actinide oxides. The standard UK fast reactor recommendation for $(U,Pu)O_2$ is retained:

$$L(T) = \begin{cases} L(293)(0.997141 + 9.80004 \times 10^{-6}T - 2.70446 \times 10^{-10}T^2 + 4.39012 \times 10^{-13}T^3) & T < 923\,K \\ L(293)(0.996521 + 1.17876 \times 10^{-5}T - 2.42851 \times 10^{-9}T^2 + 1.218676 \times 10^{-12}T^3) & T > 923\,K \end{cases} \quad (1)$$

*Diluents*

Table 1 gives the expressions used to calculate the linear thermal expansion of diluent materials. For MgO, Shaffer [3] gives some mean thermal expansions up to 2 073 K; these agree very well with the equation in Table 1.

**Table 1. Linear thermal expansion of diluent materials**

| Material | $f(T)$ | Reference |
|---|---|---|
| $\beta$-$ZrO_2$ | $\exp(1.2 \times 10^{-5}(T - 293)) = 0.99649 \exp(1.2 \times 10^{-5}T)$ | [2] |
| $Y_2O_3$ | $\exp(9.3 \times 10^{-6}(T - 293)) = 0.99728 \exp(9.3 \times 10^{-6}T)$ | [2] |
| MgO | $\exp(1.04 \times 10^{-5}(T - 293) + 2.581 \times 10^{-9}(T^2 - 293^2) - 2.83 \times 10^{-13}(T^3 - 293^3))$ $= 0.99674 \exp(1.04 \times 10^{-5}T + 2.581 \times 10^{-9}T^2 - 2.83 \times 10^{-13}T^3)$ | [2] |

*Density and lattice parameter*

The density of non-porous (U,Pu)O$_2$ at the reference temperature 293 K is calculated from the lattice parameter. $U_{1-y}Pu_yO_{2+x}$ has the fluorite crystal structure, with four actinide cations per unit cell. The reference density at 293 K is:

$$\rho_{293} = \frac{4}{N} \frac{10^{-3}}{a_{latt}^3} \left( (1 - y)(eM_{U235} + (1 - e)M_{U238}) + yM_{Pu} + (2 + x)M_O \right) \tag{2}$$

where $N$ is Avogadro's number, $e$ the $^{235}$U enrichment and the $M$ is the atomic weight. The lattice parameter $a_{latt}$ is adjusted from the pure UO$_2$ value by factors that take into account the Pu fraction $y$, the oxygen content $(2 + x)$ and the fractional burn-up $b$:

$$a_{latt} = A_1 - A_2 y - \alpha x (A_3 + A_4 y) - A_5 b \tag{3}$$

where $A_1 = 547$ pm, $A_2 = 74$ pm, $A_3 = 30.1$ pm, $A_4 = 11$ pm, $A_5 = 3.9$ pm, $\alpha = 1$ for hypostoichiometric fuel $(x < 0)$ and $\alpha = 0.5$ for hyperstoichiometric fuel $(x > 0)$.

*Lattice parameters for minor actinides*

All the actinide dioxides (MO$_2$) have the face-centred cubic (FCC) crystal structure. The cubic unit cell has side $a$, its volume is just $a^3$ and it contains four cations. The sesquioxides (M$_2$O$_3$) have the hexagonal close-packed (HCP) structure, characterised by lattice parameters $a$ (in the basal plane) and $c$ (along the axis). The volume of the primary unit cell is $a^2c\sqrt{3}/2$, but this contains just two cations; to facilitate comparison with the FCC structure, it is convenient to define a unit cell that is twice this size.

From the lattice parameters and densities of minor-actinide oxides given in [4], we may deduce the lattice parameters; the difference between the lattice parameter of UO$_2$ (547 pm) and the deduced values gives terms equivalent to $A_2$ in Eq. (3). To account for the deviation from stoichiometry, the minor actinides are assumed to produce the same effect as Pu, i.e. that the correction for stoichiometry is $a = a_0 - x((1 - y)A_3 + yA_4)$. Hence the recommended equation becomes:

$$a_{latt} = 547 - 3.6c_{Np} - 7.4c_{Pu} - 9.3c_{Am} - 11.2c_{Cm} - \alpha x(30.1c_U + 11(c_{Np} + c_{Pu} + c_{Am} + c_{Cm})) - A_5 b \quad \text{pm} \tag{4}$$

where $c_M$ is the atom fraction of actinide M and the other terms are as in Eq. (3).

*Lattice parameters for diluents*

From the given densities and molecular weights of the diluent compounds and their recommended thermal expansions [2], we may calculate the lattice parameter for the pure substances. Applying Vegard's law (linear interpolation) to lattice parameters rather than the densities, we may deduce the factors $A_X$ to go into the analogue of Eq. (3), such that:

$$a_{latt} = A_1 - \sum_M A_{2,M} c_M - \alpha x \sum_M B_M c_M - A_5 b - A_{Zr} f_{Zr} - A_Y f_Y - A_{Mg} f_{Mg} \tag{5}$$

(Note that all the $A_X f_X$ terms are *subtracted*; the M's are the actinides.) The coefficients for the diluents are $A_{Zr} = 42.2$ pm, $A_Y = 16.1$ pm, $A_{Mg} = 118.4$ pm.

## Melting points

In the European Fast Reactor (EFR) project, Philipponneau [5] recommended expressions for the solidus and liquidus temperatures of $U_{1-y}Pu_yO_{2+x}$, at burn-up $b$ that included the corrections:

$$C_{sol} = -121 c_U c_{Pu} - 340 b - 1000|x| \qquad C_{liq} = 129 c_U c_{Pu} - 340 b - 300|x| \tag{6}$$

for the effects of burn-up and O/M ratio. The expressions were based upon a melting point of 2 701 K for pure fresh $PuO_2$, as recommended by Adamson, *et al* [6]. However, a more recent assessment by Cordfunke, *et al* [7] gives the $PuO_2$ melting temperature as 2 663 ± 40 K. Using this value for $PuO_2$, taking other melting temperatures from Table 2 and applying Vegard's law, we recommend:

$$T_{sol} = 3120 c_U + 2663 c_{Pu} + 2820 c_{Np} + 2783 c_{Am} + 2838 c_{Cm} + 2983 c_{Zr} + 2712 c_Y + 3100 c_{Mg} + C_{sol} \text{ K} \tag{7}$$

$$T_{liq} = 3120 c_U + 2663 c_{Pu} + 2820 c_{Np} + 2783 c_{Am} + 2838 c_{Cm} + 2983 c_{Zr} + 2712 c_Y + 3100 c_{Mg} + C_{liq} \text{ K} \tag{8}$$

## Enthalpy and specific heat capacity

Using enthalpy differences rather than the instantaneous heat capacity at constant pressure, $c_p$ helps modelling codes to be stable and computationally efficient. Enthalpies are obtained by integrating the heat capacity expressions. The integration may span phase transitions, for example melting; if so, it must include the heat of transition. The polynomials used to calculate $H(T)$ are of the form:

$$H(T) = a_0 + a_1 T + a_2 T^2 + a_3 T^3 + a_{-1}/T \tag{9}$$

The constants $a_0$ are calculated to ensure the correct behaviour across the phase transition boundaries. The enthalpies of formation $\Delta H_f$ at 298 K for the actinide oxides are -1 074, -1 056, -932.2 and -911 kJ/mol for $NpO_2$, $PuO_2$, $AmO_2$ and $CmO_2$ respectively [4]. For $AmO_2$, the data tabulated by Cordfunke, *et al* [7] are incorrect and therefore the NEA values [8] are used.

## Diluents

Table 2 gives the coefficients (other than $a_0$) that are used in Eq. (9). $H_{trans}$ is the phase transition enthalpy; this is zero if there is merely a change in the curve used to fit the data. The $c_p$ data used to calculate these coefficients were obtained from the Scientific Group Thermodata Europe (SGTE) pure substance database [9]. Note that Table 2 uses units of J/mol rather than J/kg.

**Table 2. Data to calculate enthalpy of minor-actinide and diluent materials.**
**Bold values of $H_{trans}$ are fusion enthalpies; bold coefficients $a_i$ are for the liquid state.**

| Material | Temperature range (K) | $H_{trans}$ J mol$^{-1}$ | $a_1$ J mol$^{-1}$K$^{-1}$ | $a_2$ J mol$^{-1}$K$^{-2}$ | $a_3$ J mol$^{-1}$K$^{-3}$ | $a_{-1}$ J mol$^{-1}$K |
|---|---|---|---|---|---|---|
| β-ZrO$_2$ | 0-2 620* | $1.30 \times 10^4$ | $7.810 \times 10^1$ | $1.085 \times 10^{-16}$ | $-1.563 \times 10^{-20}$ | $-1.525 \times 10^{-7}$ |
| | 2 620-2 983 | $\mathbf{9.00 \times 10^4}$ | $8.000 \times 10^1$ | $2.013 \times 10^{-14}$ | $-1.838 \times 10^{-18}$ | $-1.026 \times 10^{-4}$ |
| | 2 983-6 000 | | $\mathbf{1.000 \times 10^2}$ | $\mathbf{-3.041 \times 10^{-16}}$ | $\mathbf{1.750 \times 10^{-20}}$ | $\mathbf{4.934 \times 10^{-6}}$ |
| Y$_2$O$_3$ | 0-1 100 | 0 | $1.140 \times 10^2$ | $1.276 \times 10^{-2}$ | $-2.999 \times 10^{-6}$ | $1.627 \times 10^6$ |
| | 1 100-2 100 | 0 | $1.981 \times 10^2$ | $-3.499 \times 10^{-2}$ | $7.377 \times 10^{-6}$ | $2.186 \times 10^7$ |
| | 2 100-2 550 | $5.40 \times 10^4$ | $3.941 \times 10^2$ | $-1.003 \times 10^{-1}$ | $1.557 \times 10^{-5}$ | $1.542 \times 10^8$ |
| | 2 550-2 712 | $\mathbf{8.10 \times 10^4}$ | $1.600 \times 10^2$ | $5.089 \times 10^{-13}$ | $-4.854 \times 10^{-17}$ | $-2.286 \times 10^{-3}$ |
| | 2 712-6 000 | | $\mathbf{2.000 \times 10^2}$ | $\mathbf{-4.612 \times 10^{-16}}$ | $\mathbf{2.765 \times 10^{-20}}$ | $\mathbf{6.309 \times 10^{-6}}$ |
| MgO | 0-1 700 | 0 | $4.749 \times 10^1$ | $2.323 \times 10^{-3}$ | $-8.892 \times 10^{-8}$ | $1.034 \times 10^6$ |
| | 1 700-3 100 | $7.70 \times 10^4$ | $7.830 \times 10^1$ | $-9.713 \times 10^{-3}$ | $1.719 \times 10^{-6}$ | $1.710 \times 10^7$ |
| | 3 100-5 100 | | $\mathbf{8.400 \times 10^1}$ | $\mathbf{-3.886 \times 10^{-16}}$ | $\mathbf{2.234 \times 10^{-20}}$ | $\mathbf{6.868 \times 10^{-6}}$ |
| NpO$_2$ | 0-2 820$^†$ | | $6.389 \times 10^1$ | $1.835 \times 10^{-2}$ | $-2.427 \times 10^{-6}$ | $7.072 \times 10^5$ |
| PuO$_2$ | 0-2 663$^†$ | | $8.450 \times 10^1$ | $5.320 \times 10^{-3}$ | $-2.038 \times 10^{-7}$ | $1.901 \times 10^6$ |
| AmO$_2$ | 0-2 783$^†$ | | $8.474 \times 10^1$ | $5.360 \times 10^{-3}$ | $-2.720 \times 10^{-7}$ | $1.929 \times 10^6$ |
| CmO$_2$ | 0-2 838$^†$ | | $8.498 \times 10^1$ | $5.401 \times 10^{-3}$ | $-3.402 \times 10^{-7}$ | $1.956 \times 10^6$ |

\* Stabilised β-ZrO$_2$ will be used instead of α-ZrO$_2$ in reactor fuels. Hence the equation for β-ZrO$_2$ is extrapolated to low temperatures; both the α-ZrO$_2$ equation and the heat of the α→β transition are ignored. The extrapolation should not cause any problems: because the coefficients $a_2$, $a_3$ and $a_{-1}$ are so small, the specific heat (to 10 significant figures) is constant.

$^†$ Assumed to be valid up to melting temperatures given in [4]; estimated for AmO$_2$ and CmO$_2$. The same reference quotes melting temperatures of the sesquioxides as 2 358 K for Pu$_2$O$_3$, 2 478 K for Am$_2$O$_3$ and 2 533 K for Cm$_2$O$_3$.

*Actinides*

Previous recommendations used in modelling codes for the enthalpy of $U_{1-y}Pu_yO_{2+x}$ at fractional burn-up $b$ were expressed in terms of the "reduced temperature" $\tau = T/T_{melt}$, with a change at the "fast ion" transition at $\tau = 0.856$. To combine these recommendations with the literature recommendations given below for the other actinides, we rewrite the expressions in terms of $T$ as:

$$H_U(T) = 239.0T + 5.284 \times 10^{-2}T^2 + 6.705 \times 10^{-7}T^3 - 1.289 \times 10^{-8}T^4 + 4.044 \times 10^{-12}T^5 \qquad (T < T_{fi}) \text{ J/kg} \quad (10)$$
$$+ 2.153 \times 10^6/T$$

$$H'_U(T) = H_U(T_{fi}) + 6.187(T - T_{fi}) \qquad (T_{fi} \leq T \leq T_{melt}) \text{ J/kg} \quad (11)$$

where $T_{fi} = 0.856 T_{melt} = (2\,671 - 291b)$ K is the fast-ion transition temperature and we have used the Philipponneau recommendation [Eq. (6)] that 1% burn-up reduces the melting temperature by 3.4 K. Above the melting temperature, the specific heat $c_P = 485$ J kg$^{-1}$K$^{-1}$ for all the actinides.

For NpO$_2$, heat capacity data over the temperature range 10 to 315 K [10] and 350 to 1 100 K [11] have been measured. These data have been reviewed by the NEA and $c_p$ equations recommended [8]. However, there is a problem in extrapolating $c_p$ above ~1 000 K; the enthalpy drop data (350 to 1 100 K) are in very poor agreement with the (good) low-$T$ data. Therefore these data have been reassessed and a $c_p$ equation recommended that fits data less than 900 K, but increases less sharply above 1 000 K. This new fit is used to calculate the coefficients in Table 2.

The thermodynamic functions for $PuO_2$ have all been measured and assessed [8]. Table 2 gives the recommended coefficients for the enthalpy Eq. (9). Assuming a molecular weight of 0.271 kg/mol to convert values from J/mol to J/kg, the new recommendation in Table 2 gives results that are within about 200 J/mol (typically 0.1-0.5%) of the values of the previous EFR recommendation.

There are no experimental heat capacity data for $Am_2O_3$ and $AmO_2$. All the $AmO_2$ values in Table 2 are estimated from the values for $CeO_2$ and $PuO_2$.

There are no $c_p$ data for Cm oxides. The specific heats of $PuO_2$ and $AmO_2$ are very close, and hence the specific heat of $CmO_2$ is obtained by extrapolating the Pu-O and Am-O data. Table 2 lists the resulting coefficients.

*Solid solutions: Combining the specific heat recommendations*

For a fuel solid solution during melting, combining enthalpies linearly using Vegard's law is not necessarily appropriate. If the melting temperature of the solid solution is higher than that of a particular component, then the fusion enthalpy (and increased specific heat) of the component should not be added to the overall fuel enthalpy until the *solution* reaches its melting temperature. Instead, an extrapolation of the solid-state enthalpy for the component must be used. Conversely, if the component has a higher melting temperature than the fuel solid solution, then the fusion enthalpy must be recognised earlier and the liquid-state enthalpy equation applied at a lower temperature than usual.

Diluents $ZrO_2$ and $Y_2O_3$ do indeed form solid solutions with the actinide oxides, but MgO does not – instead it acts as an inert matrix in which the actinides will not dissolve. Defining melting is then even more complex, because the two components will melt at different temperatures. Whether or not the fuel is able to support a shear stress significantly affects the way that modelling codes calculate stresses and strains – there is a different calculation for molten fuel. And the shear strength of a composite that has one molten and one solid component will depend on the respective volume fractions, so the "fluid" behaviour may appear at either the lower or the higher melting temperature, depending on the amount of MgO present. However, the latent heat of fusion of the two components will appear separately at the individual melting temperatures.

The power density generated in the inert-matrix fuels is sufficiently low that they are unlikely to melt, even in off-normal conditions. So a simple solution to the complexities of the melting behaviour will suffice. Here, we recommend that:

- MgO be treated as if it formed a solid solution with the other fuel components. Eqs. (7) and (8) are used to define single solidus and liquidus temperatures for the combined "solid solution".

- The fuel fusion enthalpy $\Delta H_{fus}$ be calculated using linear interpolation (i.e. Vegard's law) on the fusion enthalpies of the component parts. For the diluents, these come from Table 2. There are no data for the minor actinides, therefore the standard EFR value [12] as follows: $2.835 \times 10^5 + 1.3265 \times 10^4 y$ J/kg, will be used for all the actinides.

- During melting (i.e. between $T_{sol}$ and $T_{liq}$), the standard linear interpolation of specific heat should have $\Delta H_{fus}/(T_{liq} - T_{sol})$ added to it, and the corresponding adjustment made for the enthalpy.

- For any pure substance that melts below $T_{liq}$, the highest-temperature solid-state enthalpy equation from Table 2 shall be extrapolated up to $T_{liq}$.

- For any pure substance that melts above $T_{liq}$, the liquid-state enthalpy equation from Table 2 (and in particular, the duly derived specific heat of the liquid state) shall be extrapolated downwards so that it applies for all temperatures greater than $T_{liq}$.

## Thermal conductivity

### Actinides

Some work [13] at the Transuranium Institute (ITU) appeared to suggest that $AmO_2$ had a lower thermal conductivity than $UO_2$ and $PuO_2$. However, two separate recent examinations of the data [14,15] concluded that the reported low conductivity was due to porosity and changing O/M ratios in the samples. The conclusion was that there was no evidence that the thermal conductivity of $AmO_2$ was different to the standard EFR recommendation [12] for $(U,Pu)O_{2+x}$:

$$\lambda = \left( c_1 \sqrt{\alpha |x| + c_5} - c_6 + c_2 b + c_3 T \right)^{-1} + c_4 T^3 \qquad \text{W m}^{-1}\text{K}^{-1} \qquad (12)$$

where $\alpha = 1$ for $x < 0$, $\alpha = 0.5$ for $x > 0$, $T$ is the temperature (K), $b$ the fractional burn-up, $c_1 = 1.32$, $c_2 = 0.38$, $c_3 = 2.493 \times 10^{-4}$, $c_4 = 8.84 \times 10^{-11}$, $c_5 = 9.31 \times 10^{-3}$, $c_6 = 9.11 \times 10^{-2}$ in appropriate units. Note that there is no effect of plutonium content here, i.e. $PuO_2$ and $UO_2$ are assumed to be the same.

There is also some ITU work [13] on the thermal conductivity of $(U,Np)O_2$ and $(U,Np,Am)O_2$. ITU measured thermal diffusivity $a$, and converted the results to thermal conductivity $\lambda$. The results should be corrected for thermal expansion and for the effect of porosity on the thermal conductivity; for thermal conductivity, the modelling code TRAFIC uses a factor $(1 - p)/(1 + 2p)$ where $p$ is the fractional porosity. After correction, the ITU data for $U_{0.5}Np_{0.5}O_2$ gives the results shown in Figure 1.

**Figure 1. Comparison of corrected ITU data for thermal conductivity
of $U_{0.5}Np_{0.5}O_2$ with EFR (TRAFIC) recommendation for $(U,Pu)O_2$**

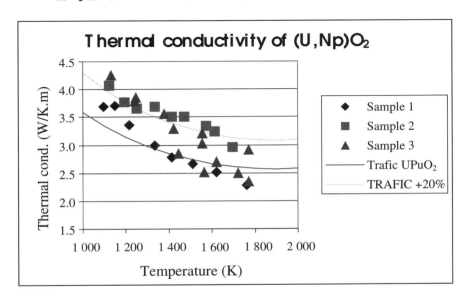

The figure also includes curves representing the standard EFR recommendation for $(U,Pu)O_2$ and the standard recommendation increased by 20%. These two curves enclose the experimental scatter quite well. We may:

- Deduce that the thermal conductivity of $U_{0.5}Np_{0.5}O_2$ is 10% ($\pm 10\%$) higher than that of pure $UO_2$.

- Hence, applying Vegard's law, recommend that the thermal conductivity of $NpO_2$ should be taken as 20% ($\pm 20\%$) more than that of $(U,Pu)O_2$ given in Eq. (12).

ITU has also reported similar work on the thermal conductivity of $(U,Np,Am)O_{2-x}$ [13,16]. Unfortunately the O/M ratio of the fuel in one experiment was unknown and for the second only a summary graph is given. However, taking both results together, the assumptions that the thermal conductivity of $AmO_2$ is the same as that of $(U,Pu)O_2$ whereas $NpO_2$ is better by 20% both look tenable.

Finally, Keller [17] recommends that the thermal conductivity of $Cm_2O_3$ be:

$$\lambda = \left(0.3542 + 1.55 \times 10^{-4} T\right)^{-1} \qquad \text{W m}^{-1}\text{K}^{-1} \qquad (13)$$

This is the standard form, with the first term representing the temperature-independent phonon scattering, and the second the temperature-dependent phonon-phonon scattering. (In the original paper the second term is given as $1.55 \times 10^{-4}/T$, but presumably the "/" is a misprint.) However, the recommendation in Eq. (12) for $(U,Pu)O_2$ includes a term for the electronic contribution to heat conduction, proportional to $T^3$. There is no such term in Eq. (13), which throws doubt on its validity at high temperatures. There is also no information on the variation with O/M ratio, or with burn-up. Comparing Eq. (13) against the $(U,Pu)O_{2+x}$ recommendation in Eq. (12) with $x = -0.5$ shows that the $Cm_2O_3$ conductivity is approximately twice that of $(U,Pu)O_{1.5}$ at 1 000 K, and equal to it at 2 000 K. Pending more data, we recommend that above 2 000 K, where the "missing" $T^3$ term would have its largest effect, $\lambda_{CmO2+x} = \lambda_{UPuO2+x}$; below 2 000 K, scale with $T$ to get:

$$\lambda_{CmO2+x}(T,x,b) = \frac{3000 - T}{1000} \lambda_{UPuO2+x}(T,x,b) \qquad (14)$$

*Diluents*

Touloukian [18] recommends the following thermal conductivity expressions:

$\beta$-$ZrO_2$         $\kappa = 3.266 - 1.666 \times 10^{-3} T - 0.351 \times 10^3 T^{-1} + 0.783 \times 10^{-6} T^2$      W m$^{-1}$K$^{-1}$    (15)

$Y_2O_3$         $\kappa = -8.898 + 5.432 \times 10^{-3} T + 7.859 \times 10^3 T^{-1} + 0.728 \times 10^{-6} T^2$      W m$^{-1}$K$^{-1}$    (16)

MgO         $\kappa = 5.247 - 17.93 \times 10^{-3} T + 14.27 \times 10^3 T^{-1} + 8.104 \times 10^{-6} T^2$                (17)

The $Y_2O_3$ equation is valid from 298 to 1 350 K. These curves are consistent with isolated data points given by Kleykamp [19] and Shaffer [3], and are recommended for modelling calculations.

*Elastic constants*

*Actinides*

There appear to be no data for minor actinides, so we will continue to use the standard TRAFIC properties [12] rewritten in the "material-independent" form:

$$E/E_0 = \begin{cases} 1 - 0.212\tau - 0.412\tau^2 & (\tau \le 0.837) \\ -5.88 + 16.238\tau - 10.244\tau^2 & (0.837 < \tau \le 0.939) \end{cases} \tag{18}$$

$$G/G_0 = \begin{cases} 1 - 0.187\tau - 0.425\tau^2 & (\tau \le 0.837) \\ -6.714 + 18.255\tau - 11.448\tau^2 & (0.837 < \tau \le 0.939) \end{cases} \tag{19}$$

where $E_0 = 2.2693 \times 10^5$ MPa, $G_0 = 8.583 \times 10^4$ MPa, $\tau = T/T_{liq}$ and $T_{liq}$ is taken as 3 120 K (the $UO_2$ value). Between 2 929 K ($\tau = 0.939$) and the liquidus temperature (or between solidus and liquidus if $T_{sol} < 2\,929$ K), $E$ and $G$ are tapered linearly to zero while $B$ remains constant. Corrections are also made for stoichiometry and porosity.

*Diluents*

Data for $\beta$-$ZrO_2$ are lacking, but for $\alpha$-$ZrO_2$ Ref. [20] quotes $E_0 = 253$ GPa. Shaffer [3] also gives some data, but they are internally inconsistent, violating the elasticity relations $G = E/2(1 + \nu)$ and $B = E/3(1 - 2\nu)$. We therefore recommend that Eqs. (18) and (19) be used with $E_0 = 253$ GPa, $G_0 = 111$ GPa (as implied by $\nu = 0.14$) and $T_{melt} = 2\,983$ K.

Ref. [20] also quotes a value for yttria $Y_2O_3$ of $E_0 = 159$ GPa. In the absence of data on Poisson ratio, shear modulus or bulk modulus, we assumed $\nu = 0.15$ (giving $G_0 = 69.1$ GPa) and recommend using Eqs. (18) and (19) for the variation with temperature.

For MgO the Shaffer data [3] are again inconsistent. Following TAPP [20], we recommend Eqs. (18) and (19) with $E_0 = 300$ GPa, $G_0 = 128$ GPa and $T_{melt} = 3\,100$ K.

Combining diluent elastic moduli with actinide moduli is a little complicated. As the material reaches its melting point, the shear modulus vanishes but the bulk modulus does not. The recommendations above all use the same polynomials (18) and (19) in the reduced temperature $\tau = T/T_{liq}$, and taper from the same value of $\tau$. Hence for the solid solution we may apply Vegard's law to obtain values of $E_0$ and $G_0$, use Eqs. (18) and (19) to get the temperature variation of $E$ and $G$, then apply the standard correction factors for porosity $p$ and O/M ratio.

**Conclusions**

This paper contains a set of recommendations that will allow preliminary modelling of advanced oxide fuels for actinide-burning reactors. However, much of the available information for diluents and minor actinides is unsatisfactory or incomplete. The most important gap is thermal properties of the minor actinides: thermal expansion, thermal conductivity, specific heat and thermal creep. We recognise that experimenting with these materials is difficult. Although some experimental work is planned, comprehensive data are unlikely to be available soon. We may have to continue to use estimates based on $(U,Pu)O_2$ data.

*Acknowledgement*

This work was funded by British Nuclear Fuels plc.

**REFERENCES**

[1]   A. Vasile, M. Boidron, S. Pillon, D. Plancq, J. Tommasi, "CAPRA-CADRA Overview Status Report", Proc. 6th Int. CAPRA/CADRA Seminar, Newby Bridge, June 2000.

[2]   M. Miyayama, "Eng. Mater. Handbook, Ceramics", S. Schneider, ed., ASM Int., Materials Park Ohio, USA (1991).

[3]   P.T.B. Shaffer, "Plenum Press Handbooks of High Temperature Materials, No. 1 Materials Index", Plenum Press, New York (1964).

[4]   J.J. Katz, G.T. Seaborg and L.R. Morss, "The Chemistry of the Actinide Elements", 2nd ed., Vols. 1 and 2, Chapman and Hall, London (1986).

[5]   J. Edwards, *et al.*, "Fast Reactor Data Manual", Fast Reactor European Collaboration report (1990).

[6]   M.G. Adamson, E.A. Aitken and R.W. Caputi, *J. Nucl. Mater.*, 130, 349 (1985).

[7]   E.H.P. Cordfunke, R.J.M Konings, G. Prins, P.E. Potter and M.H Rand, "Thermochemical Data for Reactor Materials and Fission Products", Elsevier Science, Amsterdam (1990).

[8]   R.J. Lemire (Chairman), "Chemical Thermodynamics of Neptunium and Plutonium", NEA/OECD publication, Vol. 4 (North Holland, 2001).

[9]   Landolt Bornstein, "Numerical Data and Functional Relationships in Science and Technology, Group IV: Physical Chemistry, Volume 19: Thermodynamic Properties of Inorganic Materials Compiled by SGTE", Subvolume A, Pure Substances (Springer, 1999).

[10]  E.F. Westrum, Jr., J.B. Hatcher and D.W. Osborne, "$C_p$(10-315 K) $NpO_2$", *J. Chem. Phys.*, 21, 419 (1953).

[11]  V.A. Arkhipov, E.A. Gutina, V.N. Dabretsov and V.A. Ustinov, *Radiokhimiya*, 16, 123 (1974).

[12]  R. Thetford, M.A. Mignanelli and I.J. Ford, "The Subroutines of TRAFIC", AEA Technology internal report (1993).

[13]  H.E. Schmidt, C. Sari and K. Richter, "The Thermal Conductivity of Oxides of Uranium, Neptunium and Americium at Elevated Temperatures", *J. Less-Common Metals*, 121, 621–630, (1986) (also in Proc. Actinide '85, Aix-en-Provence, September 1985). Also European Institute for Transuranium Elements Annual Report, "Programme Progress Report on Nuclear Fuels and Actinide Research", TUSR 38, p. 48, December 1984.

[14]   R. Thetford, "TRAFIC Modelling of Fuel Pins Containing Americium", AEA Technology internal report, October 1998.

[15]   K. Bakker and R.J.M. Konings, "On the Thermal Conductivity of Inert-matrix Fuels Containing Americium Oxide", *J. Nucl. Mater.*, 254, 129–134 (1998).

[16]   European Institute for Transuranium Elements Annual Report, "Programme Progress Report on Nuclear Fuels and Actinide Research", TUSR 39, June 1985.

[17]   C. Keller, "The Binary Oxides of the Transuranium Elements", Gmelin Handbook System No. 71, New Supplement Series, Vol. 4, Part B1, p. 219.

[18]   Y. Touloukian *et al.*, *Thermophys. Prop. Mater.,* Vol. 1, New York, Plenum (1970).

[19]   H. Kleykamp, "Materials Properties of Diluents for Uranium Free Actinide Compounds within the Transmutation Programme", FZK Internal Report 32.23.04, Sept. 1997.

[20]   "Thermochemical and Physical Properties database", ES Microware, Hamilton, USA.

# SOME VIEWS ON THE DESIGN AND FABRICATION
# OF TARGETS OR FUELS CONTAINING CURIUM

**J. Somers, A. Fernandez, R.J.M. Konings**
European Commission, Joint Research Centre, Institute for Transuranium Elements
P.O. Box 2340, D-76125 Karlsruhe, Germany

**G. Ledergerber**
Kernkraftwerk Leibstadt AG
CH-5325 Leibstadt, Switzerland

## Abstract

Given the shielding requirements necessary to handle curium, its separation from americium in the reprocessing step is essential to minimise material flows in the fabrication chains. The number of fuel assemblies holding americium and curium can be minimised if dedicated targets with high contents are manufactured. Given the inadequacies of powder blending and co-precipitation methods, the infiltration of porous host materials (e.g. $(Zr,Y)O_2$, $ThO_2$, $(Zr,Y,Pu)O_2$, $(Th,Pu)O_2$) is the preferred method to fabricate curium-containing targets in either Sphere-pac or pellet form. While in either case there are inherently no liquid wastes containing curium, the former reduces the number of processing steps and production scraps which would have to be recycled. There are considerable advantages in co-locating curium fabrication at the reprocessing facility, as curium in solution could be infiltrated directly in the matrix without extra precipitation, washing and calcination steps.

# Introduction

Back-end strategies for the nuclear fuel cycle are receiving ever-increasing attention. Apart from the long-term storage of spent fuel and nuclear waste and its disposal in geological repositories, emphasis is also being given to partitioning and transmutation (P&T) of long-lived fission products and minor actinides so that a reduction of the radiotoxicity of these materials can be achieved. For this purpose various fuel cycle concepts are under discussion. A combination of light water reactors (LWR) for Pu recycling and fast reactors (FR) for minor actinide transmutation is considered for this study [1]. Dedicated reactors (e.g. accelerator-driven systems – ADS) are also proposed for highly efficient minor actinide transmutation.

Regardless of the strategy contemplated, fuels incorporating minor actinides need to be developed, their relevant properties determined and in-pile behaviour tested. Higher contents of Am and Cm can be incorporated in specially designed targets. They can be used in a once-through concept if the transmutation efficiency is maximised, a condition best achieved if a uranium-free fuel is used.

A wide variety of matrix materials (metals, oxides, nitrides and carbides) to support the actinide phase can be considered. Irradiation experiments using the so-called inert-matrix mixed oxide (IMMOX) concept have been performed or are in planning stages. In the EFTTRA-T4 irradiation in the HFR Petten [2] Am oxide was incorporated in a $MgAl_2O_4$ spinel matrix, while in the ECRIX experiment [3] $AmO_2$ dispersed in a MgO matrix will be irradiated in Phénix. The CERCER fuel concept in both of these experiments utilises the (relatively) high thermal conductivity of the neutronically inert matrix. This advantage can be maintained throughout the irradiation if the actinide is dispersed in (large) particles to limit irradiation damage to the matrix itself. In a further experiment (EFTTRA-T5), the actinide phase will be incorporated in a partially yttria-stabilised zirconia (YSZ) solid solution, a concept now being tested in the Halden reactor for the incineration of plutonium [4a,4b].

Design considerations of fuels and targets have much in common with conventional fuels. Apart from the thermochemical, thermophysical and general material properties to be considered, the management of He produced during the irradiation must be addressed. In the EFTTRA-T4 and SUPERFACT [5] experiments, helium production was significantly higher than the fission gases. Two opposite solutions can be proposed. Helium retention can be best achieved by low-temperature operation of the fuel and allowance for its concomitant swelling. Satisfactory behaviour of this type of fuel under reactor transients would have to be proven, however. In contrast, He release can be accomplished by designing fuels with high porosity operating at high temperature. Here the main disadvantage lies in the reduced thermal conductivity and the necessarily larger plenum.

Given the high radioactivity of americium and especially curium, much emphasis must be given to the development of suitable methods to fabricate such fuels and targets and to handle corresponding pins and assemblies. Fabrication processes must be dust-free to avoid (or at least minimise) contamination on the surfaces of sealed containments and the production equipment therein. This is particularly important if such fabrication facilities need to be maintained and optimised on a regular basis. The consequences thereof and possible solutions meeting fuel design criteria are discussed here.

## General considerations

### *Quantities of minor actinides to be processed*

As yet no clear strategy for the transmutation of minor actinide fuels and targets has evolved, and thus the quantities of materials to be reprocessed and re-fabricated remains uncertain. Nevertheless, indicative values for various scenarios has been ascertained and discussed [6-8] (see Table 1). Present PWRs using uranium oxide (UOX) fuel would have an annual output (at 400 TW.h$_e$) of 500 and 100 kg of Am and Cm, respectively. Both of these quantities increase when Pu is recycled.

As the addition of curium in small quantities to conventional fuels would necessarily increase the mass throughput of the fabrication plant and shielding required for manufacture of assemblies and their handling, its fabrication in targets with high minor actinide content (e.g. 2 g.cm$^{-3}$) is necessary. The number of fuel assemblies (FA) and pins containing Am and Cm with this minor actinide density to be fabricated on an annual basis is estimated in Table 1. Fuel assemblies and pin dimensions typical for LWR and ANSALDO ADS design [9] are assumed. From the fabrication viewpoint, separation of Cm from Am should be considered to reduce as much as possible the number of pins containing Cm. The PUREX process now being used to separate both U and Pu from the spent fuel is being improved (e.g. DIAMEX/SANEX) to separate the minor actinides from the high-level waste (HLW) raffinate.

**Table 1. Americium and curium mass flow to storage (in kg/TW.h$_e$) for a nuclear park (400 TW.h$_e$, 60 GW$_e$) [6,7] and corresponding number of fuel pins if these nuclides are recycled for LWR and ANSALDO ADS geometries**

|  | 100%PWR UOX (open cycle – kg/TW.h$_e$) | LWR fuel 2 g.cm$^{-3}$ MA Pins/FAs | ADS fuel with 2 g.cm$^{-3}$ MA Pins/FAs |
|---|---|---|---|
| **Am** | 1.3 | 1 300/5 | 3 600/40 |
| **Cm** | 0.26 | 260/1 | 700/8 |
| **Am+Cm** | 1.56 | 1 560/6 | 4 300/48 |

### Decay of Am and Cm

The decay characteristics of selected Am and Cm isotopes are presented in Table 2 [10]. Both $^{241}$Am and $^{244}$Cm produce considerable heat by $\alpha$ decay (0.1 and 2.8 Watts.g$^{-1}$), which is higher than Pu typically used in MOX production (0.01 Watts.g$^{-1}$). This along with the higher $\gamma$ dose rates (ca. $\times$ 5 000) has important consequences for storage, transport and processing of these isotopes. In addition, $^{244}$Cm has a high spontaneous fission probability so that neutron protection must be provided.

**Table 2. Decay emissions of selected Pu, Am and Cm isotopes [10]**

| Isotope | t$_{1/2}$ (years) | Spec. act. (Bq.g$^{-1}$) | Power (Watts.g$^{-1}$) | Dose rate* (mSv.h$^{-1}$.g$^{-1}$) | Spon. fission (g$^{-1}$.s$^{-1}$) |
|---|---|---|---|---|---|
| $^{238}$Pu | 8.78E+01 | 6.34E+11 | 5.67E-01 | 1.80E+00 | 1.10E+03 |
| $^{239}$Pu | 2.41E+04 | 2.29E+09 | 1.93E-03 | 1.50E-02 | 1.00E-02 |
| $^{240}$Pu | 6.57E+03 | 8.39E+09 | 7.06E-03 | 2.60E-02 | 4.80E+02 |
| $^{241}$Pu | 1.44E+01 | 3.81E+12 | 3.27E-03 | 1.70E-01 |  |
| $^{242}$Pu | 3.74E+05 | 1.46E+08 | 1.17E-04 | 3.10E-04 | 8.05E+02 |
| $^{241}$Am | 4.33E+02 | 1.26E+11 | 1.14E-01 | 3.12E+02 | 4.80E-01 |
| $^{242m}$Am | 1.40E+02 | 3.87E+11 | 4.49E-03 | 1.20E+01 | 6.20E+01 |
| $^{243}$Am | 7.37E+03 | 7.40E+09 | 6.43E-03 | 4.40E+01 | 2.70E-01 |
| $^{244}$Cm | 1.81E+01 | 2.99E+12 | 2.83E+00 | 4.90E+00 | 4.00E+06 |

* Gamma at 1000 mm distance.

The construction of a facility to process minor actinides will thus require considerable shielding. The high specific activity could also impose constraints on the physical strength of the encasement building. Fabrication and handling processes eliminating respirable dusts with aerodynamic diameters < 10 μm (AMAD) are to be favoured, so that the associated environmental risks are minimised.

## Shielding requirements for the handling of minor actinides

The required shielding for the handling of Am-containing targets has been considered in detail elsewhere [11]. A storage facility constructed for $PuO_2$ could be used for $AmO_2$ but the mass would have to be reduced by a factor of 80. The installation of Pb (70 mm) would permit storage of the same quantity (13.2 kg). Processing and handling of Am would require corresponding shielding. For Am the shielding will also depend on the matrix material, as self-shielding in many inert matrices is less than in MOX. For neutrons, however, self-shielding by the matrix itself is not a relevant parameter.

At the Institute for Transuranium Elements (ITU) in Karlsruhe a specially shielded facility is being constructed for the fabrication of fuels and targets containing minor actinides. The shielding requirements were calculated for a variety of nuclides (based on uranium and thorium fuel cycles), so that a radiation dose rate of not more than 2 μSv/h at 1 m distance independently for photons and neutrons could be achieved. Water (500 mm) and Pb (50 mm) are required to shield neutron and gamma emissions of $^{244}$Cm (5 g) and $^{231}$Pa (10 g), respectively. With this level of shielding, significantly larger quantities of $^{237}$Np, $^{241}$Am and $^{243}$Am can be handled.

The extrapolation of these results to possible industrial fabrication of fuels and targets shows that the fabrication and pin handling for kg quantities of Cm will require even larger shielding. Therefore the number of processing steps in which Cm is handled must be minimised and simplified as much as possible. In addition, dusts must be avoided, as their accumulation would not only severely hinder operator intervention (for maintenance and decommissioning) but also potentially cause damage to sensitive equipment. Processing wastes (scraps, non-conforming materials) should also be minimised.

## Fabrication methods for the production of oxide fuels and targets containing minor actinides

Several methods can be used to produce oxide-type fuels and targets for actinide transmutation. They are reviewed below and the inherent difficulties and limitations are highlighted. Two promising innovative techniques based on infiltration of liquids into porous media are discussed in greater detail.

### *Powder metallurgy*

Powder metallurgical methods are widely used in the nuclear industry for the production of $UO_2$ and MOX fuels. In a special installation in the CEA's Marcoule facilities, powder metallurgy has been used to produce $MgO-AmO_2$ pellets for the ECRIX irradiation experiment [12]. Nevertheless, the inherent dust accumulation associated with this type of method limits its applicability for the fabrication of both Am- and Cm-containing targets.

### *Liquid to particle conversion*

Two gelation-based routes (internal and external) were developed for the production of spherical particles for high-temperature reactor (HTR) fuel kernels, Sphere-pac fuel rods and direct pressing to conventional fuel pellets [13]. Both methods require the preparation of an actinide nitrate solution with high metal contents and the dispersion of the chemically modified solution into droplets. In the internal gelation process, the droplets are solidified in a hot oil bath, which causes thermal decomposition of hexamethylenetetramine (HMTA) in the droplet to release ammonia for the gelation process. In contrast, ammonia for the external gelation process is provided by a hydroxide bath in which the dispersed droplets are collected.

The form of the resulting beads following droplet to particle conversion by gelation depends on the type of atomiser and the chemical constitution of the feed solution. A vibrating nozzle drop generator generally produces monodisperse particles with excellent sphericity, ideal for HTR or Sphere-pac applications, whereas rotating cup atomisers produce particles with a polydisperse size distribution suitable for direct pressing into pellets.

Both sol-gel processes produce excellent quality dust-free material. For minor actinide application care must be taken in the preparation of the feed solutions, especially as $\alpha$ particles can cause deterioration of the chemical additives. The external gelation process has been used to produce $(U,Pu,Am)O_2$, $(U,Pu,Np)O_2$, $(U,Np,Am)O_2$ and $(U,Np)O_2$ for the SUPERFACT irradiation experiment in Phénix [5]. The use of either of these sol-gel routes to produce fuels and targets containing Cm is detrimentally hindered not only by the higher $\alpha$ decay of Cm, but also by the difficulties that can be expected in treating and disposal of the low activity liquid wastes arising in the process.

### Pellet infiltration

The pellet infiltration process shown schematically in Figure 1 has been tested and used for the production of spinel pellets containing 11 wt.% Am for the EFTTRA-T4 irradiation experiment. The porous pellet is prepared by compaction of the matrix powder followed by calcination to remove pressing additives. The resulting green pellets are lowered into the actinide nitrate solution, which infiltrates the pores. Following its removal from the solution, the pellets are thermally treated to convert the actinide nitrate into oxide. Finally, they are sintered to give the final product, which is then characterised, before being loaded and welded in conventional fuel pins.

If the matrix powder is inactive it can be produced in conventional laboratories, and the processes involving handling of actinides are limited to the steps involving preparation of the actinide solution, infiltration, thermal treatment, sintering, characterisation and pin filling and welding. In addition there are no liquid wastes bearing actinides, and actinide scraps can be reduced further by application of rigorous quality control procedures in the selection of the green pellets to be infiltrated. Thus, the method is well suited to the production of fuels and targets containing either americium or curium. The process also offers the advantage of tailoring the distribution of the minor actinide within the pellet [14].

The main disadvantage of this method lies in the limited actinide quantity that can be infiltrated in a single step, e.g. ca. 11 wt.% in $MgAl_2O_4$ if the actinide solution concentration is 400 g/l. This could be increased by performing a second infiltration after conversion of the actinide nitrate to oxide, but obtaining a homogeneous distribution of the actinide phase throughout the pellet may prove even more elusive than for a single infiltration step. The process is also only applicable to matrix materials insoluble in nitric acid solutions (e.g. $(Th,Pu)O_2$, $(Zr,Y)O_2$) and cannot be applied for $UO_2$ or MgO.

### Porous bead infiltration

In another variation of the infiltration process [15] (see Figure 1), porous beads are prepared by internal or external gelation methods. The minor actinide is not incorporated directly in the gelation step, so that bead production can be performed in unshielded facilities for candidate materials, such as

**Figure 1. Pellet infiltration (left) and bead infiltration for the production of pellets (middle) or spheres (right) containing actinides. Steps requiring water shielding are marked.**

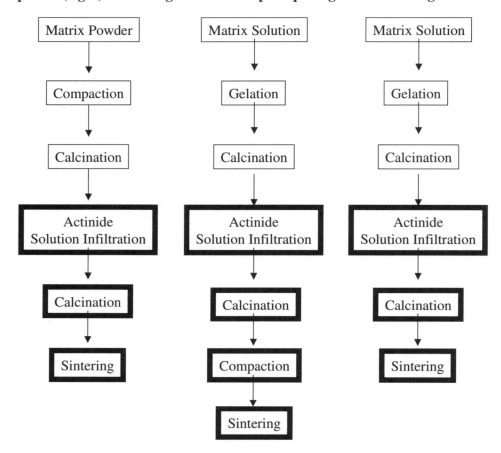

YSZ, or in lightly shielded areas, e.g. for thorium dioxide. In a further process modification, the gelation route can be used as described above in appropriately shielded facilities to prepare $(Th,Pu)O_2$ or $(Zr,Y,Pu)O_2$ matrices in the form of porous beads for minor actinide infiltration.

The beads are calcined to remove additives from the solution chemistry steps and to give the desired porosity and surface area. They are then infiltrated by the minor actinide nitrate solution in the more extensively shielded facilities. In the process now being developed, the actinide content of the final product is obtained by matching the solution concentration to the bead porosity so that only sufficient solution is added to reach the incipient wetness point of the beads. The maximum amount of material to be infiltrated is limited by two factors: (a) the bead surface area/porosity, and (b) the solubility of the actinide nitrate in aqueous media. Nevertheless, the minor actinide concentration can be increased by repeating the infiltration step following conversion of the actinide nitrate to oxide by thermal treatment. Experience has shown that at least three such infiltration/calcination cycles are possible without occurrence of undesired material agglomeration.

This bead infiltration process can be used to produce targets compacted in two different physical forms. If spheres with a polydisperse size distribution are produced in the atomisation step, they can be pressed into pellets and following measurement and inspection can be loaded into conventional fuel pins. In contrast, spheres with suitable monodisperse sizes can be mixed and loaded directly in the fuel pin and following gentle vibration to obtain higher density can be welded. This Sphere-pac concept was tested extensively both for oxide and nitride fuels in the past [16,17].

# Fabrication of targets containing curium: An assessment

With the exception of Np, powder metallurgy processes, due to their dust production, are not suited for minor actinide fabrication. The gelation route by its nature is essentially dust-free and has been used successfully in the past to produce fuels and targets containing significant quantities of Am. Nevertheless, its application for co-processing Cm-based targets or fuels must be questioned on account of the high $\alpha$ activity and heat produced, and the concomitant damage of organic components and premature release of ammonia from HMTA. Radiolysis of the aqueous media can occur and would result in major difficulties not only in the processing itself, but also in the handling and disposal of unavoidable liquid wastes. For these reasons, infiltration appears to offer the most acceptable and appropriate means to prepare fuels and targets containing Cm.

Of the possible infiltration routes, the actinide content, which can be obtained in pellet infiltration, is limited by the porosity, P, of the green pellet (typically 50%). Experience has shown that in a single infiltration step the actinide metal density reached is of the order of 0.3 g.cm$^{-3}$. This would be sufficient if Cm is added in small quantities to a conventional fuel, but not for dedicated targets.

In contrast to green pellets, beads produced by gelation are highly porous so that substantially higher quantities of minor actinides can be infiltrated. This variation of the infiltration method has been successfully used to produce $(Zr,Y,Pu)O_2$ pellets with a Pu density of 2.5 g.cm$^{-3}$. Three infiltration/ calcination cycles were used, as the Pu concentration in the infiltrant solution was limited to 200 g/l to avoid polymer formation and precipitation. For Am and Cm solution concentrations up to 400 g/l can be considered, so that the number of infiltration steps can be reduced to reach the same actinide content. The heat produced in such solutions would be considerable and appropriate cooling would be required.

If the beads are compacted into pellets, an additional fabrication step is introduced. In contrast, Sphere-pac compaction has the advantage that both the pellet compaction step and concomitant detailed dimensional and gravimetric analysis, and visual examination are not necessary. Thus, not only are two steps removed from the fabrication and control process, but also scraps from compaction and non-conforming final pellets are avoided, along with the need to meet diameter specifications.

The dust-free nature of the infiltrated spheres permits their loading in fuel pins with minimal risk of contamination of the unit. A close-packed configuration can be reached by simple vibration of the pin. The spheres produced should be monodisperse and two different sizes are required (e.g. 800 µm and 180 µm) to reach a smeared density of about 75-78%. This leaves a sufficiently large volume for FGR and He production. Both size fractions could be infiltrated with the actinide, but for fabrication simplification, the smaller phase should not necessarily be infiltrated. To compensate, the larger spheres should then contain higher actinide contents. In an additional variation, the small size fraction could be a different material, e.g. a metal, to improve the thermal conductivity so that the fuel operating temperature is reduced. Detailed calculations and comparison of the operating temperatures for both pellet and Sphere-pac CER, CERCER and CERMET fuel forms are worthy of a more detailed investigation.

From a fuel design viewpoint, the Sphere-pac concept would have sufficient porosity in the pin to accommodate swelling of the material due to accumulation of helium formed in the transmutation of americium and curium. Should the operating temperatures be sufficiently high that He and fission gases are released, the open porosity provides a path to a plenum designed for their accommodation.

Apart from the fabrication process itself, quality assurance and control measures are required and must be considered in the assessment of the appropriateness of the fabrication process (see Table 3). In general, the requirements are similar for pellet or Sphere-pac pins. Pellets, however, are selected based on measurement of dimensions and density on many or even all pellets. In contrast, yield of good spheres can be improved up to 100% without problems.

**Table 3. Quality assurance for Sphere-pac and pellet fuels containing Cm**

| Measurement | Pellet | Sphere-pac | Comments |
|---|---|---|---|
| MA solution conc. | + | + | Prior to infiltration |
| Isotopic abundance | + | + | |
| O/M | + | + | |
| Impurities | + | + | |
| Microstructure | + | – | For Sphere-pac not required |
| Density/dimensions | + | – | Large number of pellets |
| Visual inspection | + | – | Large number of pellets |
| Smeared density | – | + | NDT of all (but limited no. of) Sphere-pac pins |

Pin fabrication with spheres is easily achieved by remote technology in a shielded facility. The problem of an early cladding defect with a release of spheres from the pin early in life can be overcome by using ridged cladding material (e.g. stainless steel) which, however, might not be so favourable with regard to neutron economy. Assembly fabrication might be performed under water in a similar way as FA repair work is performed today in a nuclear power plant.

Apart from the shielding requirements necessary for the fabrication of fuels and targets containing curium, various processes would require additional engineering innovation for simplification and automation. Many of these are already available, or at an advanced stage, at MOX fuel fabrication plants, but would need to be adapted further to reduce maintenance, operator intervention and to provide extra protection for electronic equipment. Some such developments are being made in the construction of the MA-Lab at the ITU.

## Curium fuels and targets for transmutation: Logistics

The location of the reactor, reprocessing and fabrication sites also needs to be considered. In order to avoid difficult (trans)national transports, the location of all such facilities at one site would seem preferable. If not possible there are considerable advantages in the co-location of reprocessing and fabrication facilities for Cm fuel and target production at the same site. In particular the minor actinides are in liquid form at the end of aqueous reprocessing streams. If these solutions were converted directly by infiltration into suitable materials (e.g. YSZ, thoria, etc.) for use as targets or fuels at the reprocessing plant, precipitation and thermal treatment steps (as currently practised for Pu recovery) and subsequent dissolution steps would be eliminated, along with liquid waste disposal and powder packaging and transport costs.

Transportation of complete FAs could be performed in spent fuel casks, where the design of the inner basket needs to take the considerable heat production in the FA into account.

As shown in Table 1, the number of FAs containing Cm is reduced if it is separated from Am in the reprocessing steps. It would also appear advisable that Cm transmutation assemblies contain as many Cm pins as possible for ease (or at least concentrating the difficulties) of handling. Mixing of Cm pins in either UOX or MOX assemblies does not seem a suitable option, and should be avoided.

## Conclusions

The handling and fabrication of curium in fuels and targets will require the development of advanced fabrication processes to minimise the process steps and scraps produced. Furthermore, reactor safety and design criteria will not, as at present, be able to dictate the tolerances to be achieved in the fabrication steps. Grinding of such materials, to meet cladding/pellet gap specifications for example, will not be possible.

Of the processes under development at the present, the combination of sol-gel routes to produce porous beads of the host matrix (e.g. YSZ, $ThO_2$, $(Zr,Y,Pu)O_2$, $(Th,Pu)O_2$) and the infiltration of minor actinides therein offers several advantages over powder blending or direct sol-gel routes. The Sphere-pac compaction method not only requires fewer processing steps and produces less production scraps, but also has intrinsic features (i.e. porosity, chemical form of the smaller size fraction), which could prove favourable in the design of fuels and targets for transmutation of actinides.

# REFERENCES

[1]   R.J.M. Konings, J.L. Kloosterman, "A View of Strategies for Transmutation of Actinides", *Progr. Nucl. Energy*, 38, 331 (2001).

[2]   R.J.M. Konings, R. Conrad, G. Dasel, B.J. Pijlgroms, J. Somers, E. Toscano, "The EFTTRA-T4 Experiment on Americium Transmutation", *J. Nucl. Mater.*, 282, 159 (2000).

[3]   J.C. Garnier, N. Schmidt, N. Chauvin, A. Ravenet, J.M. Escleine, C. Molin, F. Varaine, C. de St. Jean, T. Philip, G, Vambenepe, G. Chaigne, "The ECRIX Experiments", Proc. of the International Conference on Future Nuclear Systems, Global'99, Jackson Hole, August 1999.

[4]   a) U. Kasemeyer, Ch. Hellwig, Y-W. Lee, G. Ledergerber, D.S. Sohn, G.A. Gates and W. Wiesenack, "The Irradiation Test of Inert-matrix Fuel in Comparison to Uranium Plutonium Mixed Oxide Fuel at the Halden Reactor, *Prog. Nucl. Energy*, 38, 309 (2001).

      b) G. Ledergerber, C. Degueldre, P. Heimgartner, M.A. Pouchon, U. Kasemeyer, "Inert Matrix Fuel for the Utilisation of Plutonium", *Prog. Nucl. Energy*, 38, 301 (2001).

[5]   J.F. Babelot, N. Chauvin, "Joint CEA/ITU Synthesis Report of the Experiment SUPERFACT1", ITU Report, JRC-ITU-TN-99/03.

[6]   M. Salvatores, "Transmutation and Innovative Options for the Back-end of the Fuel Cycle", Proc. of the International Conference on Future Nuclear Systems, Global'99, Jackson Hole, August 1999.

[7]   M. Delpech, J. Tommasi, A. Zaeta, M. Salvatores, H. Mouney, G. Vanbenepe, "The Am and Cm Transmutation – Physics and Feasibility", Proc. of the International Conference on Future Nuclear Systems, Global'99, Jackson Hole, August 1999.

[8]   P. Brusselaars, Th. Maldague, S. Pilate, A. Renard, Ph. Hemmerich, S. Janski, H. Mouney, G. Vanbenepe, "Mixed Americium-curium Target Pins Recycled in PWRs", Proc. of the International Conference on Future Nuclear Systems, Global'99, Jackson Hole, August 1999.

[9]     The European Technical Working Group on ADS, "A European Roadmap for Developing Accelerator-driven Systems (ADS) for Nuclear Waste Incineration", ENEA (2001).

[10]    J. Magill, "Nuclides 200:0 An Electronic Chart of the Nuclides on CD-ROM", *ATW*, Vol. 45 (2000), Heft 10, p. 1.

[11]    A. Renard, T. Maldague, S. Pilate, A. Harislur, H. Mouney, M. Rome, "Fuel Fabrication Restraints when Handling Americium", Proc. of the International Conference on Future Nuclear Systems, Global'97, Yokohama, October 1997.

[12]    M. Sors, M. Croixmarie, "Facilities for Preparing Actinide or Fission-product Based Targets", *Nuclear Instruments and Methods in Physics Research*, A 438, 180 (1999).

[13]    G. Ledergerber, F. Ingold, R.W. Stratton, H.P. Alder, C. Prunier, D. Warin, M. Bauer, "Preparation of Transuranium Fuel and Target Materials for the Transmutation of Actinides by Gel Co-conversion", *Nuclear Technology*, 114, 194-204 (1996).

[14]    A. Fernandez, K. Richter, J. Somers, "Fabrication of Transmutation and Incineration Targets by Infiltration of Porous Pellets by Radioactive Solutions", *J. Alloys and Compounds*, 271-273, 616 (1998).

[15]    A. Fernandez, D. Haas, R.J.M. Konings, J. Somers, "Transmutation of Actinides: Qualification of an Advanced Fabrication Process Based on the Infiltration of Actinide Solutions", *J. Amer. Ceram. Soc.*, submitted.

[16]    H.P. Alder, G. Ledergerber and R.W. Stratton, "Advanced Fuel for Fast Breeder Reactors Produced by Gelation Methods", IAEA Technical Committee, Vienna, Nov. 1987.

[17]    A. van der Linde and J.H.N. Verheugen, "Behaviour of $(U,Pu)O_2$ Sphere-pac Fuel, with Pu in the Large Spheres Only, in Low Burn-up Tests under Pressurised Water Reactor Conditions", *Nucl. Technol.*, 59, 70 (1982).

# EVOLUTIONARY AND MODULAR WATER REACTORS

*Chairs: D. Porsch, P. D'hondt, T. Downar*

# INNOVATIVE FEATURES AND FUEL DESIGN APPROACH IN THE IRIS REACTOR

**Bojan Petrović, Mario Carelli**
Westinghouse Electric Company
Science and Technology Department
Pittsburgh, PA, USA

**Ehud Greenspan, Hiroshi Matsumoto**
University of California Berkeley
Berkeley, CA, USA

**Enrico Padovani, Francesco Ganda**
Politecnico di Milano
Milano, Italy

## Abstract

The International Reactor Innovative and Secure (IRIS) is being developed by an international consortium of industry, laboratory, university and utility establishments, led by Westinghouse. The IRIS design addresses key requirements associated with advanced reactors, including improved safety, enhanced proliferation resistance, competitive electricity production cost, and improved waste management. IRIS is a modular, small/medium size (335 MWe) PWR with an integral vessel configuration. The objective has been to base its design on proven LWR technology, so that no new technology development is needed and near-term deployment is possible, yet at the same time to introduce innovative features making it attractive when compared to present PWRs. These opposing requirements resulted in an evolutionary approach to fuel and core design, balancing new features against the need to avoid extensive testing and demonstration programmes.

## Introduction

The International Reactor Innovative and Secure (IRIS) is being developed by an international consortium of industry, laboratory, university and utility establishments, overseen by Westinghouse. The consortium currently includes 18 members from eight countries. The IRIS design addresses key requirements associated with advanced reactors, including improved safety, enhanced proliferation resistance, competitive electricity production cost and improved waste management [1-3]. This design is based on proven LWR technology, so that no new technology development is needed. As a result, IRIS has potential for near-term deployment (around 2012). IRIS is a modular, small/medium size (335 MWe) PWR with an integral pressure vessel configuration (Figure 1). In contrast to loop PWRs where the primary system components are located outside the reactor pressure vessel, these components are either eliminated (large external loop pipes) or relocated (reactor coolant pumps, pressuriser, steam generators) in the IRIS integral configuration. This may be exploited to improve safety characteristics as well as to reduce the containment size. Indeed, one of the main reasons for the attractiveness of IRIS is its improved safety, which is achieved through the "safety by design" approach, as described in [3,4].

**Figure 1. Vessel layout of the 335 MWe IRIS plant**

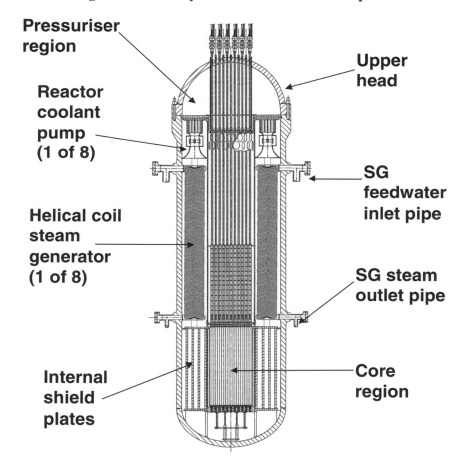

Enhanced proliferation resistance is realised through the capability to operate the plant in a straight-burn mode, which requires a long-life core design, of the order of four to eight years, or even longer [5]. This feature has a positive impact on the plant availability and capacity factors (reduced downtime for refuelling) and consequently economics, but it poses new challenges related to fuel design, neutronics and selection of core design parameters. These challenges are amplified by opposing objectives of designing a fuel with novel features, yet at the same time avoiding uncertainty and/or delay in reactor licensing and deployment that will result if extended fuel irradiation testing and higher fuel enrichment become necessary. The main fuel design decisions will be presented and discussed in this paper, along with a brief discussion of innovative, long-term research directions.

## Fuel and core design approach

One of the key requirements of the IRIS fuel and core design is to allow licensing to proceed without the need for higher than 5% fuel enrichment and extensive or extended irradiation testing, which could otherwise delay the intended near-term deployment. On the other hand, innovative features need to be implemented in order to make the design technically and economically attractive. These opposing requirements resulted in an evolutionary design, balancing between innovations and associated development and demonstration time and efforts. At the same time, it was deemed important to provide a path for future implementation of additional advanced features. This path forward is established by the variable moderation approach that is discussed below. Other fuel features are also examined in some detail.

### *Fuel cycle*

Enhanced proliferation resistance may be of special importance for certain market segments. One practical approach to promoting proliferation resistance is to make the fuel significantly less accessible by designing the core to be capable of operating in a straight-burn long-cycle mode of several years or longer. IRIS was designed from the beginning having this objective in mind, attempting to utilise advantages (and minimise drawbacks) resulting from such long cycles. Maintenance activities are optimised in such a way that opening the reactor pressure vessel (RPV) is required not sooner than every four years. Therefore, cycles of four and eight years with full fuel reload are currently being examined. The resulting improved capacity factor and reduced manpower for the maintenance and reload contribute to favourable cycle economics. Of course, a core designed for a long-cycle operation may also be operated in a more traditional (and less demanding) partial-reload shorter-cycle mode.

The eight-year cycle requires the fissile content to be increased beyond the standard enrichment (5%) for $UO_2$ fuel. This increase as well as the associated increase in the discharge burn-up would likely require irradiation testing that could delay deployment. Therefore, the following strategy is selected:

- *First core (initial fuel loading)* is designed to provide a four-year cycle utilising nearly standard and immediately licensable PWR fuel with up to 5% enriched $UO_2$, but with improved fuel utilisation.

- Core design parameters are selected so that future $UO_2$ and MOX *reload cores* with higher fissile content (and longer cycle) are feasible.

This is further elaborated in the section on the variable moderation approach.

### *Fuel utilisation*

Fuel utilisation may be improved in LWRs either via enhanced conversion [6] achieved by reduced moderation and epithermal spectrum, or via enhanced thermal neutron utilisation achieved by increased moderation. The second option is usually limited by the reactivity coefficient behaviour.

In particular, the moderator temperature coefficient (MTC) becomes positive for over-moderated lattices. Since this effect is related to soluble boron, the IRIS design aims at eliminating (or at least significantly reducing) soluble boron and retaining a well-moderated fuel lattice. For the same fuel enrichment, neutron moderation is enhanced in IRIS relative to present PWR fuel lattices. As a result, the achievable (from the reactivity standpoint) discharge burn-up is also increased; results for 5% enriched $UO_2$ fuel are given below.

A standard $17 \times 17$ PWR fuel assembly has a lattice pitch to fuel rod diameter ratio (p/d) of ~1.3, and a fuel to moderator volume ratio (Vf/Vm) of ~1.6-1.7. In the IRIS $15 \times 15$ assembly, these factors are increased to 1.4 and ~2, respectively, for 5% enriched $UO_2$ fuel, but due to the reduced soluble boron concentration, MTC remains negative. As a result, IRIS fuel of the same enrichment exhibits higher reactivity, as illustrated in Figure 2(a), which compares infinite multiplication factors corrected for neutron leakage as a function of burn-up. This represents the single-batch refuelling strategy, i.e. a straight-burn reactor operation. It may be observed from the figure that IRIS discharge burn-up is larger by ~3 500 MWd/tU (or ~9%) than that for standard fuel. Moreover, Figure 2(b) shows that a similar increase in discharge burn-up is achieved for IRIS fuel with IFBA burnable absorbers.

**Figure 2. Infinite multiplication factor (corrected for leakage) as a function of burn-up**

*(a) No burnable absorbers*

*(b) IFBA, 1 mg $^{10}B$/cm*

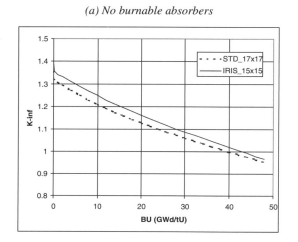
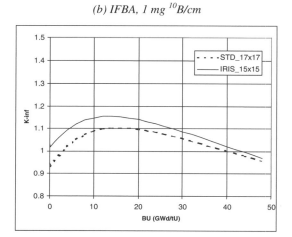

### Path forward via variable moderation approach

IRIS will utilise low-enriched oxide fuel (<20% fissile, but possibly exceeding 5%). The initial core configuration relies on 5% $UO_2$ fuel using current, licensed fabrication technology; this fuel provides a four-year core lifetime. However, for the eight-year core lifetime, $UO_2$ or MOX fuel with a fissile content in the 7-10% range is needed. A path forward is incorporated into the IRIS design that will make feasible and practical future implementation of higher enrichments and longer cycles.

Higher uranium enrichment and even more so MOX fuel require increasing the moderating ratio Vm/Vf to retain good fuel utilisation. In IRIS, the lattice pitch and fuel assembly overall dimensions are kept constant for interchangeability. At the same time the fuel rod diameter (and therefore neutron moderation) may change in the future to match a higher fissile content (7-10%) and/or MOX fuel. This is enabled through selection of fuel design parameters. The initial lattice has already somewhat increased Vm/Vf, about 2. Furthermore, it employs slightly thicker fuel rods (0.366 inches fuel pellet diameter), allowing this diameter to be reduced without leading to too-thin rods. An additional safety margin is provided by a somewhat reduced average linear heat rate (~4 kW/ft).

Fuel lattice parameters for the initial and potential future reloads are shown in Table 1. They are selected in each case so as to provide adequate neutron moderation while maintaining other parameters (e.g. pellet diameter, linear heat rate) in the desirable range. This minimises the development risk since the first core will use the proven 5% $UO_2$ fuel, while later replacements with higher-enrichment and longer-cycle $UO_2$ or MOX fuel may be introduced as they become licensed.

**Table 1. Variable moderation approach to accommodate variable fissile content**

|  | **Initial core** | **Future $UO_2$ upgrade** | **Future MOX upgrade** |
|---|---|---|---|
| **Fuel type** | $UO_2$ <5% fissile | $UO_2$ >5% fissile | MOX >5% fissile |
| **Fissile content** | 4.95% | ~8% | ~10% |
| **Core lifetime** | 4-5 years | ~8 years | ~8 years |
| **Pellet diameter** | 0.366" | 0.340" | 0.296" |
| **Clad OD** | 0.423" | 0.395" | 0.348" |
| **Lattice pitch** | 0.5922" | 0.5922" | 0.5922" |
| **p/d** | 1.4 | 1.5 | 1.7 |
| **Vm/Vf** | 2.0 | 2.5 | 3.7 |

*Burnable absorbers*

Extended cycle and associated excess reactivity impose more severe requirements on reactivity control, hence, selection and optimisation of burnable absorbers plays an important role. As compared to present PWRs, in IRIS it is necessary to hold down larger excess reactivity. Control rods will be used during the cycle in addition to burnable absorbers to provide adequate reactivity control and/or power shaping. Therefore, it is desirable to use integral burnable absorbers, rather than solid rods that would occupy control rod guide thimbles. Another observation is that the increased cycle burn-up leads to higher depletion of burnable absorber resulting in a reduced reactivity penalty as compared to present PWRs.

Several burnable absorber designs have been considered. IFBA is a thin coating of zirconium diboride on the fuel pellets. It is a well-established and well-understood burnable absorber for PWRs, and it will be used here as a reference case for comparison. Figure 3 shows the infinite multiplication factor as a function of burn-up for two different linear densities of $^{10}B$. It may be observed that there is practically no reactivity penalty for burn-ups past 30 000 MWd/tU. However, the depletion range in which the reactivity hold-down is significant is limited primarily to the first 10-15 000 MWd/tU, even though the considered IFBA loading (2 mg $^{10}B$/cm) is several times higher than that used in present PWRs. Note that such large boron loading is acceptable because IRIS fuel is designed with a significantly increased volume for fission products (and helium) gas release, i.e. its plenum length is 18-24 inches, vs. 6-8 inches typically found in PWR fuel.

Figure 4(a) examines the behaviour of integral erbium (mixed with fuel) as a burnable absorber, for several concentrations. Erbium enables larger reactivity suppression and its effect extends to a higher burn-up. Furthermore, erbium is a resonant absorber; hence, it improves the reactivity feedback behaviour. It should be noted, however, that erbium enriched in $^{167}Er$ isotope is considered in Figure 4(a). Using natural erbium would result in a reactivity penalty, as depicted in Figure 4(b). This penalty would result in ~2 500 MWd/tU burn-up reduction if all fuel assemblies contained erbium, but in a more realistic case (e.g. no absorbers at core periphery, higher burn-up in the central part) the effective core-average penalty may be reduced to 1 000-1 500 MWd/tU.

**Figure 3. Infinite multiplication factor as a function of burn-up for IFBA**

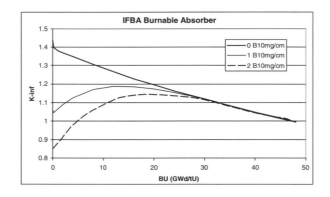

**Figure 4. Infinite multiplication factor as a function of burn-up for erbium integral absorber**

*(a) Erbium enriched in $^{167}Er$*                              *(b) Natural and enriched erbium*

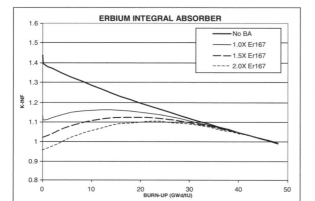
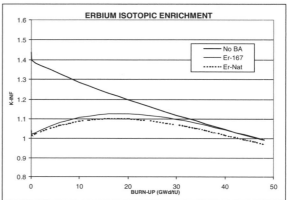

Analysis is under way to evaluate feasibility, cost and benefits of different burnable absorber options (including IFBA, Er, Gd and their combinations) but preliminary results suggest that a combination of IFBA and enriched integral erbium will be used to design the first IRIS core.

### Plutonium disposal

Separate studies were performed to assess the IRIS core characteristics from the standpoint of plutonium degradation (in $UO_2$ fuel), or plutonium burning (in MOX fuel). Considering the isotopic composition of the 5% enriched $UO_2$ fuel, the amount of fissile $^{239}Pu$ at discharge is about one quarter less in IRIS than in a standard PWR lattice, due to a more thermal neutron spectrum. Without presenting detailed results, we note that even larger effect is obtained for MOX fuel, due to the possibility to further adjust the moderation ratio by changing the fuel rod diameter. This provides a largely increased moderation ratio, Vm/Vf = 3.7 (see Table 1), and IRIS may be utilised for effective plutonium disposal through once-through depletion of MOX fuel.

### Summary of IRIS fuel characteristics

The initial IRIS fuel utilises $UO_2$ enriched up to 5%, in a square lattice $15 \times 15$ arrangement similar to present PWRs, except that neutron moderation (and fuel utilisation) is somewhat increased.

(We note that core design utilising $17 \times 17$ assembly is also possible and provides similar characteristics.) Complex combinations of burnable absorbers (e.g. IFBA and enriched erbium) are needed to provide adequate reactivity control. Longer plenum facilitates achieving higher burn-up by reducing internal pressure related issues.

The initial core utilises 5% enriched $UO_2$ fuel that enables four-year cycle operation. This fuel relies in all major features on the present and demonstrated fuel technology, with no associated testing or licensing issues. The plant may continue to operate in the four-year cycle mode for as long as desired or required. When the extended burn-up is demonstrated for $UO_2$ and MOX fuel with 7-10% fissile content, a decision may be made (based on economic analysis) whether to switch to this fuel and the eight-year cycle. To facilitate transition, fuel assembly parameters have been selected such that the core is compatible with the increased reload enrichment via the variable moderation approach discussed above (i.e. it will maintain good fuel utilisation).

## Long-term development

Finally, one aspect of the current R&D efforts, pursued vigorously by the university members of the IRIS consortium, involves evaluating more radical innovations, considered for implementation on a longer time scale, such as:

- Very tight light water lattice.

- Novel fuel rod geometry.

- Use of heavy water.

These features (individually or in some combination) are aimed to achieving epithermal neutron spectrum and a high conversion ratio, to further extend cycle length.

Additionally, the use of thorium-bearing fuel is considered in general, and in particular in tight lattices, to improve void reactivity characteristics.

These long-term research efforts are not the focus of this paper, and they are described in more detail elsewhere. Moreover, it should be emphasised that the IRIS design being developed for deployment around 2012 does not depend on this long-term research, but is instead based on evolutionary improvements of the demonstrated LWR technology, as described in previous sections.

## Conclusions

To enable near-term deployment (around 2012) of the IRIS reactor based on the current PWR fuel technology, while at the same time offering an attractive system with enhanced performance, the IRIS fuel and core design aims at an optimum balance of innovative, evolutionary and standard features. The initial core relies on proven 5% $UO_2$ fuel to facilitate timely licensing and avoid the need for fuel irradiation testing. At the same time, a path has been identified which leaves the design open for future core upgrades and cycle extension from four to eight years via a variable neutron moderation approach.

# REFERENCES

[1] D.V. Paramonov, M.D. Carelli, K. Miller, C.V. Lombardi, M.E. Ricotti, N.E. Todreas, E. Greenspan, K. Yamamoto, A. Nagano, H. Ninokata, J. Robertson, F. Oriolo, "IRIS Reactor Development", Proc. 9th Int. Conference on Nuclear Engineering (ICONE-9), Nice, France, 8-12 April 2001, ASME (2001).

[2] M.D. Carelli, B. Petrović, H. Garkisch, D.V. Paramonov, C. Lombardi, L. Oriani, M. Ricotti, E. Greenspan, T. Lou, "Trade-off Studies for Defining the Characteristics of the IRIS Reactor", Proc. 16th Int. Conference on Structural Mechanics in Reactor Technology (SMiRT 16), Washington, DC, 12-17 August, IASMiRT (2001).

[3] M.D. Carelli, L. Conway, B. Petrović, D. Paramonov, M. Galvin, N. Todreas, C. Lombardi, F. Maldari and L. Cinotti, "Status of the IRIS Reactor Development", Proc. Int. Conf. on the Back-end of the Fuel Cycle (Global'2001), Paris, France, 9-13 September 2001, paper #104, (2001).

[4] M.D. Carelli, L.E. Conway, G.L. Fiorini, C.V. Lombardi, M.E. Ricotti, L. Oriani, F. Berra and N.E. Todreas, "Safety by Design: A New Approach to Accident Management in the IRIS Reactor", IAEA-SR-218/36, IAEA Int. Seminar on Status and Prospects of Small and Medium Sized Reactors, Cairo, Egypt, 27-31 May 2001, IAEA (2001).

[5] B. Petrović, E. Greenspan, J. Vujić, T-P. Lou, G. Youinou, P. Dumaz and M. Carelli, "International Collaboration and Neutronic Analyses in Support of the IRIS Project", *Trans. Am. Nucl. Soc.*, 83, 186-187, ANS (2000).

[6] "High Converting Water Reactors", Y. Ronen, ed., CRC Press (1990).

# CORE CONCEPT FOR LONG OPERATING CYCLE SIMPLIFIED BWR (LSBWR)

**Kouji Hiraiwa, Noriyuki Yoshida, Mikihide Nakamaru, Hideaki Heki**
Isogo Nuclear Engineering Centre
Toshiba Corporation, Japan

**Masanori Aritomi**
Research Laboratory for Nuclear Reactor
Tokyo Institute of Technology, Japan

## Abstract

An innovative core concept for a long operating cycle simplified BWR (LSBWR) is currently being developed under a Toshiba Corporation and Tokyo Institute of Technology joint study. In this core concept, the combination of enriched uranium oxide fuels and loose-pitched lattice is adopted for an easy application of natural circulation. A combination of enriched gadolinium and 0.7-times sized small bundle with peripheral-positioned gadolinium rod is also adopted as a key design concept for 15-year cycle operation. Based on three-dimensional nuclear and thermal hydraulic calculation, a nuclear design for fuel bundle has been determined. Core performance has been evaluated based on this bundle design and shows that thermal performance and reactivity characteristics meet core design criteria. Additionally, a control rod operation plan for an extension of control rod life has been successfully determined.

## Introduction

Increases in nuclear plant capacity have been promoted to take advantage of economies of scale while simultaneously enhancing safety and reliability. As a result, many nuclear power plants currently play an important role in electric power generation. Presently, the next generation reactor with 1 700 MWe is under development. On the other hand, modular-type small or medium-sized reactors are anticipated to be candidates for the next generation reactor, due to their smaller investment risk and the competitiveness of power generation cost derived from the modular design.

The reactor core concept evaluated here is LSBWR [1]. A schematic view of the LSBWR is shown in Figure 1. LSBWR has a new BWR core concept optimised for long cycle operation, a small power output and a simplified BWR configuration.

The purpose of this paper is to show the design concept of long cycle operation for the fuel bundle and core and to demonstrate the core nuclear and thermal performance.

## Core design consideration for long cycle operation

One way to achieve the super-long operating cycle (over 15 years) is the adoption of a high conversion core, which is attainable with a hard neutron spectrum [2]. Some designs employ a combination of a tight lattice core and plutonium MOX fuel to attain hard spectrum [3]. Instead of hardening neutron spectrum, we adopt the combination of medium-enriched uranium oxide fuels and loose-pitched lattice bundle, because this configuration facilitates natural circulation for core cooling. To realise this idea, various design challenges must be resolved.

Figure 2 shows the relationship among these challenges. Two lines indicate the $k_\infty$ with or without burnable poison using loose-pitched $UO_2$ lattice. Increased initial poison worth and extension of poison life are required for suppression of reactivity for a long cycle operation. The increase of control rod worth and extension of control rod life are also required to control the excess reactivity for super-long operation.

Several core design challenges exist with regard to the extension of reactivity lifetime. These include, for example, realisation of long-life burnable poison, suppression of large initial reactivity and preventing positive void reactivity coefficient.

For the realisation of long-life burnable poison, isotope-enriched gadolinium is a powerful candidate. Isotope-enriched gadolinium is enriched with the $^{157}Gd$ isotope. For example, it is known that isotope-enriched gadolinium is created though a laser separation process. Isotope-enriched gadolinium is an essential technology for long cycle operation, because uranium enrichment should be increased a large amount to compensate the reactivity penalty caused by the non-enriched gadolinium currently used in core design. The reactivity penalty is caused by negative reactivity, which itself is due to the existence of even-numbered gadolinium isotopes and a reduction of uranium inventory.

To suppress large initial reactivity and attain a negative void reactivity coefficient, the location of fuel rods that include gadolinium (Gd rods) should be carefully selected. Peripheral location of Gd rods is an effective measure for their reduction, because thermal neutron flux at the peripheral position is higher than at the inner position. Furthermore, peripheral positioning is preferable to prevent positive void coefficient, as the sensitivity of void fraction becomes smaller.

An excess reactivity might be required to be adequately large compared with current BWR design, because estimation of error of the reactivity should be considered. For example, the estimation error in the reactivity comes from analytical error which accompanies long-range estimation and comes from anticipated operation condition change such as reduction in natural circulation flow.

Control rod worth might be required to be enlarged so as to suppress large excess reactivity. To this end, an effective measure could be control rod combined with a smaller bundle. This is because control rod worth increases greatly in smaller bundle design. Other countermeasures to increase control rod worth such as enriched boron are not as effective when compared with smaller bundle.

A BWR control rod is replaced every five or six years, in consideration of rod worth reduction, therefore extension of control rode life is important in long cycle operation. There are several methods for extension of control rode lifetime. One is to increase the poison content, another is to increase the number of control rods. Fortunately, the number of control rods becomes larger in combination with smaller bundle, therefore smaller bundle design is preferable for long cycle operation.

## Core design and core performance evaluation

### Bundle and core design

Fuel bundle design specifications are summarised in Table 1 and Figure 3. The active length of the fuel rod is set at 2.0 m in consideration of the pressure drop of the core, the value of which is shorter than that of SBWR. Bundle pitch is 0.7 times compared to the current BWR bundle. As a result, the area of the fuel bundle cell is half that of current BWRs.

Core design specification for LSBWR is summarised in Table 2. Bundle and control rod location are shown in Figure 4. Thermal output and cycle length are 900 MW and 15 years, respectively, based on the basic design of LSBWR [1]. A downward insertion control rod mechanism is utilised. The value of power density is the same as for SBWR.

Bundle nuclear characteristics are evaluated using lattice design code. Core nuclear and thermal performances are evaluated by 3-D simulator.

The bundle design is shown in Figure 5, and the bundle loading pattern is shown in Figure 6. The bundle design has been determined from various lattice calculations and 3-D core calculations. Since bundle shuffling is not performed, two types of fuel bundle – centre and peripheral – are needed to suppress residual gadolinium worth in the peripheral bundle. The number of Gd rod has been determined to flatten excess reactivity. In the peripheral bundle, the number of Gd rods and the concentration of $Gd_2O_3$ are decreased compared with the centre bundle. The average uranium enrichment is estimated to be 18 wt.% in both the centre and peripheral bundles. Maximum enrichment is less than 20.0 wt.%, which is preferable in several processes of enriched uranium handling, such as conversion, fabrication and transportation. All Gd rods are located at peripheral positions according to the design consideration. Additionally, isotope-enriched gadolinium is applied to the Gd rod. The enrichment of [157]Gd in total gadolinium is assumed to be 80 wt.%.

The control rod pattern is shown in Figure 7. This figure summarises the control rod position in each exposure step, and the control rod residence time in operating period is also summarised as a table. The control rod operation plan has been successfully determined so that the maximum residence time is kept less than six years.

*Reactivity performance*

Void reactivity is shown in Figure 8, and a void coefficient is shown in Figure 9. Although void reactivity has a positive value at small exposure, void coefficient has negative value even at beginning of life (BOL) because of neutron leakage. The reason why the absolute value of coefficient at BOL is less than half of end of life (EOL) is that the residual concentration of gadolinium as a thermal absorber is larger than EOL.

Excess reactivity and cold shutdown margin are shown in Figures 10 and 11, respectively. The maximum value of excess reactivity throughout the operation is 3%dk, the value of which was considered to be larger than ordinary nuclear design on BWR based on the design consideration. The shutdown margin satisfies the design limit of 1.0%dk.

*Thermal performance*

Maximum linear heat rate is shown in Figure 12. Maximum linear heat rate in the operating period is 28 kw/m, which satisfies both operating limit and design target.

Bundle peaking factor and local peaking factor are shown in Figures 13 and 14, respectively. The value of these peaking factors is almost the same level as for ordinary BWR.

## Conclusion

An innovative core design concept for LSBWR is proposed. LSBWR assumes 15-year continuous operation, natural circulation cooling and an upper entry control rod system. In addition, the following key technologies are adopted:

- Small 0.7 times bundle coupled with cruciform control rod.

- Isotope-enriched gadolinium.

- Peripheral positioning of Gd rods.

The design study has been performed on a fuel bundle and a core. A bundle nuclear design is determined based on three-dimensional calculation and core performance has been evaluated.

- Two types of bundle design are needed for the reduction of residual gadolinium worth.

- Void reactivity coefficient remains negative throughout the operation.

- Maximum enrichment needed for 15-year operation has been estimated below 20 wt.%.

- Maximum linear heat rate satisfies the design criteria.

In addition to the above results, a control rod operation plan has been successfully determined so that the maximum residence time is kept less than six years. This means that control rod combined with small bundle is an effective measure for extension of control rod life.

# REFERENCES

[1]  H. Heki, M. Nakamaru, T. Shimoda, K. Hiraiwa, K. Arai, T. Narabayashi and M. Aritomi, "Long Operating Cycle Simplified BWR", Proc. of the 9th Int. Conference on Nuclear Engineering, ICONE-9, Nice, France, 8-12 April 2001.

[2]  "High Conversion Water Reactors", Yigal Ronen, ed., CRC Press (1990).

[3]  M.D. Carlli, D.V. Paramonov, C.V. Lombardi, M.E. Ricotti, N.E. Todreas, E. Greenspan, J. Vujic, R. Yamazaki, K. Yamamoto, A. Nagano, G.L. Fiorini, K. Yamamoto, T. Abram and H. Ninokata, "IRIS, International New Generation Reactors", Proc. of the 8th Int. Conference on Nuclear Engineering, ICONE-8, Baltimore, USA, 2-6 April 2000.

**Table 1. Specification of fuel bundle design**

| | |
|---|---|
| **Active fuel length** | 200 cm |
| **Bundle pitch** | 10.9 cm |
| **Outer water gap width** | 1.3 cm |
| **Channel box thickness** | 0.13 cm |
| **Fuel rod array** | $7 \times 7$ |
| **Number of fuel rods** | 46 |
| **Number of water rods** | 3 |
| **Fuel rod pitch** | 1.26 cm |
| **Fuel rod diameter** | 1.00 cm |
| **Fuel pellet diameter** | 0.85 cm |
| **Fuel pellet stuck density** | 95% T.D. |
| **$^{157}$Gd concentration in Gd** | 80% |
| **Control rod thickness** | 0.7 cm |

**Table 2. Specification of core design**

| | |
|---|---|
| **Thermal output** | 900 MW |
| **Operating cycle length** | 15 years |
| **Fuel bundle number** | 956 |
| **Power density** | 40 kW/l |
| **Bundle array** | $34 \times 34$ |
| **Control rod number** | 225 |
| **Control rod insertion direction** | Upper entry |
| **Reactor dome pressure** | 7.2 MPa |
| **Feedwater temperature** | 215°C |
| **Core flow rate** | 12 000 t/hr |
| **Cycle exposure** | 111 GWd/t |

# Figure 1. Schematic vertical view of reactor vessel

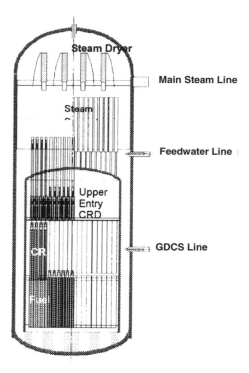

# Figure 2. Design challenges in each core design area

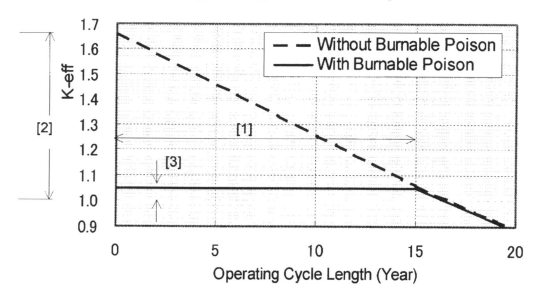

| Category | Burnable poison | Control rod | Bundle/core design |
|---|---|---|---|
| Extended reactivity life [1] required by 15-year operation. | Long-life burnable poison and suppression of large initial reactivity [2]. | Extension of reactivity life [1] required by over 15-year operation. | Limitation on maximum uranium enrichment (<20.0 wt.%). |
| Large excess reactivity [3] required by long-term flexibility and operability. | – | Large excess reactivity [3] required by long-term flexibility and operability. | – |

**Figure 3. Fuel bundle horizontal view**

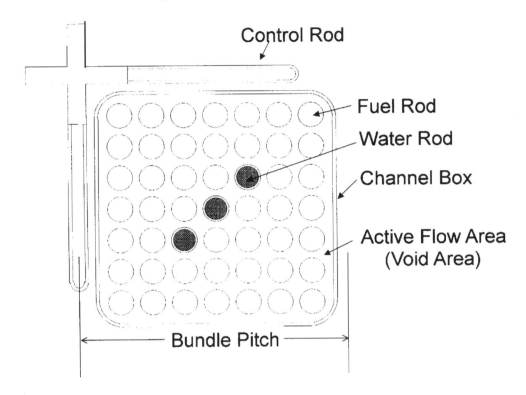

**Figure 4. Fuel bundles and control rod locations**

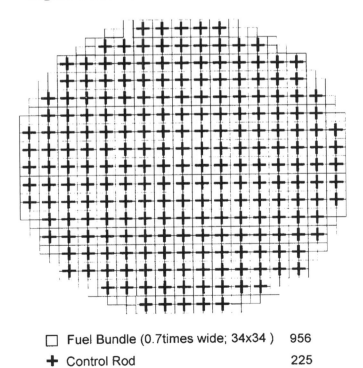

□ Fuel Bundle (0.7times wide; 34x34 )    956
✚ Control Rod                            225

## Figure 5. Bundle nuclear design

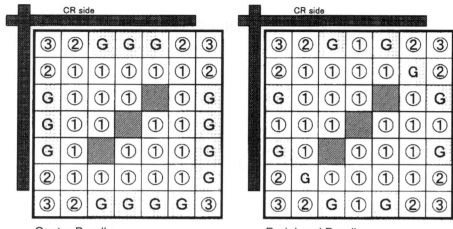

Centre Bundle     Peripheral Bundle

| ①~③ | UO$_2$ rod | $^{235}$U/U    ①: 19.9 wt.% ② 15.0 wt.% ③ 8.0 wt.% <br> (axial enrichment distribution: uniform) |
|---|---|---|
| G | UO$_2$ + Gd$_2$O$_3$ rod | Gd$_2$O$_3$ 3.0 wt.%~14.0 wt.%, $^{235}$U/U 15.0 wt.%~19.9 wt.% <br> (axial distribution Gd$_2$O$_3$: 2 or 3 region, UO$_2$: uniform) |
| (water rod) | Water rod | |
| Bundle average enrichment | | 17.9 wt.% (centre bundle), 18.0 wt.% (peripheral bundle) |

## Figure 6. Bundle loading pattern

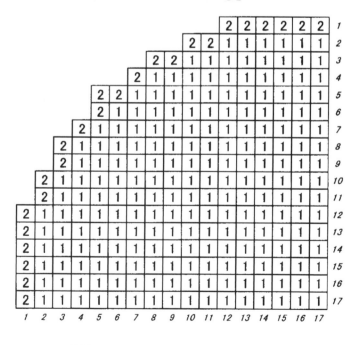

| 1 | Centre Bundle | 856 |
|---|---|---|
| 2 | Peripheral Bundle | 100 |

# Figure 7. Control rod insertion pattern and residence time for each CR

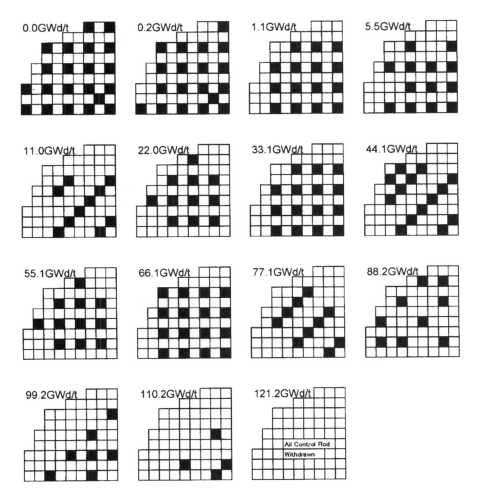

☐  CR position fully withdrawn (surrounded by 4 bundles)

■  CR position fully or patially inserted (surrounded by 4 bundles)

**Total residence time for each control rod (year)**

|     |     |     |     |     |     | 0.0 | 0.0 | 0.1 |
|-----|-----|-----|-----|-----|-----|-----|-----|-----|
|     |     |     |     | 0.0 | 4.5 | 0.0 | 1.5 | 0.0 |
|     |     | 1.6 | 1.5 | 4.5 | 3.0 | 3.7 | 0.0 | 6.0 |
|     |     | 1.5 | 4.5 | 4.5 | 3.0 | 0.0 | 4.5 | 3.0 |
|     | 0.0 | 4.5 | 4.5 | 4.5 | 0.0 | 6.0 | 4.6 | 4.5 |
|     | 4.5 | 3.0 | 3.0 | 0.0 | 4.5 | 4.5 | 4.5 | 0.0 |
| 0.0 | 0.0 | 3.7 | 0.0 | 6.0 | 4.5 | 6.0 | 0.0 | 6.0 |
| 0.0 | 1.5 | 0.0 | 4.5 | 4.6 | 4.5 | 0.0 | 4.6 | 4.6 |
| 0.1 | 0.0 | 6.0 | 3.0 | 4.5 | 0.0 | 6.0 | 4.6 | 4.5 |

## Figure 8. Void reactivity

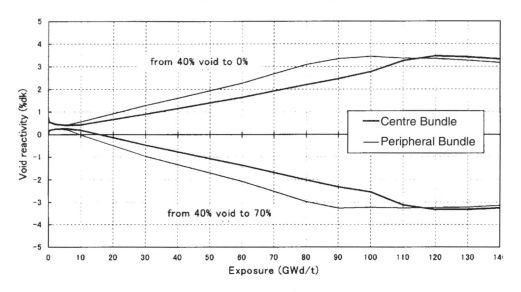

## Figure 9. Void reactivity coefficient

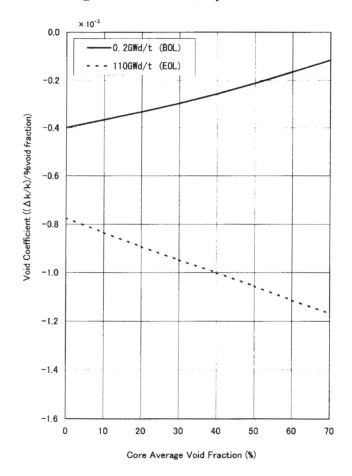

## Figure 10. Excess reactivity

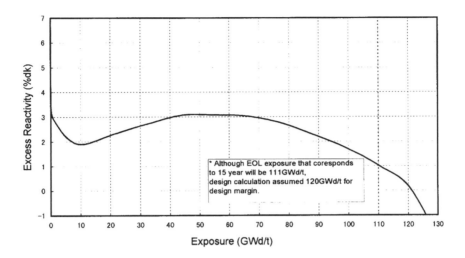

## Figure 11. Cold shutdown margin

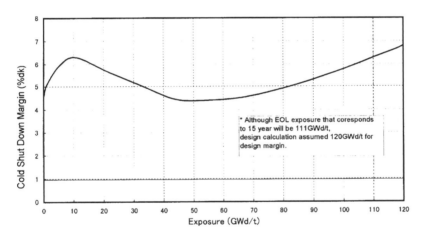

## Figure 12. Maximum linear heat rate

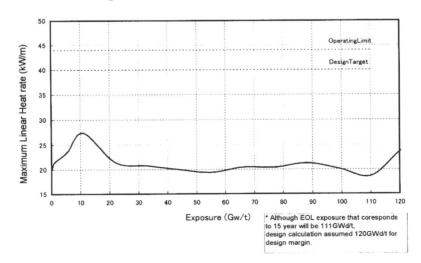

203

## Figure 13. Bundle peaking factor

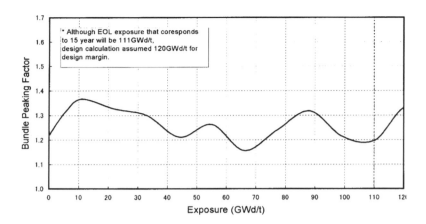

## Figure 14. Local peaking factor

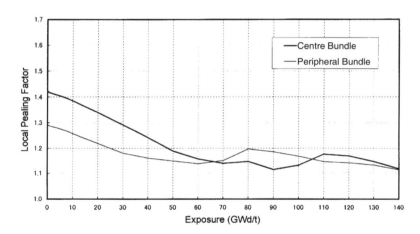

# CORE DESIGN STUDY ON REDUCED-MODERATION WATER REACTORS

**Hiroshi Akie, Yoshihiro Nakano, Toshihisa Shirakawa, Tsutomu Okubo, Takamichi Iwamura**
Department of Nuclear Energy Systems
Japan Atomic Energy Research Institute (JAERI)

## Abstract

The conceptual core design study of reduced-moderation water reactors (RMWRs) with tight-pitched MOX-fuelled lattice has been carried out at JAERI. Several different RMWR core concepts based on both BWR and PWR have been proposed. All the core concepts meet with the aim to achieve both a conversion ratio of 1.0 or larger and negative void reactivity coefficient. As one of these RMWR concepts, the ABWR compatible core is also proposed. Although the conversion ratio of this core is 1.0 and the void coefficient is negative, the discharge burn-up of the fuel was about 25 GWd/t. By adopting a triangular fuel pin lattice for the reduction of moderator volume fraction and modifying axial Pu enrichment distribution, it was aimed to extend the discharge burn-up of ABWR compatible type RMWR. By using a triangular fuel lattice of smaller moderator volume fraction, discharge burn-up of 40 GWd/t seems achievable, keeping the high conversion ratio and the negative void coefficient.

## Introduction

Light water reactors are considered to be one of the main electric power sources for at least several decades in the 21st century. For the effective utilisation of uranium resources with well-experienced LWR technology, a conceptual design study of reduced-moderation water reactors (RMWRs) has been carried out at the Japan Atomic Energy Research Institute (JAERI) in co-operation with electric power companies.

The RMWR concept is based on a tight-pitched MOX-fuelled lattice which can realise a conversion ratio of more than 1.0. The tight fuel lattice, however, makes it difficult to keep the moderator void reactivity coefficient negative. Therefore in the design study of RMWRs, it has been aimed to achieve both a conversion ratio of more than 1.0 and a negative void coefficient. With some technical approaches to reduce the moderator void coefficient, several RMWR core concepts based on both BWR- and PWR-type cores have been proposed [1,2]. The outline of these core concepts will be summarised only briefly in this paper.

One of the RMWR concepts proposed is the 1 350 MWe class ABWR compatible core without blanket, the fuel assemblies of which have square cross-sections with square fuel pin lattice [1]. Although the conversion ratio of this core is 1.03 and the void coefficient is negative, the discharge burn-up of the fuel is 26 GWd/t. The aim is to improve the burn-up characteristics of this ABWR compatible core. The current status of the improvement is mainly presented here.

## Outline of RMWR core designs

There are several proposed RMWR core concepts based on both BWR- and PWR-type cores. All the RMWR concepts realised a conversion ratio more than 1.0, and a negative void reactivity coefficient through the use of unique technical approaches.

### High conversion ratio BWR-type core

The high conversion ratio BWR type core [1] realises as high a conversion ratio as 1.1 with a triangular fuel lattice of the pin diameter of 14.5 mm and the pin pitch of 15.8 mm and a high core averaged moderator void fraction of 70%. A negative void reactivity coefficient is achieved with a very flat core seed region of about 20 cm height. To obtain the core thermal output of 3 188 MW, the core diameter is very large, about 7 m, and such a flat core is doubly stacked. An internal blanket region of about 30 cm height and lower and upper blanket regions of 20-35 cm height are also considered, and the total core height is about 1.2 m. The discharge burn-up of the MOX seed part is 66 GWd/t and the reactor operation cycle length is 14 EFPM under 4.5-batch refuelling conditions.

### Long operation cycle BWR-type core

This type of core design [1] aims at a long operation cycle, and achieved a 22 EFPM cycle length through four-batch refuelling. The active core height is 1.6 m so as to avoid a very large core diameter, and upper and lower blanket regions of 50 cm height each are placed. By introducing special void tube assemblies, a negative void coefficient is attained without significantly reducing the core height. The void tube assembly consists of a void can filled with stagnant steam, Hf γ-heater at the bottom of the tube and narrow coolant flowing channel surrounding the void can. In total, 61 void tube assemblies are arranged in the core with 252 fuel assemblies. The core diameter is 5.9 m. The fuel pin diameter and fuel pin pitch of the triangular fuel lattice are 11.9 mm and 13.2 mm, respectively. The average void fraction in the core is 60%. The thermal power of the core is 3 926 MW, the discharge burn-up of the MOX seed part is 61 GWd/t and the conversion ratio is 1.04.

## ABWR compatible type core without blanket

The ABWR compatible core [1] has the same core diameter of 5.4 m and the same core thermal power of 3 926 MW as ABWR. The fuel assemblies have square cross-sections, and the radial size and pitch are also the same as for the ABWR assembly. The fuel pin lattice in the assembly is a $9 \times 9$ square one. In order to reduce the moderator to fuel volume ratio (Vm/Vf), the pin diameter is increased to 15.17 mm, and the core averaged moderator void fraction is about 50%. For the negative void reactivity coefficient, the axially heterogeneous core concept was adopted. There are three low-enriched regions of 5.4 wt.% fis. Pu and three high-enriched regions of 8.2 wt.% fis. Pu. From the core bottom, low-enriched regions and high-enriched regions are placed in turn, and the top of the core is a high-enriched region. The heights of the low-enriched regions are 50, 30 and 40 cm, respectively from the core bottom to the top, and heights are 60, 30 and 50 cm for high-enriched regions from the bottom to the top. The total core height is 2.6 m and there is not a blanket region in the core. The conversion ratio is 1.03, the operation cycle length is 12.6 EFPM with 4.3-batch refuelling and the discharge burn-up is 25.6 GWd/t.

## High conversion ratio PWR-type core with heavy water coolant

In a PWR-type core without any void formed in the moderator, it is difficult to achieve a high conversion ratio. In the high conversion ratio PWR-type core [2], the conversion ratio becomes as high as 1.11 by introducing heavy water as a moderator. The MOX seed fuel assembly consists of a triangular fuel lattice of 9.5 mm pin diameter and 10.5 mm pin pitch. There are two types of seed assemblies with and without an axial internal blanket of 50 cm high. The seed height is 160 cm in the assembly without the internal blanket, and is 55 cm $\times$ 2 in the assembly with the internal blanket. The upper and lower blanket height is 35 cm for the both seed assemblies. These seed assemblies of hexagonal cross-section are arranged with radial blanket assemblies with a checkerboard-type distribution. The fuel pin diameter and the pin pitch in the radial blanket assembly are 17.7 and 18.7 mm, respectively. The radial blanket height is 115 cm and 22.5 cm shorter than the seed height both from the top and bottom of the seed region. The void coefficient is kept negative with this axial and radial heterogeneous core. The core diameter is about 4.7 m and the thermal output is 2 900 MW. The discharge burn-up of the MOX seed part is 55 GWd/t for the assembly with the internal blanket and 46 GWd/t for the assembly without the internal blanket. The cycle length is 15 EFPM by three-batch refuelling.

## PWR-type core with light water coolant

A conversion ratio of 1.0 also seems feasible in the light water moderated PWR-type core with a seed/blanket-type assembly [2]. In the concept of seed/blanket-type assembly, the MOX seed fuel sub-bundle is surrounded by blanket fuel rods. The seed fuel pin diameter is 9.5 mm and the triangular lattice pitch is 10.5 mm, and blanket pin has 14.4 mm diameter and 15.0 mm lattice pitch. In the blanket layer, there are also $ZrH_{1.7}$ rods to reduce void coefficient. The seed fuel is divided into two axial zones of 1 m height to improve the void coefficient. There are 50 cm high axial internal blanket and 25 cm high upper and lower blankets. From the assembly burn-up calculations, it is possible to achieve the conversion ratio of 0.98 and the cycle length of 18 EFPM under three-batch refuelling conditions in this seed/blanket-type assembly. By taking the effect of the radial blanket into account, the conversion ratio of 1.0 can be expected. The discharge burn-up of the seed region is 45 GWd/t.

# Improvement of the burn-up characteristics of ABWR compatible type RMWR

To improve the burn-up characteristics of the ABWR compatible type RMWR, a triangular fuel pin arrangement in the ABWR square assembly is considered for the reduction of the Vm/Vf value. In addition, the core averaged void fraction is increased to about 55%. The void reactivity increase in the tighter-pitched fuel lattice is to be avoided by modifying the axial Pu enrichment distribution. By reducing the enrichment in low-enriched regions, the void reactivity coefficient can be improved. In this case, on the other hand, the axial power peaking factor becomes larger. It was attempted to optimise these parameters.

## Parametric survey by one-dimensional core burn-up calculation

The pin diameter and the pin pitch were considered to be 10 mm and 11 mm, respectively, and a new ABWR compatible assembly model with 188 fuel pins is considered (Figure 1). One-dimensional core burn-up calculations were carried out to determine the axial Pu enrichment distribution. The Pu enrichment in the high-enriched region (*Eh*) was surveyed up to 20 wt.% fis. Pu, and the enrichment in low-enriched region (*El*) from 1 to a few % fis. Pu. For the reduction of power peaking due to a larger enrichment difference than the original ABWR compatible core, four layers (instead of three layers) of high- and low-enriched regions were considered. The core height of the high-enriched region (*Hh*) and the low-enriched region (*Hl*) were set to be 20 cm and 35 cm, respectively. At the core bottom is a low-enriched region and the top of the core is a high-enriched region. The total core height is 2.2 m.

**Figure 1. ABWR compatible assembly model with triangular fuel lattice**

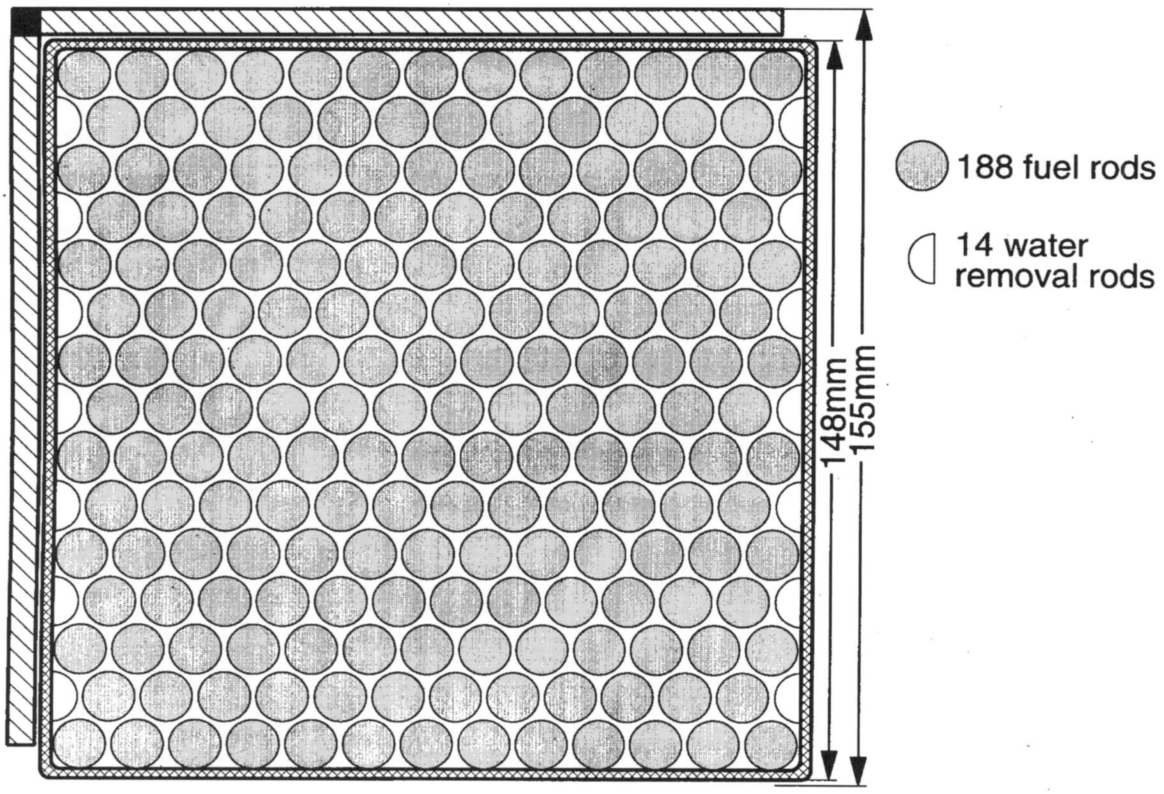

148mm
155mm

◯ 188 fuel rods

◖ 14 water removal rods

One-dimensional core burn-up calculations were performed using the SRAC95 and COREBN95 code systems [3] with JENDL-3.2 nuclear data [4]. In the parametric calculations, the target discharge burn-up was tentatively set to be 40 GWd/t. With the assumption of four-batch refuelling, the burn-up period of 40 GWd/t is divided into four cycles, and the following core characteristics were evaluated:

- Effective multiplication factor at the middle of the third cycle [k(2.5c)], which indicates the multiplication factor at EOC of a four-batch refuelling core.

- Instantaneous conversion ratio at the end of the second cycle [CR(2c)], to represent burn-up integrated conversion ratio.

- Void reactivity at the middle of the third cycle [VR(2.5c)]. The void reactivity at EOC, which is worse than that at BOC, can be estimated. Here, the SRAC/COREBN system cannot perform the coupled iteration calculation of neutronics and thermal-hydraulics. The void reactivity here is evaluated as a reactivity when the number density of hydrogen is reduced by 0.9 times in all the moderator regions but bottom reflector.

- Maximum heat flux at cladding surface at the end of the first cycle [q″(1c)]. The power peaking factor is largest at BOL, but is usually suppressed by control rod, burnable poison, etc., at BOL.

The target values of these items were 1.0 for k(2.5c) to reach 40 GWd/t discharge burn-up, 1.0 for CR(2c) and a negative value for VR(2.5c). The value q″(1c) is a measure if rod burnout takes place. From the critical heat flux experiments, it seems the heat flux must not exceed 1.0 MW/m$^2$ at worst. Taking into account the core radial power distribution and the local power peaking within assemblies, q″(1c) was aimed to be less than about 0.85.

The results of the parametric calculation are summarised in Figure 2 for the enrichment $Eh$ = 15, 16, 17 and 18 wt.% fis. Pu. In Figure 2(a), the enrichment $El$ for k(2.5c) = 1, namely the enrichment to obtain the discharge burn-up of 40 GWd/t, can be interpolated for each $Eh$ from the k(2.5c) values denoted by the symbol of a solid circle. Then the conversion ratio values (solid triangles) can also be interpolated for this enrichment, and is shown in the figure as the thin solid line denoted as "CR for k = 1.0". It can be seen that a conversion ratio of 1.06-1.1 can be achieved in this enrichment region. In the same manner as the conversion ratio, the void reactivity and the heat flux are also interpolated and shown in Figure 2(b). In this figure, VR(2.5c) and q″(1c) each have contrary dependencies on $El$. When VR(2.5c) is negative, q″(1c) becomes as high as 1 MW/m$^2$. It is difficult to achieve both a negative void reactivity and a maximum heat flux of less than 0.85 MW/m$^2$. Either the void reactivity or the power peaking must be improved.

### Modification of one-dimensional core specification

Improvement of the power distribution flattening was first investigated by increasing the height of high-enriched region $Hh$ and reducing the low-enriched region height $Hl$. The calculated void reactivity and heat flux are summarised in Figure 3 for $Hh$ = 25 cm and $Hl$ = 30 cm in the same manner as in Figure 2(b). In this case, q″(1c) is 0.85 MW/m$^2$ when VR(2.5c) is 0.0; CR(2c) is 1.08, and the aimed core performance is just realised.

For further improvement, the gap between pins was next increased from 1 mm to 1.3 mm to obtain a larger Vm/Vf value and smaller (more negative side) void reactivity. In this case, pin diameter is decreased to 0.97 mm with the same pin pitch of 11 mm. Figure 4 shows only two points of

## Figure 2. Parameter survey calculation results for ABWR compatible core RMWR with high-enriched region of 20 cm × 4 and low-enriched region of 35 cm × 4

*(a) Multiplication factor and conversion ration*　　　*(b) Void reactivity and heat flux for k(2.5c) = 1*

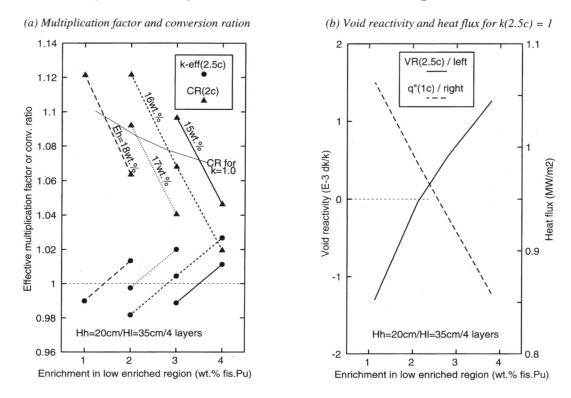

## Figure 3. Void reactivity and heat flux for k(2.5c) = 1 in the core of high-enriched region height of 25 cm

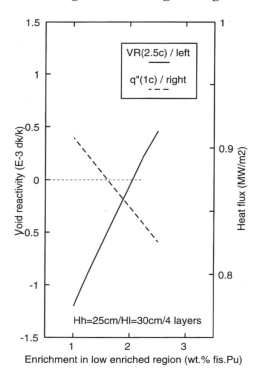

**Figure 4. Void reactivity and heat flux in the core of pin gap of 1.3 mm**

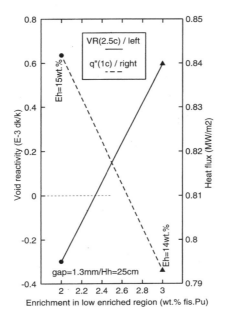

calculated VR(2.5c) and q″(1c) for the core of 1.3 mm pin gap, *Hh* of 25 cm and *Hl* of 30 cm. It is seen that the void reactivity is negative and the maximum heat flux is less than 0.85 MW/m$^2$ in the case of *Eh* = 15 wt.% fis. Pu and *El* = 2 wt.% fis. Pu. The estimated core characteristics for this core are k(2.5c) = 1.002, CR(2c) = 1.042, VR(2.5c) = -2.99 × 10$^{-4}$dk/k and q″(1c) = 0.842MW/m$^2$.

### *Three-dimensional core burn-up calculation*

From the parametric survey calculations in the previous section, we have obtained an ABWR compatible RMWR one-dimensional axial core specification to meet with the aim to increase the discharge burn-up up to 40 GWd/t. To confirm the performance of the core obtained by the one-dimensional calculation, the three-dimensional core burn-up calculation was performed.

The calculation was performed with SRAC95, ASMBURN95 and COREBN95 code systems [3] and JENDL-3.2 nuclear data. The core specifications are as follows. The pin diameter and the pin pitch of the triangular lattice are 9.7 mm and 11 mm, respectively. The square assembly of channel box outer size of 148 mm × 148 mm contains 188 fuel pins. The assemblies are arranged in the core with 155 mm assembly pitch. The core diameter is about 5.4 m without radial blanket. The core height is 2.2 m with four layers of high-enriched regions of 25 cm height and four layers of low-enriched regions of 30 cm height. The enrichment in the high-enriched region is 15 wt.% fis. Pu, and in low-enriched region is 2 wt.% fis. Pu. The core averaged void fraction is about 55%. The total core thermal power is 3 926 MW. The three-dimensional core burn-up calculation is performed by simulating four-batch assemblies refuelling.

Burn-up dependent effective multiplication factor and instantaneous conversion ratio of the equilibrium cycle are shown in Figure 5. The operation cycle length is 442 EFPD, which corresponds to 39 GWd/t discharge burn-up, and the burn-up averaged conversion ratio is 1.002. The discharge burn-up is well estimated in the one-dimensional calculations, but the conversion ratio becomes smaller than predicted. The radial power peaking in the core is large, because the optimisation of the assemblies loading and shuffling pattern is not made in this calculation.

**Figure 5. Effective multiplication factor and instantaneous conversion ratio obtained by 3-D calculation in the equilibrium cycle of ABWR compatible RMWR (pin gap 1.3 mm, high-enriched region height 25 cm and low-enriched region height 30 cm)**

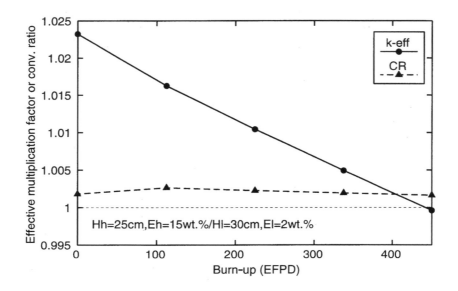

## Conclusion

For the effective utilisation of uranium resources with well-experienced LWR technology, JAERI has proposed a reduced-moderation water reactor (RMWR) concept. There are several types of design of RMWRs based on both BWR and PWR. All the RMWR concepts simultaneously attain a conversion ratio of 1.0 and a negative void reactivity coefficient. In order to achieve these core performances, each core design adopts specific technical approaches.

An ABWR compatible type RMWR concept is the simplest RMWR core design with the same radial core and assembly sizes as ABWR, with no blanket region and without any new components such as void tube assembly, heavy water moderator or seed/blanket-type assembly. In the original design, the conversion ratio of 1.0 and the negative void reactivity coefficient were achieved, but the discharge burn-up was about 25 GWd/t. By using a triangular fuel lattice of smaller moderator to fuel volume ratio (Vm/Vf) in the square assembly and reconsidering the axial Pu enrichment distribution, as a result of the parametric survey calculation on the axial Pu enrichment distribution, discharge burn-up of 40 GWd/t seems achievable, while still maintaining the conversion ratio of 1.0 and the negative void coefficient. The proposed axial heterogeneous core specification is four layers of 14-15 wt.% fis. Pu-enriched regions of 25 cm high and four layers of 2 wt.% fis. Pu-enriched regions of 30 cm high. The gap of 1.3 mm between pins seems preferable from the viewpoints of void reactivity coefficient and power peaking reduction. Even with the gap of 1.3 mm, it is possible to achieve the conversion ratio of 1.0. Further investigation is necessary to reduce core radial power peaking and the peaking factor within a fuel assembly.

# REFERENCES

[1]  T. Okubo, *et al.*, Proc. 8th international Conference on Nuclear Engineering (ICONE-8), ICONE-8422 (2000).

[2]  K. Hibi, *et al.*, *ibid.*, ICONE-8423 (2000).

[3]  K. Okumura, *et al.*, Japan Atomic Energy Research Institute Report JAERI-Data/Code 96-015 (1996) [in Japanese].

[4]  T. Nakagawa, *et al.*, *J. Nucl. Sci. Technol.*, 32, 1259 (1995).

# THE UTILISATION OF THORIUM FUEL IN A GENERATION IV LIGHT WATER REACTOR DESIGN

**Thomas J. Downar, Yunlin Xu**
School of Nuclear Engineering
Purdue University
West Lafayette, IN 47907
USA

## Abstract

During the last several years the Department of Energy has sponsored research at Purdue University on advanced reactor designs under the Nuclear Energy Research Initiative (NERI) programme. This work has involved research in "Generation IV" advanced reactor designs such as a high conversion boiling water reactor, as well research in advanced fuel designs such a metal matrix "dispersion" fuel. The unifying theme of this research has been to take advantage of the numerous benefits of the thorium fuel cycle. The Purdue research has been performed in collaboration with Argonne and Brookhaven National Laboratories for the dispersion fuel research and the high conversion reactor research, respectively. The primary contribution to both research efforts from Purdue has been on neutronics design and analysis. This paper will focus on the neutronics design and analysis of the high conversion boiling water reactor.

## Introduction

The overall objective of the research on the high conversion boiling water reactor (HCBWR) has been to design and analyse a proliferation-resistant, economically competitive, high conversion boiling water reactor fuelled with fertile thorium oxide fuel elements. The extensive previous research on high conversion LWRs in the US, Europe and Japan over the last several years provided a valuable starting point for the design developed here. In all of these previous efforts, the HCBWR here has been characterised by a very tight lattice with a relatively small water volume fraction, which therefore operates with a fast reactor neutron spectrum and a considerably improved neutron economy compared to the current generation of light water reactors. The design objective of the work at Purdue has been to utilise a thorium fuel cycle and to achieve a high conversion of thorium to $^{233}$U as well as to reduce the national accumulated inventory of plutonium and to take advantage of the enhanced proliferation resistance of the thorium fuel cycle. The high fuel burn-ups achievable with thorium fuel in a high conversion core increase the plant capacity factor and lower the overall cost of electricity. All of these advantages are consistent with the Department of Energy's objectives for the development of Generation IV nuclear power systems and proliferation-resistant fuel cycles.

One of the concerns with a tight lattice core with a hard neutron spectrum is the potential for a positive void coefficient. Design features such as "void tubes" have been proposed in some HCBWR core designs to ensure inherent negative reactivity feedbacks. One of the major objectives of the HCBWR design work at Purdue was to obviate the need for mechanical design innovations to ensure a negative void feedback, and as will be discussed in a later section, the use of the thorium fuel made this possible.

The three-year NERI research effort on the HCBWR has completed its second year. The overarching objective of the project was to provide a proof-of-principle demonstration of the HCBWR design. During the first year of the research, all the necessary computer codes were acquired and benchmarked for tight lattice reactor analysis. This included benchmarking of the HELIOS lattice physics code for tight pitch thorium fuel lattices, as well as modifications to the RELAP5 thermal-hydraulics code for tight pitch lattices and to the PARCS neutronics code for fuel depletion and for multi-group, hexagonal geometry capability. The primary focus of activity during the second year has been on the design and safety analysis of a full core high conversion reactor model with both RELAP5 and PARCS. An equilibrium cycle core of the HCR was developed and compared to the Japanese high conversion reactor design (RMWR). At Brookhaven National Laboratory, safety analysis was performed with the RELAP5 thermal-hydraulics code and focused on two of the most limiting events for a tight pitch lattice core: the station blackout and the main steam line break. The following sections will briefly outline features of the HCBWR design and discuss continuing work.

## HCR fuel design

The Japanese RMWR high conversion reactor design [1,2] was used as the starting point for the Purdue/BNL HCR. Some of the of design features of the RMWR are compared to a conventional ABWR in Table 1.

One of the most important design characteristics of this the RMWR design is the very low volume ratio of water to fuel, which is achieved by using the hexagonal lattice structure. Another important characteristic is that the height of the core is set at 160 cm in order to provide for adequate heat removal and help ensure a negative void coefficient. An additional feature of this core is the use of "void tubes" to provide for enhanced neutron leakage during core voiding. As will be discussed later in this section these void tubes were not necessary in the Purdue/BNL HCR design because it is fuelled with $^{232}$Th.

216

**Table 1. Comparison of the HCBWR/RMWR with the ABWR**

| Item | Unit | HCBWR | ABWR |
|---|---|---|---|
| Thermal power | MW | 3 926 | 3 926 |
| Core diameter | m | 5.8 | 5.4 |
| Core height | m | 1.6 | 3.7 |
| Blanket height | m | 0.3 (upper) 0.3 (bottom) | ~0.1 (upper) ~0.1 (upper) |
| Assemblies/core | – | 313 | 872 |
| Pins/assembly | – | 469/252 | 62 (8 × 8) |
| Geometry of fuel arrangement | – | Triangle lattice | Square lattice |
| Outer diameter of fuel cladding | mm | 11.9 | 12 |
| Cladding thickness | mm | 0.4 | 0.9 |
| Fuel element pitch | mm | 13.2 | 16 |
| Volume ratio water to fuel | – | 0.5 | 3 |
| Fissile enrichment | – | 9 w/o | 3 w/o |
| Fertile fuel | – | $^{232}$Th | $^{238}$U |

Because of the tighter pitch in the RMWR/HCBWR fuel pin, the neutron spectrum of the HCBWR is harder than the ABWR. The ABWR and HCBWR neutron spectrum are compared in Figure 1. The HCBWR has a smaller neutron flux in the thermal neutron energy region and a larger neutron flux in the fast energy region. This reduces neutron moderation and thereby enhances the conversion ratio.

**Figure 1. Comparison of neutron spectra for ABWR and HCBWR**

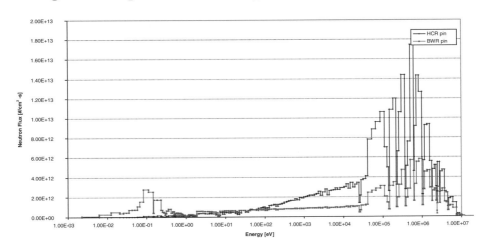

Because of the higher conversion ratio, the slope of the $k_\infty$ versus burn-up in the HCBWR is much less than that of the ABWR, as shown in Figure 2. The calculations here were performed using the HELIOS lattice physics code which has been well benchmarked for tight pitch thorium pin cells [3].

## High conversion BWR core design

This section will describe the equilibrium cycle design of the high conversion BWR. Calculations were first performed on the Japanese reduced moderation water reactor (RMWR) and then on the

**Figure 2. Comparison of fuel k∞ for HCR and ABWR**

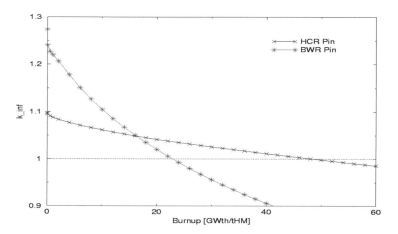

modified Purdue/BNL HCR design. The calculations on the RMWR core served the dual purpose of providing a starting point for the Purdue/BNL design, as well as validating the RELAP5/PARCS code system for high conversion reactor design. The RMWR is a research project of Japan Atomic Energy Research Institute (JAERI) and Japan Atomic Power Company (JAPCO). There are many similar neutronics characteristics between RMWR and HCBWR, such as hexagonal fuel assembly, tight lattice pitch, hard neutron spectrum. There are some published results of RMWR [1,2] that were used to compare with the HELIOS-PARCS calculated results.

### *Neutronics methods and reactor model*

PARCS is the primary code of the standard neutronics calculation procedure of Purdue University for the prediction of light water reactor core static and transient states. In order to solve the diffusion equations by PARCS, the depletion code (DEPLETOR) generates the macroscopic cross-sections at the specific core state and provides them to the PARCS. The macroscopic cross-sections at the appropriate fuel conditions can be prepared using any lattice codes such as HELIOS, CASMO, etc. Because the output format of the each lattice code is unique, an additional processing program, GENPXS, was required to provide the cross-section data in the specific format (PMAXS) that can be read by the depletion code DEPLETOR. An overview of the standard neutronics calculation procedure for the cross-section generation scheme is shown in Figure 3. A paper was presented at the American Nuclear Society M&C Meeting in Salt Lake City describing these methods [4].

**Figure 3. The standard neutronics calculation procedure of Purdue University**

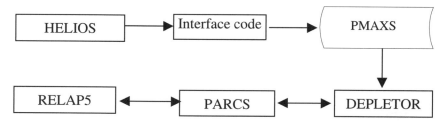

The deterministic transport code, HELIOS-ver1.6, was used to evaluate the cross-sections for fuel assemblies and the two-dimensional core characteristics with 49-group cross-sections. After computation with the HELIOS code, the post-processor code (ZENITH) generated multi-group (8 groups) assembly homogenised cross-sections for the standard neutronic design procedure of Purdue. A second paper presented at the Salt Lake ANS meeting describes the multi-group hexagonal nodal method developed for the PARCS code in order to analyse tight pitch lattices [5].

RELAP5 [6] is a well-known thermal-hydraulic code to solve for the core and system temperature/fluid distribution for a given neutronics power shape. The design parameters from RELAP5 include such items as fuel temperature, moderator temperature, moderator density and void fraction. The T/H model used here was to represent each neutronics fuel assembly with a separate thermal-hydraulics channel. A schematic of the model is shown in Figure 4. Several modifications to RELAP5 were made at BNL specifically for analysing tight pitch lattices [7]. These will be reported in a separate paper.

**Figure 4. RELAP5 HCR thermal-hydraulic model**

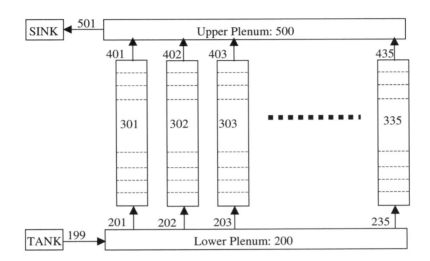

### Purdue/BNL HCR core design

*General description of RMWR core*

The RMWR core geometry is shown in Figures 5 and 6. The main design parameters of the reactor core are the same as those shown in Table 1.

**Figure 5. Horizontal cross-section of RMWR**

## Figure 6. Vertical cross-section of RMWR

The data presented in this subsection are adopted from Refs. [1,2]. Some of the details of the RMWR core design (e.g. reactivity control) were not available and therefore a "best guess" was used as a basis of comparison for the RMWR with the Purdue/BNL HCR.

*HCR core configuration*

An equilibrium core was designed with four fuel batches as shown in Figure 7.

## Figure 7. Equilibrium four-batch core (1/12 core symmetry)

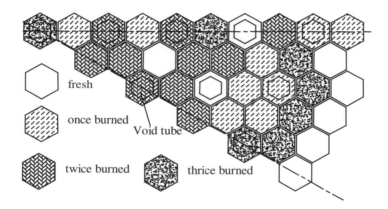

There are two kinds of fuel assemblies in this core, normal fuel assemblies and assemblies with void tubes. The assemblies with void tubes are replaced with normal fuel assemblies for a special case of the RMWR to demonstrate the positive void coefficient and for the Purdue/BNL design which does not contain void tubes. Additionally, for the Purdue/BNL design all fuel assemblies are replaced by Th-Pu fuel assemblies.

*Fuel assembly design*

There are three kinds of fuel assemblies: normal U-Pu fuel assembly, assembly with void tube and Th-Pu fuel assembly. The HELIOS calculation model used for the normal fuel assemblies is shown as Figure 8 and the assembly with void tube is shown as Figure 9. There are three different enrichments of fuel rods loaded in both types of assemblies. The highest enriched (10%) rods are placed in inner region, the middle enriched (6%) rods are placed along the sides and the lowest enriched (4%) rod are placed at corners. The enrichments of the fuel rods in each assembly are listed in Table 2.

Figure 8. Normal fuel assembly

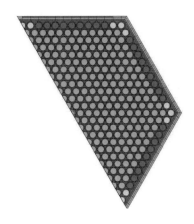

**Figure 9. Assembly with void tube**

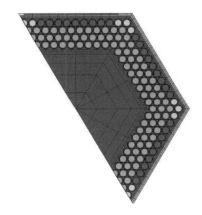

**Table 2. The enrichments of the rods in each fuel assembly**

| Fuel rod type | Normal fuel assembly | | Assembly with void tube | |
|---|---|---|---|---|
| | Number of rods | Enrichment (%) | Number of rods | Enrichment (%) |
| Highest enriched | 396 | 10 | 180 | 10 |
| Middle enriched | 54 | 6 | 54 | 6 |
| Lowest enriched | 18 | 4 | 18 | 4 |
| Total | 468 | 9.3 | 252 | 8.71 |

The U-Pu fuel consists of oxide of natural uranium and reactor-grade plutonium. The Th-Pu fuel consists of oxide with natural thorium and weapons-grade plutonium. The fissile enrichments of U-Pu fuel and Th-Pu fuel are the same.

### HCR calculation results

*Depletion results of the four-batch RMWR core*

Neither the exact loading pattern of JAERI's RMWR core is known nor some of the details of the assembly design. Therefore the results and comparisons shown here are preliminary and will be

221

updated subsequent to further collaboration with the Japanese researchers. The depletion calculation the equilibrium cycle uses average batch burns of 25, 50, 75 GWd/t for the once, twice, thrice burned assemblies, respectively. A cosine axial power shape is assumed to provide axial burn-up distribution.

The primary depletion results of the four-batch core are shown in Table 3. All the results are calculated with all control rods withdrawn (ARO). The axial power shape for BOC (0 GWd/t), MOC (12.5 GWd/t) and EOC (25 GWd/t) are shown in Figure 10. The radial burn-up and power distributions at BOC/EOC are shown in Tables 4-6.

### Table 3. Depletion results of the four-batch RMWR core

| Burn-up (GWd/t) | $K_{eff}$ | Power peaking | Average burn-up of active core | Maximum burn-up | Average void fraction (%) |
|---|---|---|---|---|---|
| 0.0 | 1.084751 | 1.995 | 35.1 | 87.8 | 40.1 |
| 2.5 | 1.073765 | 1.971 | 37.6 | 91.2 | 39.77 |
| 5.0 | 1.064456 | 1.863 | 40.1 | 94.7 | 39.89 |
| 7.5 | 1.055722 | 1.778 | 42.6 | 98 | 39.94 |
| 10.0 | 1.047487 | 1.713 | 45.1 | 101.2 | 39.94 |
| 12.5 | 1.039710 | 1.662 | 47.6 | 104.4 | 39.93 |
| 15.0 | 1.032289 | 1.621 | 50.1 | 107.4 | 39.89 |
| 17.5 | 1.025159 | 1.588 | 52.6 | 110.4 | 39.85 |
| 20.0 | 1.018371 | 1.565 | 55.1 | 113.4 | 39.81 |
| 22.5 | 1.011865 | 1.548 | 57.6 | 116.4 | 39.77 |
| 25.0 | 1.005599 | 1.532 | 60.1 | 119.4 | 39.73 |

### Figure 10. Axial power profile of four-batch HCR core

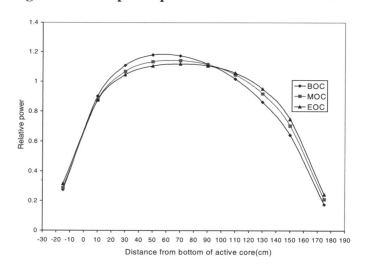

### Table 4. Relative radial power distribution at BOC

```
0.7587 1.3931 0.7756 1.5020 0.8211 1.1018 0.7611 0.9306 0.5261 0.7061
        1.2689 1.3348 1.6504 1.2776 1.1623 1.1014 0.7934 0.9373
            0.8341 1.3141 0.8675 1.2102 0.5856 0.7563 0.8043
                1.2504 1.2297 1.0197 0.7666 0.8759
                    0.5358 0.7688 0.9052
                        0.9115
```

## Table 5. Radial power distribution at EOC

```
0.6784 1.2144 0.6899 1.2683 0.7242 1.0848 0.7558 1.0442 0.5967 0.7363
       1.1370 1.1721 1.3555 1.1814 1.1618 1.1647 0.9406 1.0151
              0.7263 1.1972 0.7948 1.2199 0.6553 0.8954 0.8327
                     1.1978 1.2228 1.1188 0.9158 0.9367
                            0.6068 0.9152 0.9593
                                   0.9611
```

## Table 6. Radial burn-up distribution at EOC (GWd/t)

```
103.158 54.130 79.074 56.038 80.779 99.276 32.187 72.405 48.248 40.946
        76.557 77.730 34.351 77.445 76.166 51.069 94.534 23.045
               80.989 77.994 34.958 52.902 50.918 93.890 19.353
                      77.469 53.157 49.767 94.299 21.738
                             98.214 94.341 22.360
                                    22.430
```

The depletion results of the four-batch HCR core show that with the average enrichment ~9.2% fuel, the cycle length is around 36.5 months (25 GWd/t). This is larger than the cycle length given by the Japanese given in [1] which is 22 months with 9.8% fissile enrichment, and [2] which is a two-year cycle with 15% fissile enrichment. But the core average burn-up at EOC, 60 GWd/t, is similar to the data given in [1], which is 57 GWd/t. The void reactivity coefficients were then analysed for three cases: the RMWR with void tube, the RMWR without void tube and the core with the same geometry as the RMWR but with Th-Pu fuel instead of U-Pu fuel.

*Void reactivity coefficient analysis*

The results show that the void reactivity coefficient of the RMWR without void tubes is positive, as shown in Figure 11. The void tubes have the well-known negative reactivity effect on the void reactivity coefficient by introducing neutron leakage. The core with Th-Pu fuel has a larger negative void reactivity coefficient throughout the core life.

## Figure 11. Void coefficients as a function of core average burn-up

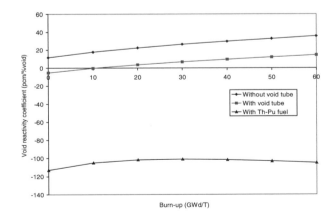

The results suggest that by using thorium fuel it is possible to design a HCBWR which has negative void reactivity without the use of void tubes. It is also possible to increase the core height to that of a normal BWR size since the large leakage is not necessary to obtain a negative void reactivity coefficient. This will have the beneficial economic impacts of providing a more efficient fuel cycle and the ability to use standard core specifications.

*Reactivity worth of the control system*

The reactivity control requirements for the RMWR and HCBWR-Th are estimated in Table 7 and shown graphically in Figure 12.

**Table 7. The control reactivity requirements for the RMWR and HCBWR-Th**

| Components | HCBWR-Th | | RMWR | |
|---|---|---|---|---|
| | **Reactivity** | **Coefficient** | **Reactivity** | **Coefficient** |
| Moderator density(-40%~40%void) | 0.068 | 85 pcm/%void [1] | 0.0192 | 24 pcm/%void [2] |
| Moderator temperature(300~550 K) | 0.005 | 2 pcm/K [3] | 0.005 | 2 pcm/K [3] |
| Fuel temperature(300~900 K) | 0.012 | 2 pcm/K [3] | 0.012 | 2 pcm/K [3] |
| XE/SM | 0.005 | | 0.005 | |
| Shutdown margin | 0.05 | | 0.05 | |
| Burn-up | 0.05 | | 0.05 | |
| Total | 0.19 | | 0.1412 | |

[1] Estimated depends on assembly calculation result.
[2] From Ref. [1].
[3] From normal LWR.

**Figure 12. The HCR reactivity requirement**

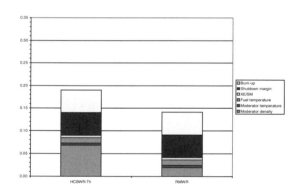

The reactivity of control rods for the RMWR and for the HCBWR-Th are estimated with the assembly calculation at cold zero power; the results are shown in Table 8 and in Figure 13.

**Table 8. The reactivity control rod worths for the HCR**

| Rod states | HCBWR-Th | | RMWR | |
|---|---|---|---|---|
| | **K** | **Worth** | **K** | **Worth** |
| No rod | 1.281351 | | 1.326621 | |
| 1 B$_4$C rod/ three assemblies | 1.213173 | 0.043858463 | 1.272925 | 0.031797447 |
| 1 Hf rod/ three assemblies | 1.219947 | 0.039281457 | 1.278851 | 0.028157133 |
| 1 Hf rod/ assembly | 1.067107 | 0.156686864 | 1.163573 | 0.105627007 |

# Figure 13. Reactivity worth of the control system

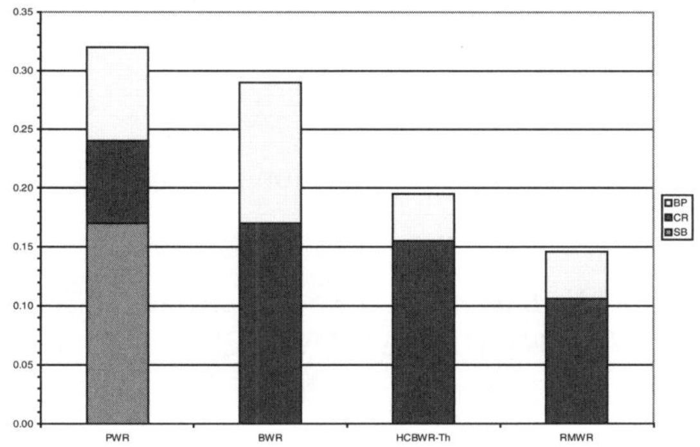

The results show that with a single control rod per assembly, the worth of control rod system can satisfy the shutdown reactivity requirements.

## Summary and continuing work

A preliminary design has been presented of a thorium-fuelled high conversion boiling water reactor. During the final year of this research project, work will continue on the optimisation of the Purdue/BNL HCR design. For example, the core height will be increased because it appears sufficient margin is available for the station blackout event. Safety analysis will also be performed on other limiting transients such as the main steam line isolation valve closure event using both the RELAP5/PARCS and TRAC-M/PARCS coupled codes.

## REFERENCES

[1] Tsutomu Okubo, *et al*., "Conceptual Designing of Reduced-moderation water Reactors (1) – Design for BWR-type Reactors", Proc. ICONE-8. p. 715, Baltimore MD, USA, 2-6 April 2000.

[2] Takamichi Iwamura, *et al*., "Research on Reduced-moderation Water Reactor (RMWR)", JEARI-research 99-058.

[3] T.J. Downar, "Feasibility Study of a Plutonium-thorium Fuel Cycle for a High Conversion Boiling Water Reactor", ANS Winter Meeting, Washington, DC, November 2000.

[4] Y. Xu, "A Fuel Cycle Analysis Capability for the US NRC Neutronics Code PARCS", ANS M&C 2001, Salt Lake City, USA, September 2001.

[5] H.G. Joo, "A Multi-group Hexagonal Nodal Method", ANS M&C 2001, Salt Lake City, USA, September 2001.

[6]     "RELAP5/Mod3 Code Manual, NUREG/CR-5535, June 1995

[7]     U.S. Rohatgi, Brookhaven National Laboratory, private communication, September 2001.

*Additional references*

T. Yokoyama, R. Yoshioka, Y. Sakashita, "A Study on Breeding Characteristics of Fast Spectrum BWR", private communication.

S. Aoki, A. Inoue, M. Aritomi and Y. Sakamoto, "An Experimental Study on the Boiling Phenomena Within a Narrow Gap@", *Int. J. Heat Mass Transfer*, Vol. 25, No. 7, pp. 985-990 (1982).

# THORIUM FUEL IN LWRs: AN OPTION FOR AN EFFECTIVE REDUCTION OF PLUTONIUM STOCKPILES

**Dieter Porsch**
FRAMATOME ANP GmbH
Erlangen, Germany

**Dieter Sommer**
Kernkraftwerk Obrigheim GmbH
Obrigheim, Germany

## Abstract

An option for the re-use of plutonium in LWRs is its utilisation with thorium as carrier. Core design studies performed for modern PWRs demonstrated their capability of being operated with exclusively Th/Pu fuel without major changes in the fuel assembly design or the safety-related reactor systems. For a commercial introduction of Th/Pu fuel in LWRs it is essential to extend the available qualification basis to higher plutonium concentrations and significantly higher exposures. As an important contribution to the database an irradiation test programme with Th/Pu fuel has been launched at the Obrigheim power station (KWO) in Germany. The initial characterisation of the test fuel rod, the measurements planned during refuelling and post-irradiation examinations with isotopic analyses is the basis for a later exploitation of the experiment. The test programme is partially funded by the EC through the 5th R&D Framework Programme.

## Introduction

World-wide, the vast majority of commercially operated light water reactors (LWRs) use uranium-based fuel. In some countries the plutonium generated in that operational mode is recycled as uranium/plutonium mixed oxide fuel (MOX).

An alternative re-use of plutonium in LWRs is its utilisation with thorium as carrier. In the past, research institutes and industry in Germany and other countries have investigated the options of the thorium-based nuclear fuel cycle in LWRs, PHWRs and HTRs in detail. The conclusion from those studies was that the use of thorium fuel offers the potential for improved resource utilisation. This holds true particularly for advanced reactor concepts specifically designed for thorium application. Even for present PWRs, however, advantages can be anticipated in the case of a thorium-based fuel cycle [1-6].

The once-through thorium fuel cycle with plutonium as the initial fissile material reveals the potential for significant plutonium reduction rates and for enhanced proliferation-resistant characteristics of the spent fuel. The degradation of the plutonium composition and the reduced solubility of thorium-based fuel render a misuse of the spent fuel even more unattractive. This is a feature in particular appealing for plutonium whose source is the dismantlement of nuclear warheads.

## Status of operational experience in LWRs

Operational experience and results from post-irradiation examinations on LWR thorium/plutonium fuel were available from an irradiation programme at the Lingen power plant (BWR, Germany) from 1971 to 1977 [7]. The total plutonium content in the rods was 2.6 w/o with a fissile content of about 86%. The discharge exposure of the fuel rods was in the range of about 20 MWd/kg. The experience gained from that irradiation programme was a valuable basis for verifying the appropriate accuracy of cross-sections and the applicability of spectral codes for the design of Th/Pu fuel assemblies.

## Thorium fuel cycle options for existing LWRs

Extensive core design studies performed for modern PWRs of the 193 fuel assembly type demonstrated their capability of being operated with exclusively Th/U or Th/Pu fuel without major changes in the fuel assembly design or the safety-related reactor systems.

Most likely, the introduction of thorium-based fuel in existing LWRs will start from uranium-based cycles. This is feasible with an appropriate fuel assembly design. Th/Pu assemblies for transition cycles, i.e. loaded adjacent to uranium fuel assemblies, are to be designed similar to MOX assemblies with two or more enrichment zones. Transition cycles can be designed taking economical considerations into account, while still meeting safety requirements with U, U/Pu, Th/U and Th/Pu fuel assemblies simultaneously present in the core.

Investigations of reactor cores loaded exclusively with thorium fuel concluded in a very similar accident behaviour compared to common U and U/Pu loaded cores. In comparison to U or U/Pu, thorium fuel mainly affects reactivity coefficients. Table 1 shows end of cycle reactivity coefficients of comparable U, "all MOX" and "all Th/Pu" cores. The moderator temperature coefficient is mainly determined by the plutonium isotopes. No major differences occur compared to U/Pu cores. The boron worth is further reduced. Using enriched boron, a common procedure in several of the actual PWRs, can, if required, compensate for the resulting somewhat higher demand on soluble boron. The integral

**Table 1. Examples of reactivity coefficients for different cores (end of equilibrium cycle)**

|  | Unit | Uranium core | All MOX core | All Th/Pu core |
|---|---|---|---|---|
| Moderator temperature coefficient | pcm/°C | -52 | -64 | -61 |
| Inverse boron worth | ppm/% Δρ | -102 | -172 | -185 |
| Integral Doppler | % Δρ | 1.4 | 1.4 | 1.0 |

Doppler seems to be slightly smaller in the case investigated. The possibility of higher Pu concentrations together with thorium offers the potential of concentrating the Pu to fewer fuel assemblies and of an improved utilisation of the provided fuel without reprocessing.

The recycling mode, with mixed cores of uranium- and thorium-based fuel, and the all-thorium mode are possible for existing PWRs. Both offer the potential for a more effective disposition of high quality plutonium, but also medium- and high-enriched uranium compared to current recycling modes.

## Irradiation experiment in Obrigheim power station (KWO)

For a commercial introduction of thorium fuel in LWRs the qualification basis has to be extended to higher plutonium concentrations and significantly higher exposures. The irradiation of lead test assemblies and the follow-up of their power history, subsequent post-irradiation examinations with isotopic analyses and theoretical benchmarks for thorium fuel assemblies and thorium-loaded cores are essential for improving the confidence in the design tools.

As an important step, among others, an irradiation test programme with thorium/plutonium fuel has started in the actual 32$^{nd}$ cycle of the Obrigheim power station (KWO) in Germany. KWO is a commercially operated PWR. Table 2 shows some basic characteristics of the plant.

**Table 2. Characteristic data of Obrigheim power station (KWO)**

|  | Unit |  |
|---|---|---|
| Thermal power | MW | 1 050 |
| No. of fuel assembly positions | – | 121 |
| No. of fuel assemblies | – | 97 |
| No. of steel dummy assemblies | – | 24 |
| Fuel assembly type | – | 14 × 14-16 |
| Fuel rod pitch* | cm | 1.43 |
| Fuel assembly pitch* | cm | 20.1 |
|  |  |  |
| Core average moderator temperature | °C | 297 |
| Average linear heat generation rate | W/cm | 214 |
| Active core height* | cm | 270 |

\* At room temperature, 20°C.

The test programme is partially funded by the EC in through the 5$^{th}$ R&D Framework Programme. The first phase of the test suite covers three irradiation cycles of the test rod and includes inspection programmes during each outage for refuelling.

The irradiation programme is initiated and co-ordinated by KWO. FRAMATOME ANP, as a subcontractor, is responsible for the mechanical, thermo-hydraulic and neutronic design of the test rod. Figure 1 shows a schematic sketch of the test rod. The irradiation test is planned as close as possible at representative commercial LWR conditions. This facilitates the direct utilisation of the experience gained from the experiment without further interpretation.

### Figure 1. Schematic sketch of the test rod

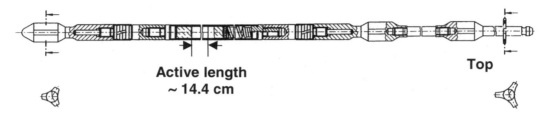

The structural parts of the test rod were manufactured by FRAMATOME and supplied to the Institute for Transuranium (ITU) in Karlsruhe. Fuel manufacturing and completion of the fuel rod including welding of the second end plug were the responsibility of the ITU. The specification of the fuel and the qualification of the manufacturing process were derived from the experience gained in the joint Brazilian/German programme [2] and provided by FRAMATOME.

The length of the pellet column of the test rod is about 14.4 cm. Spacers are designed to centre the rod in its insertion position and are responsible for defined thermo-hydraulic conditions at the probe level. The pellet diameter and the outer fuel rod diameter of 9.5 mm are consistent with the majority of commercial LWRs.

The sample is inserted into a guide tube of a regular MOX fuel assembly (average fissile Pu concentration of 3.8 w/o in natural uranium) with an average burn-up of 15.1 MWd/kg after one in-core cycle. The neutronic characteristic of the irradiation environment is therefore representative for PWR MOX neutron spectra under commercial power reactor conditions. Figure 2 shows the schematic MOX fuel assembly design and indicates possible locations for the test rod. The test rod is attached to a plugging device especially developed for irradiation tests and facilitates handling during reactor shutdown.

### Figure 2. Design of the carrier assembly for the test rod

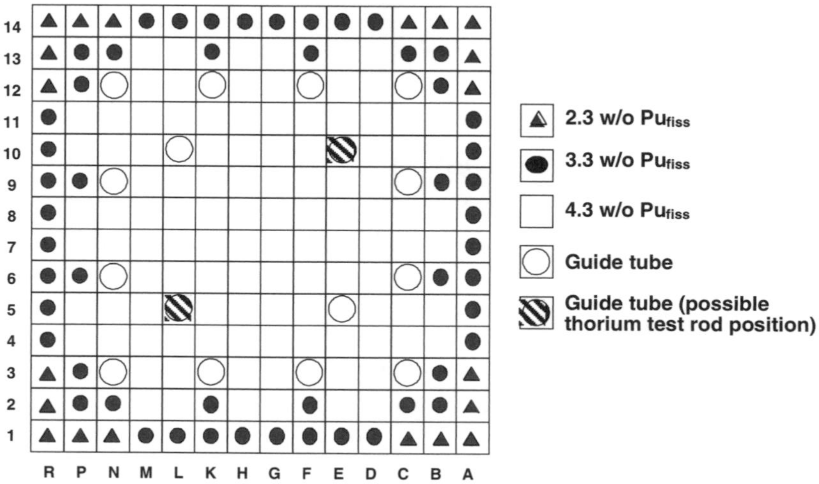

The thermo-hydraulic conditions at the active probe level limit the possible linear heat generation rate to about 200 W/cm. The minimum DNBR was not supposed to be lower than values common for KWO core designs. To meet these design criteria, the plutonium content had to be limited. Design calculations resulted in a fissile plutonium content of 3 w/o ($\approx$ 3.3 w/o $Pu_{tot}$) for the high-quality plutonium (> 90 w/o fissile content) provided. The design calculations were performed with SAV90, the FRAMATOME standard neutronic design procedure used for KWO. The code applied for the test rod design was Th-FASER, a version of the spectral code FASER for thorium treatment, and for the fuel assembly design calculations (performed in 10 energy groups) the code applied was MULTIMEDIUM, a two-dimensional multi-group transport code.

The carrier MOX fuel assembly is placed at the central core position in Cycle 32. Under actual conditions the maximum linear heat generation rate of the probe is 194 W/cm and under conservative assumptions the minimum DNBR is determined to be 2.24. This is about the same value that was calculated for the regular core without test rod and meets the design target. Figure 3 shows the axial power density distribution of the carrier MOX assembly and the axial position of the thorium segment. The expected burn-up of the probe after the first cycle is 12.4 MWd/kg.

**Figure 3. Axial power profile and position of the test rod**

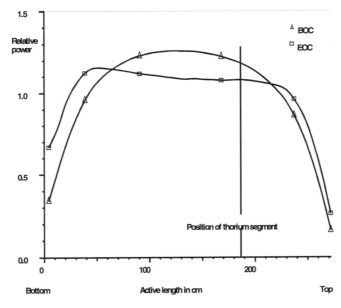

The test rod remains in this fuel assembly for two cycles. Thereafter it will be transferred to a second MOX fuel assembly for further irradiation. A continuation of the experiment is planned for up to four cycles. The final burn-up of the segment after four irradiation cycles in these two different MOX assemblies is expected to be about 40 MWd/kg and is therefore close to discharge exposures of the actual commercial fuel. The initial characterisation of the fuel rod, the measurements planned for each refuelling period of the plant and post-irradiation examinations with isotopic analysis are the basis for a later evaluation of the operational fuel rod behaviour.

This experiment will provide extensive information on the operational behaviour of Th/Pu fuel and the measurement of the isotopic composition of the spent fuel is an important contribution to the qualification of the cross-section database and design codes.

## Characteristics of spent high quality plutonium

The resulting degradation of the plutonium used for the experiment is also representative for weapons-grade material. Figure 4 shows the reduction rate for the plutonium used in the experiment compared to its use in U/Pu fuel. At a burn-up of 50 MWd/kg the total amount of plutonium is reduced to about 20% of the initially inserted material, whereas in conventional MOX fuel about 60% is still to be dealt with. A similar picture holds true for the fissile content. Less than 10% of the residual plutonium is fissile when used in Th/Pu fuel, compared to ≈ 40% when used in U/Pu fuel. Figure 5 compares the degradation of the plutonium quality for the two types of fuel. The plutonium quality (ratio of fissile to total plutonium) in the thorium-based fuel is reduced to ≈ 40%, in the U/Pu fuel the resulting quality is about equivalent to the actual commercial level.

### Figure 4. Comparison of plutonium reduction in U/Pu and Th/Pu fuel

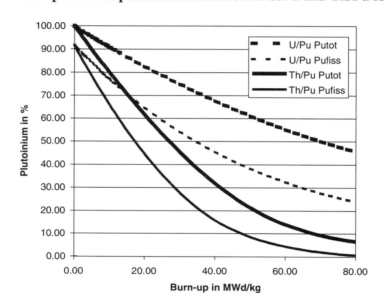

### Figure 5. Plutonium quality vs. burn-up in U/Pu and Th/Pu fuel

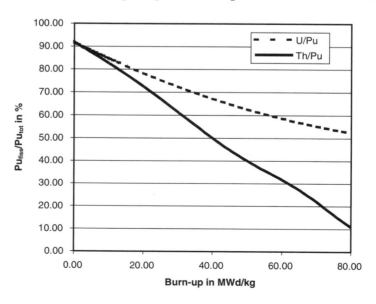

The neutronic characteristics of the uranium generated during irradiation of Th/Pu fuel with $^{233}$U as a very valuable energy source predestine that spent fuel for reprocessing. On the other hand it complicates the composition of the uranium, with its relatively high ratio of $^{232}$U (and its decay products) and $^{234}$U, handling of the separated material and contributes to an increased proliferation resistance of the fuel at higher burn-up levels.

The radioactivity of the spent fuel is slightly lower for Th/Pu fuel than for U/Pu fuel for at least the first 25 years after discharge. After that time it becomes higher. The γ-power of spent Th/Pu fuel exceeds the U/Pu level after less than five years and the gamma energy is shifted to higher intensities. For about two years the decay heat of Th/Pu is of about the same level as U/Pu; after that it is significantly higher.

## Conclusion

The utilisation of thorium is seen as an option for effectively reducing stockpiles of high-quality plutonium, but also of highly-enriched uranium. Another significant aspect in the discussion is the development of a more proliferation-resistant fuel cycle. Thorium fuel is equally capable of serving this aspect.

The irradiation experiment in progress demonstrates the status of thorium fuel cycle development. It contributes to the qualification basis for the mechanical and neutronic design of thorium fuel rods for a LWR environment.

Thorium fuel, based on developed technology, will be applicable for modern reactors within a relatively short period of time. On a longer-term perspective the utilisation of the available thorium resources world-wide could contribute to a sustainable supply of energy.

## REFERENCES

[1]    M. Peehs, G. Schlosser, "Prospects of Thorium Fuel Cycles in a Standard Pressurized Water Reactor", Siemens Forsch.- u. Entwickl.-Ber. Bd. 15 (1986), Nr. 4.

[2]    "Program of Research and Development on the Thorium Utilization in PWRs", German/ Brazilian Co-operation in Scientific Research and Technological Development, Final report (1979-1988), 1988.

[3]    A. Radkovski, "Using Thorium in a Commercial Nuclear Fuel Cycle: How to do it", *Nuclear Engineering International*, January 1999.

[4]    X. Zhao, M. Driscoll, M. Kazimi, "Rationale for Reconsidering the Thorium Cycle in Light Water Reactors", *Tran. Am. Nucl. Soc.*, Boston, June 1999, pp. 43-44.

[5]   S. Herring, P. MacDonald, "Advanced, Lower-cost, Proliferation-resistant Uranium-thorium Dioxide Fuels for LWRs", *Tran. Am. Nucl. Soc.*, Boston, June 1999, pp. 45-46.

[6]   A. Morozov, A. Galperin, M. Todosow, "A Thorium-based Fuel Cycle for VVERs & PWRs – A Non-proliferative Solution to Renew Nuclear Power", *Nuclear Engineering International*, January 1999.

[7]   Welhum, P. Pohl, "Isotopic Analysis on $PuO_2$-$ThO_2$ Fuel Irradiated in Lingen BWR Power Plant", European Applied Research Reports, Vol. 2, No. 6 (1981).

# PWRs USING HTGR FUEL CONCEPT WITH CLADDING FOR ULTIMATE SAFETY

**Yoichiro Shimazu*** (Hokkaido Univ.)**, Hiroshi Tochihara** (EDC)
**Yoshiei Akiyama** (MHI)**, Kunihiro Itoh** (NDC)
* Graduate School of Engineering, Hokkaido University
Kita 13, Nishi 8, Kita-ku, Sapporo 060-8628, Japan

## Abstract

The growth in population and energy demand not only for electricity but also for ordinary sources of heat for human lives on the one hand and the global climate crisis on the other hand, it has become clear that a consensus of available energy resources in accordance with sustainable development must be established. With regard to this fact, the effective use of nuclear energy is indispensable.

From this point of view, an innovative PWR concept has been studied that uses carbon-coated particle fuels moderated by graphite as that of HTGR, but cooled by pressurised light water. The aim of this concept is to take both the best advantages of fuel integrity against fission-product release and the reliability of PWR technology based on long operational experience. In this paper the current status of the evaluation of nuclear characteristics of the reactor and an outline of the plant is discussed.

# Introduction

The growth in population and energy demand not only for electricity but also for ordinary sources of heat for human lives on the one hand and the global climate crisis on the other hand, it has become clear that a consensus of available energy resources in accordance with sustainable development must be established. With regard to this fact, the effective use of nuclear energy is indispensable.

From this point of view, an innovative PWR concept has been studied that uses carbon-coated particle fuels moderated by graphite as that of HTGR, but cooled by pressurised light water. The aim of this concept is to take both the best advantages of fuel integrity against fission-product release and the reliability of PWR technology based on long operational experience. Similar studies have already been reported upon [1,2]. The present concept is different from previous studies from the viewpoint of the quantity of the carbon/graphite moderator. The quantity of graphite for moderation is selected so that it has sufficient thermal capacity but negative void reactivity coefficient. The fuel is clad by zircaloy lest it should contact with steam or air in case of loss of coolant accident (LOCA) or other accidents. The power density is lower than that of the current PWRs in order to maintain higher thermal margin during operation. Only control rods control the transient reactivity without soluble boron. Burnable absorbers direct long-term reactivity and suppress extra reactivity to obtain the required shutdown margin with minimum control rods. The reactor lifetime can be expected to endure for more than a few years without refuelling with an enriched uranium oxide of 5 w/o.

With such features the reactor can be operated with very high safety margins such that the maximum fuel temperature during LOCA is about 1 000°C lower than the limiting temperature of the coated-particle fuel and no operator action is required for a few hours in case of LOCA even without a passive residual heat removal system. Thus it can be envisaged that construction of a plant with this type of reactor would be acceptable even near densely populated areas. It may further be suitable for district heating or desalination.

In this paper the current status of the evaluation of nuclear characteristics of the reactor and an outline of the plant is discussed. The manufacturing process and area for the future investigation is also presented along with an application of this plant for district heating.

# Basic concept of the reactor system

It is a well-known established fact that carbon- and/or silicon-coated particle nuclear fuel for HTGR is quite resistant for fission-product release even at elevated temperatures. The graphite moderator is also helpful for the reduction of the fuel temperature in case of LOCA because of the large heat capacity. A longer time interval is thus available for emergency cooling, and one may even go so far as to eliminate the emergency cooling system. This is quite a favourable feature for a safer reactor system.

On the other hand, PWRs have been operated safely for many years in many countries. Thus the system characteristics and the safety features have been investigated in detail. In other words the reliability of the PWR system is quite high.

When these two features can be combined to design a reactor system, it can be a quite safe system based on the actual operating experience. This is the basic concept of the present reactor system. In order to realise this concept we have to show some basic features required for the reactor and the fuel. They are: 1) the void coefficient should be negative, 2) the configuration of fuel element can be established and 3) the reactor can be designed with acceptable operational characteristics. These features are discussed in the following sections.

## Nuclear characteristics

We have investigated basic nuclear characteristics of the reactor concept by performing unit cell calculations varying volume ratios of uranium, carbon and the coolant. The nuclear code used for the calculations is SRAC-95 [3], which has been developed by Japan Atomic Energy Research Institute. It has various calculation modules for geometries of a unit cell, an assembly and a reactor core of 1-D to 3-D. Neutronic calculation models such as collision probability, $S_N$ and diffusion are integrated.

In the calculation the fuel enrichment is assumed to be 5 w/o, which is the maximum value presently licensed for fuel manufacturing in thermal reactors. We also assume the use of fuel cladding. One reason for this is that the compatibility of the coated-particle fuels with the coolant under operational condition is not well known. Another reason is to avoid oxidation of graphite in case of LOCA when the fuels enter into contact with high temperature steam or air. It is known from the calculation that stainless steal cannot be used as the cladding material because of its high neutron absorption in a thermal reactor. Thus we assume zircaloy cladding. The calculation results are shown in Figure 1. From these results, it can be seen that the void coefficient can be negative. Sufficient reactivity can also be obtained for actual operation.

The main parameters of the fuel configuration are the ratios of uranium to carbon and water to the uranium. As can be seen, the void coefficient becomes more negative as the ratio of carbon to $UO_2$ is decreased. From the viewpoints of reactor safety and controllability we keep the void coefficient as negative as possible. However, the more graphite moderator we use, the more we can increase the heat capacity of the reactor, and thus we increase security by slowing the heating rate in case of LOCA. The behaviour of the multiplication factor is also dependent on the ratio of hydrogen to heavy metal (H/M). When the H/M increases the multiplication factor also increases. When H/M is increased the multiplication factor increases, however, so does the reactor size. We have to compromise the nuclear characteristics and economical demerit. In this study we selected the carbon to $UO_2$ ratio as 9 based on the HTGR fuel design. The H/M is selected as 10, which resulted in a fuel assembly design with a hexagonal lattice.

In the selection of the fuel rod diameter, we simply assumed the $UO_2$ loading per fuel rod is assumed to be identical to that of the typical PWR fuel rod, which resulted in a fuel diameter of 25.9 mm and a cladding thickness of 1.5 mm.

## Reactor operational characteristics

Based on the results above we proceeded to design a reactor. The thermal output has been chosen to be 50 MWt. As explained later, our objective of the development is effective and widespread usage of nuclear energy. From this point of view we intend to deploy the system in the areas as close as possible to the cities making full use of its ultimate safety. A good example would be an energy source of district heating system. For such a case the power of 50 MWt is sufficient. The average coolant temperature is assumed to be 250°C, which is also selected as the heat source for district heating.

In Figure 2 a plane view of the reactor is shown. We use three types of fuel assemblies, one with only fuel rods, one with integrated burnable absorber fuel rods and one with control rod guide tubes. Gadolinia is used as the absorber and the loading in each fuel assembly is optimised to reduce the control rod requirement and flat behaviour of the multiplication factor during core burn-up. We found that it can be possible to use a few types of gadolinia loading, as discussed below. The number of control rods is determined so as to satisfy the cold shutdown capability. The optimisation of the burnable poison resulted in reducing by three the number of control rods in comparison with that when one type of gadolinia loading is adopted. As the excess reactivity is controlled uniquely with control rods, the flattening of the multiplication behaviour during the operating cycle helps to simplify the control rod programming.

The burn-up characteristics of the core are also shown in Figure 2. We can expect an operating period of about three full power equivalent years, which corresponds to about four years of actual operation when the capacity factor of an actual plant is taken into account. In Figure 2 we have shown four behaviours of effective multiplication factor during the cycle with or without burnable poison. When we use only one type of gadolinia-loaded fuel assemblies, the flatness of the multiplication factor is not good enough. When we use three types of gadolinia-loaded fuel assemblies, the flatness is much improved. As can be seen in Figure 2 the heavier gadolinia loading on the inside of the reactor is better than in the outside from the viewpoint of core lifetime. The behaviour of the peaking factor is also flat and acceptable.

The reactor's reactivity coefficients are as follows. The void coefficient is about -140 pcm/%void and the Doppler coefficient is about -5 pcm/°C (pcm = $10^{-5}$ $\Delta k/k$), as expected from the previous calculation. With such a reactivity coefficient a quite stable reactor response against operational disturbance is obtained.

## Fuel configuration

The TRISO particle fuel is a particle of uranium dioxide coated by three layers of silicon carbide and pyro-carbon. It has been used in HTGRs and proved to be failure-free up to 2 000°C through irradiation tests [4]. In the present study, this feature of TRISO particle fuel is made maximum use of.

The concept of the fuel rod in the present study is shown in Figure 3. The TRISO particle fuels are packed into a zircaloy tube with graphite. The fuel assembly consists of 37 fuel rods or 31 fuel rods and 6 control rod thimbles. In order to obtain the appropriate H/M ratio, the fuel rods are arranged in a hexagonal lattice with a rod gap of 1 mm. They are encircled by strong-back rods, which provide sufficient stiffness and strength for fuel handling. The subcriticality of the single fuel assembly submerged in water is assured by the fact that the effective multiplication factor is 0.83 for the most reactive fuel assembly. However, the fuel design concept must be experimentally established in the future.

Fuel rods with integrated burnable absorber are also used. In a fuel assembly, 12 fuel rods with the burnable poison are assumed in this study. The distribution of the fuel rods with burnable poison in a fuel assembly is shown in Figure 4. The distribution is determined in order to obtain a quick burning of the poison. Thus we can suppress large excess reactivity at the beginning of cycle and reduce the reactivity penalty of the poison as much as possible at the end of cycle.

The reactor parameters of the present study are listed in Table 1.

## Safety features

Fuel rod temperature is estimated based on the thermal conductivity of HTGR fuels. A maximum heat flux of 22.6 kW/m is assumed, which corresponds to the peaking factor of about 2.5. The maximum fuel temperature is obtained as 350°C. This is far below the limiting temperature of the TRISO particle fuel. This is due to the good thermal conductivity and low power density along with the low coolant temperature. We will try to investigate the possibility of integrating the Vipac fuel concept in the future for more economical manufacturing.

**Table 1. Reactor parameters**

| | |
|---|---|
| **Thermal output** | 50 MW |
| **Average linear heat rate** | ~9.1 kWw/m |
| **Average coolant temperature** | 250d°C |
| **Core equivalent diameter** | ~1.8 m |
| **Core height** | ~1.8 m |
| **Number of fuel assembly** | 85 |
| **Number of control rods** | 24 |
| **Loading of heavy metal** | ~2.5t |
| **Operating life** | >3 EFPY |
| **Fuel rod diameter** | 29 mm |
| **Fuel rod pitch** | 30 mm |
| **Cladding thickness** | 1.5 mm |
| **Number of rods/assembly** | 37 |
| **Fuel** | TRISO |
| **Enrichment** | 5 w/o |
| **Cladding material** | Zircaloy |
| **Lattice type** | Hexagonal |

In case of LOCA, which will be the most limiting in terms of evaluation of safety, the fuel temperature increase is quite slow due to the large heat capacity of the carbon moderator. Even if the adiabatic heat conduction is assumed, it will take more than two hours for fuel cladding temperature to reach 1 200°C, which is the limiting temperature of zircaloy cladding. This is long enough for some effective counter actions to be taken. The safety feature is thus greatly improved.

In the present study a passive residual heat removal system is adopted. This system consists of a gravity-driven emergency water tank, natural two-phase circulation loop and a heat pipe system. Based on a preliminary evaluation, the capacities of these heat removal systems are 500 kW and 10k W, respectively. The emergency water fills the reactor and the natural two-phase circulation loop continues to cool down the reactor temperature. The heat pipe system cools down the air in the containment vessel. The system does not require any active driver for the residual heat removal. It will enhance the safety feature. The study of the safety analyses of various events and accidents is ongoing.

## Example of application

One of the applications of such an ultimately safe nuclear energy source will be district heating. An example of such a system is operating in Sapporo, Hokkaido in Japan [5]. It supplies heat using hot water for air conditioning and hot water in the offices covering an area of about 100 ha. The outlet temperature of the system is 200°C and the inlet temperature is 180°C. The system has five conventional boilers – two coal burners of 29 MWt, two oil burners of 46 MWt, one natural gas burner of 46 MWt – and one refuse-derived fuel (RDF) burner of 31 MWt. The plant exhausts 27 000 tonnes of carbon dioxide annually.

The typical daily power pattern of the plant is shown in Figure 5 and the yearly power pattern is shown in Figure 6. As can be seen from Figure 5 the heat demands change largely with time. This will result in high heat cycle fatigue for the plant devices and also lead to low capacity factor of the plant. In order to solve these problems we plan to integrate a reservoir tank, which can absorb the large daily variation of heat demands. Then the nuclear power can be operated in as constant a manner as possible.

The maximum size of such a tank would be 7 400 cubic meters. This tank will also be useful to improve the yearly capacity factor from 30% to 70%. Using such a tank, the maximum heat supply of 50 to 55 MWt is sufficient.

## Conclusion

We have investigated the feasibility of designing a reactor system of ultimate safety, which is established by combining the high integrity of TRISO particle fuel against the release of fission products and the high reliability of PWRs based on long operating experiences in the world. The feasibility study has shown that such a reactor system is feasible and the safety features can be expected. Thus it can be constructed near urban area to be used, for example, as a district heating plant.

We will continue our study to refine the design and to analyse safety quantitatively.

## REFERENCES

[1]  M. Kim, *et al.*, "Nuclear Feasibility of Carbon-coated Particles Fuels in PWRs", ANS Trans., Vol. 75, p. 362 (Nov. 1996).

[2]  M. Kim, *et al.*, "Use of Carbon-coated Particle Fuels in PWR Assemblies", ASN Trans., Vol. 77, p. 396 (Nov. 1997).

[3]  K. Okumura, SRAC'95, JAERI-Data/Code 96-015 (1996).

[4]  T. Ogawa, *et al.*, "A Model to Predict the Ultimate Failure of Coated Fuel Particles During Core Heat-up Events", *Nucl. Technol.*, 96, 314-322 (1991).

[5]  Direct contact with the plant owner.

# Figure 1. Result of cell calculation

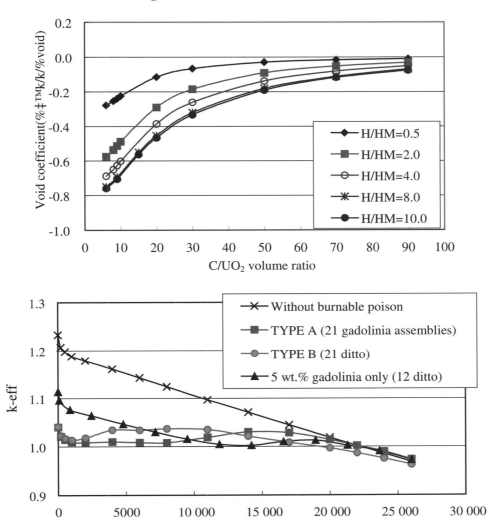

# Figure 2. Core plane view and burn-up characteristics

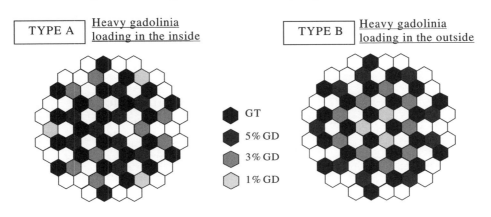

## Figure 3. Concept of TRISO fuel rod and assembly

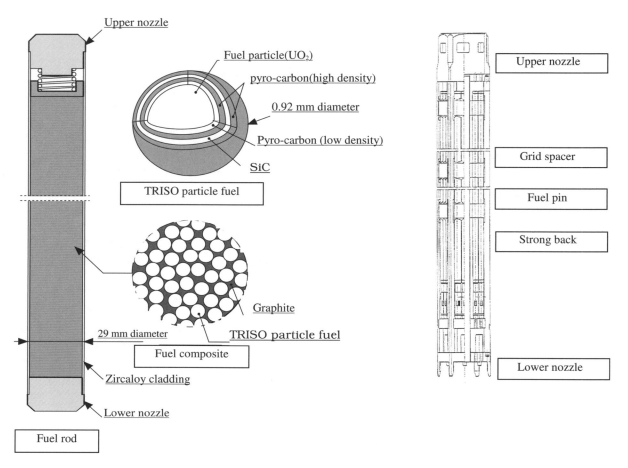

Upper nozzle

Fuel particle(UO₂)
pyro-carbon(high density)
0.92 mm diameter
Pyro-carbon (low density)
SiC

TRISO particle fuel

Graphite

29 mm diameter

TRISO particle fuel

Fuel composite

Zircaloy cladding

Lower nozzle

Fuel rod

Upper nozzle

Grid spacer

Fuel pin

Strong back

Lower nozzle

## Figure 4. Distribution of the fuel rods with burnable poison in an assembly

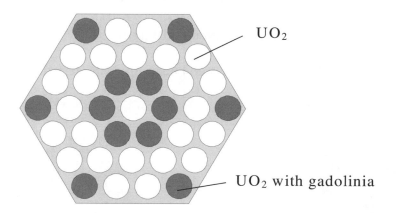

UO₂

UO₂ with gadolinia

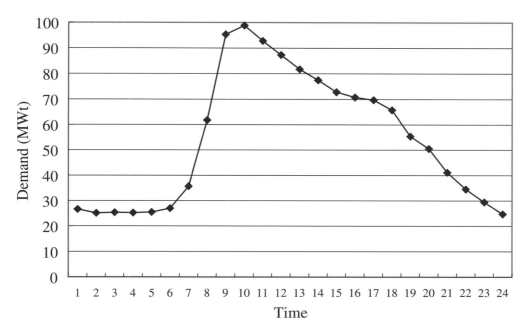

**Figure 5. Hourly power demand (Jan. 1998)**

**Figure 6. Monthly power demand (1998)**

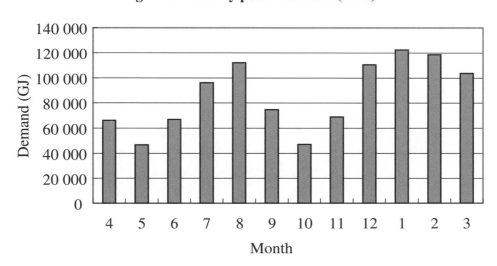

# FEASIBILITY OF PARTIAL LWR CORE LOADINGS WITH INERT-MATRIX FUEL

**U. Kasemeyer,**[*1] **Ch. Hellwig,**[1] **D.W. Dean,**[2] **R. Chawla,**[1,3] **G. Meier,**[4] **T. Williams**[5]
[1]Paul Scherrer Institute, CH-5232 Villigen, PSI, Switzerland
[2]Studsvik Scandpower, Inc., 1087 Beacon St., Suite 301, USA
[3]Swiss Federal Institute of Technology, CH-1015 Lausanne
[4]Kernkraftwerk Gösgen-Däniken AG, CH-4658 Däniken, Switzerland
[5]Elektrizitäts-Gesellschaft Laufenburg AG, CH-8953 Dietikon, Switzerland

## Abstract

Three types of plutonium-containing cores have been compared, each comprising of four different stages of plutonium deployment in an actual 1 000 MWe pressurised water reactor (PWR). In a first step, core-follow calculations for four real-life cores with increasingly larger mixed-oxide (MOX) loadings were validated against measured plant data. In a second step, all MOX assemblies were substituted by optimised Pu-Er-Zr oxide, inert-matrix fuel (IMF) assemblies. Finally, core loadings with IMF have been designed and considered which contain, on the average, the same amounts of plutonium as the four partial MOX loadings. From the latter, more realistic IMF loadings, the IMF rods with the highest power ratings were identified. Fuel behaviour calculations were then performed for these rods employing models partly validated via recent data from the comparative IMF/MOX irradiation test currently under way at Halden. Based on the various results obtained, conclusions have been drawn regarding IMF rod designs most likely to yield (in partial IMF core loadings) fuel behaviour similar to that of $UO_2$ fuel.

[*] Corresponding author: Tel: +41 56 310 2046, Fax: +41 56 310 2327, E-mail: uwe.kasemeyer@psi.ch

# Introduction

The world plutonium inventory is steadily growing due to the unavoidable production of plutonium in current light water reactors (LWRs) and the de-allocation of weapons-grade plutonium from dismantled nuclear arms [1]. Accordingly, efforts are being made to increase the Pu consumption in LWRs by using new Pu-containing inert-matrix fuels (IMFs).

Today, MOX fuel is widely used in LWRs in the form of partial core loadings [2]. The maximum amount of MOX fuel is typically limited to about 40% of the core inventory. As a result, the plutonium consumption does not exceed the amount of plutonium which is produced from the uranium present in the core. In addition, there are some disadvantages of using MOX fuel. Usually the MOX assemblies are more expensive than $UO_2$ assemblies, the time the MOX assemblies have to spend in the wet storage pool is longer and the use of soluble boron which is enriched in $^{10}B$ is sometimes demanded by nuclear safety inspectorates to achieve the specified shutdown margins.

Usually, the amount of spent fuel to be reprocessed and hence the amount of plutonium to be brought back into the core are fixed. Doing this in the form of a once-through uranium-free IMF could represent a useful complementary strategy to the currently practised single recycling of plutonium as MOX. With a steadily growing number of investigations, IMFs for LWRs have been an important research topic in recent years. Thus, collaborative international efforts have led to two long-term irradiation experiments being started in 2000 [3,4]. While the irradiation test in Petten is dedicated to the investigation of different IMF concepts, that in Halden concentrates on the comparison of MOX with IMF of a particular solid-solution type, viz. plutonium dissolved in a matrix of yttria-stabilised zirconia (Pu-Er-Zr oxide) [5]. It is this particular type of IMF which has been considered in the current comparative investigation of partial IMF and MOX loadings in an actual 1 000 MWe pressurised water reactor (PWR).

For this study, a number of real-life cycles of the plant, including cycles with partial MOX loadings, have first been modelled using the Studsvik Core Management System (CMS) [6,7,8]. Comparisons of the calculated results with measured data in terms of the boron let-down curves and detector signal distributions showed good agreement for both $UO_2$ as well as MOX loadings. CMS has then been employed for investigating the behaviour of partial core loadings with IMF relative to those with MOX, two different strategies being considered in this context. In the first case, every MOX assembly (in the cycles considered with partial MOX loadings) was substituted by an IMF assembly. Following this set of comparisons, core loadings with IMF have been designed which contain, on the average, the same amount of plutonium as the cores with partial MOX loadings. Because there is about 50% more plutonium in IMF than in MOX fuel, these cores were loaded with about 30% less IMF assemblies than MOX assemblies and the free positions were filled with $UO_2$ assemblies. From the latter, more realistic IMF loadings, the IMF rods with the highest power ratings were identified and their power history used to perform fuel behaviour calculations. Partial validation of the fuel modelling carried out in this context has been made possible via recent data from the above-mentioned IMF/MOX irradiation test at Halden [3].

The main points of the neutron physics and core behaviour comparisons have been the boron let-down curve (reflecting the cycle length) and the power distribution. The principal parameters compared in the fuel behaviour investigations are the fuel temperature and the fission-gas release.

## Geometry and materials

IMF rods of different enrichments were used to reduce power peaking within the IMF assembly. A sketch of the optimised IMF assembly (similar to the used MOX assemblies) is shown in Figure 1, with fuel rod and lattice geometry being kept the same as for UO$_2$ and MOX fuel.

### Figure 1. Quarter view of the IMF assembly used

☐ *Corner pins Pu/Er/Gd: 0.60/0.40/0.30 cm$^3$*
▨ *Border pins Pu/Er: 0.75/0.30 cm$^3$*
■ *Inner pins Pu/Er: 0.98/0.25 cm$^3$*

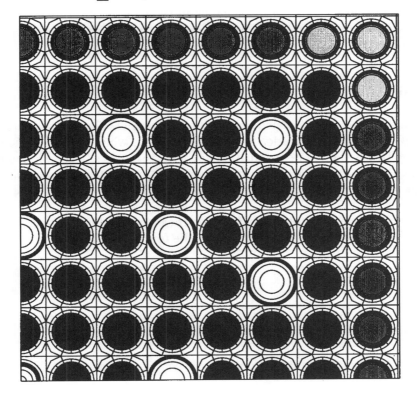

The plutonium isotopic composition assumed for the IMF was that of the MOX employed, viz.

| $^{238}$Pu | $^{239}$Pu | $^{240}$Pu | $^{241}$Pu | $^{242}$Pu |
|---|---|---|---|---|
| 1.5 | 60.2 | 25.6 | 7.4 | 5.3 |

As mentioned earlier, an actual 1 000 MWe PWR was modelled, the core containing 177 assemblies, arranged with a pitch of 21.56 cm and having an active core height of 358 cm. The average core conditions employed are as follows:

| | |
|---|---|
| Total thermal power | 3 002 MW |
| Mean power density | 102 MW/m$^3$ |
| Inlet temperature | 291.5°C |
| Outlet temperature | 324.8°C |
| Water flow | 15 981 kg/s |
| Pressure | 154.0 bar |

## Core-follow calculations

All calculations were done using the CASMO-4 (lattice) and SIMULATE-3 (nodal diffusion) codes of CMS [9]. The cross-section library employed was based on the JEF-2.2 data file, CMS having been earlier benchmarked with it for MOX applications [10-13]. In addition, recent neutronics measurements with IMF and MOX rodlets in a $UO_2$ lattice carried out in the PROTEUS facility have been analysed with CASMO-4 using the same library, satisfactory agreement being obtained between measured and calculated power distributions for both types of Pu fuel [5,14].

The first calculations were done for several real-life, 100% uranium-fuelled cycles. With the exception of the initial uranium cycle ("jump-in" core), all cycles showed a good agreement between measured and calculated data. Thus, for example, in all cases the RMS of the radial detector reaction rate distributions was lower than 2% and the soluble boron values agreed within 30 ppm. All subsequent calculations employed $UO_2$ data derived for these 100% uranium cores.

## Partial MOX core loadings

The four real-life cycles considered here with partial MOX loadings are termed MOX-1 to MOX-4. Eight MOX assemblies were loaded in cycle MOX-1, while in each of the three subsequent cycles, 20 fresh MOX assemblies were added. Four MOX assemblies, loaded in the third cycle, were unloaded in cycle MOX-4, so that the latter contained a total of 64 MOX assemblies. Figure 2 shows the comparison between measured and calculated boron let-down curves for MOX-4.

### Figure 2. Comparison between measured and calculated boron let-down curves for the fourth MOX cycle

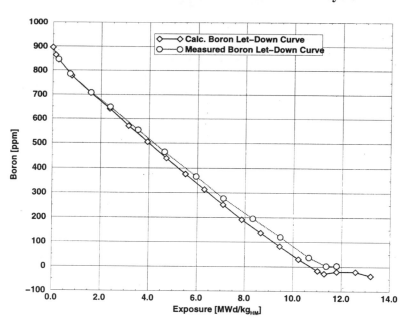

In all the partial MOX loadings, as well as in the 100% $UO_2$ cycles, the soluble boron values were predicted quite well at beginning of cycle (BOC), while SIMULATE underestimated the reactivity at end of cycle (EOC) by similar amounts. Because this behaviour was found to be independent of the cycle plutonium content, the differences currently reported between partial IMF and MOX loadings can clearly be attributed to differences in the burn-up behaviour of the corresponding cores.

The comparison of the detector reaction rate distributions showed slightly larger spreads for the partial MOX loadings than for the 100% $UO_2$ cycles, the RMS being 2 to 3%. Considering that the detector response of the MOX assemblies is about half that of the $UO_2$ assemblies, this is still a good agreement between calculation and measurement.

## One-to-one MOX substitution by IMF assemblies

As mentioned earlier, all MOX assemblies were replaced in a first step by IMF assemblies. In order to see the differences in core behaviour directly, the IMF assemblies were placed at the same positions as the MOX assemblies.

The differences between the two kinds of partial loadings have been found to be small in the first cycle, the boron let-down curves for which are compared in Figure 3. Due to the presence of burnable poisons within IMF, the reactivity of the core with a partial IMF loading is lower at BOC than that of the core with MOX assemblies. Because the highest power peaking was seen every time at BOC, most of the comparisons have been made for 6 equilibrium full power days (EFPD). The differences in the power peaking are small for the first cycle. In the IMF case, the maximum power in the $UO_2$ assemblies is greater by 1.3%, while the maximum power difference between an IMF and a MOX assembly is about 20%. Figure 4 shows the relative power distribution of the first core with partial IMF and MOX loadings, as well as the difference for each assembly.

The differences in assembly power decrease with increasing burn-up. The evolution of the relative power fraction of a MOX and an IMF assembly are compared in Figure 5, through all the four cycles considered. It is seen that the main differences occur during the first cycle, the power fractions being very similar in the subsequent cases.

**Figure 3. Comparison between boron let-down curves of the first PWR
core with IMF and MOX partial loadings (8 assemblies in each case)**

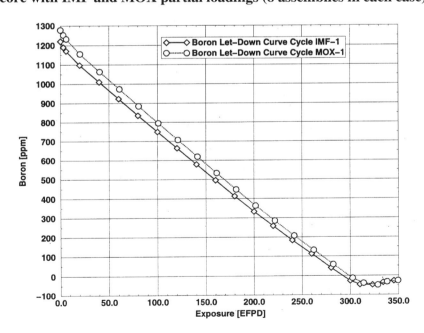

**Figure 4. Comparison between relative power distributions at BOC in the first core with partial loadings (8 assemblies in each case) of IMF (1st line) and MOX (2nd line)**

| | | | | | | | |
|---|---|---|---|---|---|---|---|
| 1.135 | 1.485 | 1.399 | 1.000 | 1.284 | 0.767 | 1.409 | 0.407 |
| 1.112 | 1.466 | 1.403 | 1.010 | 1.266 | 0.726 | 1.313 | 0.374 |
| 2.104 | 1.298 | -0.266 | -0.984 | 1.399 | 5.638 | 7.340 | 8.797 |
| 1.486 | 1.419 | 1.314 | 1.141 | 0.746 | 0.912 | 1.385 | 0.382 |
| 1.467 | 1.414 | 1.348 | 1.204 | 0.752 | 0.870 | 1.295 | 0.352 |
| 1.302 | 0.348 | -2.479 | -5.274 | -0.850 | 4.864 | 6.949 | 8.534 |
| 1.399 | 1.315 | 0.940 | 1.152 | 1.180 | 1.254 | 1.248 | 0.277 |
| 1.403 | 1.349 | 1.039 | 1.443 | 1.228 | 1.212 | 1.175 | 0.256 |
| -0.259 | -2.480 | -9.566 | -20.183 | -3.870 | 3.417 | 6.190 | 8.041 |
| 0.999 | 1.140 | 1.151 | 0.940 | 1.369 | 1.224 | 0.473 | |
| 1.009 | 1.204 | 1.443 | 1.029 | 1.380 | 1.186 | 0.447 | |
| -0.973 | -5.273 | -20.182 | -8.649 | -0.811 | 3.222 | 5.725 | |
| 1.283 | 0.745 | 1.172 | 1.367 | 1.356 | 1.246 | 0.266 | |
| 1.265 | 0.751 | 1.218 | 1.378 | 1.328 | 1.200 | 0.252 | |
| 1.415 | -0.821 | -3.824 | -0.793 | 2.113 | 3.896 | 5.442 | |
| 0.767 | 0.913 | 1.255 | 1.223 | 1.247 | 0.391 | | |
| 0.726 | 0.871 | 1.213 | 1.185 | 1.200 | 0.373 | | |
| 5.661 | 4.904 | 3.459 | 3.256 | 3.908 | 4.865 | | |
| 1.409 | 1.390 | 1.258 | 0.475 | 0.270 | | | |
| 1.313 | 1.300 | 1.185 | 0.449 | 0.256 | | | |
| 7.356 | 6.964 | 6.191 | 5.748 | 5.452 | | | |
| 0.407 | 0.386 | 0.279 | IMF | | | | |
| 0.374 | 0.356 | 0.258 | MOX | | | | |
| 8.814 | 8.538 | 8.049 | %-Diff. | | | | |

**Figure 5. Relative power fraction comparison over all four cycles between a MOX and an IMF assembly**

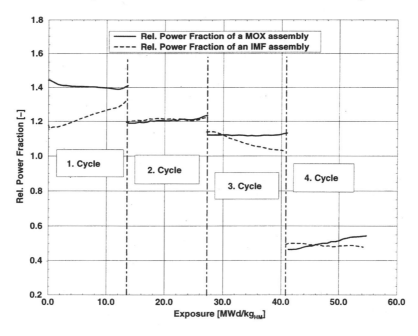

Because of the lower relative power in IMF, compared to MOX, during the first cycle, the power in the other UO$_2$ assemblies increases. Especially for cores with a large fraction of IMF assemblies, replacing MOX without core design optimisation results in the power peaking in the remaining UO$_2$ assemblies becoming too high. Figure 6 compares the radial power peaking at BOC in the case of the fourth cycle (36% IMF assemblies).

**Figure 6. Comparison between relative power distributions at BOC in the fourth core with partial loadings (64 assemblies in each case) of IMF (1st line) and MOX (2nd line)**

| | | | | | | | |
|---|---|---|---|---|---|---|---|
| 0.818 | 1.216 | 1.097 | 1.183 | 0.980 | 0.925 | 1.643 | 0.469 |
| 0.801 | 1.288 | 1.179 | 1.513 | 1.097 | 0.894 | 1.474 | 0.399 |
| 2.209 | -5.564 | -6.943 | -21.821 | -10.622 | 3.421 | 11.440 | 17.468 |
| 1.218 | 1.295 | 1.313 | 0.947 | 1.214 | 1.256 | 1.497 | 0.423 |
| 1.290 | 1.308 | 1.389 | 1.078 | 1.465 | 1.276 | 1.317 | 0.353 |
| -5.573 | -0.977 | -5.482 | -12.101 | -17.123 | -1.540 | 13.688 | 19.629 |
| 1.101 | 1.311 | 1.066 | 1.208 | 1.215 | 1.278 | 0.846 | 0.277 |
| 1.183 | 1.387 | 1.189 | 1.495 | 1.314 | 1.167 | 0.749 | 0.218 |
| -6.968 | -5.445 | -10.371 | -19.175 | -7.580 | 9.573 | 12.886 | 27.043 |
| 1.184 | 0.944 | 1.200 | 1.061 | 1.128 | 1.489 | 0.508 | |
| 1.514 | 1.076 | 1.488 | 1.106 | 1.045 | 1.287 | 0.469 | |
| -21.825 | -12.221 | -19.342 | -4.047 | 8.003 | 15.705 | 8.258 | |
| 0.981 | 1.205 | 1.196 | 1.108 | 1.560 | 0.810 | 0.288 | |
| 1.097 | 1.458 | 1.302 | 1.043 | 1.338 | 0.711 | 0.215 | |
| -10.623 | -17.351 | -8.086 | 6.165 | 16.614 | 13.982 | 34.174 | |
| 0.925 | 1.244 | 1.259 | 1.472 | 0.817 | 0.331 | | |
| 0.895 | 1.270 | 1.158 | 1.280 | 0.708 | 0.252 | | |
| 3.415 | -2.017 | 8.738 | 15.025 | 15.369 | 31.336 | | |
| 1.643 | 1.482 | 0.811 | 0.501 | 0.287 | | | |
| 1.475 | 1.311 | 0.737 | 0.466 | 0.214 | | | |
| 11.430 | 13.091 | 9.931 | 7.537 | 34.156 | | | |
| 0.470 | 0.418 | 0.272 | IMF | | | | |
| 0.400 | 0.351 | 0.216 | MOX | | | | |
| 17.453 | 19.001 | 25.656 | %-Diff. | | | | |

In all cases except the first considered cycle, the core reactivity at EOC is greater with partial IMF loadings than with MOX. Figure 7 compares the boron let-down curves of the fourth cycle (64 MOX or IMF assemblies). In this case, because the boron let-down curve is significantly flatter with IMF than with MOX, the fourth cycle could run about 9 EFPD longer with IMF. However, as mentioned above, a simple one-to-one substitution of MOX assemblies by IMF is not possible because of the significantly higher power peaking which results for the $UO_2$ assemblies.

**Figure 7. Comparison between boron let-down curves of the fourth core with IMF and MOX partial loadings (64 assemblies in each case)**

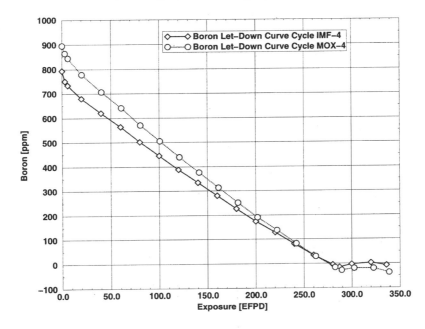

On the other hand, with an IMF assembly containing about 50% more plutonium than a MOX assembly, it is only necessary to substitute 70% of the MOX by IMF assemblies if the same amount of plutonium is to be loaded in the core. The remaining MOX assemblies can then be substituted by UO$_2$ assemblies, and this is what is considered in the following section.

### Equivalent plutonium substitution with IMF assemblies

Following the first set of comparisons, core loadings with IMF have been designed which contain, on the average, the same amount of plutonium as the cores considered with partial MOX loadings. Because of the core quarter-symmetry, the number of loaded fresh assemblies should be divisible by four. Eight IMF assemblies were loaded the first cycle, while in each of the three following cycles 12 fresh IMF assemblies (instead of 20) were loaded so that the fourth cycle contained 44 IMF assemblies. The remaining MOX assembly positions were filled with UO$_2$ fuel. While the core design of the first core was the same as in the previous set of cases, the loading schemes for the other three cores were adjusted to yield a maximum relative radial nodal power of about 1.5 and maximum IMF pin power values of about 420 W/cm.

As in the first set of comparisons, the reactivity at EOC could be increased in each case, except for the first partial IMF core. This can be seen in Figure 8, which compares the boron let-down curves for the fourth cycle with partial MOX and IMF loadings. In the present calculations, the cycle length in all cases was in fact kept the same as in the original MOX cores and no additional burn-up was accumulated. The advantage, however, was that fewer fresh UO$_2$ assemblies needed to be loaded to get the same cycle length. If additional burn-up had been aimed at, the cycle length could have been increased by about 20, 40 and 25 EFPD, respectively, for the considered second, third and fourth cycles with partial IMF loadings.

**Figure 8. Comparison between boron let-down curves of the fourth core with equivalent plutonium (44 IMF, 64 MOX assemblies)**

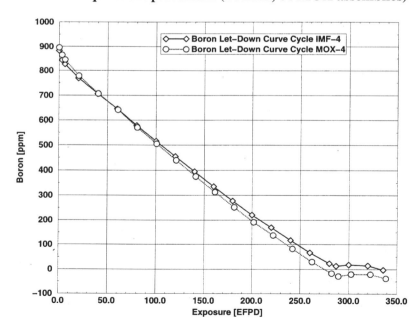

As mentioned above, the maximum relative radial nodal power fraction was adjusted in each core with partial IMF loading to be about 1.5. The highest relative radial power fractions were found at BOC in the third cycle. Figure 9 shows the comparison between partial MOX and IMF loadings in this particular case.

**Figure 9. Comparison between relative power distributions at BOC in the third core with 32 IMF (1st line) and 44 MOX (2nd line) assemblies**

```
0.985   1.304   1.256   0.863    1.067    0.993   1.526   0.455
0.975   1.301   1.079   1.471    1.274    1.001   1.419   0.399
1.051   0.225   16.401  -41.295  -16.217  -0.794  7.518   14.064

1.303   1.336   1.296   1.073    0.867    1.087   1.481   0.428
1.299   1.253   1.284   1.090    1.416    1.118   1.293   0.343
0.265   6.645   0.960   -1.550   -38.788  -2.811  14.513  25.069

1.257   1.300   1.268   0.958    1.207    1.242   1.271   0.296
1.079   1.284   1.280   1.446    1.078    1.149   0.828   0.218
16.547  1.270   -0.931  -33.776  11.907   8.061   53.458  35.816

0.867   1.084   0.966   1.202    1.324    1.433   0.494
1.470   1.091   1.445   1.256    1.063    1.348   0.430
-41.074 -0.706  -33.165 -4.323   24.549   6.336   14.664

1.066   0.875   1.217   1.337    1.519    0.684   0.227
1.273   1.415   1.070   1.062    1.477    0.880   0.240
-16.278 -38.161 13.675  25.966   2.843    -22.257 -5.491

0.988   1.098   1.282   1.461    0.691    0.260
1.001   1.119   1.157   1.349    0.881    0.303
-1.254  -1.932  10.817  8.365    -21.491  -14.268

1.523   1.500   1.307   0.520    0.232
1.419   1.304   0.851   0.425    0.240
7.353   14.991  53.634  22.343   -3.605

0.454   0.436   0.304    IMF
0.399   0.347   0.221    MOX
13.925  25.569  37.402   %-Diff.
```

The highest pin power value of a $UO_2$ fuel rod was found to be 512 W/cm, while the maximum value for an IMF rod was 434 W/cm. The axially averaged pin power history of the hottest $UO_2$ and IMF rods was used to investigate fuel behaviour characteristics using an extended version of the fuel performance code TRANSURANUS [15,16]. The results of this study are presented in the next section. Before doing so, it is useful to compare the effectiveness of the two different equivalent Pu strategies in terms of plutonium reduction.

### Plutonium reduction

Around 2/3 of the initial plutonium remains in the discharged MOX assemblies, while in the case of an IMF assembly the corresponding fraction is only 2/5. Table 1 summarises the results for the plutonium consumption achieved with IMF and MOX fuel.

**Table 1. Fractions (wt.%) of initial plutonium remaining at discharge and the total destroyed plutonium amounts for the two considered fuel types**

|  | $Pu_{tot}$ | $Pu_{fiss}$ | Destroyed Pu |
|---|---|---|---|
| Case | wt.% of initial Pu | | kg per assembly |
| MOX | 66.2 | 35.7 | 10.4 |
| IMF | 41.2 | 16.4 | 25.8 |

253

On average, 16 MOX or 11 IMF assemblies would be discharged at each stage of an equilibrium cycle. Hence, about 167 kg plutonium would be destroyed using MOX, while 284 kg plutonium would be destroyed with the use of IMF. Table 2 summarises the overall plutonium balance (i.e. including the $UO_2$ assemblies) for a single equilibrium cycle stage. It can be seen that the plutonium consumption of a core with a partial IMF loading would be four times as large as that in the corresponding case with MOX.

### Table 2. Comparison of overall plutonium balances with partial MOX and IMF loadings for a single equilibrium cycle stage

*All values are given in kg*

|  | MOX | IMF |
|---|---|---|
| **Loaded Pu[a]** | 493 | 483 |
| **Discharged Pu[b]** | 462 | 359 |
| **Pu balance** | -31 | -124 |

[a] Pu in 16 MOX or 11 IMF assemblies.
[b] Pu in 44.25 discharged assemblies (on average).

## Comparison between IMF and $UO_2$ fuel behaviour

In order to be able to model the fuel performance of IMF, the TRANSURANUS code has been modified by PSI for this new type of fuel [17]. This was done using the following material data for yttria-stabilised zirconia or – if available – for the fabricated fuel: thermal expansion, yield stress, Young's modulus, emissivity, melting temperature, specific heat and density. Since no data have been found for a creep strain correlation for yttria-stabilised zirconia, the $UO_2$ correlation was adopted for IMF (the creep strain correlation is of relatively minor importance for the calculations). The thermal conductivity was derived from measurements carried out with fuel samples at ITU, Karlsruhe, up to temperatures of 1 620°C.

There is not enough data currently available to establish a fission-gas release behaviour model for IMF. Nevertheless, there are indications from the Halden irradiation test, as well as from other related experiments, that the fission-gas release behaviour could be quite similar to that of $UO_2$. Therefore, the same model has been used for IMF. Other models for fuel irradiation behaviour (relocation, swelling, densification) have been adjusted to fit the in-pile data recently made available from Halden. As only three IMF rods (with slightly different design and instrumentation) are being irradiated in this test, and the burn-up achieved till now is only about 4.5 MWd/kg$_{HM}$ (MOX-equivalent burn-up), the validation base for the chosen assumptions is rather limited. The calculation results currently reported should not, therefore, be viewed as accurate predictions of IMF behaviour, but rather as a contribution to fuel rod design optimisation on the basis of the irradiation data available to date.

### *Pin power histories*

The CMS-calculated power histories for the hottest IMF rod and the hottest $UO_2$ rod, as well as the corresponding axial profiles, have been used as input in TRANSURANUS (each rod was virtually divided into 20 axial slices for calculation). The power histories employed are given in Table 3, the differences indicated being essentially due to the use of a burnable absorber and the absence of any plutonium breeding in the case of IMF. Because an axial averaged pin power history from CMS was used in TRANURANUS a slightly lower peaking factor than with CMS has been calculated for the hottest IMF rod by TRANURANUS.

## Table 3. Simplified power histories for the hottest rods, UO₂ and IMF, used as input for TRANSURANUS

| | UO$_2$ peak power [kW/m] | UO$_2$ avg. power [kW/m] | IMF peak power [kW/m] | IMF avg. power [kW/m] |
|---|---|---|---|---|
| Begin of 1$^{st}$ cycle | 49.7 | 40.0 | 23.7 | 19.9 |
| End of 1$^{st}$ cycle | 37.2 | 34.4 | 37.7 | 32.0 |
| Begin of 2$^{nd}$ cycle | 42.2 | 34.4 | 40.2 | 33.7 |
| After 80 EFPD | Interpolation | | 41.5 | 35.6 |
| End of 2$^{nd}$ cycle | 30.6 | 28.0 | 33.1 | 32.3 |
| Begin of 3$^{rd}$ cycle | 34.8 | 28.0 | 24.0 | 20.1 |
| End of 3$^{rd}$ cycle | 24.0 | 22.0 | 18.5 | 15.7 |
| Begin of 4$^{th}$ cycle | 27.3 | 22.0 | 17.9 | 15.0 |
| End of 4$^{th}$ cycle | 17.5 | 16.0 | 16.5 | 14.0 |

It is seen that the power history of the UO$_2$ rod decreases steadily from each BOC to EOC. The power history of the IMF rod increases first, due to Gd consumption and then shows a turnaround after 80 days in the second cycle.

### Geometrical input data

The geometry of the cladding of the IMF rod was the same as in the UO$_2$ case. The design of the pellets was modified in two respects. Firstly, the pellet outer diameter was slightly increased, i.e. the gap size slightly decreased, in accordance with the different (reduced) swelling and relocation behaviour of IMF observed in the Halden experiment. Secondly, a central hole of 3 mm diameter was introduced in order to lower the fuel centre temperature. For comparison reasons, a calculation was also done for an IMF rod without a central hole. It should be mentioned that the hole diameter was not optimised, but was rather a compromise between fabrication feasibility (the hole should be fabricated by pressing, not drilling) and residual pellet mass. The pellet mass decreases by just 11%, so that the plutonium content of the IMF needs to be enhanced only by this amount to achieve the same power generation. Such a moderately higher plutonium content is not expected to alter the material properties of this IMF type significantly.

### Results of the fuel behaviour calculations

Various results of the TRANSURANUS calculations for the UO$_2$ and IMF rods are presented in Tables 4 and 5. Graphical representations of the main results as function of burn-up are given in Figures 10 and 11.

## Table 4. TRANSURANUS results for the hottest pins at end of irradiation and under hot full power conditions

| | Unit | UO$_2$ rod | IMF rod (annular pellet) | IMF rod (full pellet) |
|---|---|---|---|---|
| **Average rod burn-up** | [MWd/kg$_{HM}$] | 60.7 | 56.3* | 50.4* |
| **Max./avg. fuel temp.** | [°C] | 942/640 | 638/589 | 1 154/679 |
| **Fission gas production** | [cm$^3$] | 4 134 | 3 514 | 3 511 |
| **Fission gas release** | [cm$^3$/%] | 431/10.4 | 458/13.0 | 717/20.4 |
| **Hot/cold plenum pressure** | [MPa] | 11.6/4.4 | 9.4/4/3.3 | 13.0/5.1 |

* MOX-equivalent burn-up.

**Table 5. Maximum values of various parameters during irradiation**

| | Unit | Specification limit | UO₂ rod | IMF rod (annular pellet) | IMF rod (full pellet) |
|---|---|---|---|---|---|
| Fuel centre temp. | [°C] | 2 005 | 1 921 | 1 647 | 2 103 |
| Avg. fuel temp. | [°C] | – | 995 | 989 | 1 134 |
| Time of max. value | [–] | – | BOC of 1st cycle | After 80 EFPD of 2nd cycle | |
| Fuel outer temp. | [°C] | 553 | 549 | 439 | 483 |
| Clad inner temp. | [°C] | – | 389 | 403 | 403 |
| Clad outer temp. | [°C] | 347 | 349 | 349 | 349 |
| Coolant temp.ᵃ | [°C] | 344 | 344 | 344 | 344 |
| Rod pressure | [MPa] | – | 12.4 | 10.1 | 14.1 |

ᵃ Coolant outlet temperature in hottest channel limited by boiling point at operational pressure of 15.4 MPa.

**Figure 10. Graphical representations of the main modelling results for the hottest UO₂ rod**

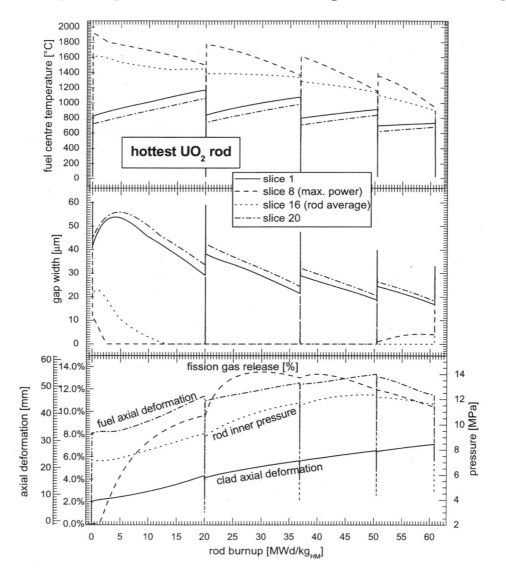

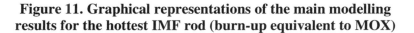

**Figure 11. Graphical representations of the main modelling results for the hottest IMF rod (burn-up equivalent to MOX)**

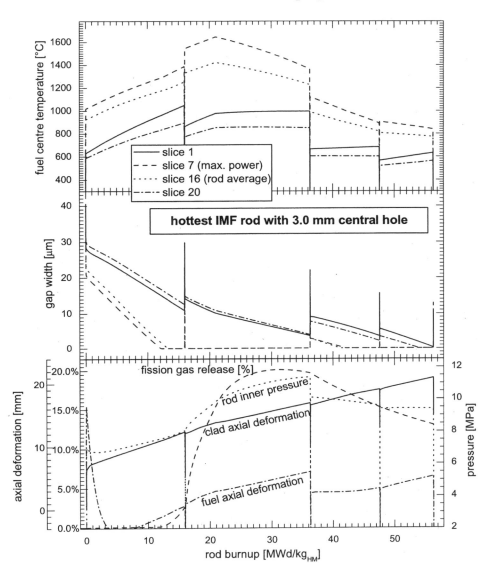

The hottest UO$_2$ rod nearly reaches the specified limiting fuel centre temperature according to the present calculations. The gap closes during the first cycle in most of the slices. In the hotter slices (slices 5 to 14), cladding lift-off can be observed during the fourth cycle. The rod inner pressure is slightly reduced by this cladding lift-off, while the fuel axial deformation is de-coupled from the clad axial deformation. The relative fission-gas release has its peak value during the second cycle. After that, it is decreasing due to the steadily increasing amount of fission gas produced.

The hottest IMF rod (annular pellet) reaches its maximum temperature after 80 days of the second cycle due to the corresponding power history (as mentioned earlier, this peak is caused by the consumption of the burnable poison). The gap closes, also in this case, during the first cycle for most of the slices. There are no signs of cladding lift-off later. The greater free volume due to the central hole is more than sufficient to compensate for the increased fission-gas release, compared to UO$_2$. The fuel axial deformation is significantly smaller than that found for UO$_2$. It is reduced by the strong IMF re-sintering observed in the Halden test, as well as by the modified (i.e. strongly reduced) swelling model for IMF. The fission-gas release in the IMF rod is higher than in the UO$_2$ rod despite the lower fuel centre temperature. This can be explained by the larger fuel volume (in the IMF case) which runs at elevated temperatures, triggering fission-gas release. The central hole lowers the fuel centre temperature, but more fuel "sees" the maximum temperature in an annular pellet. Additionally, the lower thermal conductivity and the different radial power profile result in a temperature profile with a larger fuel volume at higher temperatures.

### *Discussion of fuel behaviour modelling results*

Although relatively large uncertainties are associated with the use of the current models for IMF (based largely on yttria-stabilised zirconia), this first set of modelling results delivers useful information pertaining to the technical feasibility of this new fuel type. It indeed appears possible to employ Pu-Er-Zr oxide as IMF in a current-day PWR. The introduction of a central hole seems necessary in the fuel rod design to avoid too-high fuel centre temperatures. However, if the maximum power were decreased by 5 to 10%, further core design optimisation could even permit the use of full IMF pellets.

The present uncertainties associated with the fission-gas release model for IMF are particularly significant. However, as the gap is closed in the hot axial region, there is little effect of these uncertainties on the maximum fuel temperature. Their influence on the rod inner pressure is somewhat compensated by the larger free volume resulting from the central hole. The large uncertainties associated with the swelling model have a negligible influence on the thermal behaviour of the IMF rod.

## Conclusions

From the currently presented investigations, the following conclusions can be drawn:

- An equivalent Pu replacement of MOX assemblies in a 36% partial loading by IMF (and UO$_2$) assemblies is possible with simple core optimisation and yields the following benefits:

  - A four times larger, overall plutonium reduction.

  - Better use of UO$_2$ fuel due to the resulting longer cycle length.

- Partial loadings with higher fractions of IMF assemblies appear to be more difficult due to the increased power peaking which results in the UO$_2$ assemblies.

- Fuel performance calculations for an annular IMF rod design delivered results within currently specified limits for UO$_2$.

- Maximum fuel centre temperature in the case of full IMF pellets is about 100°C above the limit specified for UO$_2$ but could be lowered by further core design optimisation.

- The relatively large, present uncertainties in fission-gas release and fuel swelling models for IMF should not significantly influence the reported maximum fuel temperature results. Further data from the ongoing IMF irradiation test at Halden, of course, remain crucial.

# REFERENCES

[1]    J-S. Choi (IAEA), in "Conference Report of the International Conference on the Future of Plutonium", N. Numark, A. Michel, eds., Brussels, Belgium, October 2000.

[2]    P. Ledermann, D. Grenèche, Proc. of the International Conference in Future Nuclear Systems GLOBAL'99, Jackson Hole, Wyoming, USA, Aug./Sep. 1999.

[3]    U. Kasemeyer, Ch. Hellwig, Y-W. Lee, G. Ledergerber, D.S. Sohn, G.A. Gates, W. Wiesenack, "The Irradiation Test of Inert-matrix Fuel in Comparison to Uranium Plutonium Oxide Fuel in the Halden Reactor", *Prog. Nucl. Energy*, 383-4, 309-312 (2001).

[4]    R.P.C. Schram, K. Bakker, H. Hein, J.G. Boshoven, R.R van der Laan, T. Yamashita, G. Ledergerber, F. Ingold, "Design and Fabrication Aspects of a Plutonium-incineration Experiment Using Inert Matrices in a Once-through-then-out Mode", *Prog. Nucl. Energy*, 383-4, 259-262 (2001).

[5]    R. Chawla, C. Hellwig, F. Jatuff, U. Kasemeyer, G. Ledergerber, B.H. Lee, G. Rossiter, "First Experimental Results from Neutronics and In-pile Testing of a Pu-Er-Zr Oxide Inert-matrix Fuel" ENS TOPFUEL 2001, Stockholm, Sweden, 27-30 May 2001.

[6]    M. Edenius, K. Ekberg, B.H. Forssen, D. Knott, "CASMO-4, a Fuel Assembly Burn-up Program; User's Manual", StudsvikSOA-951, Studsvik of America, Newton, MA, Sep. 1995.

[7]    T. Bahadir, "CMS-link User's Manual", StudsvikSOA-9704, Studsvik of America, Newton, MA (1997).

[8]    A.S. DiGiovine, J.D. Rhodes, "SIMULATE-3 User's Manual", StudsvikSOA-9515, Studsvik of America, Newton, MA (1995).

[9]    M. Edenius, *et al.*, "Core Analysis: New Features and Applications", *Nuclear Europe Worldscan*, No. 3/4, p. 35, March/April 1995.

[10]   M. Edenius, D. Knott, K. Smith, "CASMO-SIMULATE on MOX Fuel", Proc. Int. Conference on Physics of Nuclear Science and Technology, Vol. 1, p. 135, Long Island, NY, October 1998.

[11]   Dave Knott, *et al.*, "New Cross-section Libraries for CASMO-4 Based on JEF-2 & ENDF/B-6", Proc. Int. Conference on Physics of Nuclear Science and Technology, Vol. 1, p. 51, Long Island, NY, October 1998.

[12]   M. Edenius, D. Knott, K. Smith, "CASMO-SIMULATE on MOX and Advanced Designs", Technical Meeting of Fuel Assembly and Reactor Physics and Calculation Methods Groups of the German Nuclear Society (KTG), p. 27, Karlsruhe, Germany, February 1998.

[13]  M. Mori, M. Kawamura, K. Yamate, "CASMO-4 SIMULATE-3 Benchmarking against High Plutonium Content Pressurized Water Reactor Mixed-oxide Critical Experiment", *Nuclear Science and Engineering*, Vol. 121, p. 41, September 1995.

[14]  R. Chawla, P. Grimm, P. Heimgartner, F. Jatuff, G. Ledergerber, A. Lüthi, M. Murphy, R. Seiler, R. van Geemert, "Integral Measurements with a Plutonium Inert Matrix Fuel Rod in a Heterogeneous Light Water Reactor Lattice", *Prog. Nucl. Energy*, 38/3-4, 359-362 (2001).

[15]  K. Lassmann, "TRANSURANUS: A Fuel Rod Analysis Code Ready for Use", *J. Nucl. Mater.*, 188, 295-302 (1992).

[16]  K. Lassmann, A. Schubert, J. van de Laar, C.W.H.M. Vennix, "Recent Developments of the TRANSURANUS Code with Emphasis on High Burn-up Phenomena", IAEA Technical Committee Meeting on Nuclear Fuel Behaviour Modelling at High Burn-up, Lake Windermere, UK, June 2000.

[17]  Ch. Hellwig, U. Kasemeyer, "Modelling the Behaviour of Inert Matrix Fuel with TRANSURANUS-PSI", Enlarged Halden Group Meeting, Lillehammer, Norway, March 2001.

# CEA STUDIES ABOUT INNOVATIVE WATER-COOLED REACTOR CONCEPTS

**P. Dumaz, A. Bergeron, G.M. Gautier, J.F. Pignatel, G. Rimpault, G. Youinou**
Commissariat à L'Énergie Atomique, Nuclear Energy Division

## Abstract

The first part of the paper concerns the lowering of the operating point of a standard 900 MWe PWR. The primary and secondary pressures have been decreased by about a factor two compared to the reference. For the core design, the critical heat flux increase resulting from the pressure reduction leads to larger margins. It is confirmed that the temperature reduction can ease the fuel burn-up increase. Calculating loss of coolant accidents only, it is shown that the emergency injection system could be simplified.

The second part of the paper is about the calculation of an integral PWR operating in natural circulation. It is shown that a one-batch long-life core is achievable. The thermal-hydraulic study shows the good resistance of such a concept to loss of coolant transients. In the last part of the paper, preliminary results obtained during the evaluation of a super-critical pressure concept are discussed.

## Introduction

In the beginning of the 90s, the CEA achieved a detailed evaluation of innovative reactors proposed up to that time (AP600, SIR, PIUS, etc.). These evaluations led the CEA to launch a significant reactor innovative programme. Since 1999, the main objective of this programme is to study gas-cooled reactor technology (from the HTR up to a fast reactor with an integrated fuel cycle).

As far as water-cooled concepts are concerned, most of the studies were completed in the 90s, but there is still a significant interest for this type of coolant. The three research subjects were:

- The search for improvements for the "standard" three- or four-loop PWR considering limited modifications of the design only (free soluble boron PWR [1], over-moderated cores [2], etc.).

- The study of innovative PWRs having small and medium outputs.

- The study of PWRs with better capabilities regarding uranium utilisation and the minimisation of wastes.

In this paper, three studies related to the following three subjects are presented: the decrease of the PWR operating point (pressure, temperature), the PWR with an integrated design for small reactor outputs and super-critical pressure concepts (which are supposed to be able to obtain fast neutron spectra; in fact here, only results concerning thermal concepts are presented).

## The low-pressure PWR

The interest of decreasing the operating point of a "standard" PWR (i.e. the 900 MWe with three loops) comes from the quite low decrease of the thermal efficiency with the primary pressure. From 155 bar to 85 bar, the expected variation of the net efficiency would be 33% to about 30% (Figure 1). On the other hand, a significant reduction of the following items is expected:

- The weight and the required performance of all major components and systems (including the safety classification).

- The energy stored in the primary and secondary circuit (containment size).

- The corrosion problems (lower temperature).

**Figure 1. Estimation of the net efficiency**

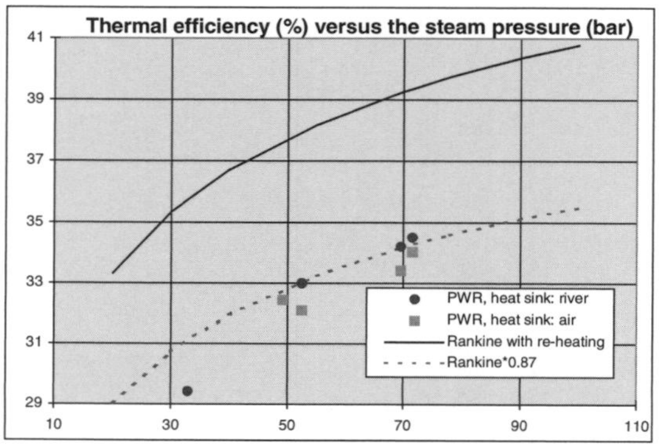

Considering the French 900 MWe reactor, a systematic analysis of the decreasing operating point has been carried out running calculations for both the low pressure and the normal pressure. The objective was to determine the potential interests of decreasing the operating point having a well-known reference. The purpose was not to design a new three-loop low pressure PWR reactor. It is expected that this design option could be chosen in more innovative PWR concepts. The following has been studied:

- The neutronic and thermal-hydraulic core design.

- The fuel element design.

- The release and deposition of corrosion products.

- The architecture of the safety systems.

These technical studies were supplemented by an economic evaluation. A primary pressure of 85 bar was chosen in order to have a secondary pressure of about 30 bar, then a net efficiency of 30%. A consistent set of reactor parameters was obtained using the COPERNIC code (Table 1). The purpose of COPERNIC is to ease the study of innovative reactor concepts. Taking into account some very general specifications, COPERNIC will provide a consistent set of data describing a preliminary reactor design. Then, this preliminary design can be more precisely studied using more sophisticated computer codes. COPERNIC was initially developed for innovative PWRs (both loop and integral types), and has now been extended to other reactor concepts, particularly gas-cooled nuclear reactors.

**Table 1. COPERNIC results, comparison with the standard 900 MWe PWR**

| | | Standard | Low pressure |
|---|---|---|---|
| **Thermal power** | **(MW)** | 2 785 | 2 785 |
| **Electrical power** | **(MW)** | 940 | 856 |
| **Primary pressure** | **(bar)** | 155.0 | 85.0 |
| **Core inlet temperature** | **(°C)** | 286.5 | 241.1 |
| **Core outlet temperature** | **(°C)** | 324.9 | 282.6 |
| **Vessel outlet temperature** | **(°C)** | 322.4 | 280.9 |
| **Vessel flow rate** | **(kg/s)** | 13 498 | 13 996 |
| **Linear heat rate** | **(kW/m)** | 17.9 | 17.9 |
| **Secondary pressure** | **(bar)** | 58.0 | 28.8 |
| **Pressuriser** | **(m$^3$)** | 39.4 | 26.9 |
| **NSSS mass** | **(t)** | 1 921 | 1433 |
| **Specific mass (t/MWe)** | | 2.045 | 1.673 |
| **Containment** | **(m$^3$)** | 49 449 | 38 653 |

*Core design*

From the neutronic point of view, the slight increase of the moderation ratio, 10% (due to the water density increase), is the most significant parameter variation. This does not greatly affect the core neutronic design. Using the APOLLO-2 computer code [3], a three-batch $UO_2$ core with an 18-month fuel cycle was calculated (discharge burn-up of 45 GWd/t). APOLLO-2 is a transport code, the V2.4 version was used in association with its 172-group nuclear data library CEA93.V4 (92 groups in the fast and epithermal energy range and 80 groups in the thermal energy range), derived from the JEF-2.2 database.

The [235]U enrichment required decreases from 4.21% to 3.97%. This is explained by the reactivity increase due to the better moderation and the slight reduction of the neutron leakage (decrease of the migration area).

As far as the reactor control is concerned, there are not significant variations from 155 bar to 85 bar. The penalising point for the low-pressure case is the increase of required burnable poison due to a lower limit of the soluble boron concentration (1 200 ppm instead 1 700 ppm).

The core thermal-hydraulic design is favoured by the increase of the critical heat flux at 85 bar (Figure 2 [4]). Larger margins are then expected for the departure from nucleate boiling ratio (DNBR) in steady state and in loss of flow transients. Using the FLICA-IV computer code, a three-dimensional two-phase flow code devoted to core analyses [5], both cases were calculated. Two critical heat flux correlations were used: W3 and Groeneveld [4]. The different corrective factors (bundle geometry, grids) were not taken into account (at low and standard pressures). In Table 2, the minimum DNBRs are indicated, and it is confirmed that the margins are larger in the low-pressure case.

**Figure 2. Critical heat flux (Groeneveld) for different pressures at a mass flux of 3 000 kg/m$^2$/s**

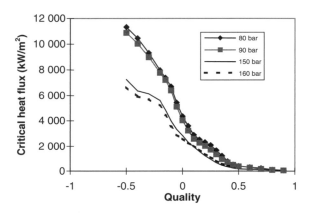

**Table 2. Minimum DNBR**

|  | W3 | | Groeneveld | |
|---|---|---|---|---|
|  | Low pressure | Standard pressure | Low pressure | Standard pressure |
| **Steady state** | 3 | 2.6 | 4 | 3.25 |
| **Loss of primary pumps** | 2.6 | 2.09 | 3.5 | 2.35 |

*Fuel design*

The primary pressure decrease has some effects on the fuel element design. It was expected that a high burn-up could be obtained more easily with a low primary pressure. First of all, quite clearly, the lower primary temperature reduces the cladding corrosion, one of the issues limiting the fuel burn-up. This was confirmed using the COCHISE V1B-1 computer code [6].

The fuel thermo-mechanical behaviour was studied using the METEOR computer code [7]. Assuming an initial internal rod pressure of 25 bar, the cladding compressive stress is lower for the low primary pressure case leading to a much later gap closing. The internal pressure reaches the primary coolant pressure at about 50 GWd/t instead 75 GWd/t for the standard PWR case calculation.

To obtain similar results, it was proposed to reduce the initial gap pressure or to increase the plenum volume. Assuming an initial pressure of 10 bar, a better behaviour is obtained, and the coolant pressure is reached at 80 GWd/t. The evaluation of the fuel-cladding interaction is much less easy; on the one hand, the later gap closing should be in favour of a better resistance to FCI, and on the other hand, the lower cladding temperature tends to increase the cladding embrittlement. Further studies would be necessary to settle this issue.

### Release and deposition of corrosion products

The primary pressure decrease also requires some modifications in the water chemistry in order to optimise the release and the deposition of corrosion products. Using the PACTOLE computer code [8], it was found out that without chemistry modifications, the product deposition on the core was much too high. The suitable solution seems to be an increase in pH, which leads to a reduction of the core deposition with some increase of the corrosion product deposition in the steam generators (a sensitive component regarding the minimisation of staff doses).

### Safety systems simplification

A potential advantage of decreasing the primary pressure is the simplification of the safety injection systems. A limited analysis of LOCAs using the CATHARE code [9] shows that it would be possible to remove the high-pressure injection system and the accumulators.

### Preliminary economic evaluation

This preliminary analysis had a quite limited objective: to run the SEMER computer code which is under development at CEA [10]. Furthermore, it is worth recalling that the low-pressure three-loop reactor considered was not an optimised reactor design but only a possible design option. Only the main circuits, components and systems which depend on the pressure (primary piping, reactor vessel, safety injection system, pressuriser, primary pump, steam generators, steam circuit, EFWS, containment) were taken into consideration. For the considered components, a significant reduction of cost is obtained: about 200 MF. Compared with the overall reactor investment cost (about 7 000 MF), this does not seem very significant.

## The integrated concepts

The integrated design means that the steam generators are installed within the reactor vessel (Figure 3). The pressuriser and the primary pumps could also be installed in the vessel. A first very good example of this type of concept is the Safe Integral Reactor (SIR) project [11] studied at the end of the 80s. In this project, the power output was 320 MWe for a 640 MWe twin unit station. No new nuclear reactors being ordered in the UK, this project did not have the opportunity to be finalised by a construction.

In the framework of the different initiatives of the USDOE (NERI, Generation IV International Forum), this type of PWR concept is again a subject of deep interest [12]. The CEA study is about a small reactor (about 50 MWe) operating in natural circulation at full power. Following the USDOE guidelines, it was attempted to consider a one-batch core with the longest possible fuel cycle (15 years).

## Figure 3. Schematic of an integrated PWR

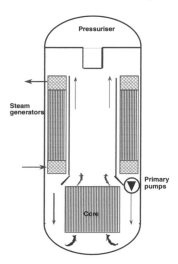

Using the COPERNIC computer code, a preliminary design analysis was achieved in order to obtain a first set of reactor parameters (reference case in Table 3). The main imposed parameters to COPERNIC were:

- A thermal power of 150 MW.

- A primary pressure of 155 bar.

- A core height over diameter of 0.7.

- A core with an hexagonal tight lattice (pitch over rod diameter, P/D, of 1.1 which gives a Vmoderator/Vfuel of about 0.7 taking into account all the core water) and "standard" fuel rod dimensions.

## Table 3. Results of the COPERNIC analysis

| | | Reactor designation | | | |
|---|---|---|---|---|---|
| | | Reference | PP increase | NCR decrease | Standard PWR |
| Thermal power | (MWth) | 150 | 150 | 150 | 2 785 |
| Electrical power | (MWe) | 45.6 | 45.9 | 45.9 | 940 |
| Natural circulation ratio | $(\Delta(\rho.g.h)/\Delta Ptot)$ | 100% | 100% | 40.3% | 1.3% |
| Primary pressure | (bar) | 155 | 175 | 155 | 155 |
| Core inlet temperature | (°C) | 258 | 260 | 260 | 286.5 |
| Core outlet temperature | (°C) | 320 | 327.6 | 320 | 322.4 |
| Fissile length | (m) | 1.35 | 1.25 | 1.02 | 3.66 |
| Linear heat rate | (kW/m) | 4 | 5.1 | 9.3 | 17.9 |
| Secondary pressure | (bar) | 48 | 50 | 50 | 58 |
| Number of steam generators | | 8 | 8 | 8 | 3 |
| Vessel height | (m) | 14 | 14 | 14.7 | 12.5 |
| Specific mass | (t/MWe) | 9.72 | 8.01 | 6.24 | 2.05 |

As far as the full natural circulation is concerned, a low linear heat rate (4 to 5 kW/m) must be considered and consequently a high value of the specific mass (about 9 t/MWe) is obtained. The size of the vessel body is about the size of a standard 900 MWe PWR. The primary pressure increase improves the natural circulation and then allows a reduction of the specific mass. Adding some forced convection (natural circulation ratio of 40%), one can observe a significant increase of the linear heat rate (>> factor 2) then a reduction of the specific mass. It is worth mentioning the required pumping power is quite low (3 kW), but with a very low discharge head and a high volumetric flow rate.

*Neutronic analysis*

The feasibility of one-batch long-life cores has been investigated using the APOLLO-2 computer code. The burn-ups and cycle lengths that could be reached were evaluated, taking into account different neutronic design parameters:

- Two oxide fuels, $UO_2$ (10, 15 and 20% of $^{235}U$) and MOX (15, 20 and 25% of plutonium coming from 41 GWd/t $UO_2$ fuel).

- Two plutonium isotopic compositions due to different cooling times, 3 years and 20 years. The fissile contents are respectively 65.3% and 63.5%.

- Six moderating ratios ranging from a P/D of 1.1 to 1.7.

- Three specific fuel powers, 12 W/g, 20 W/g and 30 W/g, which mean discharge burn-ups from 60 to 150 GWd/t (for 15 years).

- Two coolants, light water and heavy water.

In order to reach the 15-year objective with a one-batch MOX core, the Pu content must be between 22.5 and 25%. The lattice must be tight (P/D = 1.1), the specific power low (12 W/g) and the neutron leakage below 10% of the absorption (Figure 4). The average burn-up at discharge is about 60 GWd/t.

**Figure 4. Impact of the specific power for the MOX fuel
(3 years old, 20% Pu, P/D = 1.1, light water)**

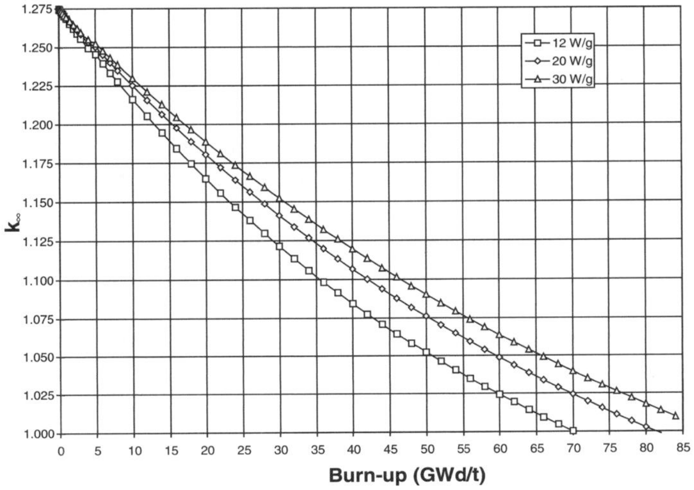

267

For UO$_2$ cores, more configurations fulfil the objective. A 10% enriched UO$_2$ core with an intermediate lattice (P/D = 1.4) can reach about 50 GW.d/t (i.e. 13 years with 12 W/g) if the neutron leakage level is kept below 7.5%. A 20% enriched UO$_2$ core with a tight lattice (P/D = 1.1) can reach 90 GWd/t (14 years with 20 W/g) if the neutron leakage level is kept below 7.5%. Limiting the neutron leakage is important in order to increase the cycle length. This, however, might be incompatible with the necessity to maintain a negative void coefficient. For both UO$_2$ and MOX fuels, the required contents are above the capability of existing fuel fabrication plants.

Rough estimations of the void coefficient have been achieved to determine the maximal core size. Tight lattices with high fissile contents are considered. The core height over diameter ratio, H/D, was assumed to be 0.7 (like in COPERNIC). The respective dimensions of the different cores having a zero void reactivity coefficient after irradiation and for a whole core voidage, i.e. the most penalising situation, are evaluated using APOLLO-2 with a one-dimensional model. The core reactivity ($k_{eff}$) is written, in a classical way, as the product of the $k_\infty$ and the non-leakage probability:

$$k_{eff} = k_\infty \cdot P_{nonleakage} = \frac{k_\infty}{1 + M^2 B^2}$$

where the $\frac{\Delta k_{eff}}{k_{eff}}$ is decomposed into $\frac{\Delta k_\infty}{k_\infty}$ and $\frac{\Delta P_{nonleakage}}{P_{nonleakage}}$ which allows to evaluate the purely spectral contribution ($k_\infty$) and the contribution related to the migration of the neutrons (non-leakage term). One had to solve $\frac{\Delta k_{eff}}{k_{eff}} = 0$ after irradiation and for 99.9% voidage. In order to take into account the effect of the flux redistribution between the nominal and the voided configurations, it is assumed that $B_{nom}^2 = 1.5 \cdot B_{geom}^2$ and $B_{void}^2 = B_{geom}^2$.

The MOX cores must be short: between about 80 cm and 90 cm for the light-water cases, and between about 40 cm and 60 cm for the heavy-water cases. The UO$_2$ core heights are between 110 cm and 130 cm (20% $^{235}$U), and between 140 cm and 180 cm (15% $^{235}$U). These are smaller cores than the core size obtained with COPERNIC to ensure 100% natural circulation at 150 MWe.


*Thermal-hydraulic analysis*

The use of the CATHARE thermal-hydraulic computer code had to confirm the sizing obtained by COPERNIC and make a preliminary analysis of some transients.

For the steady state, the CATHARE-calculated vessel flow rate is 415 kg/s, about 7% lower than the COPERNIC value (445 kg/s). The agreement is also very good for the temperatures. For the hot assembly channel, CATHARE predicts some boiling with a maximum void fraction of 25% (but we are still well below the DNB). Some LOCA transients were calculated without taking into account any safety injection system. Two break diameters (2.56 cm and 10 cm), located at the elevation of the steam generator outlet lines, were considered. In both cases, the core temperatures are close to the saturation temperature. For the "large" break, the primary pressure reaches the containment pressure after 2 500 s. For the small break, after six hours of calculation, the primary pressure is about 15 bar, the coolant inventory being still very high. This shows the good resistance to LOCAs of this concept.

## The super-critical pressure concepts

The super-critical pressure concepts are water-cooled reactors operating above the critical point (221 bar, 374°C). These concepts were evaluated during the 50s and 60s by Westinghouse and General Electric without any construction achieved due to concerns about the economic competitivity and the materials issue at high pressure and temperature. For about ten years, new studies have been published, by the Kurchatov Institute [13], AECL (CANDU-X programme [14]) and the University of Tokyo [15]. These studies can benefit from the significant progress made in conventional fossil power plants using super-critical pressure steam.

Among these new concepts, those of the University of Tokyo seem quite attractive because they could be seen as simplified BWRs (no re-circulation pumps, separator-dryer) with a PWR size vessel. They are based on a once-through direct steam cycle, the water enters the reactor as liquid (about 300°C) and exits as high temperature steam (about 500°C), providing a high thermodynamic efficiency (about 44%). Many versions have been proposed by the University of Tokyo [15,16], from fast neutron spectra to thermal spectra. CEA started to study these concepts to assess their real fast spectrum capabilities and to identify the key technological issues which must be overcome. Some parametric neutronic calculations were run showing the difficulties to obtain a positive regeneration gain and a negative void coefficient. A European programme has also been launched (5th Framework Programme of the European Union), the High Performance Light Water Reactor (HPLWR [17]), mainly devoted to the analysis of a thermal concept, the SCLWH-H [15]. CEA made thermal-hydraulic and neutronic evaluations of this core concept which consists of a hexagonal tight lattice of $UO_2$ fuel rods (axial and radial zoning of the $^{235}U$ enrichment) and water rods with descending water (Figure 5, Table 4).

### *Thermal-hydraulic evaluations of the SCLWR-H design*

The FLICA computer code has been used to perform assembly sub-channel analysis. Supercritical water properties as well as heat transfer and deteriorated heat flux correlations were added to the FLICA code. A cosine axial power was assumed.

## Figure 5

*a) Schematic of the SCLWR-H assembly [15]*  *b) Enrichment distribution (1/12 of the assembly)*

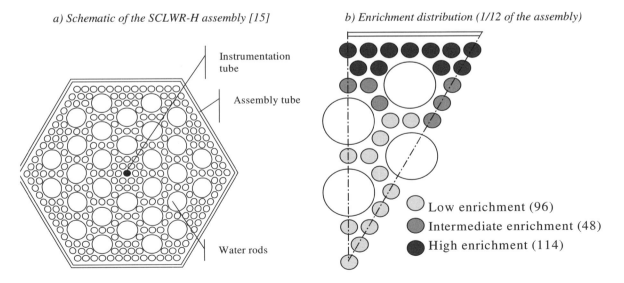

**Table 4. University of Tokyo thermal concepts**

|  | SCLWR | SCLWR-H |
|---|---|---|
| **Electric power, MW** | 1 013 | 1 570 |
| **Thermal efficiency, %** | 40.7 | 44.0 |
| **Pressure, bar** | 250 | 250 |
| **Cladding** | Stainless steel | Nickel alloy |
| **Number of assemblies** | 163 | 211 |
| **Fuel rods per assembly** | 258 | 258 |
| **Water rods per assembly** | 21 | 21 |
| **Inlet/outlet temperature, °C** | 324/397 | 280/508 |
| **Feedwater flow rate, kg/s** | 2 313 | 1 816 |

The results indicated substantial impact of the heat transfer coefficient on the axial temperature profile of the cladding, the Dittus-Boelter correlation leading to the highest cladding temperature (Figure 6). Calculations showed large temperature differences (500°C) between the centre of the assembly and the water rod. This is explained by the peculiar geometry of the water rod wrapper that generates regions of low velocity (Figure 7). The radial variations of the axial velocities within the assembly are high, up to a factor of 3, which cause high temperature variations in the assembly. Therefore, the cladding temperatures would exceed the current design criterion (620°C). The geometry of the sub-channel must be modified.

**Figure 6. Cladding temperatures (Dittus-Boelter)**

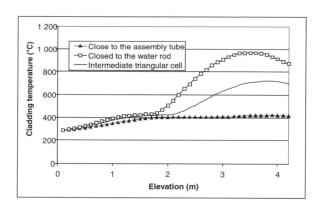

**Figure 7. Axial mass velocities**

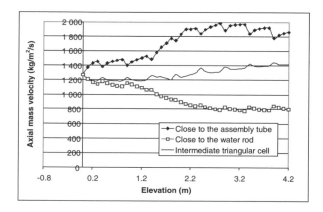

## Neutronic evaluations of the SCLWR-H design

In a first series of 2-D subassembly calculations performed at beginning of cycle (BOC), the temperature of water rods was assumed to be constant along the core (350°C). For the coolant channel, four temperatures were considered: 330°C, 370°C, 400°C and 480°C. These calculations put in light the influence of the two assembly parameters: the coolant density and the three axial and radial zones. The results obtained (Table 5, Figure 8) are consistent with the other HPLWR partner results. They clearly justify the axial zoning. On the other hand, the radial zoning seems to be insufficient to limit the local power peaking factors, especially in the upper part of the core where the coolant density is very low. The moderation distribution in the subassembly is not uniform and justifies the three zone enrichments. Different solutions are currently being investigated to improve this distribution: use of solid moderator, increase of the water gap between subassemblies, etc. This study also pointed out that the average core moderation ratio is about 20% lower than in a standard PWR, which explains the high enrichment used. Sensitivity calculations showed the huge neutronic penalty due to the nickel alloy cladding.

**Table 5. $k_{eff}$ of the assembly**

| Coolant temperature | Average enrichment water density | 4.72%, lower core (3.81, 4.72, 5.49%) | 5.16%, mid core (4.16, 5.16, 6.00%) | 6.0%, upper core (4.87, 6.04, 7.02%) |
|---|---|---|---|---|
| 330°C | 0.6807 | 1.17811 | | |
| 370°C | 0.5405 | 1.16242 | 1.18711 | |
| 400°C | 0.1665 | | 1.11792 | 1.15736 |
| 480°C | 0.0960 | | | 1.13939 |

**Figure 8. Neutron spectra in different locations of the assembly (mid-core at a temperature of 370°C)**

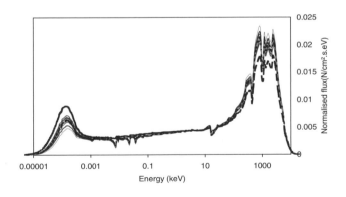

For the Doppler reactivity coefficient, when the fuel temperature is dropped to 293°K while other regions are kept at the same temperature, the coefficients obtained vary between -2.36 pcm/°C and -3.04 pcm/°C. For the water density coefficients, considering a 1% decrease in water coolant density, the values obtained ranged from -0.0909 to -0.2453. Considering a 99% decrease in the coolant density (voided assembly), the density coefficients vary between -0.1732 and -0.2598. At first look, these coefficients could be regarded as satisfactory, but the water density increase could be a more likely event and could lead to a faster transient than in standard PWRs.

## Conclusions

The three studies presented demonstrate the CEA's capabilities to calculate and evaluate innovative water-cooled concepts. Today, the main remaining ongoing study is the super-critical pressure concept. A future significant activity will be the evaluations conducted in the framework of the "Generation IV International Forum", many proposals made being water-cooled reactors.

## REFERENCES

[1]  G.L Fiorini, *et al.*, "PWR Safety Systems: Consequences of the Total or Partial Elimination of Soluble Boron on Plant Safety and Plant Architecture", *Nucl. Tech.*, Vol. 127, pp. 239-258, Sept. 1999.

[2]  R. Girieud, *et al.*, "A 100% MOX Core Design using a Highly Moderated Concept, GLOBAL'97, Yokohama, Japan, Oct. 1997.

[3]  S. Loubiére, *et al.*, "APOLLO-2, Twelve Years Later", M&C'99, Madrid, Spain (1999).

[4]  D.C. Groeneveld, *et al.*, "The 1995 Look-up Table for Critical Heat Fluxes in Tubes", *Nucl. Eng. and Design*, Vol. 163, pp. 1-23 (1996).

[5]  I. Toumi, *et al.*, "FLICA-IV: A Three-dimensional Two-phase Flow Computer Code with Advanced Numerical Methods for Nuclear Applications", *Nucl. Eng. and Design*, Vol. 200, pp. 139-155 (2000).

[6]  A. Giordano, *et al.*, "COCHISE: code de prévision de la corrosion externe des gaines d'assemblages combustibles", *Journal de physique IV*, Volume 11, pp. 1-151 (2001).

[7]  C. Struzik, *et al.*, "High Burn-up Modelling of $UO_2$ and MOX Fuel with METEOR/ TRANSURANUS 1.5C", ANS Light Water Fuel Performance Meeting, Portland, USA (1997).

[8]  D. Tarabelli, *et al.*, "Prediction of Light Water Reactor Contamination using the PACTOLE Code", ICONE8, Baltimore, USA (2000).

[9]  F. Barré, M. Bernard, "The CATHARE Code Strategy and Assessment", *Nucl. Eng. and Design*, Vol. 124, pp. 257-284 (1990).

[10]  S. Nisan, J-L. Rouyer, "SEMER: A Simple Calculational Tool for the Economic Evaluations of Reactor Systems and Associated Innovations", ICONE9, Nice, France (2001).

[11]  Safe Integral Reactor, program summary, ABB Combustion Engineering Power, July 1990.

[12]  M. Carelli, *et al.*, "Status of the IRIS Reactor Development", Global 2001, Paris, France, 9-13 Sept. 2001.

[13] V.A. Silin, *et al.*, "The Light Water Integral Reactor with Natural Circulation of the Coolant at Supercritical Pressure B-500 SKDI", *Nucl. Eng. and Design*, Vol. 144, p. 327 (1993).

[14] S.J. Bushby, *et al.*, "Conceptual Designs for Advanced High-temperature CANDU Reactors", ICONE8, Baltimore, USA (2000).

[15] K. Dobashi, *et al.*, "Conceptual Design of a High-temperature Power Reactor Cooled and Moderated by Supercritical Light Water", ICONE6, Nice, France (1998).

[16] K. Kitoh, *et al.*, "Pressure and Flow-induced Accident and Transient Analyses of a Direct Cycle Supercritical Pressure Light-water-cooled Fast Reactor", *Nucl. Tech.*, Vol. 123, p. 233, Sept. 1998.

[17] G. Heusener, U. Muller, T. Schulenberg, D. Squarer, "A European Development Program for a High Performance Light Water Reactor (HPLWR)", SCR-2000, The University of Tokyo, Japan, Nov. 2000.

# BARS: BWR WITH ADVANCED RECYCLE SYSTEM

**Kouji Hiraiwa, Yasushi Yamamoto, Ken-ichi Yoshioka, Mitsuaki Yamaoka**
Toshiba Corporation, Isogo Nuclear Engineering Centre, Japan

**Akira Inoue, Junji Mimatu**
Gifu University, Japan

## Abstract

A study is being evolved concerning the neutronic and thermal characteristics of a fast spectrum BWR core with a tight fuel lattice for an innovative fuel cycle system known as BARS (BWR with an Advanced Recycle System), the aim of which is Pu multi-recycling and minor actinide (MA) burning. The BARS core has unique characteristics: its neutron spectrum is very hard through tight fuel lattice with fuel pin gap of 1.3 mm and it has a neutron streaming channel to keep negative void reactivity on top of one-third of the fuel bundles. Therefore, the neutronic benchmark tests and the thermal-hydraulic tests were planned to measure void and Doppler reactivity effects, neutron streaming and the thermal-hydraulic performance of the tight lattice bundle. Through our core design study and benchmark tests, we will make clear the feasibility of the conceptual design of BARS core.

## Introduction

The LWR will play an important role in nuclear power generation for a longer time than was previously expected due to the delay of commercialisation of the LMFBR. It is one of the promising candidates for recycling various actinides such as U, Pu and minor actinides (MAs) because it is well demonstrated. In particular, BWR has an advantage in breeding or high conversion of fissile plutonium as boiling in water decreases the amount of moderator to induce a fast neutron spectrum. The fast spectrum is also advantageous in multi-recycling of Pu because the production of higher isotopes is suppressed. Moreover, it helps to burn fuels with recovered MAs and low decontaminated fission products (FPs) such as rare earths (RE), which endows the recycling system with nuclear proliferation resistance.

Taking into account the advantage of the fast spectrum BWR, an innovative fuel cycle system named BARS (BWR with an Advanced Recycle System) is proposed as a future fuel cycle option whose aim is to enhance the utilisation of uranium resources and reduce radioactive wastes [1,2,3]. As shown in Figure 1, in the BARS, the spent fuel from conventional LWRs is recycled as a MOX fuel for a BWR core with the fast neutron spectrum through oxide dry-processing and vibro-packing fuel fabrication.

**Figure 1. Concept of BWR with Advanced Recycle System (BARS)**

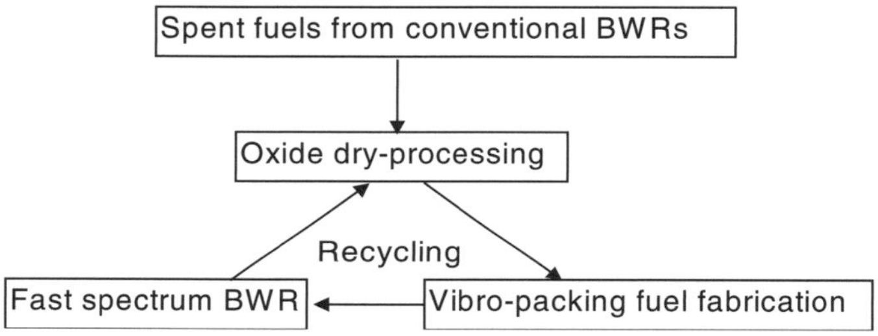

The fast neutron spectrum is obtained through triangular tight fuel lattice, which tends to shift coolant void reactivity toward the positive. Therefore, neutronics design efforts were made to keep it negative under the condition of target core characteristics. As a result, a core concept with a neutron streaming channel aiming at negative void reactivity was proposed.

Verification of the neutronics calculation method is required since the BARS core has the fast neutron spectrum through tight fuel lattice and reactivity effect due to neutron streaming phenomena as shown in the next section. Therefore, a programme of benchmark tests has been established to verify the neutronics calculation method. In the tight lattice core, the gap between the rods must be narrow, about 1.3 mm, so as to achieve a high conversion ratio. As the gap between the rods becomes narrower, the critical power performance may be worse due to the smaller flow area. Because there are few critical power data for such a tight lattice, critical power measurement tests should be performed in order to establish a database. In this paper, the BARS core design is summarised and the programme of the measurement tests is presented for verification of both the neutronics calculation method and thermal characteristics.

## BARS core description

The BARS core concept is summarised below. Table 1 displays typical BARS core and fuel specifications. To achieve a fast neutron spectrum, a tight lattice fuel assembly was adopted with a

**Table 1. Specifications of BARS core**

| Item | Unit | BARS core | Conventional ABWR |
|---|---|---|---|
| Power | MWe | 1 356 | 1 356 |
| Core equivalent diameter | m | 5.2 | 5.2 |
| Core height | m | 1.6; normal fuel<br>0.8; partial fuel | 3.7 |
| Number of fuel assemblies | – | 208; total<br>132; normal fuel<br>76; partial fuel | 872 |
| Number of fuel pins | /assembly | 658 | 60 |
| Pin lattice type | | Triangle | Rectangular ($8 \times 8$) |
| Pin diameter | mm | 11.2 | 12.3 |
| Cladding thickness | mm | 0.3 (SUS) | 0.86 (Zr) |
| Pin gap | mm | 1.3 | 4.0 |
| Pin pitch | mm | 12.5 | 16.3 |
| Bundle pitch | mm | 317 | 155 |
| Ratio of flow area to fuel area | – | 0.49 | 3.1 |

water to fuel volume (W/F) ratio of about 0.5. It is well known that the void reactivity coefficient in a fast spectrum BWR core has the tendency to be positive. Thus, a new core concept was proposed in order to improve the void reactivity coefficient under the restriction of core diameter by adopting a neutron-streaming channel as described in Table 1.

Figure 2 shows a core profile of the BARS core in a large BWR plant (reactor thermal output of 3 926 MWt, core diameter of 5.2 m, core height of 1.6 m). The diameter of the BARS core should be approximately the same as that of conventional BWR cores to suppress the plant construction cost. The height is less than half that of the conventional BWR, and partial fuel assemblies whose active fuel length is about half that of the normal fuels are arranged by one-third of the whole core, as shown in Figure 3.

**Figure 2. Vertical view of BARS core**

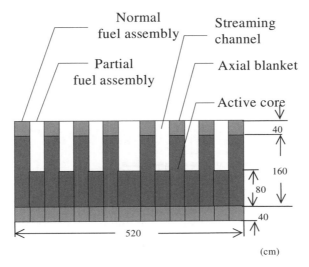

277

**Figure 3. BARS core layout**

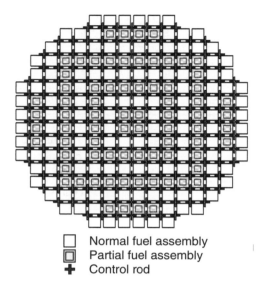

☐ Normal fuel assembly
▣ Partial fuel assembly
✚ Control rod

When void fraction increases, the streaming channel at the upper part of the partial assembly will enhance axial leakage of neutrons which have leaked out through the side of the normal assemblies and the top of the fuel bundle of the partial assemblies as shown in Figure 4. The cavity can in the streaming channel not only provides a streaming path for the leaked neutrons from the fuel but also suppresses softening of the neutron spectrum by expelling water in the channel.

**Figure 4. Vertical view of fuel assemblies**

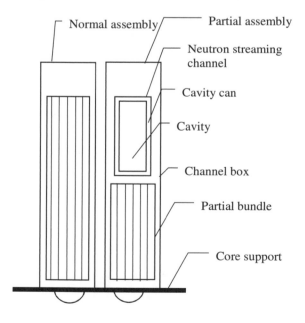

Figure 5 shows a horizontal cross-sectional view of a fuel assembly. The follower above the control rods and water removal plates attached on the outer side of channel box expels surplus water and helps to decrease W/F ratio. These concepts as well as a large size channel box (~30 cm) enable a W/F ratio as low as 0.49 in the BARS core. The average void fraction is designed to be about 60%.

## Figure 5. Horizontal cross-section of fuel assembly

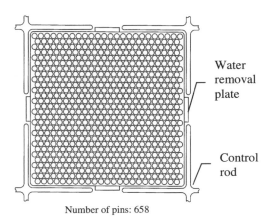

Water removal plate

Control rod

Number of pins: 658

## Characteristics of BARS core

Table 3 shows neutronic characteristics of the core loaded with MOX fuel accompanied with no MAs and REs. The burn-up calculation was done by using the conventional diffusion burn-up code SRAC [4], taking into account the fuel bundle heterogeneity. The void reactivity coefficient was analysed by the continuous-energy Monte Carlo code MVP [5]. The cross-section library used was based upon JENDL-3.2, which is attached to each code.

### Table 3. Neutronic characteristics of BARS (equilibrium cycle)

| Item | Unit | BARS |
|---|---|---|
| Refuelling scheme | – | 1 year × 4 batches |
| (1) Burn-up reactivity swing | %Δk/k | 1.6 |
| (2) Breeding ratio | – | 1.04 |
| (3) Core average discharge fuel burn-up | GWd/t | 44 |
| (4) Maximum linear heat rate | W/cm | 400 |
| (5) Void reactivity coefficient | $10^{-4}$ Δk/k/%void | -0.1 |
| (6) Heavy metal inventory | t | 132 |

Load fuel: MOX without rare earths.
Pu fissile/Pu total = 0.575, U enrichment = 0.02.

The operation cycle length is one year, and there are four refuelling batches. The Pu enrichments of load fuel were determined as shown in Table 4 and Figure 6 such that criticality is attained at the end of the equilibrium cycle (EOEC). The enrichment of outermost assemblies has been set higher than the other fuels aiming at radial power flattening. The enrichment is varied in the axial direction; enrichment located in the lower region of fuel assemblies near the axial core centre is about 20% to 30% lower than those above and below the region. The breeding ratio is 1.04 with an average discharge burn-up of 44 GWd/t.

### Table 4. Pu enrichment distribution (wt.%)

|  | Low enrichment zone | Other zones |
|---|---|---|
| Inner assembly | 11 | 14 |
| Outermost assembly | 11 | 17 |

**Figure 6. Axial Pu enrichment distribution**

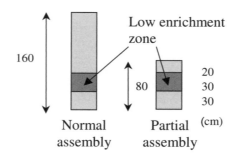

The void reactivity coefficient is negative through the equilibrium cycle. It was found that axial enrichment distribution leads to more negative void reactivity than uniform distribution, as described below. Since there is no fuel in the upper half of the partial fuel assembly and the neutrons tend to leak from the streaming channel, the neutron flux in the upper half of the normal assembly is lower than that in the lower half. By placing a low enrichment zone at the lower half of fuel assembly, it is possible to flatten the axial neutron flux distribution. Therefore, the neutron streaming effect by void fraction changes remarkably, increasing due to the increase of the neutron flux in the upper half of the normal assembly adjacent to the streaming channel. The axial enrichment distribution is also advantageous from the viewpoint of power flattening.

As the next step, core design is evolving for the core loaded with MOX fuel accompanied with MAs and REs recycled though the use of the oxide dry processing method.

## Benchmark tests on NCA

The BARS core has unique characteristics: its neutron spectrum is different from conventional BWRs and LMFBRs and it utilises neutron-streaming phenomena that are difficult to accurately evaluate. Therefore, a programme to perform benchmark tests in the Toshiba Nuclear Critical Assembly (NCA) has been established to verify the neutronics calculation method. The NCA is a slightly enriched, uranium-fuelled, light-water-moderated critical assembly, which has been utilised to verify both LWR design codes and the specific fuel design.

One purpose of the tests is to verify the method in a fast spectrum core within the limitation of the use of slightly enriched uranium fuel. The important feature to be measured is the $^{238}$U reaction rate because it occupies the main part or one of the main parts in the evaluation of the conversion ratio and reactivity coefficient (void and Doppler reactivity coefficients). The neutron streaming effect is also an important feature to be measured. The evaluation accuracy of the effect is not so sensitive to the difference of the neutron spectra between uranium- and MOX-fuelled cores because it is strongly dependent on the collision reaction rate. Another purpose is to verify the new method to evaluate the reactivity coefficients based upon the measured data of modified conversion ratio [6,7], aiming at applying the method to the measurement in a MOX-fuelled test core in the future. The new method is very advantageous because it can evaluate the characteristics of the tight lattice zone only, eliminating the effect of the driver core zone composed of a conventional LWR lattice.

So far, basic characteristics such as conversion ratio, neutron spectrum and power distribution were measured in the NCA test core with the tight lattice test zone [8,9]. New tests are now being prepared for the measurement of reactivity coefficients and the neutron streaming effect. One of the fuel rod arrangements of the test core is shown in Figure 7. The basic characteristics mentioned above

**Figure 7. One of the fuel rod arrangements of the test core**

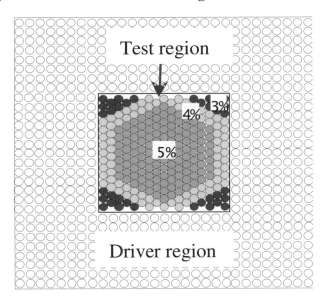

are measured in the central test zone with the tight fuel lattice surrounded by the driver zone with a conventional LWR fuel lattice to attain criticality. The test region for the streaming effect measurement is shown in Figure 8. The neutron streaming effect through the central void region is simulated with changing the water volume fraction by changing the diameter of the void tube outside the central region. Figure 9 illustrates the test device to measure the Doppler coefficient.

**Figure 8. Test region for streaming effect measurement**

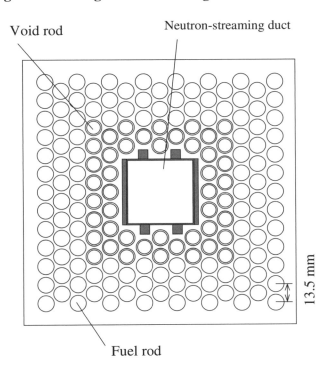

**Figure 9. Test rod for Doppler coefficient measurement**

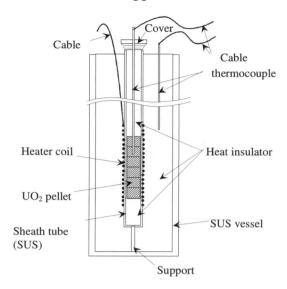

**Thermal-hydraulic test**

In the BARS core, a tight lattice fuel must be adopted because the conversion ratio needs to be improved to about 1. However, as a fuel rod gap becomes narrower, thermal-hydraulic performance, especially at critical power, becomes worse. Therefore, the thermal power of the BARS reactor core could be influenced largely by thermal-hydraulic performance of the tight lattice fuel. In the previous section, the fuel is designed as a triangular lattice whose rod gap is 1.3 mm. In such a tight lattice bundle, there are very little critical power test data and critical power correlations applicable to critical power prediction.

Thus, three tests are planned:

1)  Visualisation test.

2)  High-pressure thermal-hydraulic test.

3)  Counter-current flow limitation (CCFL) test.

*Visualisation test*

The purpose of the visualisation test is to investigate the boiling transition behaviour in the narrow gap bundle. It is hard to visualise the two-phase flow in the rod bundle under BWR operating conditions. Therefore, it is planned to perform the two single-channel tests under atmospheric conditions. One is the unheated test for measuring the physical quantities of the two-phase flow in narrow channel, i.e. liquid film thickness, velocity and so on.

The other is the heated test to visualise the liquid film behaviour just before the boiling transition occurs. The heater rods are made of glass, whose surface is coated by $SnO_2$. Because the boiling transition behaviour will be investigated, these test sections have the same flow geometry as shown in Figure 10.

**Figure 10. Flow geometry of the test channel for visualisation test**

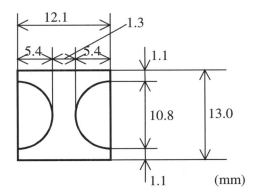

## High-pressure thermal-hydraulic test

The purpose of the thermal-hydraulic test is to create a database of the critical power performance of the tight lattice bundle whose rod gap is about 1.3 mm. In a previous paper [3], it was reported that the critical power test was performed with the tight lattice bundle and Arai's correlation could be applied to predicting critical power of tight lattice bundle. However, that previous critical power test was performed with a longer heating length than that of the actual BARS core. Therefore, the thermal-hydraulic tests were planed to enhance the thermal-hydraulic database for the tight lattice bundle, to verify the applicability of Arai's correlation on the BARS core and to develop a more accurate correlation for predicting critical power. Moreover, the two-phase flow instability test, transient boiling transient test and pressure drop test were planed to design the BARS core from the thermal-hydraulic point.

Figure 11 shows a cross-sectional view of the test assemblies. As shown in Figure 11, it is planned to make two types of the test bundle. One is the seven-rod test bundle with a hexagonal channel box, the other is a 14-rod test bundle with a rectangular channel box. The purpose of the seven-rod bundle test is to survey the rod gap effect and heating length effect on the critical power and pressure drop. On the other hand, the purpose of the 14-rod bundle test is to check the critical power for various radial power distributions.

**Figure 11. Cross-sectional views of the test bundle for the high-pressure thermal-hydraulic test**

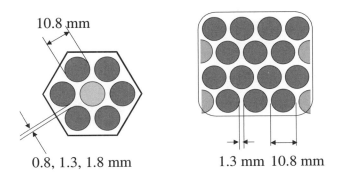

Figure 12 shows the system diagram of the test facility. This test facility is usually called BEST (Toshiba BWR experimental loop of stability and transient). This loop has a capability of testing under BWR operating conditions.

**Figure 12. Flow diagram of test facility (BEST)**

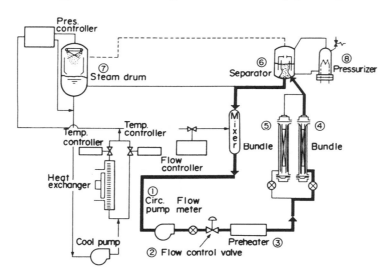

On critical power test, pressure, inlet water temperature and flow rate were set to programmed level first. Then, the bundle power is raised step by step by very small magnitudes. Critical power is defined as a power when the rod surface temperature jumps by 14°C from the temperature under nucleate boiling conditions. Test conditions were planned to be as follows.

| | |
|---|---|
| Pressure | 1~8 MPa |
| Mass flux | 500 ~ 2 000 kg/m²s |
| Inlet subcool | 20~80 kJ/kg |

### Counter-current flow limitation (CCFL) test

On the tight lattice bundle, because the flow area becomes narrower than that of the usual BWR fuel, it is a concern that a little spray water could fall into the core when a loss of coolant accident (LOCA) occurs. Therefore, the purpose of this test is to investigate the CCFL characteristic and to make a CCFL correlation for the tight lattice bundle. It is planned that the correlation based on this CCFL test be applied to the LOCA analysis. Figure 13 shows the CCFL test section and cross-sectional view of the tight lattice test bundle. The conditions that the water is unable to fall down into the test bundle a are defined CCFL condition. In the CCFL test, the water spays to the test bundle from the upper plenum. On the other hand, the vapour is blown into the test section under the test bundle.

## Conclusion

A study is being evolved on neutronic and thermal characteristics of a fast spectrum BWR core with a tight fuel lattice for an innovative fuel cycle system known as BARS (BWR with an Advanced Recycle System), the aim of which is Pu multi-recycling and the burning of minor actinides (MAs). The BARS core has unique characteristics: its neutron spectrum is very hard through tight fuel lattice with fuel pin gap of 1.3 mm and it has a neutron streaming channel to keep negative void reactivity on top of one-third of the fuel bundles. Thus, benchmark tests were planned to measure void and Doppler reactivity effects, and neutron streaming effect at the Toshiba Nuclear Critical Assembly (NCA) in order to verify the neutronics calculation method. Because there are few critical power data for such a

**Figure 13. CCFL test section and cross-sectional geometry of the test bundle**

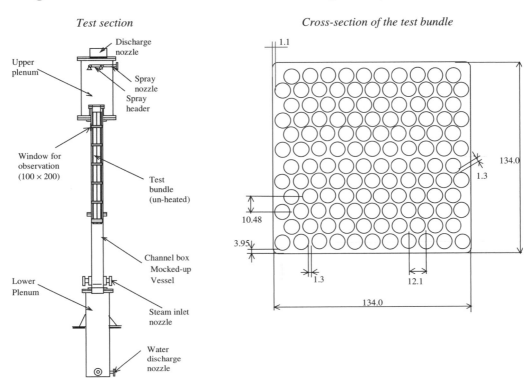

tight lattice, a critical power measurement test was planned to get the database. Not only critical power test, but also the transient BT test, CCFL test and visualisation test were planned to design the tight lattice core at the thermal-hydraulic points and investigate CHF mechanism when the rod gap becomes extremely narrow. Through our core design study and benchmark tests, we will determine the feasibility of the conceptual design of BARS core.

## REFERENCES

[1]   T. Yokoyama, *et al.*, "Study on Breeding Characteristics of Fast Spectrum BWR", International Conference on the Physics of Nuclear Science and Technology, Islandia Marriott Long Island, Long Island, NY, 5-8 October (1998).

[2]   Y. Sakashita, *et al.*, "Core Characteristics of Breeding BWR for BARS (BWR with Advanced Recycle System)", 7th International Conference on Nuclear Engineering (ICONE-7), Tokyo, Japan, 19-23 April (1999).

[3]   M. Yamaoka, *et al.*, "Study on Fast Spectrum BWR Core for Actinide Recycle", 9th International Conference on Nuclear Engineering (ICONE-9), Nice, France, 8-12 April (2001).

[4]     K. Okumura, *et al.*, "SRAC95: General Purpose Neutronics Code System", JAERI-Data/Code 96-015 (1996) (in Japanese).

[5]     T. Mori, *et al.*, "MVP/GMVP: General Purpose Monte Carlo Codes for Neutron and Photon Transport Code Based on Continuous Energy and Multi-group Methods", JAERI-Data/Code 94-007 (1994).

[6]     K. Yoshioka, *et al.*, "Determination of Void Coefficient from Modified Conversion Ratio Measurements", Proc. of 2000 Fall Meeting of the Atomic Energy Society of Japan, G41 (2000) (in Japanese).

[7]     I. Mitsuhashi, *et al.*, "Doppler Coefficient Measurement Method in an LWR Pu Lattice Employing Modified Conversion Ratio", Proc. of 2000 Fall Meeting of the Atomic Energy Society of Japan, G42 (2000) (in Japanese).

[8]     S. Miyashita, *et al.*, "Critical Experiments of Low Moderated Tight Lattice Core: (1) Measurements of Neutron Flux and Modified Conversion Ratios", Proc. of 1999 Fall Meeting of the Atomic Energy Society of Japan, E27 (1999) (in Japanese).

[9]     K. Yoshioka, *et al.*, "Critical Experiments of Low Moderated and Tight Lattice Core: (2) Analysis with MCNP Code", Proc. of 1999 Fall Meeting of the Atomic Energy Society of Japan, E28 (1999) (in Japanese).

# ADVANCED FUEL CYCLE FOR LONG-LIVED CORE OF SMALL-SIZE LIGHT WATER REACTOR OF ABV TYPE

**A. Polismakov, V. Tsibulsky, A. Chibinyaev, P. Alekseev**
Russian Research Centre "Kurchatov Institute"
Kurchatov sq., 123182, Moscow, Russia
Phone: +7-095-196-73-77, Fax: +7-095-196-37-08, E-mail: polismakov@dhtp.kiae.ru

## Abstract

The paper presents the results of advanced fuel cycle optimisation for a long-lived core for a small PWR of ABV-type though selection of fuel composition and the use of advanced burnable absorbers.

## Introduction

Advanced small reactors could be considered an important component of the future nuclear power structure. Approaches to design of the core and primary cooling system should account for operational experience of transport reactors, but should be revised for safety enhancement. To simplify reactor control, monitoring and maintenance, the safety of small reactors should mainly be provided by passive systems and inherent properties of core design and materials.

One of the major safety requirements for a small reactor operation is the reduction of reactivity accident probability and consequences. Reactivity to be compensated by control rods at the beginning of the fuel cycle in existing transport pressurised water reactors (PWR) ranges from about 8 up to $20\%\Delta(1/k)$. Technological difficulties concerning the introduction of reactivity control by boric acid in the primary coolant and a low core breeding ratio are the main reasons for such high reactivity margins for burn-up.

The paper presents results of advanced fuel cycle optimisation for a long-lived core of a small PWR of ABV-type through selection of fuel composition and the use of advanced burnable absorbers.

## ABV reactor for "Volnolom" floating nuclear power plant

The design of the "Volnolom" floating nuclear power plant (FNPP) with two ABV reactor facilities (Figure 1) was developed in the beginning of the 90s [1]. The ABV reactor of 38 MWt and 12 MWe can produce 12 Gcal/h. In particular, the cost of using the ABV reactor for heat delivery to the Arctic Sea cost is considered. Design lifetime of the FNPP is about 10 years.

### Figure 1. ABV reactor facility general view [1]

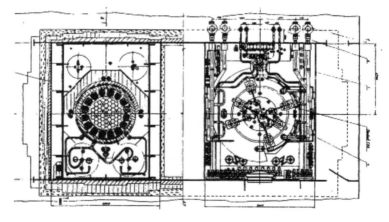

The ABV core consists of 55 hexagonal fuel assemblies (FAs). The FA design is a shortened variant of the VVER-440 FA design. The layout of FAs accepted as the reference for calculation is shown in Figure 2. The reference core layout consists of FAs of the same type.

## ABV-type reactor fuel cycle duration

To avoid the intermediate refuelling at the site the ways were studied to obtain the fuel cycle close to FNPP life time (10 years). On the basis of data related to Ust-Kamchatsk region (RF), the load factor of two-unit ABV-type facility was estimated close to 0.6. Thus, about 2 190 effective full-power days (EFPD) was considered as a target of the study.

## Figure 2. Reference ABV-type fuel assembly layout

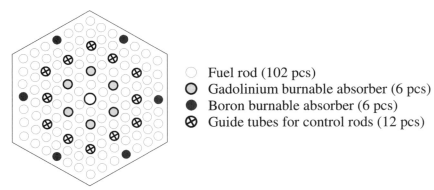

○ Fuel rod (102 pcs)
◎ Gadolinium burnable absorber (6 pcs)
● Boron burnable absorber (6 pcs)
⊗ Guide tubes for control rods (12 pcs)

As a first step, fuel cycle duration was studied for a small reactor of ABV-type charged with various fuel compositions. All core and primary circuit parameters except fuel type remain constant.

The following fuel compositions were studied: Zr-Al-based inert-matrix fuel, uranium dioxide, fuel particles. Inert-matrix fuel U-Zr-Al with 17% enrichment by $^{235}$U was considered as a reference option to be optimised. Fuel enrichment of 21% by $^{235}$U was set as the upper limitation in this study. The concentration of uranium grains in inert and graphite matrices was chosen with maximum respect for the requirements of fuel structural stability.

The main result of the study is as follows: a fuel cycle duration of about 10 years can be achieved with the use of uranium dioxide fuel with fuel enrichment of 8% by $^{235}$U (Figure 3).

## Figure 3. k-effective of ABV-type reactor vs. operational time for various fuel compositions

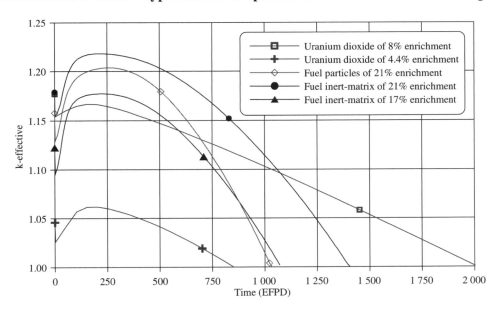

The next step of the study was an attempt to minimise reactivity swing with burn-up in order to enhance reactor safety parameters. Reactivity swing in the considered fuel cycles changes from 17 up to 22%$\Delta(1/k)$ (Figure 3). Such a high reactivity margin should be compensated by control rods. Thus, reactor control and diagnostics become more complex. In addition, the probability of an accident with high positive reactivity insertion due to control rod ejection becomes higher.

For these reasons the next stage of the study was aimed at modification of the ABV-type core layout in order to reduce reactivity swing with burn-up, while maintaining or extending fuel cycle duration.

## Modification of ABV-type core layout

If fuel pellets of VVER-1000 were used in the ABV-type core, loading of $^{235}$U and $^{238}$U would reach 2 000 kg. The average fuel burn-up for a fuel cycle of 2 000 EFPD, in this case, would equal 40 MWd/kgU. As the value found for fuel enrichment (8%) seems rather high from the viewpoint of the use of technologies developed for serial fuel rods, therefore, an increase in the fuel loading was chosen as the measure for extending the fuel cycle duration.

As seen from Figure 3, efficiency of boron and gadolinium burnable absorbers in the reference ABV-type core layout is not enough for compensation of reactivity swing with burn-up. In fact, due to low concentration of gadolinium and boron in burnable absorbers, they burn in the first half of the fuel cycle, and then only reserve a useful place, not influencing neutron balance. The low efficiency of burnable absorbers in the reference core variant results in a large number of control rods, which is used for compensation of excess reactivity. These control rods are also require space, which could be used for location of fuel rods.

Thus, the following measures were studied so as to increase fuel loading and simultaneously enhance the burnable absorber efficiency:

- increasing the number of FAs from 55 up to 61 pcs (see Figure 4);

- using advanced burnable absorbers located in the fuel (gadolinium) and on the fuel pellet surface (zirconium diboride film);

- replacing a part of the guide tubes for control rods by fuel rods due to the decrease of reactivity swing with burn-up.

**Figure 4. ABV-type small reactor core optimised layout**

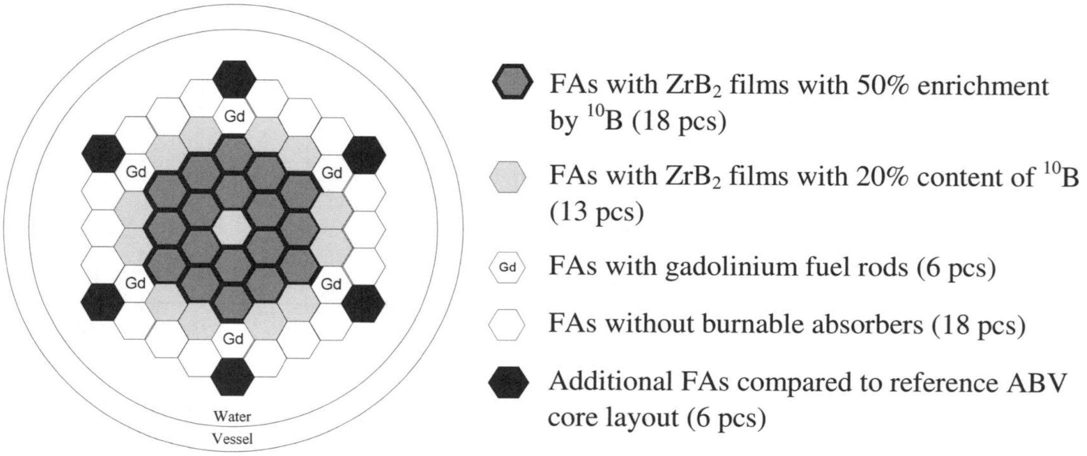

FAs with ZrB$_2$ films with 50% enrichment by $^{10}$B (18 pcs)

FAs with ZrB$_2$ films with 20% content of $^{10}$B (13 pcs)

FAs with gadolinium fuel rods (6 pcs)

FAs without burnable absorbers (18 pcs)

Additional FAs compared to reference ABV core layout (6 pcs)

As a result of a large number of optimisation calculations, the most acceptable option for fuel loading was obtained. $ZrB_2$ films 20 μm thick were used in 31 FAs both with natural content of $^{10}B$ (20%) and with 50% enrichment by $^{10}B$ (see Figure 4). Six FAs contain no zirconium diboride, but 12 fuel rods contain 8 wt.% of natural gadolinium. Twenty-four (24) outer-row FAs have no guide tubes for control rods, as these are replaced by fuel rods.

The chosen option for fuel loading is rather effective from the viewpoint of minimisation of reactivity to be compensated by control rods. Dependencies of k-effective on time for the reference and optimised core layouts are shown in Figure 5.

**Figure 5. k-effective of ABV-type reactor vs. operational time for various core layout options**

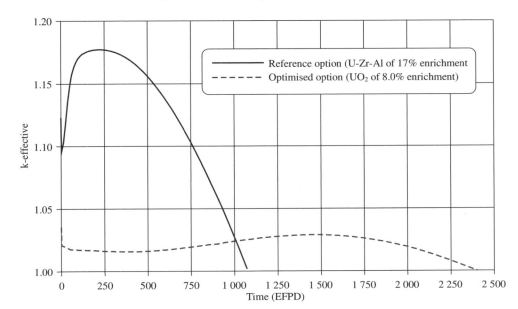

Loading of uranium isotopes increases up to 2 568 kg, while average fuel burn-up is close to 37 MWd/kgU for a fuel cycle length of 2 450 EFPDs. Average burn-up is the highest in the central FA and equals 41 MWd/kgU. FA power peaking factor decreases from 1.5 in the reference core layout to 1.4 in the optimised option.

As seen from Figure 4, burnable absorbers in the optimised ABV-type core layout are located in the central region of the core. Thus, in beginning of cycle the power peak is located near the core periphery. Then, with fuel burn-up the maximum of the power field shifts to the core centre. Due to this effect the reactivity value to be compensated by control rods is reduced by 5-8%Δ(1/k) in the beginning of cycle.

## Conclusions

The paper presents results of advanced fuel cycle optimisation of a long-lived core for a small PWR of ABV-type through selection of fuel composition and the use of advanced burnable absorbers. ABV is a marine PWR project of 38 MWt at 15 MPa with natural circulation of primary coolant and inert-matrix fuel U-Zr-Al.

The replacement of inert-matrix fuel by uranium dioxide and the use of $ZrB_2$ and gadolinium as burnable absorbers in the core central zone allowed to obtain the following parameters for the fuel cycle of the ABV-type reactor:

- The time of operation at nominal power level is 2 450 effective days.

- The reactivity to be compensated by control rods is about $3\%\Delta(1/k)$.

- Fuel assembly power peaking factor does not exceed 1.4 during the whole fuel cycle.

- Fuel assembly peak and average fuel burn-up are 41 and 37 MWd/kg U, respectively.

The reduced value of reactivity to be compensated by control rods allows for significant safety enhancement by reducing the possible consequences of control rod ejection accident.

## REFERENCE

[1]  Yu. Panov, V. Polunichev, K. Zverev, "Use of Reactor Plants of Enhanced Safety for Sea Water Desalination, Industrial and District Heating", in IAEA-TECDOC--923: *Non-electric Applications of Nuclear Energy*, Proc. Advisory Group Meeting on Non-electric Applications of Nuclear Energy, Jakarta (Indonesia), 21-23 Nov. 1995.

# A NEW APPROACH FOR THE SYSTEMS DEDICATED TO TRANSMUTATION: THE REACTOR WITH COMPENSATED BETA

**Bruno Bernardin**
Commissariat à l'Énergie Atomique, France

## Abstract

Hybrid systems are often presented as a solution with a future to incinerate nuclear waste. Indeed, the presence of minor actinides in fuel degrades certain neutron parameters significant for safety – like the effective beta – and makes difficult the realisation of a critical core dedicated to transmutation. The operation of a core in subcritical mode eliminates these problems. However, the design of a hybrid system of industrial size is rather complicated due to the power of the accelerator. This paper proposes a new approach which tends to solve the problem upstream, i.e. to compensate for the intrinsic weakness of beta. The principle rests on the use of an adequate coupling between the accelerator and the neutron flux so as to simulate additional delayed neutrons. Thus, the level of subcriticality of the core can be very small and a commercial accelerator could be used. The whole of the system behaves as a critical reactor having sufficient beta.

## Summary

In France, 80% of electricity is of nuclear origin, and the continuation of this means of production will rest, amongst other things, on the capacity to control waste. Our nuclear capacity is relatively young and *a priori*, to concentrate the incineration of waste with long life in some dedicated reactors appears more logical than to modify the whole of the park and the fuel cycle. This process is described as a "double strata" strategy.

Waste to be incinerated includes minor actinides and some fission products. However, a high concentration of these elements in a reactor core has consequences detrimental to the physical parameters which control the behaviour of the reactor and determine its safety. A solution to circumvent these bad effects consists of making these reactors operate on a subcritical level, with an external contribution of neutrons. The accelerator-driven system (ADS) implements this solution in which the auxiliary neutrons are produced by a source of spallation placed at the centre of the core and supplied with a beam of protons produced by an external accelerator.

The association of an accelerator to a reactor produces an acceptable over cost in the frame of the double strata strategy (the ADS would account for approximately 5% of the park). On the other hand, the feasibility of the accelerators necessary to consider industrial incinerators is, today, very far from being acquired. The difficult points relate to obtaining a beam of protons of several tens of MW and a level of reliability limiting the spurious shutdowns to a few events a year.

These observations led to the proposal for a different approach, designated "reactor with compensated beta" (RCB). Contrary to the ADS for which the low value of beta is circumvented by a margin of significant subcriticality, the proposal rests on a principle which aims at compensating for the weakness of beta. Practically, the RCB looks like an ADS because it includes a core, a target producing external neutrons and an accelerator. The difference is that the subcriticality of the core is very small ($k_{eff} = 0.997$) and that the accelerator is controlled by the evolution of the neutron flux present in the core, according to a law simulating the concentration of delayed neutrons. Thus, the core being slightly subcritical makes it possible to enormously reduce the power of the accelerator. The dynamic control of the neutron source according to an adequate law confers on the external neutrons the property of delayed neutrons. So, even if the core itself is subcritical and has a low effective beta, the whole system has the behaviour and the characteristics of a traditional critical reactor. The apparent beta of this system includes the intrinsic beta due to the fuel, increased by the value of the level of subcriticality of the core. Obviously, the RCB must be associated with the concept of cores not being able to melt, so as to avoid all risks of criticality in the event of accidents.

## Introduction

Currently two means are privileged to consider the transmutation of long-lived wastes. They concern, on the one hand, the use of dedicated critical reactors and, on the other hand, the use of subcritical systems supplied with an external neutron source. The concepts based on the direct spallation of minor actinides by a beam of particles or on the fission of this waste by neutrons resulting directly from a target of spallation are not retained. Indeed, the incineration of a weighable quantity of waste by these methods would require beams with unrealistic intensities. The more plausible options, critical reactor or ADS, raise nevertheless many theoretical and practical difficulties, at a point such that it is not possible to decide today in favour of one or the other. It is realised that a significant programme of study with the aim of evaluating and comparing these options must be carried out and that the risk of failure is not negligible. One thus re-examines, from a strictly need-based point of view, if there are different solutions to approach the transmutation of long-lived waste.

## Recalls on the current orientations – advantages and disadvantages

### *Dedicated critical reactors*

The introduction of minor actinides into traditional reactors (PWR or FBR) is possible in small quantities (a few %). Beyond that, degradations affecting certain significant parameters for safety become unacceptable (lowers Doppler coefficient and fraction of delayed neutron beta).

The objective to concentrate waste in a limited number of installations (double strata strategy) led to the study of dedicated reactors for which the tolerance for a stronger concentration of waste would be improved. It seems that this is possible to a certain extent by altering the drawing of the core (to increase Doppler) and the composition of fuel (to increase beta). Nevertheless, the room for manoeuvrability is weak and it is known that the most critical parameter will remain beta.

The step envisaged consists of determining, from a strict safety point of view, the minimal value of beta necessary for a critical reactor, and then to define a core which respects this value within a suitable margin (in playing on the composition of fuel and on the capacity of incineration).

In truth, this exercise will not be simple to realise and could easily be compromised by the safety authorities.

On the assumption that acceptable beta could be obtained, the dedicated reactor, incinerator of waste, will have to be satisfied nevertheless with beta appreciably weaker than that of the traditional fast reactors. Even if it respects safety requirements, it is located in withdrawal of these reactors with respect to certain types of accidents. It is an uncomfortable situation for a new generation of reactor. Apart from the problem of beta, the dedicated critical reactor presents the attraction of simplicity if one compares it with an equivalent ADS.

### *Subcritical hybrid systems*

The idea of using an ADS as a waste incinerator precisely responds to the difficulty in preserving acceptable beta in a critical reactor containing minor actinides. Let us recall that this constraint on the beta is dependent on the need of maintaining, for a critical reactor, a period (which depends on beta) higher than the time-constants of phenomenon ensuring the stability of the system (thermal feedback, dilation, system of regulation). In addition, the weaker the beta, the easier it is to imagine a situation leading to a critical prompt state. In a subcritical reactor, this constraint obviously does not exist, it behaving as a simple amplifier of the external source $\left( A = -\dfrac{1}{\rho} \right)$.

The fact of having a margin of subcriticality constitutes *a priori* an asset of the ADS with respect to reactivity accidents In fact, the demonstration of this advantage and its quantification in the case of serious accidents remains to be made. For the moment, this supposed interest prevailed and resulted in pushing this logic completely. Today, levels of $k_{eff}$ ranging between 0.9 and 0.95 are usually evoked. This choice makes it possible, for example, to eliminate control rods and to assign the care of completely ensuring the piloting and the compensation of burn-up to the spallation source. On the other hand, the realisation of the target and the accelerator will have to attain very high requirements in terms of power and controllability.

The following disadvantages can also be noted:

- The response times to variations of source or reactivity are very short, causing fast transients of power. The use of a pulsated beam can pose problems of stress of materials and fuel.

- Contrary to the case of the critical reactor, thermal feedbacks have a limited effect in the ADS. The more significant the level of subcriticality, the more this effect is reduced. This phenomenon explains why the primary loss of flow is more penalising for an ADS than for a critical reactor. Let us note that there is nevertheless a parry with this incident if one takes the precaution of supplying the accelerator with the energy provided by the system, the loss of the coolant then leading automatically to the stopping of the accelerator.

## Report

Today, neither of the two solutions studied have a decisive advantage. They both still require heavy comparative studies, without a guarantee that one between the two will lead to a proposal for a coherent project. Raising the question of a possible third solution thus appears legitimate.

## Proposal

The proposal emerges from the following observations:

- A critical reactor could be suitable if a beta of about 350 pcm (as for the FBR) were ensured.

- A critical reactor can be assimilated to a subcritical reactor of beta pcm supplied with the source consisting of the delayed neutrons.

- It is possible to simulate the existence of an additional group of delayed neutrons through the presence of an external source.

One can thus imagine the design of a very slightly subcritical reactor ($k_{eff}$ = 0.997) fed by a neutron spallation source of low power. Binding the power of this source to the neutron power of the reactor through a relation of the type which governs the precursors of delayed neutrons transforms the external source into an internal one and is equivalent to adding a new group of delayed neutrons. The characteristics of this group of delayed spallation neutrons can be adjusted (constant decrease, relative fraction) so as to lead to the value of the total beta desired. In this light, the system behaves like a critical reactor and is controlled as such.

The kinetic equations of the point model for a core without source and with only one group of delayed neutron precursors are written:

$$\frac{dn}{dt} = \frac{\rho - \beta}{\Lambda} \cdot n(t) + \lambda_1 C_1(t) \tag{1}$$

$$\frac{dC_1}{dt} = \frac{\beta}{\Lambda} \cdot n(t) - \lambda_1 C_1(t)$$

where $n(t)$ is the number of neutrons, $C_1(t)$ is the concentration of delayed neutron precursors, $\rho$ is the reactivity, $\beta$ is the fraction of delayed neutrons, $\Lambda$ is the mean lifetime of neutrons and $\lambda_1$ is the decay constant of the delayed neutron precursors.

The equations above characterise a critical core when ρ is equal to zero.

The following equations characterise a subcritical core supplied with an external neutron source as in a traditional ADS:

$$\frac{dn}{dt} = \frac{[\rho - \Delta\rho] - \beta}{\Lambda} \cdot n(t) + \lambda_1 C_1(t) + S \tag{2}$$

$$\frac{dC_1}{dt} = \frac{\beta}{\Lambda} \cdot n(t) - \lambda_1 C_1(t)$$

When the control rods are positioned such that ρ is equal to zero, Δρ represents the level of subcriticality of the core. The external neutron source $S$ makes it possible to maintain a stationary flux in the core.

One can write for the source:

$$S = \frac{Z\varphi^*}{Q} \cdot I \tag{3}$$

where $I$ is the current of protons of the beam, $Q$ is the load of the proton ($1.6 \cdot 10^{-19}$ C), $Z$ is the number of neutrons produced by proton and $\varphi^*$ is the importance of the source.

Now, to obtain the reactor with compensation of beta (RCB), a specific constraint on the kinetic behaviour of the source must be introduced:

$$\frac{dS}{dt} = \lambda_2 \frac{\Delta\rho}{\Lambda} \cdot n(t) - \lambda_2 S(t) \tag{4}$$

where $\lambda_2$ is a constant data. In stationary working, Eq. (2) is not modified and $S$ is equal to:

$$\frac{\Delta\rho}{\Lambda} \cdot n$$

However, if power changes for any reason, $S(t)$ follows $n(t)$ with a delay just like the concentration in precursors of delayed neutrons.

Changing Δρ into Δβ in Eq. (2) and $S(t)$ into $\lambda_2 C_2(t)$ in Eq. (4) gives the following system:

$$\frac{dn}{dt} = \frac{\rho - (\beta + \Delta\beta)}{\Lambda} \cdot n(t) + \lambda_1 C_1(t) + \lambda_2 C_2(t) \tag{5}$$

$$\frac{dC_1}{dt} = \frac{\beta}{\Lambda} \cdot n(t) - \lambda_1 C_1(t)$$

$$\frac{dC_2}{dt} = \frac{\Delta\beta}{\Lambda} \cdot n(t) - \lambda_2 C_2(t)$$

This system describes the behaviour of a reactor becoming critical when ρ is equal to zero with an additional group of delayed neutrons and consequently increased beta.

It is simply enough to evaluate $n(t)$ thanks to a measurement of the neutron flux and to deduce in real time $S(t)$ according to Eq. (4). Then the accelerator is driven so that the current of the beam is equal to:

$$I(t) = \frac{Q}{Z\varphi^*} \cdot S(t) = \frac{Q}{Z\varphi^*} \cdot \lambda_2 C_2(t) \tag{6}$$

This manner of binding the external neutron source to the neutron flux of the moment thus makes it possible to send delayed neutrons into the core. Although the core is slightly subcritical and has weak beta, the whole of the system behaves like a critical reactor having sufficient beta to guarantee a good controllability.

Indeed, it should be recalled that the reactors are controllable thanks to the presence of the delayed neutrons which lengthen their stable period:

- Without delayed neutrons the period would be very short: $T = \dfrac{\Lambda}{\rho}$.

- With delayed neutrons the period is: $T = \dfrac{(1-\beta) \cdot \Lambda + \sum\limits_i \dfrac{\beta_i}{\lambda_i}}{\rho} \neq \dfrac{\sum\limits_i \dfrac{\beta_i}{\lambda_i}}{\rho}$.

As an example, let us take the case of a core containing minor actinides and provided, consequently, an intrinsic beta of 100 pcm.* To reach good stability, it is necessary to add about 250 pcm to obtain a usual beta total of 350 pcm.

The decay constant of the group of delayed neutrons coming from the spallation source ($\lambda_2$) will be taken at about 0.08 s$^{-1}$. This value is representative of the various values of natural groups. The level of subcriticality of the core ($\Delta\rho$) under operation, is selected to be equal to the deficit of beta to compensate, that is to say 250 pcm in this example.

This two values $\lambda_2$ and $\Delta\rho$ are introduced into the computer solving Eq. (4) to calculate $S(t)$. Knowing the characteristics of the accelerator and the spallation target, it is easy to control the proton beam so as to obtain the desired source of delayed neutrons $S(t)$.

## Discussion

### Type of reactor proposed

This system is equivalent to a critical reactor with compensated beta. The weakness of intrinsic beta, due to a high proportion of minor actinides in the fuel, is compensated by the fictitious addition of a group of additional delayed neutrons. This operation is carried out while using, on the one hand, a core whose weak subcriticality simulates the fraction of delayed neutrons to add and, on the other hand, an external neutron source which evolves according to the power, in the manner of the concentration of a group of precursors of delayed neutrons.

---

* *Nota bene*: 1 pcm = $10^5 \times \dfrac{\Delta k}{k}$.

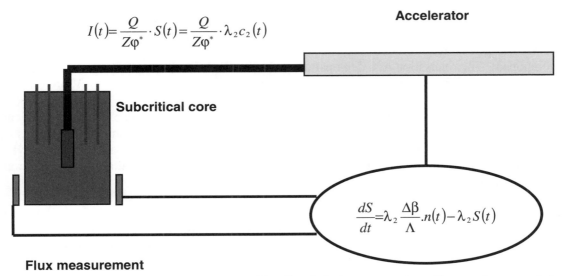

$$I(t) = \frac{Q}{Z\varphi^*} \cdot S(t) = \frac{Q}{Z\varphi^*} \cdot \lambda_2 c_2(t)$$

**Accelerator**

**Subcritical core**

$$\frac{dS}{dt} = \lambda_2 \frac{\Delta\beta}{\Lambda} \cdot n(t) - \lambda_2 S(t)$$

**Flux measurement**

**Feedback between neutron flux and source intensity**

Although conceived like an ADS, this type of reactor – including a subcritical core, an external neutron source consisting of a spallation target and a proton accelerator – behaves like a traditional critical reactor. The feedback established between the source and the neutron power by means of a fictitious tank of precursors of delayed neutrons ensures the stability of the ADS to be minimally subcritical and "transforms it" into a critical reactor.

This approach is necessary because, just as the operation of a traditional critical reactor is not possible with too-weak beta, the operation of a traditional hybrid reactor with a very low level of subcriticality is also not possible because of the risk of a drift towards the critical state. Indeed, a neutron amplifier is unstable by nature (ineffective feedbacks), its viability conditioned by the existence of a margin of subcriticality such that it guarantees in all circumstances the maintenance of a subcritical state. If the system became critical, the source would cause a linear increase in power with a slope: $S \cdot \dfrac{\Lambda}{\Lambda + \sum\limits_i \dfrac{\beta_i}{\lambda_i}}$.

*Advantages*

The RCB allows:

- To rediscover the effectiveness and the interest of feedbacks specific to the critical reactors.

- To rediscover a better, flattened distribution of flux which can possibly be adjusted with the control rods and with different zones of enrichments. The problem of sharp flux, characteristic of the amplifiers of energy with low multiplicative coefficient, disappears.

- To decrease the requirements of the accelerator. The maximum power of the source is reduced by a factor of 20 to 30 compared to a conventional ADS. For example, with a level of subcriticality of 300 pcm, an RCB of 3 000 MW would require a beam intensity of only 6.5 mA with protons of 1 GeV. This increase of the source power could make it possible to reduce the relative size of the target or to consider, with more chance of success, the realisation of a solid target which would eliminate the problems involved in the use of lead-bismuth. Lastly, although a significant challenge remains concerning the accelerator, the reduction by a factor of 20 to 30 of the power requirements seriously diminishes the constraints which pressed on this component. The realisation of a demonstrator of 100 MW becomes, consequently, possible in the short term, with a beam current requirement limited to 0.36 mA with 600 MeV protons.

- To reach a good level of safety. The safety of an RCB is improved compared to the critical reactor equivalent by the fact that in addition to the means of traditional shutdown systems is added the possibility of erasing, by turning the beam, a significant part of the delayed neutrons. This possibility, to stop more quickly and with a reinforced reliability the neutron power, can be advantageous for certain types of incidents.

- To maintain the behaviour of a usual reactor. The piloting and the behaviour of a RCB will be identical to those of a critical reactor. The criticality approach, the power increase, the compensation of the burn-up, the control of the radial profile of flux and possibly the follow-up of load will be carried out by absorbing rods.

- To discover convergences with gas-cooled fast reactors. This type of reactor is extremely close to a gas FBR with an annular core. It is even possible to consider during the life phases of operations in traditional critical mode (without spallation source) and phases of operations in incinerator mode (with spallation source and minor actinide loading). This reasoning obviously abounds in the direction of the realisation of a common demonstrator FBR and incinerator which could be the subject of a start-up by stage.

### Problems encountered

Compared to a traditional reactor for which beta results from an intrinsic fission process, reactor RBC has a beta which results partly from a computing process external of which reliability is not perfect. There is thus a risk that the relation desired between the intensity of the source and the power of the reactor is disturbed or disappears, with for consequence the possibility that the system will behave as an ADS. However, the operation of an ADS with a $k_{eff}$ of about 0.997 presents potential dangers and will have thus to be detected to lead to an emergency stop. It will also be necessary to be vigilant with regard to the transients of sources relating to the insertion of the maximum intensity of the accelerator during operations for intermediate power.

Concerning the first point, the reliability of the chain including the measurement of the power, the calculation of the variable of the accelerator and its command can be guaranteed on a desirable level by the traditional processes (redundancies of measurements and the hardware, operation at 2/3). The same applies to the control of the correct operation of this chain and the release of an emergency stop.

Concerning the second point, one can imagine a limitation of the current of the beam according to the consigned power and a threshold of emergency stop of the type of intensity of the beam/primary flow. A component of passive safety can be introduced by supplying a small specific alternator which would be dedicated to the accelerator. This would be dimensioned so that for a given level of operation, the power available would not allow a significant and fast increase in the intensity of the beam. In the event of loss of coolant of the primary circuit, the accelerator and thus all the installation would stop without other intervention. It should be noted that the stop of the accelerator beam must start an emergency stop though a fall of the control rods to guarantee the maintenance of the cold shutdown. Indeed, the cooling of the core introduces a reactivity higher than the initial margin of subcriticality and could lead, without this precaution, to a critical state.

# Conclusion

The transmutation of long-lived wastes could advantageously be carried out in dedicated reactors if the decrease in value of beta pulled by the presence of minor actinides could be compensated for. The use of the hybrid systems would constitute an interesting alternative if one could function with a low level of subcriticality in order to minimise the power of the accelerator and the target. This point is crucial today because it conditions the importance of the over-cost compared to a traditional reactor and thus the viability of the ADS.

It has been shown that it is possible to reconcile these two approaches by pointing out that an ADS functioning on a low level of subcriticality $\Delta\rho$ and having a source controlled according to a law describing the concentration of an additional group of precursors of delayed neutrons was equivalent to a critical reactor having beta increased by $\Delta\rho$. The resulting system, that it has been proposed to name "reactor with compensated beta" (RCB), presents the behaviour of a critical reactor and has the related advantages. It also has potentialities specific to the ADS such as the possibility of eliminating, by cut of the beam, a fraction of the precursors of delayed neutrons at an emergency stop or the possibility of inserting passive safety by supplying the accelerator with a dedicated alternator. In addition, the reduction in the requirements with regard to the level of the neutron source makes an experiment like MEGAPIE directly representative of a target for a system of 1 000 MW.

In return, the simulation of the existence of an additional group of delayed neutrons needs operations which will require monitoring and control to be included in the safety system. However, this aspect is not different from what already exists in many forms for regular reactors.

The system suggested also seems to be a variation of the concept of a gas-cooled reactor with an annular core. It militates in the direction of a common realisation of a demonstrator FBR gas RCB.

# FAST SPECTRUM REACTORS

**Chairs: P. Alekseev, H. Sekimoto**

# A SIMPLIFIED LMFBR CONCEPT (SFR)

**D.V. Sherwood, T.A. Lennox**
NNC Ltd.
Booths Hall, Chelford Road, Knutsford Cheshire, WA13 8QZ, UK
Tel: +44 (0)1565 633800, Fax: +44 (0)1565 843837, E-mail: david.sherwood@nnc.co.uk

## Abstract

NNC Ltd. is developing concepts that could provide the basis for the next generation of advanced fast reactors. One such concept is the simplified fast reactor (SFR), which incorporates a number of specific means of simplifying an LMFBR. A reduced-temperature steam cycle has been adopted (relative to EFR) to simplify the steam generation system and to reduce the reactor operating temperatures below the creep range of the structural materials. The fuel-handling system has been based on the use of a single removable roof plug to which the upper internal structure (UIS) is attached. A nitrogen-filled fuel-handling vault is provided over the roof area in which all the fuel-handling operations are carried out remotely. The secondary sodium circuit is eliminated by the use of Cu-bonded steam generators (SGs) in which the heat transfer occurs via three solid boundaries which separate the primary coolant from the water/steam.

## Introduction

Following the cessation of the EFR project, NNC Ltd. is assessing concepts that could provide the basis for the next generation of advanced fast reactors. Wherever possible these concepts utilise not only the existing knowledge gained through the operation of DFR and PFR and the extensive R&D programme for commercial-size fast reactors, but also the knowledge gained through the operation of AGRs and PWRs.

The SFR concept reflects the views of NNC's former Managing Director, Derek Taylor, who believed that existing designs were unnecessarily complicated as a consequence of unrealistic design targets and that considerable simplification was essential. Therefore, although the concept is largely based on NNC's extensive experience in developing commercial LMFBRs (CDFR and EFR), many of the original design bases have been re-addressed to simplify the concept.

In particular the SFR concept is characterised by:

- simplification (of arrangement, parameters, materials, design, fabrication, construction and operation);

- utilisation of existing knowledge and proven technology wherever appropriate (FR & LWR);

- application of innovative ideas where simplification and reduced cost are apparent;

- improved inspectability, maintainability and repairability;

- safety comparable to (or better than) existing LMFBR designs;

- reduced costs;

- high reliability;

- flexible size.

## Simplification aims

A number of specific means of simplifying an LMFBR have been brought together in SFR. A reduced-temperature steam cycle has been adopted (relative to EFR) to simplify the SG design and to reduce the reactor operating temperatures to below the creep range of the structural materials. Wherever possible, the reliability of welds has also been raised by reducing the number of components immersed in the sodium and by developing flow paths and structures of simpler shape. Gas/sodium interfaces are also avoided close to the welds.

Other aims include simplification of the fuel-handling system, the avoidance of leaving the fuel-handling equipment in the primary circuit during operation and the minimisation of the need for leak jackets on sodium pipework and vessels. However, the most significant simplification aim is the suppression of the secondary circuit.

## Basic SFR plant concept

SFR consists of a very compact, loop-type reactor as shown in Figure 1. The secondary sodium circuit has been eliminated and all the vessels are located in individual vaults containing an inert-nitrogen atmosphere. A large plug in the reactor roof is removable for simplified fuel handling and inspection that is carried out remotely in a dedicated vault above the reactor roof. Separate SG buildings are not required, resulting in a very compact plant layout.

### Figure 1. SFR heat transport system

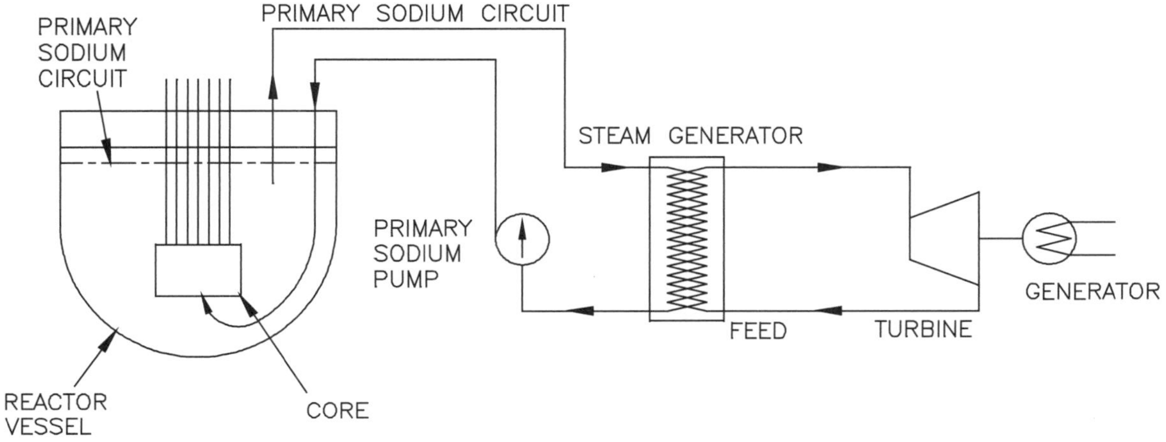

The basic SFR plant parameters are listed in Table 1, where they are compared with EFR.

### Table 1. Comparison of SFR and EFR reactor parameters

| Parameter | SFR | EFR |
|---|---|---|
| Thermal power | 3 600 MW (but could be modularised) | 3 600 MW |
| Electrical power (net) | 1 350 MW (approx.) | 1 470 MW |
| Core inlet temperature | 310°C | 395°C |
| Core outlet temperature | 495°C | 545°C |
| SG inlet temperature | 495°C | 525°C |
| SG outlet temperature | 310°C | 340°C |
| Steam temperature | 460°C | 490°C |
| Feedwater temperature | 210°C | 240°C |
| Diameter of primary vessel | 12.0 m | 17.2 m |
| Number of secondary loops | 6 | 6 |

## *SFR site layout*

The elimination of the secondary sodium circuits and ancillaries and the housing of the SGs within the secondary containment building (i.e. the elimination of separate SG buildings), all result in a very compact site layout for SFR, as shown in Figures 2-4. The building costs are hence reduced.

## Figure 2. Plan view of SFR secondary containment building

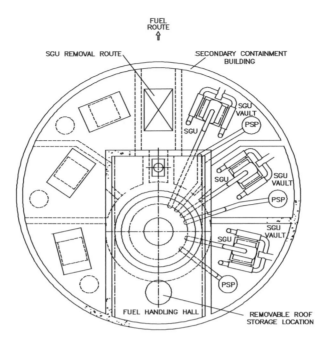

## Figure 3. Elevation on steam generator vault

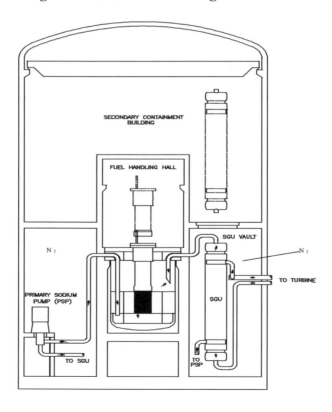

**Figure 4. Elevation of fuel handling hall**

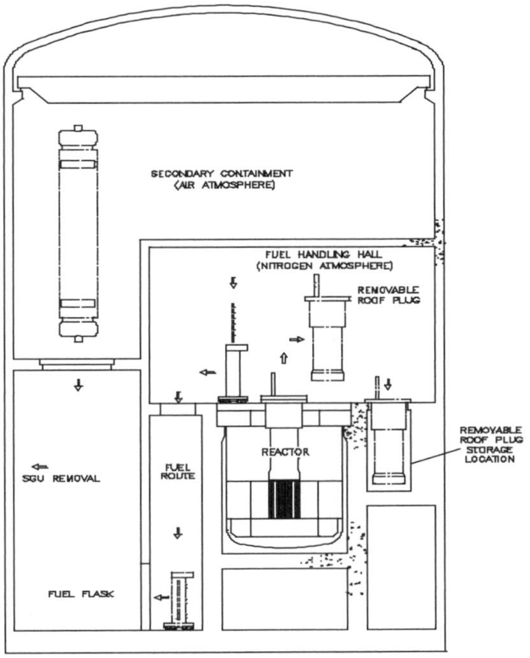

## Core and fuel

Currently the fuel is based on the classic LMFBR design, but with a reduced core outlet temperature and a smaller gas plenum. However, the elimination of the wrapper is being investigated (possible through the adoption of lower operating temperatures) to simplify fuel fabrication and reprocessing and hence reduce fuel cycle costs. Bottom-supported absorber rods are also being investigated to simplify the UIS, as is a suspended core support with the aim of simplifying inspection and the improvement of welded joint integrity.

## Primary vessel

The primary vessel is designed to have a removable roof plug to which the UIS, control rod drives, etc., are attached. This allows excellent access for refuelling and inspection of the reactor internals. The removable roof plug also eliminates the need for rotating shields during refuelling, so that the roof area becomes very uncluttered. The vessel is surrounded by a nitrogen atmosphere and located in a lined concrete vault. The potential for the elimination of the leak jacket is being investigated. The vessel internal layout is arranged to minimise diameter, avoid thermal striping, improve inspectability and reduce costs.

## Cooling system

SFR operates at temperatures lower than those conventionally adopted in a FR to improve structural integrity by keeping below the creep range of the materials. This also allows a more cost-effective SG system to be adopted and perhaps a longer plant life. The SG system used incorporates an innovative Cu-bonded concept that eliminates the need for a secondary sodium circuit and its ancillaries.

Each SG is located in its own vault, surrounded by a nitrogen atmosphere (Figure 2). This allows most leak jackets to be eliminated. In all cases the use of under-sodium welds is minimised.

### Copper-bonded steam generator

In an LMFBR, a secondary sodium coolant circuit and an intermediate heat exchanger (IHX) are normally used to ensure adequate safety margins in the event of a sodium/water reaction resulting from a leak in the SG. An alternative method of ensuring that a sodium/water reaction in a SG does not threaten the safety and integrity of the reactor is to provide multiple solid boundaries (through which the heat can pass) between the primary coolant and the steam/water. The failure of one or more of these boundaries would avoid a reaction provided that at least one boundary remained intact. Major repairs are therefore unlikely to be required, increasing the availability of the plant.

The principle of the Cu-bonded SG is shown in Figure 5. It consists of two separate sets of tubes closely integrated to form a bundle. One of the tube sets contains water/steam and the other the sodium coolant. The two sets of tubes are separated by a solid Cu layer that mechanically bonds to both sets. Therefore three boundaries, comprising the two tube walls and the Cu matrix, separate the primary sodium from the water/steam.

#### Figure 5. Arrangement of rectangular tubes in a copper-bonded SG

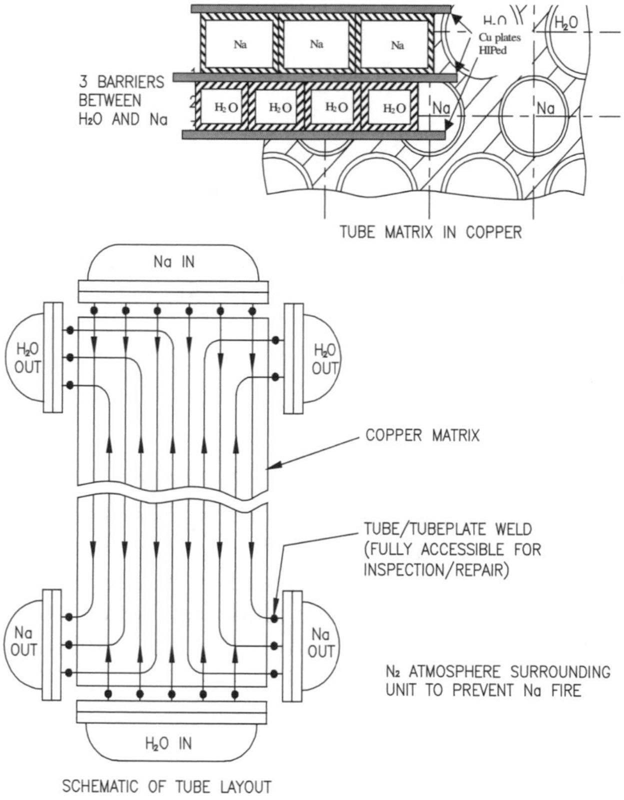

The tubes can be arranged in a variety of ways within the Cu matrix, e.g. straight tubes, serpentine tubes or coiled tubes, in order to optimise the efficiency of the heat exchange and the overall dimensions and cost of the SG. However, in order to prevent leakage along the tube/Cu interface should a tube fail and to maximise heat transfer, it is preferable that the tubes and Cu matrix be mechanically bonded. This procedure can be achieved through the use of HIPing. It is the development of this manufacturing process that has made an economic design of a Cu-bonded SG now possible.

The advantages of the Cu-bonded SG can be summarised as:

- removal of the secondary sodium circuit and ancillaries;

- the provision of three separate boundaries between the sodium and steam;

- economic manufacture is possible using the HIP process;

- internal inspection of all tubes is possible by removal of end covers;

- decay heat removal can be incorporated into the tube matrix if required;

- location in a nitrogen atmosphere eliminates the sodium fire risk;

- reduced cold trap duty;

- high reliability.

Disadvantages are few but include:

- large SG mass (although this is greatly reduced if rectangular tubes are used and is offset by the elimination of the secondary circuit components);

- tube replacement being difficult but very unlikely to be necessary.

This Cu-bonded SG concept can also be applied to conventional LMFBRs.


### Fuel handling

In an LMFBR the control rod drives, thermocouples, etc. are usually incorporated into the UIS, which also serves to distribute the cooling flow to the reactor vessel outlet ducts. However, the UIS covers a large part of the reactor core, including all the fuel subassemblies and therefore must be moved out of the way to allow refuelling to commence. Most LMFBR refuelling schemes incorporate single or twin rotating plugs. These add to the initial reactor cost and their constant repositioning during refuelling makes the process slow. In addition the presence of the UIS hampers in-service inspection and repair and is itself difficult to repair.

The SFR refuelling scheme involves the provision of a single, removable plug on which the UIS is mounted. During refuelling the plug and the UIS are lifted clear off the roof and placed in a parked position. Once the core is exposed a railed-mounted manipulator-type charge machine is brought into operation via the opening in the roof. A nitrogen-filled fuel-handling vault is provided over the roof area in which all the fuel handling operations are carried out remotely. Once removed from the reactor, the irradiated fuel is placed in a sodium-filled container which is lowered into the irradiated fuel store via a valve system. New fuel is placed in the fuel-handling vault prior to the refuelling campaign, via a similar valve system.

## Economic perspective

Although SFR is based largely on well-developed technology, it incorporates a number of innovative features which, following a period of detailed design and R&D, offer the potential for significant reduction in capital cost. Specific savings that can be readily identified include:

- a significant reduction in the building size;

- the elimination of secondary sodium circuit and associated ancillaries;

- the elimination of separate SG buildings;

- the elimination of rotating plugs;

- reduced inspection and maintenance costs;

- high plant reliability and availability.

Reducing the complexity and size of LMFBRs is an essential step in the process of achieving a competitive commercial FBR. The SFR concept is aimed specifically at achieving this goal.

## SFR status

The current status of the SFR design is that the simplification aims and preliminary design concept have been identified, allowing the principle components to be outlined.

The core and fuel are currently based on existing LMFBR technology although alternative options are also being considered.

A Cu-bonded SG utilising circular tubes has already been studied in some detail. Further development of the design is currently being undertaken based on the use of rectangular tubes in order to greatly reduce the mass of material contained in the matrix and hence the costs.

## Conclusions

The SFR concept embodies the many benefits of existing knowledge but is further advanced by incorporation of the philosophy of simplification and the addition of some innovative ideas. This philosophy reduces the complexity of the arrangement, the severity of the operating parameters and simplifies the choice of the materials, the design, the fabrication, the construction and the operation. In particular:

- The secondary sodium circuit and associated ancillaries are eliminated.

- The structural integrity and reliability of the systems and components are improved.

- The inspectability/maintainability/repairability are improved.

- The costs are reduced and the reliability greatly improved.

# THE DESIGN OF THE ENHANCED GAS-COOLED REACTOR (EGCR)

**H.M. Beaumont, A. Cheyne, J. Gilroy, G. Hulme, T.A. Lennox, R.E. Sunderland**
NNC Ltd.
Booths Hall, Chelford Road, Knutsford Cheshire, WA13 8QZ, UK
Tel: +44 (0) 1565 633800, Fax: +44 (0) 1565 843837, E-mail: richard.sunderland@nnc.co.uk

**D.P. Every**
BNFL plc
Springfield Works, Preston, Lancashire, PR4 0XJ, UK
Tel: +44 (0) 1772 762482, Fax: +44 (0) 1772 762470, E-mail: denis.p.every@bnfl.com

## Abstract

A preliminary concept for the fast spectrum enhanced gas-cooled reactor (EGCR) has been studied as a possible Generation IV reactor, based upon the developed technologies of the liquid-metal fast breeder reactor (LMFBR) and the UK's advanced gas-cooled reactor (AGR). EGCR is a $CO_2$-cooled fast reactor and comprises a single reactor with a thermal power of about 3 600 MW in a single pre-stressed concrete pressure vessel. Enhanced passive safety features are provided and a debris tray included.

The conceptual design and its principal features are described in this paper. The engineering studies to arrive at the conceptual design are summarised and the differences from the AGR reactor are highlighted. Considerable simplification is possible in comparison to the LMFBR and the AGR. The design is supported by the extensive and successful operation of the AGR's which now regularly achieve very high availability.

## Introduction

The long-term interest in fast reactors (FRs) continues as always to be for the most efficient utilisation of fuel reserves. The development of FRs using liquid sodium as the coolant has established the system's viability. However the need for such fuel utilisation is not pressing, and without an economic advantage over alternative established nuclear systems and some way to go to demonstrate high reliability of the current LMFR designs, there is no incentive to proceed to build.

Today, the choice of coolant for the fast reactor is being reconsidered in light of changes to the requirements and priorities that led to the selection of sodium as the preferred fast reactor coolant in the early days of fast reactor development. The scarcity of highly-enriched uranium and plutonium, leading to the adoption of very high power density cores to minimise fissile inventory, and high breeding gain to allow the rapid introduction of fast reactors into the nuclear park, is no longer an issue. On the other hand, today's requirements place a strong emphasis on the improvement of plant inspectability and maintainability, the elimination of chemical hazards associated with the coolant and the reduction of one of the major safety concerns, the coolant void effect. Further, there is a current interest in exploring particular advantages of the fast reactor for its Pu-burning flexibility, to manage the growing Pu stockpile and to irradiate minor actinides and fission products to reduce the long-term toxicity of nuclear wastes at the back end of the fuel cycle. For a commercial fast reactor, it is considered that the most that may be needed is a sufficient breeding capability, using the Pu coming from thermal reactors to launch the FRs, and modest breeding/burning flexibility to manage Pu stocks.

A consequence of this new environment is that fast reactor concepts are being reviewed world-wide, with both critical and subcritical reactors considered, and with gas and lead alloy coolants, as well as sodium, included in the investigations. Due to the ability of the gas-cooled fast reactor to respond positively to all of the changed fast reactor requirements described above, and also to its potential for further innovative development and for commercial competitiveness, NNC and BNFL are pursuing the development of EGCR as a Generation IV reactor concept in the UK. Fuji Electric is also co-operating in Japan. EGCR is a fast reactor, cooled by carbon dioxide, which uses the developed technology from the UK advanced gas-cooled reactor (AGR) and the LMFBR core and fuel technology.

With regard to the management of nuclear materials, alternative systems are increasingly being considered as part of the CEA-led CAPRA (Pu management)/CADRA (waste incineration) projects. In this context NNC and BNFL have also been studying gas-cooled fast reactors as dedicated incinerators of minor actinides and long-lived fission products. These aspects are considered elsewhere [1].

## The EGCR conceptual design

The basic ETGBR concept proposed in the 1970s [2] had a thermal power of 1 680 MW, but alternatives were proposed as part of a future development programme with thermal powers up to 3 320 MW. These future concepts relied on significantly higher primary circuit gas pressure to achieve the higher ratings. The EGCR 1500 concept [3] achieves a core thermal power of 3 600 MW whilst retaining the primary circuit gas pressure of the existing AGRs (42 bar). The initial choices for EGCR were based on the aim of meeting the EFR performance and economic parameters. A number of preliminary studies were performed to assess important aspects of the concept including core physics, in-service inspection and repair, safety and economics.

Thermal-hydraulic studies of the EGCR subassembly (SA), core and primary circuit have shown that, in order to meet the AGR pressure drop constraints and the fuel pin performance constraints in a gas-coolant environment, the concept required optimisation. In particular an increased fuel pin pitch was required compared with liquid-metal cooling.

The overall reactor arrangement is shown in Figure 1, a twin reactor station in Figure 2. A reactor net output in the region of 1 400+ MW(e) is envisaged.

## Figure 1. EGCR concept

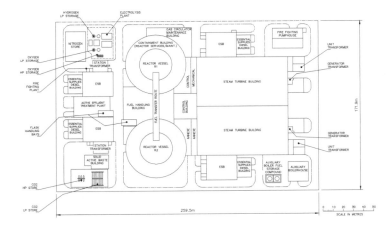

## Figure 2. EGCR twin station layout

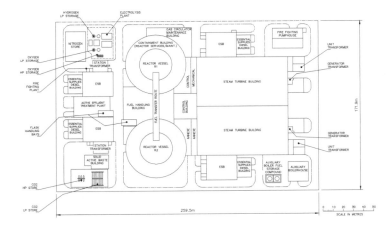

The reactor design is based on Heysham 2/Torness AGRs and utilises a single cavity pre-stressed concrete pressure vessel (PCPV) to house the core structure, twelve steam generators and eight gas ($CO_2$) circulators. The proposed steam/water circuit employs steam to steam reheat, in preference to the gas/steam reheat used on AGR. This enables the EGCR boilers to have additional heat transfer capacity by extending their tube/platen length, utilising the equivalent space which was occupied by the reheater units on AGR. Fuel and radial breeder assemblies use the latest technology developed for LMFBRs. A central removable handling machine is included to transfer all core components. All control rods will be actuated from below the core. A circular containment building is provided.

For a twin reactor layout maintenance and service facilities can be shared. This would be similar to the twin AGR arrangement as shown in Figure 2, but would have an output of ~2 800 MW(e), which shows that a very compact site layout is feasible. Reactor services essential for reactor operation will be located within the individual containment building of each reactor. Major portions of the refuelling system can, however, be common, particularly the external fuel store. A common service facility for circulator maintenance is envisaged similar to the AGR arrangements.

Segregation and separation both within and external to the containment buildings will follow AGR practices which involve dividing each containment building into quadrants and providing separation and fire protection between each quadrant. External to the building the approach to separation is maintained by positioning the four individual essential services and essential supplies buildings remote from one another but still in close proximity to the containment buildings. For EGCR all buildings up to and including the essential services/supplies will be located on the seismic raft.

Compared with AGR, the EGCR is much simplified. The AGR flow baffle dome is eliminated. The AGR graphite moderator is absent (reduced core size, reduced requirements for coolant conditioning) and the large number of penetrations in the AGR roof is eliminated. The AGR nitrogen injection system (for emergency shutdown) is also eliminated.

### Fuel and core design

The preliminary core consists of two enrichment zones with a thermal output of 3 600 MW(th) from 550 subassemblies (Figure 3).

The inner region of the core contains 334 subassemblies and the outer region 216. A standard LMFBR, PE16 clad, fuel pin is proposed with a 1 200 mm column of MOX fuel. This arrangement provides the potential for a high burn-up (>20% ha peak). However to achieve this burn-up the clad must resist a damage dose of the order of 230 dpa-NRT(Fe). The pins are contained within a hexagonal wrapper. Reactivity control is based on 24 control and shutdown and nine diverse shutdown absorber rods. A five-batch refuelling system with an interval of 24 months between shutdowns is proposed for this core. High breeding cores, with breeding gain in excess of 0.2, can be achieved with the addition of axial breeder and up to two rows of radial breeder.

Control rod absorbers will be actuated via mechanisms located within the vessel bottom slab (Figure 1). Mechanism handling for maintenance purposes will be undertaken from within the room below the bottom slab. Control rod absorbers will be handled via the pantograph charge machine following de-latching from their mechanisms. The 24 control rods operate via screw-drive, electromagnet mechanisms below the core, which provide a downward motion from above the core. The nine shutdown rods operate via an electromagnet and incorporate a passive shutdown feature.

### Pressure vessel

The reactor vessel (Figure 1) is a vertical concrete cylinder (as for the AGR) with helical multi-layer pre-stressing tendons in the walls arranged so that no tendons are required across the top and bottom slabs. Circumferential galleries at the top and bottom of the vessel provide access for tendon re-tensioning

## Figure 3. Fuel subassembly and core layout of EGCR

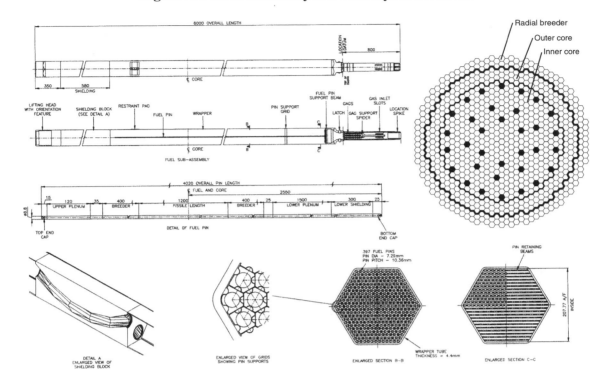

and inspection. The vessel inner surface is lined with steel to provide a leak-tight membrane. The liner is also insulated and cooled (as for the AGR) to maintain concrete temperatures within limits. Ties are provided between the liner and the concrete to prevent inward deflection of the liner.

The heat transfer system comprises four parallel circuits; each circuit contains three boiler units and two gas circulators. Each circuit is located in a 90° sector of the reactor.

The $CO_2$ at a pressure of nominally 42 bar flows upwards through the subassemblies, where it is heated to about 525°C, and downwards through the boilers. The gas is driven round the reactor circuit by gas circulators which draw $CO_2$ from the boiler outlets and discharge to a plenum below the core support structure and diagrid at about 250°C.

The gas inlet temperature and steam outlet temperatures in the boiler are consistent with the EFR parameters and are chosen to suit the selected materials.

### Boilers and decay heat removal

The boilers are located in the annular inter-space between the PCPV and the cylindrical gas flow baffle which surrounds the core. The rectangular units have been rotated through 90° with respect to their orientation in the AGR in order to provide a better reactor layout and minimise the PCPV diameter. The units are supported on the underside by beams which are suspended from the inner cylinder and the vessel liner. Water feed and main steam feed are connected to the boiler sections through separate penetrations.

317

Any circuit of three boilers can be taken out of service while the reactor is on load without the hot gas being drawn through it. Gas flow through the shutdown boilers is limited by closure of the inlet vanes of the two stationary circulators within the circuit.

The boiler (steam generator unit – SGU) is a 300 MW(th) unit with a platen-style tube bundle based on the AGR, but with increased tube length, utilising the space equivalent to that occupied by the re-heaters in AGR. This SGU features a plain (un-finned) tube, manufactured in a single material 9% Cr, 1% Mo steel. This will improve their structural integrity which, together with the good inspectability, maintenance and reparability established for AGRs allows consideration of very long lifetimes (~60 years).

For decay heat removal (DHR), two independent and diverse types of decay heat removal system are proposed. The first system uses dedicated decay heat boilers and forced circulation with a limited number of gas circulators operating at low speed. These decay heat boilers are fed by an independent feed system – the decay heat boiler feed system (DHBFS). The DHBFS automatically starts following a reactor trip signal. The steam/water from the decay heat boilers is passed to the dump condenser, which is initially vented to atmosphere. Longer-term, the system is fully re-circulatory. The heat from the dump condenser is then transferred to atmosphere by forced draught cooling towers. The DHBFS system provides long-term DHR without external power. The second system is put into operation if either de-pressurisation of the reactors occurs or if the DHBFS fails. This system utilises the main boilers fed by the emergency boiler feed system (EBFS). With this system natural circulation through the core is effective when the reactor is pressurised.

### *Circulators*

The eight gas circulators are of the AGR centrifugal type driven by electric motors. Coolant flow is controlled by variable inlet guide vanes. Each circulator, complete with motor and vane control gear, is a totally enclosed unit located in a horizontal penetration at the bottom of the vessel wall. The motor runs under full coolant pressure. The whole assembly of motor impeller and guide vanes can be sealed off from the reactor circuit after de-pressurising the reactor. After the internal seal is made, the outer pressure closure may be removed and the circulator assembly replaced. A special area is provided for maintenance and subsequent pre-installation testing.

### *Fuel-handling route*

Refuelling will be undertaken with the reactor shutdown. An in-reactor fuel-handling machine will be located in a central penetration within the reactor roof, and will cover the complete core and the discharge position at the core boundary. The machine will be removed from the reactor during power operation, with a separate sealed roof plug providing the pressure boundary when the reactor is at power.

### Core design for optimum thermal-hydraulic performance

To enable the effect of core design on thermal-hydraulic performance to be examined, a numerical model of the core, referred to as the GATHER model, has been assembled. The GATHER model consists of a set of algebraic equations that relate the fundamental thermal-hydraulic characteristics to overall core design parameters. The principal parameters which govern the design of the core are the dimensions of the core, the pressure drop necessary to drive the required coolant flow through the core and the maximum temperature which can be allowed in the fuel cladding. The model is based on a simplification of the core geometry. It is assumed that the core is constructed from a collection of identical fuel pins in a triangular arrangement aligned with the direction of flow. The presence of subassembly wrappers and control sub-channels is neglected and the fuel is assumed to occupy the full cross-section of the pins.

Figures 4 and 5 show contour plots in the plane of mean pin linear rating and pitch-to-diameter ratio for the maximum can temperature and the core pressure drop based on the solutions obtained. Figure 4 shows that maximum pin surface temperature increases with both linear rating, because more powerful pins become hotter, and pitch-to-diameter ratio because more widely spaced pins lead to lower velocities and hence lower heat transfer coefficients. Figure 5 shows that core pressure drop also increases with linear rating. Higher velocities are required to cool the more powerful pins for the same core temperature rise. Core pressure drop falls as pitch-to-diameter ratio increases because the flow passages become less resistive.

**Figure 4. Maximum can temperature**

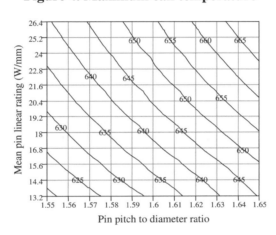

**Figure 5. Core pressure drop**

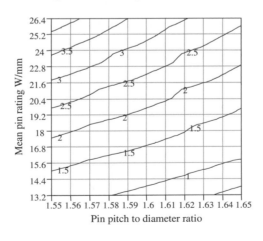

As a consequence of these studies the pressure drop in the core has been reduced while the maximum pin temperature has been maintained by means of a reduction in mean pin linear rating and an increase in the pin spacing. This optimised design satisfies the limits on pressure drop and pin temperature with a slightly larger core diameter than the original design. This occurs because there are more pins due to the lower mean linear rating and the pins are more widely spaced. The options proposed by the GATHER model are then examined using a more detailed model based on the code VIPRE01.

## Severe accident conditions

### Transient analysis

To investigate the performance of the EGCR primary circuit under accident conditions three severe accident transients, including a de-pressurisation transient, have been calculated in addition to the normal reactor trip. The transients have been assessed with the GASBET code, which is a modified version of the BESBET code originally developed at NNC as a whole plant model for LMFBR transients. A significant amount of remodelling of BESBET has been necessary partly to accommodate the change from a liquid to a gaseous coolant but mostly to allow for the elimination of the secondary circuit.

The new gas-cooled fast reactor primary circuit model consists of the core, the hot plenum, the boilers, the circulators and the diagrid. GASBET also includes a model of the steam/feedwater system. The new model will be extended to handle non-symmetric conditions at the next stage of its development.

A typical fault transient is that of de-pressurisation. A de-pressurisation of the primary vessel is assumed to occur through a breach area of $0.03$ m$^2$, which is the maximum PCPV penetration. When the circulator inlet pressure falls to 27 bar the reactor is tripped. The circulators are assumed to trip 10 s after the reactor and after a further ten seconds it is assumed that half the circulators become unavailable. The speed of the remaining circulators is assumed to be increased to compensate. As the pressure falls circulator speed is assumed to increase automatically according to the density/pressure relationship used for the AGR circulators. The flow rate stabilises at about 12% of full flow until the available circulators reach full speed at about 940 s. Thereafter the flow rate falls while the system pressure continues to fall.

The preliminary analysis is shown in Figure 6. The leak is assumed to start at 100 s. Before the reactor trip, which is initiated at 430 s (330 s after the leak started), the core outlet and plenum temperatures rise to a peak of 657°C due to the fall in reactor flow rate. The peak clad temperature rises to a maximum of 796°C in the period before the reactor trip. Following the trip it falls quickly to around 300°C and then rises again to 431°C before declining on a longer time scale. It would therefore seem possible to trip the reactor sooner either with a higher threshold on the pressure measurement or by tripping the reactor on high core outlet temperature. These transient studies will be used to further optimise the core safety and performance.

### Debris tray for hypothetical core disruptive accident

Unlike the LMFBR, EGCR does not have a potentially rapid, energetic, coolant phase-change driven, reactivity fault. Hence any potential core disruptive accident in EGCR has a very low probability. Nevertheless, since hypothetical core disruptive accidents are a topic of discussion for fast reactors, a debris tray installed below the core inside the PCPV has been studied. The debris tray (Figure 7) consists of a ceramic crucible, a heavy metal layer, a sacrificial material layer and an associated external cooling system on the surface of the PCPV liner. This cooling system is very similar in structure to the PCPV liner cooler. The debris tray is sized and arranged to manage the whole core at maximum decay power.

**Figure 6. Temperatures during the de-pressurisation transient**

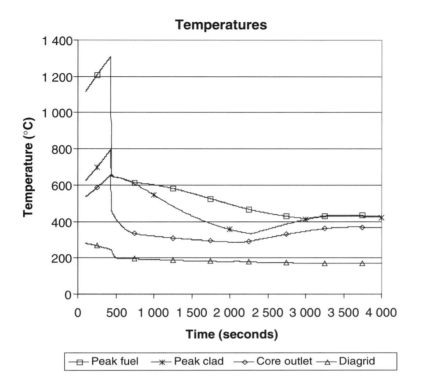

**Figure 7. Debris tray for EGCR**

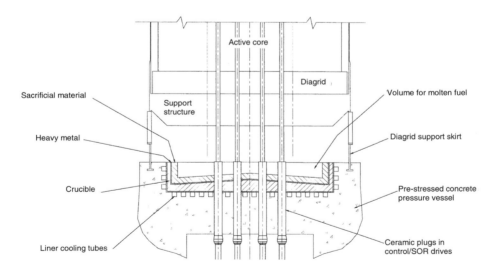

In a postulated severe accident, if the fuel melted, the molten fuel would relocate easily to the debris tray because of the wide gaps between the fuel pins.

First, the molten fuel is cooled to below its melting temperature by melting and vaporising the sacrificial material. After fuel solidification, cooling water in the dedicated cooling system passively cools the fuel long-term. The heavy metal seals the gaps in the ceramic crucible and the absorber rod sleeves. Although the geometry of the debris tray is "safe by shape" absorber material in the debris tray mixes with the molten fuel to assure no re-criticality. The PCPV integrity is maintained. Preliminary analysis confirms that the total core inventory of molten fuel on the debris tray can be adequately cooled both short and long-term.

## Cost reduction and development potential

Several cost reduction items have been studied, but particular attention has been given to a reduction in the size of the PCPV (the external dimensions of the PCPV for the AGR are 46 m height and 32 m diameter). The decay heat boilers in EGCR have been moved to the ceiling of the PCRV in order to reduce the vessel height, and the upper structures of the PCPV have been simplified. The boiler size and arrangement have also been optimised. The diameter of the core and the radial shield have been evaluated and the preliminary internal diameter of the PCPV settled at 17 m. Consequently the outer diameter of the PCPV has been reduced to 25 m. The reactor building size has also reduced. As EGCR has about twice the power output of the AGR within a much smaller PCPV the construction cost is much less. With further development, other potential cost reduction measures such as local containment and innovative boiler design have been identified.

The reactor pressure and core outlet temperature initially adopted for EGCR are based on the conditions experienced in AGR and LMFBR. However there is the prospect of using a higher temperature and pressure in the future to enhance the power output or reduce the PCRV size. Direct cycle can also be considered.

Following a preliminarily study of the construction schedule for AGR and EGCR it has been determined that EGCR can have a significantly shorter construction period than AGR because it has much simpler internal structures. These studies confirm that EGCR has the potential to achieve a cost similar to that of an LWR.

Whilst a large reactor concept has been presented here, the basic concept shows sufficient flexibility to be readily adaptable to other core sizes or modular concepts.

## Conclusion

The EGCR represents a commercial fast reactor utilising demonstrated AGR and LMFBR technology. The preliminary analysis confirms the capability of this concept to meet the appropriate economic and safety requirements and with further development has the potential for additional cost and safety improvements. The concept also offers the flexibility to manage plutonium stocks in the short term while providing a longer-term breeding capability.

*Acknowledgements*

The ECGR study was carried out in conjunction with Fuji Electric and JNC, whose support is gratefully acknowledged.

# REFERENCES

[1]  H.M. Beaumont, G. Hulme, J.T. Murgatroyd, R.E. Sunderland, E.K. Whyman, S.J. Crossley, "Flexibility of $CO_2$-cooled Fast Reactors for Plutonium and Minor-actinide Management", ARWIF 2001, Chester, October 2001.

[2]  W. Kemmish, "Gas-cooled Fast Reactors", *Nuclear Energy*, Vol. 21, Nr. 1 (1982).

[3]  M. Nakano, D. Sadahiro, H. Ozaki, S.D. Bryant, A. Cheyne, J.E. Gilroy, G. Hulme, T.A. Lennox, R.E. Sunderland, H.M. Beaumont, M. Kida, M. Nomura, "Conceptual Design Study for the Enhanced Gas-cooled Reactor (EGCR)", GLOBAL 2001, Paris, September 2001.

# THE FLEXIBILITY OF $CO_2$-COOLED FAST REACTORS FOR PLUTONIUM AND MINOR ACTINIDE MANAGEMENT

**H.M. Beaumont, R.E. Sunderland, T.A. Lennox, J.T. Murgatroyd, E.K. Whyman, G. Hulme**
NNC Ltd.
Booths Hall, Chelford Road, Knutsford Cheshire, WA13 8QZ, UK
Tel: +44 (0) 1565 633800, Fax: +44 (0) 1565 843837, E-mail: heather.beaumont@nnc.co.uk

**S.J. Crossley**
BNFL plc
Springfields Works, Nr. Preston, Lancashire, PR4 OXJ, UK
Tel: +44 (0) 1772 762216, Fax: +44 (0) 1772 762470, E-mail: steven.j.crossley@bnfl.com

## Abstract

In recent years there has been increased interest in gas-cooled fast reactor systems for the management of plutonium and minor actinides. In this context NNC and BNFL have been involved in the investigation of a number of different gas-cooled fast reactor systems ranging from a conventional, plutonium-burning design based on existing technology to a dedicated minor-actinide-burning system. The potential of a gas-cooled fast breeder reactor has also been demonstrated. The studies have considered core design and optimisation, core performance, safety parameters and preliminary transient studies. This paper reviews the main concepts and core designs considered to date.

## Introduction

In recent years there has been increased interest in gas-cooled fast reactor systems for the management of plutonium and minor actinides. Gas-cooled fast reactors (GCFR) have a number of advantages over liquid-metal-fuelled concepts. There are obvious safety, economic and technical advantages when using a relatively benign, readily available and optically transparent, gaseous coolant which is compatible with both air and water, compared to sodium which reacts explosively with water and requires special handling and disposal. The absence of a significant positive coolant void reactivity effect in gas-cooled cores compared to LMFRs allows the potential for loading greater quantities of minor-actinide isotopes. Loading large quantities of minor-actinide isotopes into sodium-cooled cores has been shown to lead to unacceptably large positive coolant void effects. An additional attractive feature of gas cooling over sodium cooling, in terms of minor-actinide incineration, is the harder neutron spectrum of gas-cooled cores. Gas-cooled fast reactors therefore offer considerable flexibility in core design.

Drawing on the extensive UK experience gained in the successful design and operation of the $CO_2$-cooled advanced gas reactors (AGRs) coupled with the experience gained from the design of gas-cooled and liquid-metal fast reactors, a number of different gas-cooled fast reactor cores which range from a conventional, plutonium-burning design based on existing technology to a dedicated minor-actinide-burning system, have been investigated. The potential of a gas-cooled fast breeder reactor has also been demonstrated. This paper reviews the main, representative concepts and core designs considered to date concentrating on the core physics and safety aspects particularly in terms of plutonium and minor-actinide management. In this context the advantages of gas cooling over sodium cooling are highlighted. Three basic designs for commercial sized reactors with a thermal power output of 3 600 MW are considered in some detail:

(i) a conventional core design, optimised for plutonium burning, which utilises conventional MOX fuel pellets;

(ii) a dedicated minor-actinide-burning concept, fuelled mainly with minor actinides and only sufficient plutonium to obtain criticality;

(iii) a core design optimised for plutonium breeding.

The studies have considered core design, core performance, safety parameters, and preliminary transient studies and have demonstrated the significant flexibility of gas-cooled fast reactors. The cores have not been specifically optimised for thermal-hydraulics and hence the performance characteristics should be regarded as preliminary.

## Plutonium-burning GCFR cores

In the 1970s and 1980s a gas-cooled fast reactor concept, ETGBR [1], based on the contemporary existing technology was investigated in some detail, combining the experience of the early advanced gas-cooled thermal reactors (AGR) and LMFR technology. This basic ETGBR concept had a thermal power of 1 680 MW but alternatives were proposed with thermal powers up to 3 320 MW. NNC has reviewed and updated the ETGBR concept [2,3] for a 3 600 MW(th) GCFR burner core based on more recent AGRs and the EFR LMFR core and subassembly technology. The studies supporting this concept are summarised below.

The GCFR reactor design concept is based on the Hinkley Point B and Heysham 2/Torness AGR2s which utilise a single cavity pre-stressed concrete pressure vessel to house the core structure, steam generators and gas circulators. A number of preliminary studies were performed assessing the important aspects of the concept including thermal-hydraulics, core physics, in-service inspection and repair, safety and economics. These are discussed in some detail in Ref. [2].

## Initial core physics assessment

An initial core physics assessment was carried out with the aim of demonstrating that an acceptable size of a gas-cooled fast reactor core could be designed such that a power output of 1 450 MWe can be achieved. This initial core was not specifically aimed at plutonium burning. A possible fuel pin and subassembly concept was established. An assessment of performance and enrichment levels for a core capable of achieving a 20% ha peak fuel burn-up and of absorber rod worths was carried out.

The core has two enrichment zones, with 334 SAs in the inner region and 216 in the outer region, resulting in a total of 550 fuelled SAs. Conventional, metal-sheathed, LMFBR pins fuelled with MOX are included in the core. The main core design parameters are shown in Table 1. The pins are clad in the high burn-up, high damage dose cladding material PE16 which was developed in the UK. An initial thermal-hydraulic assessment of this adopted SA/pin concept and core, along with a suitable primary circuit have shown that the adopted pin design would require some further optimisation to meet the AGR pressure drop constraints along with fuel pin performance constraints in a gas coolant environment. For this initial core it was found that a significantly greater plutonium enrichment in the outer core region was necessary compared to that in the inner core. A further, more detailed analysis was carried out with the aim of improving this ratio.

### Table 1. Main core design parameters for the initial Pu-burning GCFR core

| Parameter | Value | Parameter | Value |
|---|---|---|---|
| Reactor thermal output | 3 600 MW | Wrapper inside A/F | ~167 mm |
| Nominal electrical output | 1 400 MW | No. control rods | 24 |
| $CO_2$ gas pressure | 42 bar | No. diverse shutdown rods | 9 |
| Active core height | 1 500 mm | Pu enrichment inner region | 19.76% |
| Number of fuelled SAs | 550 | Pu enrichment outer region | 31.26% |
| No. inner-region SAs | 334 | Cycle length | 334 EFPD |
| No. outer-region SAs | 216 | Number of cycles | 5 |
| No. fissile pins per SA | 169 | Peak pin burn-up | 20% h.a |

## Detailed design

The detailed core design GCPu00 was established based on the pin and subassembly concept of the initial studies. To achieve improved plutonium burning rates dilution has been incorporated into the initial core concept while retaining as conventional a core as possible. Two methods of incorporating dilution into the core were considered, these being the use of un-fuelled pins within the fuel SAs and the incorporation of diluent SAs containing empty steel pins. Dilution in the form of diluent subassemblies was adopted so that the same design could be used for the fuelled subassemblies in both the inner and outer cores. The diluent subassemblies were modelled with 169 inert pins in the fuel region with the above and below core structure identical to those of the fuel subassemblies. Both solid steel pins and empty fuel pins were assessed and found to be equally effective, indicating that the dilution effect is primarily due to geometry, rather than to neutron absorption in the steel. This is not surprising, given the relatively hard neutron spectrum in a GCFR, and allows considerable flexibility in the design of the diluent subassemblies. However, it also imposes geometric limits on the size of the dilution effect that can be achieved with steel diluents alone.

The number and arrangement of diluent subassemblies in the core was optimised to achieve as flat a radial power shape as possible with an enrichment ratio as close as possible to unity. Both the number and the arrangement were chosen such that no diluent should be adjacent to another diluent or to either type of absorber rod. For each fuelled subassembly replaced by a diluent, an additional fuelled subassembly was added to the outside of the core to keep the total number of fuelled pins in the core approximately constant.

It was found that with steel diluents alone, it was not possible to achieve exactly equal enrichments with an acceptable power shape – even with almost all the available positions for diluents occupied, the dilution effect was not sufficient. Consequently the diluent arrangement and the enrichment ratio were optimised simultaneously to achieve a compromise between power shape and enrichment ratio. The boundary between the inner and outer cores was also varied and it was found that if the outer core was made wider, the enrichment ratio could be made closer to unity, whilst still balancing the inner and outer core peak powers. However, this was at the cost of reducing the power of the outer ring of subassemblies relative to the peak.

A compromise was reached for which the enrichment ratio was 0.9, the inner and outer core peak powers were essentially balanced and the minimum power in the outer ring was 0.6 times the peak. It may be possible to achieve an enrichment ratio of exactly one if absorbent diluents are used, but given the stated intention of focusing on conventional technology, it was decided to continue with steel diluents in this study. The optimised core layout is shown in Figure 1.

## Figure 1. GCPu00 core layout

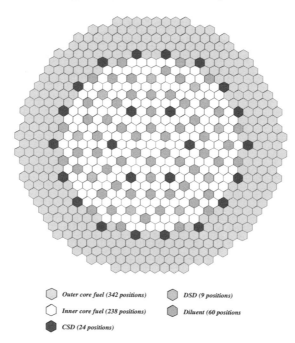

Outer core fuel (342 positions)   DSD (9 positions)

Inner core fuel (238 positions)   Diluent (60 positions)

CSD (24 positions)

A 120° core sector model which explicitly represents the batch refuelling was used to assess the core physics performance. Diffusion theory calculations were carried out in 33 neutron energy groups using the European Fast Reactor code scheme, ERANOS, along with the ERALIB1 cross-section libraries (which are based on JEF-2.2). An iterative optimisation procedure was carried out with the aim of achieving a 20% peak burn-up at end-of-life (EOL) and a calculational $k_{eff}$, after applying correction factors [4] to allow for modelling approximations, of close to unity.

The main core parameters and results from this study are summarised in Table 2. The results are compared to those of a Pu-burning adaptation of the EFR core design studied recently by NNC, the EFR CD 9/91 active core region [5] and the CAPRA 04/94 reference core [6] which was designed as a high-enrichment Pu-burner optimised for 20% h.a burn-up.

**Table 2. Comparison of the GCPu00 core design and performance with LMFRs**

| Core parameter | GCPu00 core results | EFR CD 9/91 (core region only) | Pu burning adaptation of EFR CD | CAPRA 04/94 Pu burning core |
|---|---|---|---|---|
| Reactor thermal output | 3 600 MW | 3 600 MW | 3 600 MW | 3 600 MW |
| Active core height/No. fuelled SAs | 1.5 m/580 | 1.0 m/387 | 1.0 m/387 | 1.0 m/366 |
| No. SAs (inner/middle/outer region) | 238/0/342 | 207/108/72 | 207/0/180 | 150/0/216 |
| Wrapper inside A/F | ~167 mm | ~174 mm | ~174 mm | ~168 mm |
| No. fissile pins per SA | 169 | 331 | 331 | 336 |
| Initial Pu inventory | 13.16 te | 8.8 te | 9.69 te | 9.233 te |
| No. control/diverse shutdown rods | 24/9 | 24/9 | 24/9 | 24/9 |
| Pu enrichment inner region | 24.89% | 18.33% | 20.8% | 43.0% |
| Pu enrichment intermediate region | n/a | 22.36% | n/a | n/a |
| Pu enrichment outer region | 27.62% | 26.87% | 26.1% | 44.7% |
| Cycle length/number of cycles | 338 EFPD/6 | 340 EFPD/5 | 340 EFPD/5 | 285 EFPD/3 |
| Peak pin burn-up | 18.3% ha | 20% ha | 20.2% | 20% ha |
| Peak clad damage (dpa NRT Fe) | 167 | 190 | 202 | 123 |
| Coolant void reactivity | 341 pcm | 2 100 pcm | 2 558 pcm | 1 564 pcm |
| Doppler constant | -598 pcm | -650 pcm | -730 pcm | -455 pcm |
| Prompt neutron lifetime | $7.5 \times 10^{-7}$ sec | $4.1 \times 10^{-7}$ sec | $4.7 \times 10^{-7}$ sec | $8.4 \times 10^{-7}$ sec |
| Total β-effective | 337 pcm | 362 pcm | 343 pcm | 324 pcm |
| Reactivity loss over a cycle | 2 911 pcm | 2 900 pcm | ~ 2 640 pcm | ~ 9 000 pcm |
| Pu/MA burning rate (kg/TWhe) | 31.81/-4.8 | 15.0/-3.0 | 21.28/-4.3 | 74/-9.7 |

The GCPu00 core is a larger core than both the CAPRA core and EFR active core region, having more fuel SAs, but fewer pins per SA and a higher active core height. The Pu inventory is over 30% greater in the GCPu00 core compared to both the CAPRA and EFR active cores. Thermal-hydraulics require an increased pin pitch in gaseous coolants compared to sodium as a result of $CO_2$ being a poorer heat transfer medium than sodium. Pin spacing is thus larger. In order to meet pin temperature constraints fuel ratings are generally lower hence enrichments are lower, and to achieve similar burn-ups to sodium-cooled cores a higher fuel inventory is required and hence a larger core. However there is considerable flexibility in the choice of core height and diameter.

The Pu enrichment in GCPu00 is designed to be much lower than the CAPRA core with the average enrichment being comparable to that of the EFR core. This lies well below the maximum enrichment of 45% Pu, which can be reprocessed using the PUREX process. The fuel residence time is

2 028 EFPD for the GCPu00 core, longer than the 1 700 EFPD of the EFR core and over twice that of the CAPRA core, for comparable fuel burn-ups. In fact the calculated burn-up of the GCPu00 core in this study is slightly lower than that of the EFR and CAPRA cores, thus indicating that the cycle length could be further increased to obtain a 20% ha burn-up. For the GCPu00 core the peak clad damage is below the accepted limit and the burn-up is thus not constrained by this parameter.

The GCPu00 Doppler constant is close to -600 pcm (-1.78$), comparable with EFR and 24% higher in magnitude than the CAPRA burner core. The coolant void reactivity is significantly lower in the gas-cooled core compared to the two sodium-cooled cores, it being of the order of 1$ compared to 4.8$ and 5.8$ for the CAPRA and EFR cores. This gives the gas-cooled core a significant safety advantage over liquid-metal-cooled cores.

The Pu burning rate is approximately 33% greater in the gas-cooled core compared to the Pu-burning adaptation of the EFR core, but around half that of the CAPRA core. There is the possibility of placing dedicated target SAs fuelled with minor actinides around the periphery of the core, providing added flexibility for this core as both a Pu- and MA-burner. This is currently being studied.

Preliminary transient analysis studies have confirmed that there are no characteristics for this core which would exclude it from further study and the transient behaviour may be acceptable. More extensive transient studies would be needed to confirm the transient behaviour.

## The GCMA-dedicated MA-burning FR

As an enveloping concept, ignoring issues relating to the availability of sufficient minor actinides (MAs) to fuel the reactor, a full-sized 3 600 MWth commercial reactor fuelled largely by MAs was chosen for study. The core plan for this study was taken from the preliminary Pu-burning core-scoping studies described previously. The core design parameters were also taken from that study as shown in Table 1. A nitride matrix, using 100% $^{15}$N for the fuel, was adopted as recommended by the CAPRA project. It is necessary to utilise $^{15}$N, rather than the more common $^{14}$N to avoid the production of the active isotope $^{14}$C. The MA nitrides were modelled in a solid solution with ~25% of inert zirconium nitride (ZrN).

Finite difference diffusion theory calculations were performed in 33 neutron energy groups with a 120° sector TRIZ model. Prior to burn-up all fuel SAs contained fresh fuel. Successive values of flux, isotopic compositions and macroscopic cross-sections were produced by a series of 100-day burn-up steps with the CSDs at zero insertion; the flux of one step being used to burn the fuel during the next step, using the newly determined concentrations. The fuel dwell time was determined on the basis of obtaining a peak clad damage of 180 dpa NRT Fe. By considering the anticipated ability of the CSD rods to control the cycle reactivity swing (with allowances for uncertainties and margins), estimates were made as to the refuelling scheme requirements. Subsequently specific measures were introduced to achieve single batch operation to the above damage limit.

The studies were carried out in three phases:

(i)   core fuelled only with a minor actinide, zirconium nitride matrix;

(ii)  core fuelled with minor actinide, zirconium nitride matrix with plutonium nitride;

(iii) core fuelled with (MA,Zr,Pu)N matrix including moderator.

## Phase 1 study: Core fuelled only with a minor actinide, zirconium nitride matrix

In the first phase a core concept was established with a suitable fuel matrix containing only MA fissile material and the capability of achieving criticality at clean core. An optimisation was carried out in a series of calculations covering a range of minor actinide to zirconium loadings. Optimum values were chosen with the aim of achieving a clean core reactivity close to zero with all rods out. The MA and Zr loadings in the inner and outer core regions were varied in order to balance the peak ratings (and if possible the peak burn-ups) in the two regions. This initial phase illustrated that a core fuelled only with MAs in a solid nitride solution with ZrN could be designed to be critical with a clean core. The reactivity rise with burn-up of about +8 000 pcm at 1 200 EFPD would be such as to require at least a two-batch cycle. As would be expected, this initial option exhibited a significantly larger net MA consumption rate than the cases with added initial Pu (described below); however the net production rate of Pu was about half the net MA consumption rate. The Doppler constant was unacceptably small.

## Phase 2 study: Core fuelled with minor actinide, zirconium nitride matrix with plutonium nitride

In the second phase the effects of adding a small amount of Pu to the fuel were investigated with a view to reducing the reactivity rise with burn-up of the Phase 1 (MA,Zr)N core. The effect of the Pu quality was also investigated. The added Pu was in the form of a nitride utilising 100% $^{15}$N to be compatible with the (MA+Zr)N fuel matrix. In adding Pu to the initial MA fuel to reduce the reactivity rise with burn-up the guiding aim was to have a minimum quantity of Pu consistent with achieving a single-batch cycle. The range of Pu concentrations considered all resulted in very low Doppler values, despite increasing the Pu content. The effect on the Doppler constant of using a poorer quality Pu was small. The MA consumption rate decreases with an increase in the Pu proportion of the total actinide (MA+Pu) component of the fuel matrix, which was offset by a comparable decrease in net Pu production. This results in an approximately constant net MA+Pu consumption rate of around 109 kg/TWhe with Pu, and 101 kg/TWhe without.

## Phase 3 study: Core fuelled with (MA,Zr,Pu)N matrix including moderator

In the third phase of the studies the effect of introducing a moderating material, with its consequent softening of the spectrum, was investigated. This spectrum change was also expected to reduce the clad damage rate, and thus to have some potential to increase the fuel life. It is also expected to improve the Doppler constant due to the softening of the spectrum.

Initial survey calculations were carried out in which $^{11}$B$_4$C as moderator was introduced within the fuel SAs. It was found however that this material did not engender any substantial improvement in the Doppler constant. A doubling of the moderator to fuel ratio compared to previous CAPRA-LMFR studies only resulted in an SOL Doppler constant of about -48 pcm. Use of hydrogenated materials such as zirconium hydride (ZrH$_2$) provide an alternative, more efficient moderator for obtaining an increase in Doppler values, and stochiometric ZrH$_2$ was chosen for study. Preliminary calculations indicated that utilising a similar ZrH$_2$ moderator pin to fuel pin ratio to that in the NNC studies for a CAPRA-LMFR [7] offers an improvement in the Doppler constant. Thus a model using moderated fuel SAs was created, in which ZrH$_2$ replaced the fuel matrix in 30 pins of the 169 pin GCFR SA design previously studied.

A series of scoping calculations resulted in a preferred case. For the preferred case the burn-up time was extended from the 1 200 EFPD of the scoping studies to a more realistic 2 000 EFPD. In the model all fuel SAs have a uniform burn time. More detailed core performance calculations were carried out for the preferred case and the results are presented in Table 3.

## Table 3. Performance of the dedicated MA core with $ZrH_2$

| Parameter | Units | Value |
|---|---|---|
| Plutonium in [MA+Pu] | % | 27.5 |
| [Pu+MA]/Zr<br>    Inner core<br>    Outer core | % | <br>60/40<br>75/25 |
| Peak burn-up @ 2 000 EFPD<br>    Inner core<br>    Outer core | % (ha) | <br>34.7<br>24.2 |
| Cycle length | EFPD | 2 000 |
| Peak clad damage @ 2 000 EFPD | dpa NRT Fe | 140 |
| Estimated EFPD to reach 180 dpa NRT Fe | | 2 251 |
| Doppler constant<br>    SOL<br>    @ 1 200 EFPD<br>    @ 2 000 EFPD | pcm | <br>-247<br>-337<br>-377 |
| Reactivity swing over 1 200 EFPD | pcm | +799 |
| Reactivity change | pcm/EFPD | 0.67 |
| MA consumption rate over 2 000 EFPD<br>Pu production rate over 2 000 EFPD | kg/TWeh | 132.2<br>23.3 |
| Total β-effective | pcm | 151 |
| Prompt neutron lifetime | sec | 3.8E-7 |
| Total rod array worth | pcm | 5 935 |
| CSD array worth (DSDs out) | pcm | 4 026 |
| Peak linear ratings<br>    SOL<br>    @ 1 200 EFPD<br>    @ 2 000 EFPD | W/cm | <br>469<br>517<br>456 |
| SOL coolant void reactivity (peak) | pcm | 454 |

The peak burn-up in the inner core (34.6% ha) is beyond the usual limits set for oxide fuel. However there is currently little experience with nitride solid solutions, and burn-up limits remain to be proven in practice. Further optimisation may balance the inner and outer burn-ups reducing the higher, inner core peak. This could also reduce the variation in the ratio of inner/outer core ratings which are discussed below.

The inner/outer ratings were well-balanced at SOL with the peak in the outer core being only 33 W/cm above that of the inner core, but at about 1 200 EFPD the peak rating occurs in the inner core and is of a much higher value. However, with continued burn-up to EOL, 2 000 EFPD, the imbalance between the inner and outer core peaks diminish, with a consequent reduction in the peak rating. The acceptability or otherwise of such rating values depends on performance, transient and fuel performance analyses, which are beyond the scope of the present study. Preliminary transient analysis studies have shown that an acceptable core response in accident scenarios can be obtained.

### Comparison of gas-cooled and sodium-cooled dedicated MA burners

This scoping study of a gas-cooled fast reactor as a dedicated MA-burner has served to demonstrate a potentially viable concept, and has provided preliminary physics, performance and safety results for a potential design. This establishes a basis for future studies and the optimisation required to further advance this concept. It has been demonstrated that, for a 1.5 m active core height, and maintaining the radial dimensions of the EFR primary circuit, a gas-cooled reactor can be designed to obtain a power output of 3 600 MW(th) using a mainly minor-actinide fuel.

Fuelling a sodium LMFR mainly with minor actinides has a number of adverse consequences on core safety, in particular on the Doppler constant, the coolant void reactivity and delayed-neutron fraction. Scoping studies have shown that a 3 600 MW(th) sodium-cooled core fuelled mainly with minor actinides is not a viable option due to the unacceptably large coolant void effect and therefore there are no available LMFR studies available with which to draw direct comparisons.

In liquid-metal-cooled cores a number of options exist for improving these safety parameters in a core fuelled mainly with minor actinides. Along with reducing the core size to enhance leakage effects and hence reduce the coolant void reactivity, these options include:

- the use of a dense liquid lead coolant;

- the inclusion of either $B_4C$ or zirconium hydride moderator to enhance the Doppler effect;

- including small amounts of fissile material in the form of plutonium or uranium.

Overall even if all these measures are implemented in an LMFR the core size is severely limited to a few hundred MWth. In contrast a dedicated MA-burning gas-cooled core is not so limited and a commercially sized core seems viable.

## EGCR breeder reactor

Recently, detailed studies have been carried out on a $CO_2$-cooled high breeding gain core concept including both radial subassemblies and an axial breeder blanket [8,9].

A number of survey calculations were carried out to optimise the thermal-hydraulic characteristics of the core and to define a fuel element design. The fuel pin diameter and pin pitch were optimised to meet temperature constraints. The core performance of the proposed design was assessed using a 120° core sector model in TRI-Z geometry in which the batch refuelling was represented explicitly.

Comparing the gas-cooled breeder design to that of a sodium-cooled breeder brings out essentially the same points as noted in the comparison of the burner cores. The gas-cooled core is larger than the sodium-cooled core, there being more fuel SAs and thus a higher initial Pu inventory. The fuel residence time is substantially greater for the gas-cooled core compared to the equivalent sodium-cooled core because of the lower ratings. The coolant void reactivity is significantly lower in the gas-cooled core compared to the EFR sodium-cooled core, it being of the order of 1.18$ compared to 5.8$ for EFR cores. This gives the gas-cooled core a significant safety advantage over the liquid-metal-cooled cores. The EGCR Doppler constant is close to -500 pcm (or -1.4$), which is 20% lower than the EFR core, but still adequate given the substantially lower coolant void effect.

The breeding gain (BG) quoted for the EFR CD9/91 core is close to zero compared to a breeding gain of +0.2 for the gas-cooled core. However the EFR variant here has less axial breeder than EGCR and was designed for a BG close to zero. EFR variants with breeding gains of up to 0.15 have been studied. Ref. [8] indicates that removing the radial breeder assemblies from the EGCR core would reduce the breeding gain to around 0.03, which is comparable with the EFR CD9/91 design.

As for the GCPu00 burning core discussed above, analysis of three representative transients for EGCR indicated a satisfactory performance with no apparent characteristics which would preclude further study.

## Conclusions

A number of advantages of using a gaseous $CO_2$ coolant compared to a liquid metal such as sodium have been highlighted. These include:

- It is ready available and relatively inexpensive.

- It is optically transparent and does not require specialist handling.

- It is relatively benign and does not react explosively with air or water.

- It has been successfully employed as a coolant in the AGR thermal reactors with high levels of availability.

- There is a significantly smaller coolant void effect which is a limiting feature in LMFRs, particularly in terms of minor-actinide burning.

Overall the three studies discussed here illustrate the flexibility of gas-cooled fast reactor cores with potential applications ranging from burning or breeding plutonium, to burning minor actinides in dedicated cores.

*Acknowledgements*

The CAPRA project is a European collaborative project led by the CEA. Useful discussions with colleagues in the CAPRA project are acknowledged. The GCPu00 plutonium burner and dedicated minor-actinide-burner study was sponsored by BNFL as part of the CAPRA project. The ECGR study was carried out in conjunction with Fuji Electric and JNC, whose support is gratefully acknowledged.

## REFERENCES

[1]    W.B. Kemmish, M.V. Quick, I.L. Hirst, "The Safety of $CO_2$-cooled Breeder Reactors Based on the Existing Gas-cooled Reactor Technology", *Progress in Nuclear Energy*, Vol. 10, No. 1 pp. 1-17 (1983).

[2]    T. Abram, D.P. Every, B. Farrar, G. Hulme, T.A. Lennox, R.E. Sunderland, "The Enhanced Gas-cooled Reactor (EGCR)", ICONE-8, Baltimore, MD, USA, 2-6 April 2000.

[3]    T.A. Lennox, D.M. Banks, J.E. Gilroy, R.E. Sunderland, "Gas-cooled Fast Reactors", ENC'98, Nice, France, Sept. 1998.

[4]     G. Rimpault, P.J. Smith, T.D. Newton, "Advanced Methods for Treating Heterogeneity and Streaming Effects in Gas-cooled Fast Reactors", M&C'99, Madrid, Spain, 27-30 Sept. 1999.

[5]     "EFR, European Fast Reactor. The Approach to Europe's Future Need for Electricity", EFR Associates brochure, Dec. 1993.

[6]     A. Languille, J.C. Garnier, P. Lo Pinto, B.C. Na, D. Verrier, J. Depliax, P. Allen, R.E. Sunderland, E. Kiefhaber, W. Maschek, D. Struwe, "CAPRA Core Studies, the Oxide Reference Option", GLOBAL'95 Versailles, Paris, France, Sept. 1995.

[7]     H.M. Beaumont, E.K. Whyman, R.E. Sunderland, D.P. Every, "Heterogeneous Minor Actinide Recycling in the CAPRA High Burn-up Core with Target Sub-assemblies", GLOBAL'99, Jackson Hole, Wyoming, USA, 29 Aug.-3 Sept. 1999.

[8]     M. Nakano, H. Sadahiro, H. Ozaki, S.D. Bryant, A. Cheyne, J.E. Gilroy, G. Hulme, T.A. Lennox, R.E. Sunderland, M. Kida, T. Imagaki, "Conceptual Design Study for the Enhanced Gas-cooled Reactor (EGCR)", GLOBAL'2001, Paris, France, Sept. 2001.

[9]     H.M. Beaumont, A. Cheyne, J. Gilroy, G. Hulme, T.A. Lennox, R.E. Sunderland, D.P. Every, "The Design of the Enhanced Gas-cooled Reactor (EGCR)", ARWIF, Chester, UK, Oct. 2001.

# RBEC LEAD-BISMUTH-COOLED FAST REACTOR: REVIEW OF CONCEPTUAL DECISIONS

**P. Alekseev, P. Fomichenko, K. Mikityuk, V. Nevinitsa, T. Shchepetina, S. Subbotin, A. Vasiliev**
Russian Research Centre "Kurchatov Institute"
Kurchatov sq., 123182, Moscow, Russia
Phone: +7-095-196-70-16, Fax: +7-095-196-37-08, E-mail: kon@dhtp.kiae.ru

## Abstract

A concept of the RBEC lead-bismuth fast reactor-breeder is a synthesis, on the one hand, of more than 40 years experience in development and operation of fast sodium power reactors and reactors with Pb-Bi coolant and, on the other hand, of large amount of R&D activities concerning the core concept for a modified fast sodium reactor. This paper briefly presents the main parameters of the RBEC reactor as a candidate for commercial exploitation with the future nuclear power structure.

## Introduction

Development of the RBEC reactor was a logical continuation of the Russian conceptual R&D direction related to advanced fast reactors based, first of all, on experience gained in operation of fast sodium reactors existent in Russia: BOR-60, BR-10, BN-350 and BN-600.

In the late 70s/early 80s at a first stage of this direction a concept was developed for the core of a fast sodium-cooled reactor with extended nuclear fuel breeding and a number of innovations [1]. Some of these innovations include: wide fuel rod lattice, fuel assemblies (FAs) without shrouds, low core hydraulic resistance, low coolant heating-up, heterogeneous U-Pu core composition with core breeding ratio (CBR) close to 1, etc. As a result of the first stage:

- Proposals were formulated by Kurchatov Institute on sodium-cooled reactor BN-500EC with coolant natural circulation, FAs without shrouds and wide fuel rod lattice.

- A technical project was prepared for an advanced BN-1600M sodium-cooled fast reactor.

- Proposals were formulated for the modification of the BN-800 core.

In the mid-80s at a second stage of the fast reactor R&D direction a conceptual project was developed for lead-bismuth fast breeder reactor RBEC, reviewed in detail in this paper. The scientific leader at the first and second stages was the Kurchatov Institute.

Currently, there are several directions for conceptual development of advanced LBFBRs in RF:

1. Concepts oriented toward use in the near future and based as much as possible on traditional decisions and technologies. Examples would include lead-bismuth coolant experience from nuclear submarines, integral layout of primary circuits, hexagonal fuel assemblies, mixed uranium-plutonium oxide fuel, intermediate circuits, steam parameters close to those used in existing reactors, etc.

2. Advanced concepts requiring additional studies and based on a number of innovations. Examples include the use of pure lead coolant, square fuel assemblies, high-density mixed nitride fuel, two-circuit scheme with elimination of intermediate circuit, supercritical steam parameters, etc.

The RBEC [2-4] and SVBR [5] reactor projects can be assigned to the first direction, while BREST-type reactor projects [6] lean toward the second direction.

## Reactor general parameters

The aim of the RBEC project [2-4] was creation of a nuclear steam-generating power plant on the basis of Russian experience in design and operation of fast reactors and liquid-metal technology. A high level of self-protection should be provided by inherent core safety properties, thermal-physical properties of lead-bismuth coolant, use of natural circulation for emergency core cooling, application of passive safety systems along with traditional active ones and qualitative factory fabrication of the equipment.

The three-circuit scheme was implemented in the reactor design of a 900 MWt, 340 MWe power unit by OKB Gidropress, RRC KI and IPPE. The design and thermal-hydraulic parameters of RBEC are based, as much as possible, on technical decisions proved in BN-type reactors cooled by sodium, and they correspond to existing experience on fuel, structural materials and technology of liquid-metal coolant. Major technological processes of the NPP equipment fabrication were chosen to be mainly based on the previously developed nuclear power technologies.

The RBEC reactor facility contains the following main systems (Figure 1):

- primary system structurally made as a monoblock unit;

- intermediate (secondary) system;

- turbine system;

- air emergency core cooling system;

- refuelling system;

- system for gas heating or emergency cooling of monoblock vessel;

- system for electric heating of secondary circuit;

- system for filling and drainage of primary and secondary coolant;

- clad failure detection system;

- system of the primary and secondary coolant technology;

- control and protection system, automatic control, etc.

**Figure 1. General view (a), top view (b) of RBEC reactor and monoblock view (c)**

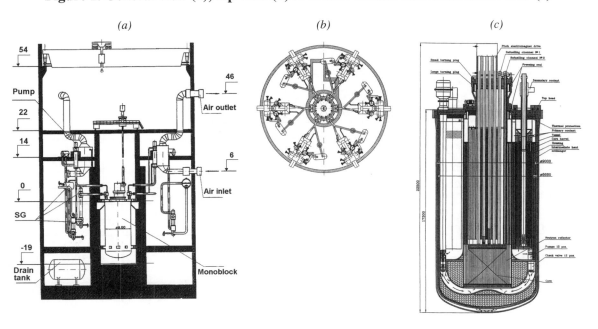

The design of main equipment was developed using the following basic principles:

- Dimensions of basic equipment (in particular, vessel and other components of the monoblock) are restricted by limits allowing for factory fabrication and transportation of equipment or components to the site.

- The unit's thermal-hydraulic parameters were chosen to solve problems of coolant technology and corrosion resistance of structural materials using heavy-metal coolant experience.

- Seismic resistance of the facility is provided for earthquakes of up to magnitude 8 on the MSK-64 scale.

The plant is placed in hermetic reinforced concrete containment which may be partly or fully installed underground in order to increase the equipment seismic stability and to create the best conditions for localisation and elimination of hypothetical accident consequences. On the basis of an estimation of the seismic stability of the monoblock vessel, the depth of the containment location, corresponding to the zero level of the monoblock support structures, was accepted in the given project. An integral layout of the primary circuit is used in the RBEC reactor. The monoblock containing the primary circuit is a vessel with double walls to prevent a loss-of-coolant accident. The gas gap between external and internal walls (70 mm) is under control and used for gas heating of the monoblock before the reactor is filled with coolant or for emergency cooling of the vessel. The monoblock unit of 9 000 mm diameter contains the core with axial and radial blankets, 12 (two for each loop) intermediate heat exchangers (IHX), 12 (two for each loop) circulation axial pumps with electric drives, thermal and neutron vessel shield. The design value of reactor vessel thickness ranges from 90 to 120 mm for different structural parts. The reactor pump has a nominal rotation speed of 985 rpm, drive power of 200 kW and head of 2.75 m. The check valves are installed at the pump outlets to prevent an inverse flow of primary coolant through the intermediate heat exchangers after trips of several pumps. After trip of all pumps (under natural circulation conditions) the check valves are kept open. Major characteristics of the RBEC are given in Table 1.

**Table 1. Major characteristics of the RBEC reactor**

| Primary circuit | |
| --- | --- |
| Coolant | Pb-Bi |
| Thermal power, MW | 900 |
| Electric power, MW | 340 |
| Number of loops | 6 |
| Number of reactor pumps | 12 |
| Total reactor flow rate, t/h | 220 000 |
| Core inlet/outlet coolant temperature, °C | 400/500 |
| Primary coolant pressure in the core, MPa | 2 |
| Helium pressure above free level of primary and intermediate coolant, MPa | 0.09 |
| Power removed by natural circulation with rated heating-up, % of rated power | 11 |
| Total power of air cooling heat exchangers, % of rated power | 3 |
| Total mass of Pb-Bi, t | ~6 500 |
| Metal consumption, t | 3 500 |
| Seismic resistance (MSK-64) | 8 |
| Fuel cycle duration, year | 4 |
| Time interval between refuellings, year | 1 |
| Annual excess fuel production (292 eff. days), kg | ~160 |
| Design lifetime of main equipment, year | 40 |

**Table 1. Major characteristics of the RBEC reactor (*cont.*)**

| Intermediate circuit | |
|---|---|
| Coolant | Pb-Bi |
| Number of intermediate heat exchangers (IHX) | 12 |
| Intermediate temperature at steam generator inlet/outlet, °C | 480/380 |
| **Turbine circuit** | |
| Coolant | Water |
| Feedwater temperature, °C | 260 |
| Generated steam pressure, MPa | 15 |
| Generated steam temperature, °C | 460 |
| Steam production, t/h | 1 580 |

## Core parameters

In-assembly heterogeneity is used in the RBEC core: 78 fuel rods with mixed uranium-plutonium oxide fuel and 42 fertile rods with depleted uranium carbide are installed in a hexagonal fuel assembly without shroud with pitch of 15.3 mm [Figure 2(b)]. FA pitch in cold conditions is 176 mm.

The RBEC reactor core [Figure 2(a)] consists of 253 hexagonal fuel assemblies. Two types of MOX fuel with different Pu content are used in fuel rods to flatten the power density radial distribution. The central low-content zone (LCZ) consists of 121 fuel assemblies with 27.5% Pu content in fuel rods. The high-content zone (HCZ) includes 132 FAs with 37.1% Pu content in fuel rods. The core is surrounded by 126 assemblies of radial blanket with fertile rods of depleted uranium carbide. One hundred ninety-two (192) assemblies of neutron reflector are installed around the core.

**Figure 2. RBEC core (a) and fuel assembly (b)**

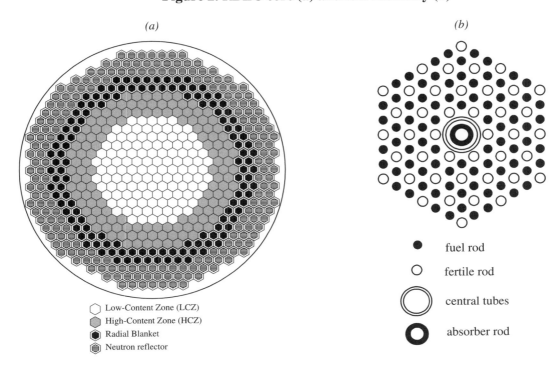

The hollow pellet of mixed uranium-plutonium oxide fuel with density of 9.03 g/cm$^3$ has an outer diameter of 7.9 mm and an inner diameter of 1.2 mm. Fuel cladding composed of 12%Cr-Si steel (EP-823 in Russian classification) has an outer diameter of 9.0 mm and a thickness of 0.45 mm. The active part of the fuel rod was determined in order to reach the CBR value close to 1, and equals 1 500 mm. The top and bottom axial blankets of 350 mm each contain the pellets of low-enriched UO$_2$ with a density of 9.5 g/cm$^3$. Thus, a column of fuel rod pellets includes 350 mm of bottom blanket, 1 500 mm of MOX fuel and 350 mm of top blanket. The pellet of low-enriched uranium carbide with a density of 12.4 g/cm$^3$ has an outer diameter of 10.7 mm and no inner hole. Cladding of the same steel as for the fuel rod has an outer diameter of 11.0 mm and a thickness of 0.45 mm. The column of fertile pellets has a height of 2 200 mm.

The fuel and fertile rods are initially filled with helium at pressure of 1 MPa. A 800-mm high gas plenum is designed in the bottom of fuel and fertile rods. The initial gas pressure and height of the gas plenum were chosen to have gas pressure inside the fuel rod below coolant pressure by the end of fuel life. Two coaxial steel tubes with diameters of 39 and 33 mm, respectively, are installed in the centre of each FA. The outer tube is used as a support structure for 10 spacer grids installed with axial steps of 312 mm. The inner tube is secured by the bayonet grips in the supporting plate and prevents the floating up of FAs. In 144 central FAs the inner tube is also used as a guide tube for absorber rods of active or passive type. The hollow pellet of B$_4$C with a density of 2.1 g/cm$^3$ and 80% $^{10}$B enrichment has an outer diameter of 20 mm and an inner diameter of 14 mm. Cladding of the same steel as for the fuel rod has an outer diameter of 23 mm and a thickness of 0.5 mm. The column of absorber pellets has a height of 1 500 mm. The main data on the RBEC core design characteristics are summarised in Table 2. Geometrical data correspond to cold (T = 20°C) conditions.

**Table 2. RBEC reactor core design parameters**

| Parameter | Value |
|---|---|
| Number of FAs in the core, including | 253 |
| Low-content zone (LCZ) | 121 |
| High-content zone (HCZ) | 132 |
| Number of FAs in radial blanket | 126 |
| Number of fuel rods in FA | 78 |
| Number of fertile rods in FA | 42 |
| Number of absorber rods in the core | 144 |
| Outer/inner fuel pellet diameter, mm | 7.9/1.2 |
| Outer/inner fertile pellet diameter, mm | 9.8/0.0 |
| Fuel rod outer diameter, mm | 9.0 |
| Fertile rod outer diameter, mm | 11.0 |
| Clad thickness, mm | 0.45 |
| Thickness of absorber rod clad, mm | 0.5 |
| Outer diameter of absorber rod, mm | 23.0 |
| Inner diameter of absorber rod, mm | 14.0 |
| FA pitch, mm | 176 |
| Fuel rod pitch in FA, mm | 15.3 |
| Height of the core, mm | 1 500 |
| Height of the top axial blanket, mm | 350 |
| Height of the bottom axial blanket, mm | 350 |
| Height of the gas plenum, mm | 800 |

## Operational parameters

A partial fuel cycle in the RBEC reactor is 292 effective days long. Replacement of 63 or 64 FAs in the core and 15 or 16 in the radial blanket is supposed in one reloading. Thus, fuel life for the radial blanket is twice as long as fuel life for the core.

About 160 kg of major fissile Pu isotopes is produced in a partial fuel cycle of 292 effective days. The power output of each reactor region is mainly determined by the fission of heavy nuclei $^{235}U$, $^{238}U$ and $^{239}Pu$, $^{240}Pu$, $^{241}Pu$, $^{242}Pu$ and by radiative capture. The maximal fuel rod linear power rating and fuel burn-up in the RBEC reactor are given in Table 3. Power peaking factors in the core and the power fraction for the separate reactor regions averaged over the whole fuel cycle are given in Table 4. The values of the breeding ratio and its components at the different burn-up stages are given in Table 5.

### Table 3. Peak fuel rod linear power rating and fuel burn-up in the RBEC reactor

|  | Peak fuel rod linear power, W/cm | | Peak fuel burn-up, % ha |
|---|---|---|---|
|  | BOL | EOL |  |
| Fuel rod | 450 | 345 | 11.6 |
| Fertile rod | 70 | 274 | 3.4 |

### Table 4. Power distribution in the RBEC reactor

| Axial power peaking factor in the core | | 1.29 |
|---|---|---|
| Radial power peaking factor in the core | | 1.20 |
| Power fraction generated in region, % | LCZ | 47.4 |
|  | HCZ | 42.9 |
|  | Total in core | 90.3 |
|  | Axial blanket | 2.9 |
|  | Radial blanket | 6.8 |
|  | Total in blankets | 9.7 |

### Table 5. Breeding parameters in the RBEC reactor

| Breeding ratio component | Fresh fuel | Equilibrium cycle |
|---|---|---|
| LCZ | 0.66 | 0.59 |
| HCZ | 0.46 | 0.42 |
| Total in core | 1.12 | 1.01 |
| Radial blanket | 0.26 | 0.26 |
| Axial blanket | 0.18 | 0.17 |
| Total in blankets | 0.44 | 0.43 |
| **Reactor** | **1.56** | **1.44** |

## Safety parameters

The effective delayed neutron fraction for the RBEC reactor is 0.38%. Prompt neutron lifetime is $5.1 \cdot 10^{-7}$ s. Components of temperature and power reactivity effects are presented in Table 6. The value of full void reactivity effect is equal to 1.15% $\delta(1/k_{eff})$. The reactivity change during the partial fuel cycle taking into account neptunium and americium effects is equal to 0.21% $\delta(1/k_{eff})$.

## Table 6. Components of temperature and power reactivity effects

| Components | Temperature reactivity coefficient, $\delta(1/k_{eff})/°C$, $10^{-5}$ | Power reactivity coefficient, $\delta(1/k_{eff})/MW$, $10^{-5}$ |
|---|---|---|
| Doppler | -1.657 | -0.724 |
| Coolant expansion | 0.152 | 0.042 |
| Radial reactor expansion | -0.650 | -0.086 |
| Axial reactor expansion | -0.164 | -0.132 |
| Total | -2.319 | -0.900 |

Two independent control and protection systems (CPS) – active and passive – are used in the RBEC reactor to meet the RF regulatory guide requirements. Each of these systems can shut down the reactor and keep it subcritical from all possible nominal and accident conditions under the assumption of a single failure of the most effective cluster. As was mentioned above, absorber rods of both active and passive CPS systems are installed in the 144 central channels of FAs of the RBEC reactor.

Active CPS consists of 72 absorber rods located in the central tubes of 72 FAs in the LCZ zone. The absorber rods of the active CPS are moved by the driving rods controlled with the use of electric motors. The driving rods of the active CPS are combined in 24 clusters, each of which controls three absorber rods. Each three-rod cluster is coupled with and controlled by one shaft. The driving rods of CPS are not coupled with the active absorber rods. In normal operation the 48 active absorber rods follow the driving rods moved by the operator and the 24 active scram rods are kept below the bottom axial blanket by the driving rods.

Under emergency conditions, all driving rods are set free and flowing up with all active rods which enter the core. The CPS design was chosen due to features of heavy coolant application and provides:

- exclusion of coupling of driving rods with absorber rods and, therefore, increase of structural reliability;

- insertion of the absorber rods into the core during refuelling, when the driving rods should be raised above the fuel assemblies;

- insertion of the absorber rods into the core during hypothetical severe accidents accompanied by damage to the vessel components, for example, in a failure of the rotating plug fasteners and flowing up of the rotating plug with shafts and driving rods.

Passive CPS consists of 72 absorber rods located in the central tubes of 49 FAs in the LCZ and 23 FAs in the HCZ. The passive absorber rods are kept in the same position below the core as the active absorber rods with the use of special triggers which are bi-metal plates made of ferritic-martensitic and austenitic grade steels with different thermal expansion coefficients. The trigger is installed in the FA central tube and via the shaft prevents flowing up of a passive absorber rod. When coolant temperature at the FA outlet exceeds 580°C, thermal deformation of the trigger reaches the critical value and leads to release of the shaft and flowing up of the passive absorber rod.

## Intermediate and turbine circuits

The secondary (intermediate) circuit is formed by six circulation loops, each connected to two IHXs. The secondary loop contains the pump, buffer tank and steam generator with multiple natural circulation and steam superheating (see Figure 3). Secondary IHX inlet and outlet are connected to air-cooled heat exchangers of an emergency cooling system. The whole facility is located in the hermetic reinforced concrete containment.

## Figure 3. Diagram of the RBEC intermediate and turbine circuits

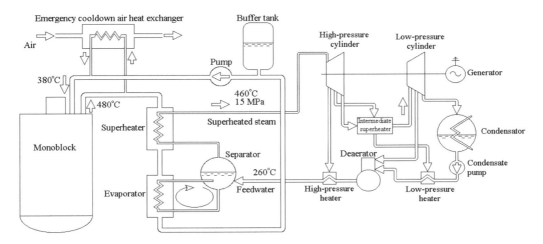

The intermediate circuit of the RBEC facility allows to exclude the following potential dangers:

1) The possibility of steam or water entrainment in the core during an emergency inter-circuit leak in the steam generator that can cause inadmissible changes in reactivity, power density and heat transfer in the core.

2) The possibility of primary circuit over-pressurisation during an emergency inter-circuit leak in the steam generator, as well as damage, in this case, of in-vessel devices as a result of hydraulic impacts, which can accompany fast evaporation of water in liquid-metal coolant.

3) The possibility of primary coolant freezing in steam generators, for example, because of steam header break and intensification of heat transfer from primary to turbine circuit. Such a scenario with coolant freezing is possible in the RBEC reactor only in secondary steam generators, thus, the primary coolant circulation is preserved and decay heat removal from the primary circuit can be organised through channels of emergency cool-down. In particular, in the RBEC reactor heat removal from the primary circuit with freezing steam generators will be organised through the IHX to coolant in non-freezing, adjacent to the IHX, sites of secondary pipelines and then through emergency air cooling heat exchangers to atmospheric air.

The IHX is a shell-and-tube once-through counter-flow heat exchanger with a thermal power of 75 MW. Primary coolant velocity is 0.7 m/s, secondary coolant velocity is 1.2 m/s. The intermediate coolant temperatures at monoblock inlet and outlet are 380°C and 480°C. The coolant flow rate through one IHX is the same as in the primary circuit, 5 092 kg/s. Pressure drop is 0.3 MPa. The pump of the secondary circuit is a centrifugal one similar to that accepted in sodium-cooled facilities. The pump drive power is 2.4 MW. The head is 16.7 m.

The RBEC steam generator design is similar to that of the steam generator used in the BN-600 reactor. The steam generator consists of an evaporator, separator and superheater. The evaporator is a counter-flow heat exchanger with multiple natural circulation and the superheater is a once-through counter-flow heat exchanger. The RBEC scheme uses feedwater heating in a large-volume separator at a pressure of 15 MPa and water saturation temperature of 613 K, significantly exceeding the lead-bismuth freezing point. The use of such a steam generator scheme in the RBEC reactor allows to significantly decrease the probability of intermediate coolant freezing in a steam generator in case of an accidental drop in feedwater temperature.

The air-cooling heat exchanger is connected to the hot and cold legs of each secondary loop. Under normal plant operating conditions, the heat exchangers are isolated from air by means of the cut-off valves. The coolant flow rate through them is kept at the minimal level to prevent the heat exchangers from freezing. Under emergency conditions with plant black-out, the electric magnets, keeping the cut-off valves closed, are de-energised. The cut-off valves are opened and natural circulation of air and coolant is established through the heat exchangers. The use of the air-cooling heat exchangers, according to the completely passive scheme, is possible too. In this case, they must be continuously in operation, leading to the external loss of about 3% of the plant thermal power.

## Conclusions

The main aim of the development of the RBEC lead-bismuth cooled fast reactor was to demonstrate the possibility to combine existent advantages of various reactor technologies in one nuclear power facility so as to improve economic and breeding parameters compared to BN-type reactors, while simultaneously demonstrating safety enhancement and environmental acceptance.

Design and technological decisions experienced in practice and checked experimentally became the basis of the RBEC design:

- *Wide fuel rod lattice.* This allows to reduce hydraulic resistance in fuel assemblies, to increase the coolant natural circulation level, to use fuel assemblies without shrouds, to decrease the fraction of steel in the core and, thus, not only to improve core breeding ratio (CBR), but also to create conditions for reducing the void reactivity effect.

- *In-assembly heterogeneity.* This allows to increase effective fuel density in the core without developing new types of mixed fuel, to obtain CBR above 1 and to increase fuel burn-up without increase of neutron fluence.

- *High-density carbide fuel.* In fertile rods with low linear rating power and burn-up, this permits not to fill the fuel-clad gap with liquid metal.

- *Parameters of turbine circuit experienced in the nuclear power.* This allows to use design decisions for steam generator and turbines checked in practice.

The use of lead-bismuth coolant at low heating-up in the core and outlet temperature allows to use 12%Cr ferritic-martensitic steels resistant against radiative swelling and radiative creep. Corrosion resistance of this fuel-clad material was checked in practice for these temperatures with the use of special technological processes of oxygen concentration maintenance in the coolant.

In our opinion, if in the near future the creation of a fast reactor with heavy-metal coolant is required to demonstrate its advantages in terms of economics, breeding properties and inherent safety, then one of the best projects from the viewpoint of feasibility and provision of mathematical modelling for normal and accident conditions will be a reactor of RBEC type.

# REFERENCES

[1]    V. Orlov, N. Ponomarev-Stepnoi, I. Slesarev, P. Alekseev, *et al.*, "Concept of the New Generation High Safety Liquid-metal Reactor (LMFR)", Proc. Int. Conf. Safety New Generation Power Reactors, USA, Seattle, May 1988.

[2]    P. Alekseev, S. Subbotin, *et al.*, "Potential Possibilities of a Three-circuit Scheme for the Enhancement of Lead-cooled Reactor Safety", Presented at ARS'94, Intern. Topical Meeting on Advanced Reactors Safety, Pittsburgh, PA, USA, April 1994.

[3]    V. Orlov, I. Slesarev, P. Alekseev, *et al.*, "Two- and Three-circuit Nuclear Steam Generating Plant With Lead-cooled Fast Reactor (RBEC)", Proc. of the Seventh All-union Seminar on Reactor Physics Problems, "Volga-1991", p. 62, Moscow, USSR, September 1991.

[4]    V. Titov I. Slessarev, P. Alekseev, *et al.*, "Three-circuit Nuclear Steam Generating Plant with Lead Coolant", Proc. of the ENS meeting, RDIPE, Moscow, USSR, October 1990.

[5]    B. Gromov, A. Dedoul, O. Grigoriev, *et al.*, "Multi-purposed Reactor Module SVBR-75/100", ICONE-8072, Proceedings of ICONE-8, 8th International Conference on Nuclear Engineering, Baltimore, MD, USA, 2-6 April 2000.

[6]    V. Orlov, A. Filin, V. Tsikunov, *et al.*, "Design of 300 MWe and 1 200 MWe BREST Reactors", Int. Conf. on Heavy Liquid Metal Coolants in Nuclear Technology, Obninsk, SSC RF-IPPE (1998).

# DESIGN AND PERFORMANCE STUDIES FOR MINOR-ACTINIDE TARGET FUELS

**T.D. Newton and P.J. Smith**
SERCO Assurance, Winfrith, Dorset, England

## Abstract

Studies have demonstrated the general neutronics feasibility of sodium- and gas-cooled fast reactor core designs for the effective management of minor-actinide stockpiles. One of the ways in which this can be achieved is by heterogeneous recycling in minor-actinide-fuelled target subassemblies. This paper presents the results of two studies. The first illustrates the change in target subassembly performance due to the consideration of reference rather than more approximate design methods. In particular, the consequences on performance due to changes in resonance self-shielding during irradiation are considered. The second study considers the optimisation of a moderated target subassembly design for the gas-cooled fast reactor. The use of a moderating material allows an increased residence time for the target in the core while retaining the existing design limit on the fuel clad exposure. However, it also introduces a number of potential difficulties including power peaking in immediately neighbouring standard fuel core subassemblies.

## Introduction

In the current phase of the CAPRA/CADRA project [1] investigations are taking place, based on previously developed sodium- and gas-cooled fast reactor reference core designs, to study the possibilities for the effective management of minor-actinide stockpiles. One of the ways in which this can be achieved is by the heterogeneous recycling of minor actinides in minor-actinide-fuelled target subassemblies. Scenarios include the loading of the target subassemblies in in-core locations, or alternatively, in ex-core locations in the first row of the radial reflector. A further extension of this concept is the inclusion of moderating material in the target subassembly design. This has the aim of extending the target residence time within the core, thus leading to improved target subassembly management where the need for reprocessing of the targets is reduced or even eliminated.

The particular performance characteristics of minor-actinide fuels require that significant consideration should be given to the methods that are employed for their analysis when compared to the treatment of more conventional fuels. The fissile inventory and composition of minor-actinide fuels can vary significantly over a relatively short time scale during irradiation. This in turn can lead to large changes in the resonance self-shielding in the fuel during its irradiation. Also, significant quantities of helium are generated in subassemblies fuelled predominantly with minor actinides. This can result in pin over-pressurisation, which has important consequences for fuel residence time and fuel pin design. The use of moderating material, while bringing the benefit of increasing residence time, also introduces a number of potential difficulties (including power peaking) in the immediately neighbouring standard core fuel subassemblies.

This paper presents the results of two studies concerning the utilisation and performance of minor-actinide-fuelled target subassemblies. The first illustrates the change in target performance due to the consideration of reference rather than more approximate design methods. The second study considers the optimisation of a moderated target subassembly design for the gas-cooled fast reactor and shows the effect of the presence of the moderating material on the core and target performance.

These studies have been performed within the context of the European CAPRA/CADRA project and are sponsored by BNFL.

## Performance issues to be considered

In order to investigate the feasibility of heterogeneous recycling using minor-actinide-fuelled target subassemblies a number of performance issues have been addressed. These are described briefly below:

- *Minor-actinide consumption rates.* A basic requirement of a minor-actinide target design is that the reactor system should be capable of achieving an equilibrium consumption of both plutonium and minor actinides. This means that both plutonium and minor actinides should be burned in proportions such that the minor actinides coming from a partitioned waste stream from a fleet of nuclear reactors are consumed, together with those produced within the fast reactor itself. The proportions required will depend on the type and mix of reactors in the fleet as well as the fuel cycle scenario being considered.

- *Minor-actinide burn-up.* Another important parameter when determining the performance of a fast reactor system is the heavy atom burn-up that can be achieved. For minor-actinide-fuelled targets this parameter influences the mass of minor actinides that will need to be recycled and re-fabricated, the mass flow to the waste stream and the fuel cycle endgame arisings. It is possible that moderating material can be successfully used within a target subassembly design to extend the target residence time within the core. The ultimate goal is to remove the requirement for reprocessing of the targets by achieving an incineration level during a single residence that is sufficiently high to give minimal waste arisings. In order to achieve this goal it is estimated that a target burn-up of between 80-90% is required.

- *Clad damage*. This is an important parameter, as it is the main limitation on the target residence time. The peak clad damage over the target lifetime should not exceed 200 dpa NRT Fe. One of the main incentives for including moderator material in the target is to soften the neutron spectrum, reducing the clad damage rate allowing higher burn-ups and longer residence times.

- *Peak linear rating*. In order to arrive at a feasible fuel pin and target subassembly design it is essential to ensure that the pin power, or peak linear rating, and hence the temperature, does not exceed safety levels and that a sufficient margin to fuel melting is allowed to accommodate any operational increases in power. For safe operation the peak linear rating should not exceed a calculated maximum limit of 500 W/cm. When considering a potential target design it is necessary to perform a thermal-hydraulic analysis to ensure that the material temperatures are within the design limits. These considerations are further complicated by the inclusion of moderator material which can cause power peaking in the immediately neighbouring core fuel subassemblies.

- *Resonance shielding*. As several of the minor actinide isotopes are relatively short-lived during irradiation the fissile inventory, and hence the flux within the target, can vary significantly over a short time scale during irradiation. This in turn can lead to large changes in the resonance self-shielding in the target fuel during its lifetime in the core. The effect on the minor-actinide isotopic transmutation rates, and also the heavy atom burn-up, can be significant.

- *Helium and fission gas production*. Significant quantities of helium are generated in subassemblies fuelled only with minor actinides. Helium arises from the alpha decay of relatively short-lived minor actinides and can result in pin over-pressurisation and consequently possible pin failure. It is important to determine the quantity of helium generated as it can have significant consequences on the target residence time and the fuel pin design. This problem is compounded by the generation of fission-product gases during irradiation.

## Core description

This paper presents the results from two studies. The first study concerns an evaluation of target subassembly performance for a high burn-up sodium-cooled fast reactor core design. The second study considers the optimisation of a moderated target subassembly design for a $CO_2$-cooled fast reactor. Both cores have a rating of 3 600 MW(th) with an assumed load factor of 80%. The reactors are assumed to have a net thermal efficiency of 40.5% and therefore have a 1 458 MW electrical output. Further details specific to each core design are given below.

### *Sodium-cooled core*

This sodium-cooled core design is based on a high burn-up version of the CAPRA 04.94M reference oxide core [2]. That reference option is based on a mixed-oxide fuel with a high plutonium content, around 45% Pu. The use of a high-plutonium-content fuel is only possible by the introduction of dilution in the form of special diluent subassemblies and the use of diluent pins within the core fuel subassemblies. Each fuel subassembly has a pitch of 18.14 cm and contains a total of 469 pins including

398 fissile pins and 82 pins containing $^{11}B_4C$ moderator. Diluent subassemblies placed within the core consist of a steel wrapper filled with empty steel pins. The core has two radial enrichment zones. For the purposes of the current study the core fuel subassemblies also include $NpO_2$ dispersed homogeneously in the core fuel matrix.

The design of the minor-actinide-fuelled target subassemblies used for this core design is based on that of the fuel subassemblies, with each target containing 469 pins. The target subassemblies do not include moderator pins. All of the pins in the target subassemblies contain isotopes of $AmO_2$ and $CmO_2$ dispersed in an inert spinel matrix which extends over the 1.0 m fissile height of the core. For both in-core and ex-core target options the required minor-actinide target loading of the target subassemblies has been optimised to produce an equilibrium consumption of plutonium and americium. For the in-core option the required number and position of the target subassemblies within the core is consistent with maintaining an adequate core power distribution. For the ex-core option a total of 78 target subassemblies are located around the edge of the core and replace the first row of radial reflector assemblies. The main core design parameters for both options are shown in Table 1.

**Table 1. Core design parameters for in-core and ex-core targets**

|  | In-core targets | Ex-core targets |
|---|---|---|
| Number of fuel S/A in the inner core | 159 | 156 |
| Number of fuel S/A in the outer core | 210 | 210 |
| Number of target S/A | 40 | 78 |
| Number of diluent S/A | 9 | 52 |
| Minor actinides per target S/A (% volume) | 12.5 | 15.5 |
| Minor actinides per target pin (% volume) | 36.7 | 40.3 |
| Inner-core enrichment (mass %) | 39.6 | 41.9 |
| Outer-core enrichment (mass %) | 41.5 | 43.2 |
| Number of cycles × cycle length (EFPD) | 6 × 220.4 | 6 × 236.7 |
| Fuel and target residence time (EFPD) | 1 322.4 | 1 420.2 |

### Gas-cooled core

The gas-cooled core design, employing a $CO_2$ coolant, is based on the technology of the advanced gas-cooled reactor, but incorporates the design features of a large liquid-metal fast reactor such as EFR [3]. The maximum plutonium enrichment in this scenario is around 30%, significantly lower than that adopted for the sodium-cooled core. The core has a fissile height of 1.5 m and contains a total of 580 fuelled subassemblies split into two enrichment zones, with 238 fuelled subassemblies in the inner core and 342 in the outer core. Each fuel subassembly has a pitch of 18.061 cm and contains 169 fissile pins. The core also contains 60 diluent subassemblies in order to achieve balanced inner and outer core peak powers.

## Calculation models and methods

### Neutronics

All the core configurations evaluated during these studies have been modelled using version 1.2 of the European Fast Reactor code scheme ERANOS along with the ERALIB1 nuclear cross-section data library [4]. Within the ERANOS code scheme the cell code ECCO uses the subgroup method to treat resonance self-shielding effects to prepare broad group self-shielded cross-sections and matrices for each material in the core model. A fine group slowing-down treatment is combined with the subgroup method within each fine group to provide an accurate description of the reaction thresholds and resonances for the heterogeneous geometry of each type of critical and subcritical subassembly. Whole core flux and depletion calculations have been performed in three-dimensional geometry for each core configuration studied.

For the first study, examining the effect on target performance due to the consideration of reference rather than more approximate design methods, both a reference and a standard calculation route have been employed. In the standard route the shielded cross-section data and spectra generated by the ECCO cell code for the clean target fuel at start of life have been used throughout the lifetime of the target fuel within the core. However, in the reference route the shielded fission and capture cross-section data, as well as the target spectrum and fluxes, has been re-calculated by ECCO at each step during the target irradiation. Whole core flux calculations have then been carried out using nodal transport theory.

For the second study, on the optimisation of a moderated target subassembly design for the gas-cooled core, the standard calculation route described above has been used. Whole core flux calculations were performed using finite difference diffusion theory. It has been shown that the use of these standard methods for a gas-cooled core requires a correction to the whole core reactivity to account for residual transport, heterogeneity and neutron streaming effects. A correction of +2.45% has been applied to all calculations of absolute core reactivity.

### *Fuel inventory*

Target fuel inventory calculations have been performed during this study using version 7B of the FISPIN fuel inventory code [5]. A link has been used to allow irradiation-dependent shielded cross-section and spectrum data generated by the ERANOS code scheme to be transferred into a format suitable for input into the nuclide libraries used by FISPIN. Calculations have then been performed using FISPIN for both in-core and ex-core options, to determine actinide and gaseous fission-product inventories as well as the quantity of helium generated in the target fuel pins.

### *Thermal-hydraulics*

To develop and optimise a moderated target subassembly design which satisfies the design constraints imposed on power and temperature, a series of thermal-hydraulics calculations have been performed. The objective of the analysis is to determine the magnitude of the gas-flow rate, and hence the imposed core pressure drop that would occur for the different arrangements of moderator and fuel considered. Consequently the efficiency of the heat transfer from the fuel to the remaining subassembly materials (clad, moderator and coolant) has been evaluated. For a potential subassembly design to be considered viable the pressure drop in the target must be the same as that in the standard core fuel subassemblies. Furthermore it has been verified that the fuel, clad and moderator temperatures are within acceptable limits.

### Target performance

For the first study, concerning the consequences on target performance due to the consideration of reference rather than design methods, results are presented for the calculated actinide masses at the end of target life for both in-core and ex-core options in Table 2.

**Table 2. Results for the calculated actinide masses
at the end of target life for in-core and ex-core options**

| | In-core (kg) | | | Ex-core (kg) | | |
|---|---|---|---|---|---|---|
| | Standard route (S) | Reference route (R) | Difference (S⇒R) | Standard route (S) | Reference route (R) | Difference (S⇒R) |
| $^{235}$U | 0.2 | 0.3 | +0.1 | 2.1 | 1.4 | -0.7 |
| $^{237}$Np | 4.4 | 3.8 | -0.6 | 21.3 | 22.9 | +1.6 |
| $^{239}$Pu | 14.9 | 20.3 | +5.4 | 45.9 | 41.7 | -4.2 |
| $^{240}$Pu | 51.9 | 52.5 | +0.6 | 257.1 | 246.4 | -10.7 |
| $^{241}$Am | 293.9 | 256.0 | -28.9 | 651.3 | 701.3 | +50.0 |
| $^{242}$Cm | 26.3 | 31.1 | +4.8 | 19.4 | 16.3 | -3.1 |
| $^{244}$Cm | 242.4 | 257.5 | +15.1 | 452.1 | 420.8 | -31.3 |

The mass of helium produced over the lifetime of the target fuel has been calculated from an integral of the alpha decay rate over time. The helium pressure in the target fuel pin has been calculated assuming that all of the helium has been released from the fuel into a plenum of cold length 90 cm. A similar calculation has been performed to determine the fission-gas pressure. The predicted helium and fission-gas pressures are given in Table 3.

**Table 3. Predicted helium and fission-gas pressures**

| | In-core (MPa) | | | Ex-core (MPa) | | |
|---|---|---|---|---|---|---|
| | Standard route (S) | Reference route (R) | Difference (S⇒R) | Standard route (S) | Reference route (R) | Difference (S⇒R) |
| Helium | 21.5 | 23.3 | +1.8 | 36.3 | 34.9 | -1.4 |
| Fission gas | 5.0 | 5.2 | +0.2 | 6.6 | 6.3 | -0.3 |

For the second study, which considers the definition and optimisation of a moderated target subassembly design for a gas-cooled fast reactor, a parametric survey has been performed to identify possible fuel and moderator contents as well as target pin dimensions which satisfy the imposed criteria on thermal and neutronic performance. The target design is intended to produce a 90% mass destruction of the minor actinides during a single core residence time while not exceeding a peak damage of 200 dpa NRT Fe. The study has concentrated on a moderated target design suitable for in-core positions. A core layout has been adopted that includes 90 in-core targets to maintain a reasonable core power profile. The target design used as a basis for this work contains a mixture of two types of pin, a fuel pin containing americium and curium oxide ($AmO_2$ + $CmO_2$) in an inert spinel matrix ($MgAl_2O_4$), and a moderator pin containing zirconium hydride ($ZrH_2$). The main results of this survey are summarised in Figure 1.

## Discussion

The results from the first study show that the performance characteristics of minor-actinide-fuelled target subassemblies are strongly influenced by the methods that are employed. The actinide masses for both the in-core and ex-core options at the end of target life are significantly modified when irradiation-dependent resonance-shielded cross-sections, spectra and fluxes are included in the calculation route. For example, for the in-core option the masses of $^{241}$Am and $^{244}$Cm are changed by 28.9 kg and 15.1 kg, respectively. This effect is even more marked for the ex-core targets where the mass of $^{241}$Am is increased by 50 kg.

## Figure 1

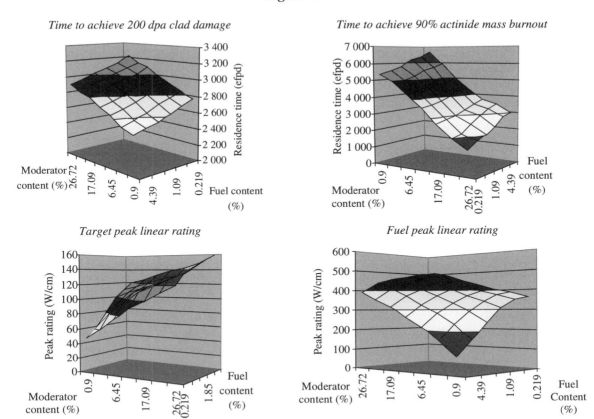

*Time to achieve 200 dpa clad damage*

*Time to achieve 90% actinide mass burnout*

*Target peak linear rating*

*Fuel peak linear rating*

It is also evident that the nature of these changes is strongly influenced by the target position, and consequently its spectrum, as the observed trends are different depending on whether in-core or ex-core targets are considered. The characteristics of the in-core targets are dominated by the surrounding core fuel, which is always in the same equilibrium state. However, there are significant modifications to the spectrum in ex-core target positions during the target lifetime. Clearly, if irradiation-dependent data, and in particular the changes in resonance shielding, are not taken into account the predicted actinide masses, and consequently the minor-actinide consumption and burn-up, will be significantly in error.

A similar situation is apparent when evaluating the helium and fission-gas production at the end of target life. The results of this study indicate that the calculated helium pressure is altered by up to 10% when irradiation-dependent resonance-shielded cross-sections are included in the calculation. Again the nature of the change is strongly dependent on the target location within the core. It can also be remarked that the pressure arising from the fission gases is significantly less, approximately one-fifth, of that arising from helium. The production of helium has the dominating influence on minor-actinide target pin design. It is evident that detailed reference methods must be taken into account during the design of minor-actinide-fuelled target pins.

From the second study the results of the preliminary parametric survey suggest that a moderated target design is feasible that avoids the requirement for multi-recycling. The target residence time required to reach the 200 dpa NRT Fe limitation on clad damage is relatively insensitive to either the fuel, or in particular the moderator content. This is a surprising result as the main benefit of using moderator material was expected to be a softer neutron spectrum, reducing the clad damage rate and therefore allowing a longer target residence time. The limited effectiveness of the moderator can be understood from Figure 2. The spectrum in the target is significantly altered, showing the effect of the moderator. However the changes to the spectrum take place in an energy range where the damage cross-section is at its lowest. The impact of the moderator on the damage rate is therefore relatively small.

**Figure 2. Demonstration of the limited effectiveness of the moderator**

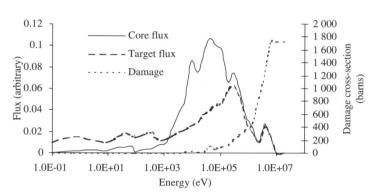

The time to achieve 90% destruction of the minor actinide content is strongly dependent on the fuel, and especially the moderator content. Only a small part of the defined parameter space exists, at low fuel and high moderator content, where a 90% mass destruction can be achieved within the 200 dpa NRT Fe clad damage limit. This corresponds to a fuel and moderator volume content within the target subassembly of approximately 2% and 23% respectively. The thermal-hydraulic analysis shows that these volume proportions can be satisfactorily distributed within the proposed target subassembly design to give an acceptable core pressure drop and material temperatures. The peak linear rating in the neighbouring core fuel subassemblies is 462 W/cm which is within acceptable limits. It can be noted however that it is necessary to include a relatively high number of target subassemblies in the core to avoid localised power peaking and achieve the equilibrium incineration rate.

On this basis a moderated target design is proposed which contains a total of 469 pins, of which 312 contain fuel and 157 contain $ZrH_2$ moderator. The inner clad/outer pin radii are 0.2198/0.2605 cm and 0.3651/0.3939 cm for the fuel and moderator pin respectively. The performance of this moderated target design has been evaluated for both in-core and ex-core options, and the main results are summarised in Table 4.

**Table 4. Performance results for the moderated target design for in-core and ex-core options**

| | In-core | Ex-core |
|---|---|---|
| Number of fuel S/A  – Inner core | 240 | 238 |
| – Outer core | 360 | 342 |
| Number of target S/A | 90 | 81 |
| Number of diluent S/A | 0 | 60 |
| Cycle length (EFPD) | 368 | 368 |
| Target residence time (number of cycles) (EFPD) | 2 944 (8) | 4 416 (12) |
| Peak/mean target burn-up (mass %) | 90.9/89.8 | 91.6/89.8 |
| Minor actinide consumption (targets) (kg/TWeh) | 6.23 | 4.38 |
| Peak linear rating (W/cm) – Targets | 134 | 82 |
| – Fuel | 462 | 466 |
| Peak clad damage (dpa NRT Fe) | 199 | 197 |

The peak clad damage and linear ratings are within the accepted limits for both the in-core and ex-core options. Due to the low flux and power at the outer edge of the core the ex-core targets are resident in the core for a significantly longer period (12 cycles) than the in-core targets (8 cycles). For the ex-core option it has been necessary to leave some of the available positions for targets vacant in order to avoid excessive power peaking in fuel subassemblies' neighbouring target positions. However, this study demonstrates that it is possible to include moderator material in target subassemblies, in either in-core or ex-core positions, without causing any additional difficulties.

The target burn-up has achieved the goal of 90% for both the in-core and ex-core options. This can be compared with the peak burn-up of 35% achieved for the in-core option with un-moderated targets for the sodium-cooled core design. As the target residence for both cores is very similar it is apparent that although the inclusion of moderator material has not greatly increased the target lifetime, it has significantly increased the burn-up that can be achieved in the same time. The minor-actinide consumption rates in the targets are 6.23 and 4.88 kg/TWeh for the in-core and ex-core options respectively. This can be compared to a total minor-actinide production rate in the reference core of 4.80 kg/TWeh, indicating that an equilibrium consumption has been achieved. However these consumption rates are much lower than those obtained for the sodium-cooled core design, 12.1 and 11.5 kg/TWeh for the in-core and ex-core options respectively. Such low consumption rates are due to the much smaller minor-actinide mass inventory in the moderated targets in the gas-cooled core. Therefore, although the requirement for target multi-recycling can be avoided with the use of moderating materials, the rate of minor-actinide consumption is rather low. The rates could be improved by increasing the proportion of actinides in the target material. However the scope for this would appear to be rather small due to the limitation imposed by the peak clad damage. Further study of these issues, and the consequences for the fuel cycle, are required.

## Conclusions

Detailed studies have been undertaken to evaluate the performance of minor-actinide-fuelled target subassemblies for the heterogeneous recycling of minor actinides. This option involves the loading of target subassemblies in either in-core locations, or alternatively in ex-core locations in the first row of radial reflector. A detailed assessment of target performance is necessary due to the impact of the significant changes in the target minor-actinide composition that take place during irradiation. The use of moderator material to increase the target residence time, and therefore reduce or even eliminate the requirement for reprocessing, also introduces a number of potential difficulties.

This paper presents and reviews the results from two studies. The results from the first study show that the calculated performance characteristics of minor-actinide-fuelled target subassemblies are strongly dependent on the methods that are employed. The predicted actinide masses at the end of target life are significantly modified when irradiation-dependent resonance-shielded cross-sections, spectra and fluxes are included in the calculation route. It is also evident that the nature and magnitude of these changes is strongly affected by the target location within the core. If accurate reference methods are not utilised the predicted actinide masses, and consequently the minor-actinide consumption and burn-up, may be significantly in error. The inclusion of irradiation-dependent data also has a noticeable effect on the calculated amount of helium and fission gas that is produced in minor-actinide target pins at the end of life. It is important therefore that these reference methods be taken into account during target fuel pin design.

The results obtained in the second study suggest that it is possible to have a feasible moderated target design that removes the requirement for multi-recycling. A moderated target design has been proposed that is capable of attaining a 90% mass destruction during a single residence while maintaining acceptable core neutronic and thermal-hydraulic performance. Although the inclusion of moderator material does not significantly increase the lifetime of the target, the burn-up that can be achieved during that time is much improved. While an equilibrium consumption of minor actinides has been achieved the consumption rates for moderated targets are much lower than those obtained for other core and target designs. These rates could be improved by increasing the proportion of actinides in the target material, although the scope for this would appear to be rather small due to the design limits on the peak clad damage. Further studies of these issues, and their impact on the fuel cycle, are required.

*Acknowledgements*

Discussions with colleagues in the CAPRA/CADRA project were extremely useful during the course of this work and are acknowledged. This study has been sponsored by BNFL.

**REFERENCES**

[1]     A. Vasile, G. Rimpault, J. Tommasi, K. Hesketh, R.E. Sunderland, W. Maschek, G. Vambenepe, "Core Physics Results from the CAPRA/CADRA Programme", Proceedings of GLOBAL'2001, Paris, France (September 2001).

[2]     H.M. Baumont, E.K. Whymann, R.E. Sunderland, D.P Every, "Heterogeneous Minor-actinide Recycling in the CAPRA High Burn-up Core with Target Subassemblies", Proceedings of GLOBAL'99, Jackson Hole, USA (September 1999).

[3]     T.D. Newton, P.J. Smith, S.J.Crossley, R.E. Sunderland, "Optimisation of the Gas-cooled Fast Reactor for Plutonium and Minor-actinide Management", Proceedings of PHYSOR'2000, Pittsburgh, USA (May 2000).

[4]     J.Y. Doriath, C.W. McCallien, E. Kiefhaber, U. Wehmann, J.M. Rieunier, "ERANOS: The Advanced European System of Codes for Reactor Physics Calculations", Proceedings of Int. Conference on Mathematical Methods and Super Computing in Nuclear Computations, Karlsruhe, Germany (April 1993).

[5]     E.B. Webster, "FISPIN for Nuclide Inventory Calculations: Introductory Guide for Version 7B", ANSWERS/FISPIN(98)03.

# APPLICATIONS OF "CANDLE" BURN-UP STRATEGY TO SEVERAL REACTORS

**Hiroshi Sekimoto**
Research Laboratory for Nuclear Reactors, Tokyo Institute of Technology
O-okayama, Meguro-ku, Tokyo, Japan

## Abstract

The new burn-up strategy CANDLE is proposed, and the calculation procedure for its equilibrium state is presented. Using this strategy, the power shape does not change as time passes, and the excess reactivity and reactivity coefficient are constant during burn-up. No control mechanism for the burn-up reactivity is required, and power control is very easy. The reactor lifetime can be prolonged by elongating the core height. This burn-up strategy can be applied to several kinds of reactors whose maximum neutron multiplication factor changes from less than unity to more than unity, and then to less than unity. In the present paper it is applied to some fast reactors, thus requiring some fissile material such as plutonium for the nuclear ignition region of the core, but only natural uranium is required for the other region of the initial reactor and for succeeding reactors. The drift speed of the burning region for this reactor is about 4 cm/year, which is a preferable value for designing a long-life reactor. The average burn-up of the spent fuel is about 40%; that is, equivalent to 40% utilisation of the natural uranium without the reprocessing and enrichment.

## Introduction

A new reactor burn-up concept CANDLE (*C*onstant *A*xial shape of *N*eutron flux, nuclide densities and power shape *D*uring *L*ife of *E*nergy producing reactor) is proposed [1,2]. In this strategy, shapes of neutron flux, nuclide densities and power density do not change along burn-up but move in the axial direction of a core with a constant velocity for a constant power operation during the whole reactor life (Figure 1). The burning motion can be either upward or downward, but in the present paper the upward motion is described so as to simplify the description. If this concept is feasible, a long-life reactor can be designed, for which the lifetime is easily set by adjusting the core axial length. The change of excess reactivity along burn-up is theoretically zero for ideal equilibrium conditions, and shim rods will not be required for this reactor. The core characteristics, such as power feedback coefficients and power peaking factor, are not changed during the operating lifetime. Therefore the operation of the reactor becomes much easier as compared to conventional reactors.

**Figure 1. Concept of CANDLE burn-up strategy**

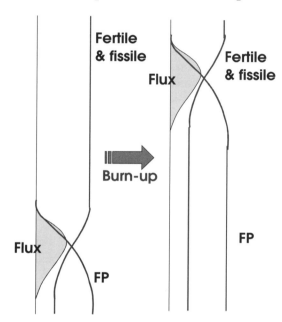

The upper part of the core is axially uniform, but the distribution of each nuclide density is complicated in the burning region. Thus, the construction of the ignition zone it is not a simple task. However, the configuration of the succeeding core is easy for this burn-up strategy. The burning zone at the end of reactor life can be used as the ignition zone of the succeeding core as shown in Figure 2.

Simple simulation is difficult to be applied without a good initial nuclide density distribution for the ignition region. Usually we do not have such distributions. It is better to find the equilibrium state at first, and to then construct ignition region by modifying the burning region. The calculation method to obtain the equilibrium state directly is given in this paper.

The CANDLE burn-up strategy can be applied to several reactors, as long as the infinite neutron multiplication factor of the fuel element of the reactor changes along burn-up in a proper way. Only fast reactor cases are presented in the present paper, though thermal reactors can offer interesting examples by introducing high fissile enrichment and burnable poisons.

**Figure 2. Concept of CANDLE burn-up and refuelling strategy**

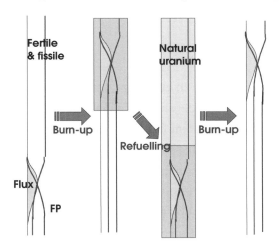

## Calculation method for CANDLE burn-up in equilibrium state

### Basic equations for CANDLE burn-up

The present work is the first step of the feasibility study for the CANDLE burn-up strategy. The purpose of this study is to find the equilibrium state which will satisfy the CANDLE requirement that distributions of nuclide densities, neutron flux and power density move axially with corresponding constant shapes and the same constant speed along burn-up for constant power operation. Once these shapes are obtained as realistic values for a given reactor design, the CANDLE burn-up strategy can be considered feasible for the design. Nuclide density distributions similar to the equilibrium ones may be used as proper initial distributions for the actual reactor as already mentioned. Many kinds of density distributions can be considered as such a distribution. However, the present study only includes finding a steady state for the CANDLE burn-up, and the study on initial conditions and practical simulations using these conditions are subjects which will be addressed in the future, and as such are not discussed in this paper.

For calculating the steady state CANDLE burn-up, a co-ordinate transformation is employed to put the burning region at rest in the transformed co-ordinate system. In this way the calculation region can be limited to a proper size of region. Otherwise, the necessary calculation region is continuously expanding, since the burning region moves steadily with the iteration of calculation. The convergence judgement is also simplified for this co-ordinate system. The actual mathematical treatment is shown below.

In the present paper a cylindrical core is considered. For an $r$-$z$ co-ordinate system, the neutron balance equation and nuclide balance equation in the core can be written as the following equations, respectively:

$$\frac{1}{r}\frac{\partial}{\partial r}rD_g\frac{\partial}{\partial r}\phi_g + \frac{\partial}{\partial z}D_g\frac{\partial}{\partial z}\phi_g - \sum_n N_n\sigma_{R,n,g}\phi_g + \sum_n N_n\sum_{g'}f_{n,g'\to g}\sigma_{S,n,g'}\phi_{g'}$$

$$+\frac{\chi_g}{k_{eff}}\sum_{g'}\sum_n N_n\nu\sigma_{F,n,g'}\phi_{g'}=0 \tag{1}$$

$$\frac{\partial N_n}{\partial t} = -N_n \left( \lambda_n + \sum_g \sigma_{A,n,g} \phi_g \right) + \sum_{n'} N_{n'} \lambda_{n' \to n} + \sum_{n'} N_{n'} \sum_g \sigma_{n' \to n,g} \phi_g \qquad (2)$$

where:

$\phi_g = \phi_g(r,z,t)$ = neutron flux in $g$-th energy group;

$N_n = N_n(r,z,t)$ = nuclide number density of $n$-th nuclide;

$D_g = D_g(r,z,t)$ = diffusion coefficient for $g$-th energy group;

$\chi_g$ = probability that fission neutron will be born in $g$-th energy group;

$k_{eff}$ = effective neutron multiplication factor;

$\sigma_{R,n,g}$ = removal cross-section of $n$-th nuclide for $g$-th energy group;

$\sigma_{A,n,g}$ = absorption cross-section of $n$-th nuclide for $g$-th energy group;

$\sigma_{F,n,g}$ = fission cross-section of $n$-th nuclide for $g$-th energy group;

$\sigma_{S,n,g}$ = slowing-down cross-section of $n$-th nuclide from $g$-th energy group;

$f_{n,g' \to g}$ = element of slowing-down matrix of $n$-th nuclide from $g'$-th energy group to $g$-th energy group;

$\sigma_{n' \to n,g}$ = transmutation cross-section of $n'$-th nuclide to $n$-th nuclide for $g$-th energy group;

$\lambda_n$ = decay constant of $n$-th nuclide;

$\lambda_{n' \to n}$ = decay constant of $n'$-th nuclide to $n$-th nuclide.

The neutron flux level is normalised by the total power of reactor. The production of FP can be evaluated using $\sigma_{n' \to n,g}$ given by:

$$\sigma_{n' \to n,g} = \sigma_{F,n',g} \gamma_{n' \to n}$$

where $\gamma_{n' \to n}$ = yield of $n$-th nuclide (FP) from neutron-induced fission by $n'$-th nuclide (actinide).

The spontaneous fission is neglected in the present study. The removal cross-section is the sum of the absorption and slowing-down cross-sections:

$$\sigma_{R,n,g} = \sigma_{A,n,g} + \sigma_{S,n,g}$$

and the absorption cross-section is the sum of fission and capture cross-sections.

The distributions of neutron flux and nuclide densities move with burn-up. Their relative shapes are constant and their positions move with a constant speed $V$ along $z$-axis for the CANDLE burn-up. For this burn-up scheme, when the following Galilean transformation [3] given by:

$$r' = r$$

$$z' = z + Vt$$

$$t' = t$$

362

is applied to Eqs. (1) and (2), then they will be changed to:

$$\frac{1}{r'}\frac{\partial}{\partial r'}r'D'_g\frac{\partial}{\partial r'}\phi'_g+\frac{\partial}{\partial z'}D'_g\frac{\partial}{\partial z'}\phi'_g-\sum_n N'_n\sigma_{R,n,g}\phi'_g+\sum_n N'_n\sigma_{n,g-1\to g}\phi'_{g-1} \tag{3}$$

$$+\frac{\chi_g}{k_{eff}}\sum_{g'}\sum_n N'_n\nu\sigma_{F,n,g'}\phi'_{g'}=0$$

and:

$$\frac{\partial N'_n}{\partial t'}=-v\frac{\partial N'_n}{\partial z'}-N'_n\left(\lambda_n+\sum_g\sigma_{A,n,g}\phi'_g\right)+\sum_{n'}N'_{n'}\lambda_{n'\to n}+\sum_{n'}N'_{n'}\sum_g\sigma_{n'\to n,g}\phi'_g \tag{4}$$

If $v = V$, the distributions of nuclide densities and neutron flux stand still and Eq. (4) becomes:

$$-V\frac{\partial N'_n}{\partial z'}-N'_n\left(\lambda_n+\sum_g\sigma_{A,n,g}\phi'_g\right)+\sum_{n'}N'_{n'}\lambda_{n'\to n}+\sum_{n'}N'_{n'}\sum_g\sigma_{n'\to n,g}\phi'_g=0 \tag{5}$$

It can be expected from this equation that the speed of the burning region, $V$, is proportional to the flux level, if the effects of radioactive decay of nuclides can be neglected.

### *Iteration scheme*

We have two equations, Eqs. (3) and (5), which should be solved simultaneously with the flux normalisation condition for obtaining the equilibrium CANDLE burn-up state. From this point the prime used for the Galilean-transformed variables is omitted for simplicity, so as not to engender any confusion in the following discussion. An iteration scheme is introduced to solve these equations. From a given flux distribution, nuclide density distributions are obtained using Eq. (5), and then from these nuclide density distributions a more accurate distribution of neutron flux is obtained using Eq. (3) with the normalisation condition. This procedure is repeated until it converges. However, in usual cases, the exact value of $V$ is unknown. Then an initial guess is introduced, and this value should be improved at each iteration stage. If the employed value of $V$ is not correct, the distributions of neutron flux and nuclide densities are expected to move along the $z$-axis. Therefore, the value of $V$ can be modified from the value of distance by which those distributions move per each iteration stage.

In order to define the position of these distributions, the centre of neutron flux distribution is introduced, which is defined as:

$$r_C=\frac{\int\phi(\mathbf{r})\mathbf{r}d\mathbf{r}}{\int\phi(\mathbf{r})d\mathbf{r}}$$

where $\mathbf{r}$ is the co-ordinate vector representing $(r,z)$ and integration is performed over the whole core. In the present paper we consider our problem more theoretical than practical. Though the height of the core is finite for the practical case, the infinite length is considered as the core height for the present study since the height is an artificial parameter and can be changed.

Next, the manner in which $V$ can be modified at each iteration step is discussed. When the velocity $V^{(i)} \neq V$ is employed, the distribution is considered to move with the velocity proportional to $V^{(i)} - V$ from the analogy of Eqs. (4) and (5). One cycle of iteration corresponds to passing the time proportional to $\Delta z/V^{(i)}$ considered from Eq. (5), where $\Delta z$ is the mesh width of the $z$-axis. Since the system in the present paper is cylindrically symmetric, only the $z$ direction should be considered. Therefore, when the $z$ co-ordinate value of $r_C$ is obtained as $z_C^{(i)}$ for the $i$-th iteration for a given velocity $V^{(i)}$, the following relation can be expected:

$$\Delta z_C^{(i)} = \alpha \left( V^{(i)} - V \right) / V^{(i)}$$

where $\Delta z_C^{(i)} = z_C^{(i)} - z_C^{(i-1)}$ and $\alpha$ is a constant. From this relation we can derive the following equation to obtain a proper estimate of $V$ for the $(i+1)$-th iteration using the results for the $i$-th and $(i-1)$-th iteration:

$$V^{(i+1)} = V^{(i)} V^{(i-1)} \frac{\Delta z_C^{(i)} - \Delta z_C^{(i-1)}}{\Delta z_C^{(i)} V^{(i)} - \Delta z_C^{(i-1)} V^{(i-1)}} \tag{6}$$

Now we have a whole iteration scheme, but it is required to use two good initial guesses of $V$, $V^{(1)}$ and $V^{(2)}$. They can be estimated from several trial calculations with different values of $V'$, where the value of $V'$ is fixed during the iteration. Here Eqs. (3) and (5) are solved repeatedly. If the height of the core is infinity, the burning region moves forever for $V' \neq V$. Though the infinite cylindrical core is considered in this paper for the sake of theoretical purity, to enable finding two good initial guesses at this calculation stage, large but finite height is treated, and a zero-flux boundary condition is set for both upper and lower core boundaries. Then, the motion of the burning region finally stops after several iterative calculations even for $V' \neq V$, since the boundary condition does not permit the burning region to pass the boundary. The iteration is converged. If the value of the attempted $V'$ is more different from $V$, then the burning region of the core continues to move until it converges and arrives closer to the boundary. The case in which the burning region stays closer to the boundary provides a smaller $k_{eff}$ value since the neutron leakage becomes larger. Therefore the $V'$ value, which gives the largest value of $k_{eff}$, should be the best candidate of the initial guess of $V$ among all trial values. By using the best two values of $V$ the iterative calculation mentioned above is begun.

## Calculation conditions and results

### Calculation conditions

The CANDLE burn-up strategy can be applied to several reactors, when the infinite neutron multiplication factor of the fuel element of the reactor changes along burn-up as follows. It starts from a value of less than unity, and increases with burn-up, becoming more than unity after a certain amount of burn-up. Its maximum value occurs at a certain value of burn-up, and then decreases with burn-up, becoming less than unity and continuing to decrease with burn-up. The spatial regions before and behind the burning region at which the neutron multiplication factor is less than unity are inevitable to fix the shape of the burning region by shifting both the front and back ends of the shape to the same direction with the same speed. The condition that the infinite neutron multiplication factor for some interval should be more than unity is also inevitable for keeping the system critical. The amount of the surplus of the infinite neutron multiplication factor above unity should be large enough to supply excess neutrons to the front region for fissile nuclide production.

To satisfy this condition an excellent neutron economy should be satisfied in the equilibrium state. Even for fast reactors it is not easy to realise this scenario when only natural uranium is charged in the front region. Only fast reactor with excellent neutron economy can realise it. A lead-bismuth-eutectic (LBE) cooled metallic fuel fast reactor, whose fuel volume fraction is 50%, is such a reactor. For this kind of neutronically excellent reactor, only natural uranium or depleted uranium may be enough for the fuel in the upper uniform region of the core. In the present study a 3 GWt LBE-cooled metallic fuel fast reactor is investigated. The parameters of the standard reactor design are shown in Table 1. The fuel and coolant are changed to other material in the present study to investigate the performance of CANDLE burn-up strategy for different reactor designs.

**Table 1. Design parameters of standard reactor design**

| Total thermal output | | 3 000 MWt |
|---|---|---|
| Core and reflector dimensions: | Core radius | 2.0 m |
| | Radial reflector thickness | 0.5 m |
| Fuel-pin structure: | Diameter | 0.8 cm |
| | Cladding thickness | 0.035 cm |
| | Pellet density | 75% TD |
| Materials: | Fuel | U-10%Zr |
| | Cladding | HT-9 |
| | Coolant | LBE |
| Fuel volume fraction | | 50% |

The group constants and their changes with respect to temperature and atomic density are calculated using part of the SRAC code system [4] along with the JENDL-3.2 nuclear data library [5]. Setting the core height infinity is impossible in the actual calculation, and it is thus set 8 m. This value is also used for the first calculation stage to find two initial guesses used for the second stage to obtain the final result with iteration.

The temperature of each region for cross-section calculations is estimated by considering the previous similar reactor calculation results [6]. Since that reactor design is considerably different from the present reactor design, the temperature distribution employed may be considerably different from the actual distribution, but it is good enough for the present study to check the possibility for performing the CANDLE burn-up strategy. After assuring the possibility of the CANDLE strategy, optimisation of the design should be performed, and in this stage thermal-hydraulic calculations should be performed to obtain accurate thermal characteristic values such as the temperature distribution.

The vacuum boundary condition is set at each outer boundary. For the first calculation stage this boundary condition is desirable for the top and bottom boundaries, since the axial movement of the burning region should be finally stopped. For the second stage the effect of the upper and lower boundary conditions should be negligibly small, since the ideal condition is the infinite length of core. When the core height for the calculation is chosen large enough so that the neutron flux and leakage may become negligibly small at the top and bottom core boundaries, the neutron flux distribution in the burning region is not affected by the change of core boundaries, and all distributions can be considered as ones for the infinite core height conditions. The effect of the boundary condition is investigated by changing the boundary condition from vacuum to reflective for the final solution in the present study.

In the present calculation 20 actinides and 66 fission products are employed. The capture cross-section of the nuclides produced by neutron capture of the nuclide at the end of the nuclide chain is assumed to be the same as the cross-section of the nuclide at the end of the chain. It is equivalent to the nuclide at the end of the chain remaining the same even after capturing neutrons.

*Obtained results and discussions*

The obtained results are shown in Tables 2 and 3. These results are obtained for the vacuum boundary condition at the top and bottom core boundaries. In order to confirm that the input core height is long enough and the employed boundary condition is proper, the boundary condition is changed to the reflective condition and the newly obtained results are compared to the original ones. The newly obtained results are the same as those shown in Tables 2 and 3. Therefore, 8 m can be considered large enough and it can be concluded that these results hold for the infinite height core case.

**Table 2. Effective neutron multiplication factor, burning velocity and spent fuel burn-up for different fuel material**

| Fuel material | Metal | Nitride | Oxide |
|---|---|---|---|
| Multiplication factor | 1.015 | 0.99 | 0.926 |
| Burning velocity (cm/s) | $1.2 \times 10^{-7}$ | $1.1 \times 10^{-7}$ | $1.5 \times 10^{-7}$ |
| Average burn-up (GWd/t) | 426 | 445 | 452 |

**Table 3. Effective neutron multiplication factor, burning velocity and spent fuel burn-up for different coolant material**

| Coolant material | LBE | He | Pb | Na |
|---|---|---|---|---|
| Multiplication factor | 1.015 | 1.035 | 1.012 | 1.006 |
| Burning velocity (cm/s) | $1.2 \times 10^{-7}$ | $1.2 \times 10^{-7}$ | $1.3 \times 10^{-7}$ | $1.2 \times 10^{-7}$ |
| Average burn-up (GWd/t) | 426 | 413 | 427 | 415 |

The LBE-cooled metallic-fuelled reactor is chosen as a reference reactor. The total thermal power output is 3 GW. The fuel is changed to nitride and oxide. The calculation results are shown in Table 2. Only metallic fuel shows the effective neutron multiplication factor more than unity. Table 3 shows the calculation results for different coolants. LBE and lead are used for the reflector for the LBE- and lead-cooled reactors, respectively, but HT9, which is used for construction material and cladding, is employed for the other cases. Helium shows a good performance as concerns neutron economy. Sodium gives the worst result from this point of view, but its neutron multiplication factor is still more than unity. The burning velocity and average burn-up do not change very much for the different coolants.

The speed of the burning region $V$ is about 4 cm/year. This is small enough to realise a long-life reactor. Even 20 years of operation requires only about 83 cm, which is less than the axial length of burning region. The average burn-up of spent fuel is about 400 GWd/t. The average burn-up shows a very high value about 40% of the inserted natural uranium. It is much higher than the LWR system, which can utilise only about 1% of the natural uranium even for the case when the reprocessing is employed for utilising its recovered plutonium. The burn-up of spent fuel for CANDLE is comparable to the FBR with the reprocessing system, whose natural uranium utilisation is usually estimated to be 60-70%. The important difference of the present system from these conventional systems is the lack of requirement for reprocessing and/or enrichment. Therefore, it should be further mentioned that the CANDLE burn-up strategy applied to a good neutron economy fast reactor is preferable not only from the viewpoint of natural resource utilisation and economics but also that of non-proliferation.

The flux distribution dangles in the radially outer region as shown in Figure 3. This is attributed to the fact that the burn-up progresses slower in this region, since the flux level is lower than in the central region. This shape is not good as regards neutron economy. It may be improved by adding fissile materials in the peripheral zone of the natural uranium region.

**Figure 3. Neutron flux distribution for CANDLE burn-up**

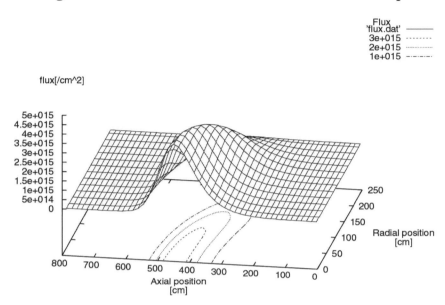

## Conclusions

The new burn-up strategy CANDLE is proposed, and the calculation procedure for its equilibrium state is presented. The power shape does not change with passing time in this strategy, and the excess reactivity and reactivity coefficient are constant during burn-up. No control mechanism for the burn-up reactivity is required, and power control is very easy. Reactor lifetime can be prolonged by elongating the core height.

This burn-up strategy can be applied to several kinds of reactors, whose maximum neutron multiplication factor changes from less than unity to more than unity, and then to less than unity. In the present paper it is applied to some fast reactors which require a fissile material such as plutonium for the nuclear ignition region of core, but only natural uranium is required for the other regions of the initial reactor and for succeeding reactors.

The drift speed of the burning region for this reactor is about 4 cm/year, which is a favourable value for designing a long-life reactor. The average burn-up of the spent fuel is about 40%. This is equivalent to 40% utilisation of the natural uranium without the reprocessing and enrichment.

Using natural uranium as the fresh fuel is an interesting possibility, as neither enrichment nor reprocessing is required after the second core. However, the neutron economy is barely marginal, and it is very difficult to realise the CANDLE burn-up under these conditions. Adding fissile materials in the fresh fuel renders realisation of the CANDLE strategy much more feasible. In this case the number of neutrons absorbed in fertile materials and fission products is decreased and the burn-up of spent fuel decreases.

The CANDLE burn-up strategy can be applied even to thermal reactors. For this case the infinite neutron multiplication factor usually decreases with burn-up monotonically during its whole history. However, by adding burnable poisons the behaviour of the infinite neutron multiplication factor can be changed such that it increases at the beginning of the core life, satisfying the condition for CANDLE burn-up. The block-type high-temperature gas-cooled reactor is a good option for this idea. The core construction of the succeeding reactor is easy, since the fuels are axially separated.

## REFERENCES

[1]     H. Sekimoto and K. Ryu, "Demonstrating the Feasibility of the CANDLE Burn-up Scheme for Fast Reactors", *Trans. American Nuclear Society*, 83, 45 (2000).

[2]     H. Sekimoto, K. Ryu and Y. Yoshimura, "A New Burn-up Strategy CANDLE", *Nucl. Sci. Engin.* (2001, to be published).

[3]     H. Goldstein, "Classical Mechanics", Addison-Wesley, Reading, Massachusetts (1950).

[4]     K. Okumura, *et al.*, "SRAC95; General Purpose Neutronics Code System", JAERI-Data/Code 96-015, Japan Atomic Energy Research Institute (1996).

[5]     T. Nakagawa, *et al.*, "Japanese Evaluated Nuclear Data Library Version 3 Revision-2: JENDL-3.2", *J. Nucl. Sci. Technol.*, 32, 1259 (1995).

[6]     H. Sekimoto and S. Zaki, "Design Study of Lead- and Lead-bismuth-cooled Small Long-life Nuclear Power Reactors using Metallic and Nitride Fuel", *Nucl. Technol.*, 109, 307 (1995).

# MOLTEN-SALT REACTORS

*Chair: W. Zwermann*

# MOLTEN-SALT REACTOR FOR BURNING OF TRANSURANIUM NUCLIDES FORMING IN CLOSED NUCLEAR FUEL CYCLE

**P.N. Alekseev, A.A. Dudnikov, V.V. Ignatiev, N.N. Ponomarev-Stepnoy,
V.N. Prusakov, S.A. Subbotin, A.V. Vasiliev, R.Ya. Zakirov**
Russian Research Centre "Kurchatov Institute", Kurchatov sq., 123182, Moscow, Russia
Phone: +7-095-196-76-21, Fax: +7-095-196-37-08, E-mail: apn@dhtp.kiae.ru

## Abstract

A concept of nuclear power technology system with homogeneous molten-salt reactors for burning and transmutation of long-lived radioactive toxic nuclides is considered in this paper. The disposition of such reactors in fuel cycle enterprises allows them to be provided with power and facilitates the solutions of problems with radwaste with minimal losses.

# Introduction

Realisation of nuclear power on large scales and for long terms is possible, if a number of conditions are met: economic efficiency, availability of resources, safety, non-proliferation of nuclear materials, ecological acceptance.

One of the main arguments for the creation of large-scale nuclear power in the 21st century is practically infinite amounts of fuel resources provided by the possibility of breeding new nuclear fuel: plutonium and $^{233}$U. Therefore, a strategy of the nuclear power development is oriented at gradual transition toward a closed fuel cycle with required (without excess and shortage) fuel breeding. The structure of the large-scale nuclear power will include reactors of various types that can be divided by their main functions:

- energy production;

- required fuel breeding;

- burning of long-lived radwaste and production of isotopes.

To engender acceptance of nuclear power for the long-term it is necessary – as of now – to pay proper attention to the elimination of the undesirable consequences of nuclear power production. Such consequences are caused by the accumulation of large amounts of long-lived radioactive toxic nuclides in reactors at various stages of the nuclear fuel cycle.

An evident proof is required to demonstrate that the potentially negative impacts of radioactive wastes emerging from nuclear power can be minimised and made to be practically harmless. This directly relates to the solution of the fuel cycle closure problem.

Ecological anxiety has resulted in the illegality of transporting any radwaste through the frontiers of a number of states. This has consequently led to an additional technical limitation: the attempt to solve the problem of spent nuclear fuel by each separate state without taking into account the installed capacity of the nuclear power sector.

Thus the problem is two-fold; not only must we propose innovative technologies and show a degree of readiness to demonstrate their efficiency, we must also try to formulate and recommend organisational steps for their implementation, as their actual creation and implementation quite probably exceed the capability of individual countries.

Today, few doubt the possibility of reliable monitored storage of radioactive nuclides with half-lives shorter than 30 years.

It is more difficult to prove the reliability of storage of long-lived radioactive toxic nuclei. An ultimate lifetime of geological disposition is considered $10^3$ years, so it is still not possible to guarantee isolation of these nuclides from the biosphere for the term of their radioactive decay (more than $10^5$ years).

Some specialists propose to transmute such nuclei into short-lived ones and, thus, to solve this problem. As applied to nuclides and within the nuclear power framework, the term "transmutation" signifies the closure of the fuel cycle on this nuclide.

A nuclear power technology system with molten-salt reactors which operates as a burner of radioactive nuclides emerging from the nuclear fuel cycle, provides electric power for enterprises and the region and produces isotopes, is thus proposed to be created at the "Mayak" enterprise. Analysis of the possible fuel compositions for MSRs shows that molten fluoride salts, which provide the necessary neutron spectrum for effective burning of radwastes and have high coefficients of electric-chemical separation between groups of elements (actinides-lanthanides), are the most acceptable.

This paper considers the return of regenerated uranium and plutonium (partly) in the power sector, while the remaining plutonium, neptunium, americium and curium are burned in molten-salt reactors to fission products with separation of them in a chemical loop of the reactor facility, the end result being the creation of a solid from of radwaste ready to be disposed of.

Conceptual technological schemes, which include LWR fuel regeneration and the preparation of radwastes for burning in a molten-salt reactor, are developed for homogeneous molten-salt reactors and solid-fuel light-water reactors. The main technological operations in these schemes are fluoridation of spent fuel with subsequent electrical-chemical separation of fuel components and fission products. Analysis of the technological schemes demonstrates an advantage of the MSR burner which consists in the fact that there is no operation of fuel rod re-fabrication in MSR [1] and, thus, loss of actinides as radwastes is potentially considerably less than for solid-fuel systems.

## Structure of future nuclear power

The amount of long-lived radwastes in an equilibrium fuel cycle of large-scale nuclear power will depend on the types and ratio of thermal and fast reactors as well as on the technology of the fuel cycle closure and the way in which the neutron balance is provided in the system.

It is necessary to begin the development of such technologies and the creation of the structures that will be required (taking into account the duration of the development). A three-component structure is possible in the future large-scale nuclear power with a closed U-Pu and Th-U nuclear fuel cycle. A system of reactors of three types is considered in this structure [2,3]. These reactors – in addition to production of energy – perform various functions:

- *Fast reactors*. Nuclear fuel breeding and provision of the necessary neutron balance in the system.

- *Thermal reactors*. Extension of the sphere of utilisation in the fuel and power balance and minimisation of Pu amounts in the equilibrium fuel cycle of the nuclear power.

- *Reactors-burners (i.e. molten salt rectors)*. Closure of the fuel cycle on MA, minimisation of long-lived radwastes (RW), production of isotopes.

The multi-component structure of nuclear power can be optimised by using the difference between the neutron properties of thermal and fast reactors, decreasing the total amount of transuranium elements in the structure. A positive neutron balance in the system of nuclear power reactors can provide not only extended nuclear fuel breeding, but also burning of the most dangerous radwastes. For these aims a special reactor-burner should be developed which operate by itself or in a subcritical regime in combination with an external neutron source.

To protect the environment from the release of long-lived radwastes (minor actinides and long-lived radioactive toxic fission products) and to solve the problems concerning full closure of the nuclear fuel cycle on long-lived radioactive toxic nuclei development and introduction in the nuclear power structure of a new element, a reactor-burner of long-lived radwastes (RW) is proposed. A molten-salt homogeneous critical reactor is chosen as such a reactor in this paper.

## Nuclear power technology complex with the molten-salt reactor for burning transuranium nuclides in the closed nuclear fuel cycle

The concept of the nuclear power technological complex with the homogeneous molten-salt reactors aimed at the burning and transmutation of long-lived radiotoxic nuclides is put forward. The placement of such a reactors at nuclear fuel cycle enterprises makes it possible to provide them with energy and facilitates the solution of the radwaste management problem with minimal losses.

The reactor based on the melting of salts, namely, the molten-salt homogeneous reactor with circulating fuel, operating as a long-lived radwaste burner in the nuclear fuel closed cycle may turn out to be an effective aid for safeguarding the environment from the long-lived radiotoxic nuclide contamination. Simultaneously, the molten-salt reactor can provide energy for the enterprise (where the spent nuclear fuel is cooled, reprocessed, where the regenerated fuel is fabricated and where the monitored and controlled short-lived radwastes are stored) and for the energy consumers surrounding the enterprise concerned.

The RRC "Kurchatov Institute", along with collaborators from OKB "Gidropress", have developed a conceptual design of the reactor installation with the molten-salt reactor (MSR) – the long-lived nuclides burner – with the main characteristics of the reactor involved being evaluated. Relying on these data the RRC "KI" and GI "VNIPIET" collaborators have substantiated the conceptual relation of the molten-salt reactor installation-burner to the RT-1 enterprise design after its modernisation.

The reprocessing of 500 t/year spent nuclear fuel from VVER-1000 type reactors with a burn-up of 50 MWt days/kg at the RT-1 enterprise is expected to annually produce the following amounts of transuranium nuclides:

- $^{238-242}$Pu – 5 330 kg/year;

- $^{237}$Np – 270 kg/year;

- $^{241-243}$Am – 230 kg/year;

- $^{243-246}$Cm – 45 kg/year.

The ultimate service period of the geological radwastes repository is reckoned to be equal to $10^3$ years. This is why it is not yet possible to ensure the isolation of radioactive nuclides from the biosphere for the entire period of their radioactive disintegration (equal to more than $10^5$).

The regenerated uranium and plutonium (partial) retrieval to nuclear power is being implemented. The left-over plutonium, neptunium, americium and curium are burned in molten-salt reactors into fission products, the latter being obtained in the reactor installation chemical loop and the solid radwastes resulting from the cycle being sent to the repository.

The power-technological complex incorporating two molten-salt reactors with a total thermal capacity equal to 5 GWt will provide the burning of up to the 600 kg/year of Np, Am, Cm and to one tonne a year of the regenerated plutonium, producing 2.2 GWe of the electrical energy. There are possibilities of a further increase in the effectiveness of burning minor actinides in the molten-salt reactors. These possibilities will be substantiated in the following stages of the research work after the verification and correction of calculation codes according to the results of the experiments made, including the transition to a subcritical operation regime with an external neutron source.

The technical-economic assessment of the complex concerned permits to hope for the economic acceptability of the application of the molten-salt reactor-burner for long-lived radwastes. The cost index of the electrical energy supplied by the system concerned is taken to be the criterion of its economic expedience. The cumulative expenditures for the energy production are diminished by the sum saved, with the savings obtained in the closing stage of the closed nuclear fuel cycle based on the existing techniques of radwaste management at the RT-1 PO "Mayak". The savings, as estimated by the GI "VNIPIET", is equal to $0.19/year transuranium nuclides.

| | |
|---|---|
| Production profitability | 58% |
| Electrical energy supply cost | 0.97 cent/kWh |
| Capital investments pay-back period | 10.8 years |

It is thus proposed to build power-technological complexes at spent nuclear fuel reprocessing enterprises with molten-salt reactors operating as burners of the radioactive nuclides emerging from the nuclear fuel cycle, generating electrical energy for the enterprise and the region and producing isotopes. The analysis of the possible fuel compositions for the molten-salt reactors shows that the most acceptable are the fluoride salts melts which provide the necessary neutron spectrums for the effective burn-up of radwastes. In addition, the fluoride salts melts have high coefficients concerning the electrochemical separation between the element groups (actinides-lanthanides).

Plutonium, neptunium, americium and curium are burned in the molten-salt reactors to the fission products, the latter accumulating in the reactor installation chemical loop and the accumulated solid wastes being sent to the repository. For the homogeneous molten-salt reactor and for the solid-fuelled light-water reactors (LWR) the conceptual technological schemes are developed. These schemes include fuel regeneration for LWR reactors and the preparation of radwastes for their burn-up in the molten-salt reactor.

The main technological procedures in these schemes are the addition of fluorine to the irradiated fuel with the subsequent electrochemical separation of the fuel components and the fission products.

The analysis of the technology involved demonstrates the advantage of the molten-salt reactor-burner (MSR-burner) over the solid-fuelled reactor systems. This advantage is that there are no procedures in the MSR for fuel rod re-fabrication and hence, the actinide losses of the radwastes may be substantially lower than in the solid-fuelled reactor systems.

The molten-salt reactor concept is of the same age as the nuclear power as a whole. It differs in principle from the traditional concept of the solid-fuelled reactor, as it allows the correction of the nuclear fuel composition without reactor shutdown. The first experimental MSR began to operate in the USA in 1951 in the framework of the programme of reactor installations for the aviation engine. In 1965 in the USA the MSRE reactor started to operate with a thermal capacity of 8 MWt, the purpose being to test the operational ability of the design's separate units, elaborate the fuel and coolant technology and to study such a type of reactor dynamic. During the four-year campaign the reactor operated successfully without the design changes using all the main kinds of fissioned fuel ($^{235}U$-$^{238}U$ and $^{233}U$-$^{232}Th$). This experiment convincingly demonstrated the possibility of creating the power MSR with a circulating fuel at the technological level of that time.

In the design projects of the Molten Salt Breeder Reactor (MSBR) and the Denaturated Molten Salt Reactor (DMSR) as the first circuit fuel carriers, salt compositions based on Li,BeF$_2$ (at a ratio of 2:1) were used. Today, the chemical and physical properties of such compositions are studied in more details, which is why the salt composition involved was used at the preliminary calculation stage for the MSR-burner.

As the transuranium concentrations are relatively low in the melts ($\leq$1% mol), the data on the properties of only the basic salt 66LiF-34BeF$_2$ may be used in the calculations of the fuel circuit with LiF-BeF$_2$-XF$_3$ (X-Pu; Np; Am; Cm), the measurement error being in the range of 10-15%.

The main technical characteristics of the reactor installation with the molten-salt reactor-burner are presented in Table 1.

A cylindrical core configuration was chosen (height – 3 m, diameter – 3 m). The reactor vessel is made of the home-made alloy of the Hastalloy HM modified type and is calculated and made for the pressure 0.5 MPa. The fluence for the fast and thermal neutrons can amount to $10^{20}$ and $5*10^{21}$ n/cm$^2$

**Table 1. Main technical characteristics of the molten-salt reactor installation**

| Characteristics | Values |
|---|---|
| Electrical power, MWe | 1 000 |
| Thermal power, MWt | 2 500 |
| Number of reactor shutdown systems | 1 – by self-drainage of the primary coolant<br>2 – by absorber rod insertion |
| Circuits number | 3 |
| **Primary circuit** | |
| Coolant | Fuel composition with the salt carrier 66LiF-34BeF$_2$ |
| Coolant melting temperature, °C | 458 |
| Coolant temperature, °C          – Core inlet | 620 |
| – Core outlet | 720 |
| Core coolant flow rate, kg/s (m/s) | 1.07·10$^4$ (5.34) |
| Gas pressure in the compensating volume (excessive), kPa | ~200 |
| Equipment layout | Integral (monoblock) |
| Circuit hydraulic resistance, kPa | 900 |
| Number of circulating pumps with electric drives | 4 |
| Electric power of one pump drive, kW | 2 000 |
| Core dimensions, m          – Diameter | 3 |
| – Height | 3 |
| Number of IHX | 12 |
| Quantity of the IHX tubes 9 × 1 in the heat exchanger | 2 977 |
| Length of the tubes heat exchanging part, m | 5.5 |
| Monoblock dimensions, m          – Diameter | 5 |
| – Height | 15 |
| Volume of materials in the monoblock, m$^3$ – Coolant | 61 |
| – Ejector with reflector | ~125 |
| – Metal (based on nickel alloy – hastalloy) | 37 |
| Number of drainage tanks | 6 |
| Volume of one tank, m$^3$ | 16 |

**Table 1. Main technical characteristics of the molten-salt reactor installation (*cont.*)**

| Secondary circuit | |
|---|---|
| Coolant | Salt 92NaBF$_4$-8NaF |
| Coolant melting temperature, °C | 385 |
| Coolant temperature, °C　　　– Steam generator inlet | 620 |
| 　　　　　　　　　　　　　　– Steam generator outlet | 470 |
| Total coolant flow rate through 8 steam generators, kg/s(m/s) | $1.1 \cdot 10^4$ (5.89) |
| Gas pressure in the compensating volume (excessive), kPa | ~200 |
| Equipment layout | Four-loop with the common points in the collectors of the heat exchangers |
| Circuit hydraulic resistance, kPa | 650 |
| Number of circulating pumps with the electric drives | 4 |
| Electric power of one pump drive, kW | 2 000 |
| Number of the drainage tanks | 8 |
| **Third circuit** | |
| Working medium | Water, steam of supercritical parameters |
| Working medium temperature, °C　　　– Steam generator inlet | 400 |
| 　　　　　　　　　　　　　　　　　– Steam generator outlet | 538 |
| Steam pressure at the steam generator outlet, MPa | 24.5 |
| Working medium flow rate, kg/s | 2 700 |
| Steam generator type | Once-through, heat exchange surface in the form of coaxial packets |
| Number of the steam generators | 8 |
| Quantity of the tubes $17 \times 3$ in one steam generator | 1 208 |
| Heat exchange surface of the tube packets of 1 steam generator, m$^2$ | 2 250 |
| Hydraulic resistance of steam generator for working medium, kPa | 3 300 |

for the 30 years of operation at 700°C. To organize the cooling of the graphite reflector with the thickness equal to 0.5 m, the share of the fuel molten salt equals 1%. The gap compensating the difference in the thermal expansion of the reactor vessel and the reflector is equal to 50 mm.

The principal hydraulic scheme has the main systems of the reactor installation with the molten salt reactor:

- primary circuit (Figure 1);

- secondary (intermediate) circuit to which belong steam generators of the steam-water circuit;

- draining and filling system of the coolant first and second circuits.

The eutectic mixture NaBF$_4$-NaF is suggested for the second circuit coolant. This mixture is cheaper than the fuel salt and its melting temperature is lower. In the steam generator a vapour with a supercritical pressure of 24.5 MPa and a temperature of 538°C are produced. The nuclear power plant full thermal efficiency is equal to about 44%.

Some auxiliary systems functionally included into the other component parts of the NPP and providing the reactor installation operation are presented on the scheme by their address indexes.

# Figure 1

*(a) Loop layout of unit*                    *(b) Integrated layout of unit*

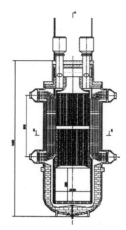

The first circuit system transmits the thermal power generated in the core volume to the intermediate circuit heat exchangers. The coolant movement along the first circuit occurs according to the following scheme.

The coolant from the core outlet plenum enters the suction header of the four circulating pumps connected in parallel. From the pumps' plenum the coolant moving downward in the space between the tubes of the intermediate heat exchanger sections enters the core inlet plenum.

The core outlet plenum and the circulation pumps' pressure tank of the first circuit are connected to the buffer tank of the first circuit through which (under the force of the operating pumps) the coolant permanent circulation is implemented in order to provide the separation of the gas fission products from the coolant. Through the coolant volume in the buffer tank the helium barbotage is performed to intensify the separation process.

The intermediate circuit heat exchanger sections are connected at the second circuit coolant inlet and outlet to the annular collectors situated outside the monoblock. The four heat exchange loops, each of which comprises a circulation pump and two steam generators connected in parallel in the secondary circuit, are connected to the annular collectors.

The first circuit coolant planned and emergency drainage is fulfilled into the drainage tanks by the gravitational forces. The coolant drainage occurs through the pipeline connecting the core inlet plenum with the drainage tanks and equipped with the "freezing" device. Each of the drainage tanks has a passive heat removal system based on the thermal tubes and the air-cooling radiators.

## Economic parameters of the nuclear technological complex

Table 2 contains the results of:

- calculations of the main economic parameters of the nuclear power technological complex (NPTC) with two molten salt reactors with an electric capacity of 1 100 MWe each (the reactor thermal power is 2 500 MWt);

- qualification of the economic expedience of the NPTC construction and operation.

## Table 2. Preliminary technical-economic indices

| | Indices | Units | Values |
|---|---|---|---|
| 1 | Number of reactor installations in the NPP | – | 2 |
| 2 | Number of hours of the installed power | hour | 8 000 |
| 3 | Operational period | years | 40 |
| 4 | Annual electric energy supply | $10^9$ kWh | 14.4 |
| 5 | Plant production efficiency in burning actinides<br>– High activity plutonium<br>– Long-lived MA | kg/year<br>kg/year | 1 000<br>550 |
| 6 | Capital investments | mln. USD | 2 726 |
| 7 | Annual operational expenditure | mln. USD/year | 252.2 |
| 8 | Tariff for purchasing the AO "Chelyabenergo" electric energy AO | cent/kWh | 2.0 |
| 9 | Cost of the selling energy | mln. USD/year | 288 |
| 10 | Annual economy in the fuel cycle of the VVER-1000 due to the transmutation of the long-lived MA | mln. USD/year | 104.5 |
| 11 | Total profit from the energy selling | mln. USD/year | 140.3 |
| 12 | Production profitability | % | 58 |
| 13 | Produced electrical energy cost | cent/kWh | 0.97 |
| 14 | Capital investments pay-back time<br>– In case of the pay-back owing to the plant profit<br>– In case of the pay-back owing to the total profit | year<br>year | 13.6<br>10.8 |

The expected capital investments for the NPTC construction and the current expenditures for its operation (the full production yield cost) are taken to be the main economic parameters in the work.

The economic expedience of constructing the nuclear power plant with the molten-salt reactors is defined by its main target – "the burning" of the most dangerous elements of the high-radiation wastes formed during the VVER-1000 reactor spent nuclear fuel reprocessing with simultaneous electrical energy production for the electrical network. The cost index of the electrical energy supplied to the consumers which is to be compared with the cost index of the electrical energy produced by other advanced reactors developed in Russia is taken to be the expedience criterion.

Due to this electrical energy production the total cost is decreased by the saved sum which may be obtained in the closing stage of the closed nuclear fuel cycle of the nuclear power plant with the VVER-1000 reactors, i.e. in the treatment of the long-lived isotopes of the transuranium elements.

The economic indices of the NPTC with the molten salt reactors are defined for its placement conditions on the PO "Mayak" territory. The expenditures saving index at the nuclear power plant with the VVER-1000 type reactors is defined according to the projects data on the treatment of the radwastes formed during the NPP spent nuclear fuel reprocessing at the RT-1 PO "Mayak" enterprise, performed by the GI "VNIPIET". All the calculations are made in 1991 basic prices. To define the economic indices in the world prices the exchange value of the rouble equal to (the USA) $1 = 1 rouble in 1991 is used.

The cost of the NPTC construction with two molten-salt reactors is defined, taking into account the following main technological components of the system:

- reactor block, including the first and second circuits filling with the eutectic melts of the fluoride compounds LiF, $BeF_2$, $NaBF_4$ and NaF (two blocks per a plant);

- steam turbine installation to generate and supply the electrical power into the network with the turbines of K-1200-2400 type (two turbines per plant);

- facilities for the chemical-technological cleaning and regeneration of the molten-salt fuel based on the pyroelectrochemical process and on the high-radiation waste reprocessing resulting in obtaining the compositions of the type of the solid monolithic natural mineral kryolite which is stored at the NPTC for 40 years (one complex for the two reactor blocks);

- auxiliary and service facilities (systems of technical water supply, energy supply, transport vehicles and communication etc.).

## REFERENCES

[1]    V.M. Novikov, V.V. Ignatiev, V.N. Fedulov, V.N. Cherednikov, "Molten-salt Reactors: Perspective and Problems", M., Energoatomizdat (1990).

[2]    S.A. Subbotin, P.N. Alekseev, V.V. Ignatiev, *et al.*, "Harmonization of Fuel Cycles for Long-range and Wide-scale Nuclear Energy System", Proc. Int. Conf. GLOBAL'95, Versailles, France, 11-14 September 1995, Vol. I, p. 199.

[3]    P.N. Alekseev, V.V. Ignatiev, S.A. Konakov, *et al.*, "Harmonization of Fuel Cycle for Nuclear Energy System with the use of Molten-salt Technology", *Nuclear Engineering and Design*, 173, pp. 151-158 (1997).

# AMSTER: A MOLTEN-SALT REACTOR CONCEPT GENERATING ITS OWN $^{233}$U AND INCINERATING TRANSURANIUM ELEMENTS

**D. Lecarpentier, C. Garzenne, J. Vergnes**
EDF-R&D, Département Physique des Réacteurs
1, Avenue du Général de Gaulle, F-92140 Clamart, France

**H. Mouney**
EDF, Pôle Industrie – Division Ingénierie et Service, Etat Major, CAP AMPERE
1, Place Pleyel, F-93282 Saint-Denis Cedex

**M. Delpech**
CEA, Direction des Réacteurs Nucléaires
Département d'Etudes de Réacteurs, Service de Physique des Réacteurs et du Cycle
CEA Cadarache, Bât. 320, F-13108 St. Paul-lez-Durance

## Abstract

In the coming century, sustainable development of atomic energy will require the development of new types of reactors able to exceed the limits of the existing reactor types, be it in terms of optimum use of natural fuel resources, reduction in the production of long-lived radioactive waste, or economic competitiveness. Of the various candidates with the potential to meet these needs, molten-salt reactors are particularly attractive, in the light of the benefits they offer, arising from two fundamental features:

- A liquid fuel does away with the constraints inherent in solid fuel, leading to a drastic simplification of the fuel cycle, in particular making in possible to carry out on-line pyrochemical reprocessing.

- Thorium cycle and thermal spectrum breeding. The MSBR concept proposed by ORNL in the 1970s thus gave a breeding factor of 1.06, with a doubling time of about 25 years.

However, given the tight neutron balance of the thorium cycle (the η of $^{233}$U is about 2.3), MSBR performance is only possible if there are strict constraints set on the in-line reprocessing unit: all the $^{233}$Pa must be removed from the core so that it can decay on the $^{233}$U in no more than about ten days (or at least 15 tonnes of salt to be extracted from the core daily), and the absorbing fission products, in particular the rare earths, must be extracted in about fifty days.

With the AMSTER MSR concept, which we initially developed for incinerating transuranium elements, we looked to reduce the mass of salt to be reprocessed in order to minimise the size and complexity of the reprocessing unit coupled to the reactor, and the quantity of transuranium elements sent for disposal, as this is directly proportional to the mass of salt reprocessed for extraction of the fission products. Given that breeding was not an absolute necessity, because the reactor can be started by incinerating the transuranium elements from the spent fuel assemblies of current reactors, or if necessary by loading just as much $^{235}U$ as is needed, we aimed to optimise core design to obtain self-generation (breeding factor = 1), while relaxing the constraints on the reprocessing unit as far as possible. This implies doing away with $^{233}Pa$ extraction and minimisation of the mass of salt to be reprocessed daily (no more than a few hundred kg for a reactor with the power of the MSBR: 1 GWe). This has the added advantage of relaxing the design requirements concerning reprocessing loss rates, which could remain appreciably the same as those obtained with the hydrometallurgy reprocessing techniques ($10^{-3}$), while guaranteeing excellent performance in terms of reducing the quantity of long-lived radionuclides sent for disposal.

We also examined the possibility of a mixed thorium-uranium composition, calculating the maximum uranium fraction that would sustain self-generation. This thorium-uranium fuel has the two-fold advantage of making better use of natural resources, while helping combat proliferation by lowering the concentration of fissile isotopes in the uranium.

## Introduction

The AMSTER molten-salt reactor concept, which has been under development by EDF for the past three years, is part of a broader examination of the development of sustainable nuclear power. This implies two vital and essential constraints:

- the regeneration of fertile fuel, failing which natural fissile material resources would be rapidly exhausted;

- minimal production of long-term radiotoxic waste.

A logical approach incorporating these two constraints considerably limits the number of candidates competing for the title "reactor of the future". We also explain how the extremely particular fuel cycle of molten-salt reactors gives them a decisive edge over the usual solid-fuel reactors. Having said this, the AMSTER concept makes maximum use of the technologies already developed at the Oak Ridge laboratory. We will see how the above constraints led us to resize the reprocessing installation initially intended for the Oak Ridge MSBR project so as to adapt it for the AMSTER concept. Finally, the calculation methods we have developed for studying the molten-salt reactor fuel cycle are briefly described. They are applied to the calculation of two particular AMSTER configurations, one incinerating on PWR transuranium waste on a thorium support, with the other being a thorium-based breeder reactor, producing minimal transuranium waste.

## Description of the AMSTER concept

AMSTER is a continuously reloaded, graphite-moderated molten-salt critical reactor, using a $^{232}$Th support.

### General presentation

Critical molten-salt reactors were extensively studied in the 60s and 70s. Research was carried out at the Oak Ridge National Laboratory, where an 8 MWh prototype, the Molten-salt Reactor Experiment (MSRE) was successfully operated between 1965 and 1969. This experiment was followed by a 1 GWe project, the Molten-salt Breeder Reactor (MSBR) [1], on which the overall AMSTER design is based. The aim at the time was breeder reactors. Today, this type of reactor is again of interest to specialists, due to its incinerating capacity. Figure 1 shows the basic layout for this type of reactor.

**Figure 1. Schematic diagram of molten-salt reactor**

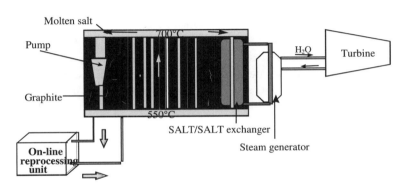

The core of a molten-salt reactor consists of an array of graphite hexagons. Each hexagon contains a hole through which the salt circulates. The diameter of the salt hole is 8 cm for a hexagon, 13 cm on a side. We used salts of the same type as those used in the MSBR project. The composition adopted for the thorium support fuels is $70LiF-15BeF_2-15ThF_4$. For our studies, we kept the MSBR design, with 48.4 $m^3$ of salt (30.4 $m^3$ in the active part of the core), weighing about 150 tonnes, and half or just over being heavy nuclides dissolved in the salt.

## Operating principle

When the salt enters the array, it becomes critical and heats up. It enters at a temperature of about 550°C and exits at 700°C. The core inlet temperature is determined by the melting temperature of the salt, which itself depends on its composition. (~500°C for the salts chosen in our study), and the exit temperature is determined by the strength of the materials other than the graphite (hastelloy).

Once heated, the salt is entrained by pumps and passes through salt/salt exchangers to heat a secondary salt. The thermal energy is then extracted from the secondary salt via a steam generator which feeds a turbine. For this study we consider a global thermal efficiency of 44%.

## Fuel salt recycling and feeding

On leaving the core, an on-line reprocessing unit takes a small fraction of the fuel for reprocessing, so as to extract the fission products (FP) which are poisoning the core. This reprocessing is accompanied by injection into the salt of new nuclei, $^{235}U$, $^{232}Th$, $^{238}U$ or transuranium elements (TRU), to replace the heavy nuclei already fissioned. Thus the salt processing unit includes the cycle front-end (salt enrichment) and back-end (fission-product extraction). An important asset of molten salts is the fact that they are immediately compatible with pyrochemical reprocessing. However, it is necessary to extract all heavy nuclides from the salt before being able to extract the FP; uranium is first extracted in $UF_6$ form, then the TRU are extracted in a salt/liquid-metal exchanger. Given a suitable separation factor in this elementary operation (about 10), and by using consecutive stages, we expect that it would be possible to ensure that the salt contains a residue of about $10^{-3}$ times the initial mass of TRU (or even less). Then the thorium and the fission products (FP), except for the long-lived fission products to be incinerated, are extracted from the salt with no need for a high separation capacity. The residual TRU in the salt are extracted with the FP, which can be vitrified and stored in the same way as fission products today. Finally, all the heavy nuclides are dissolved into the purified salt in order to be sent back into the core. In this way, they are kept in the reactor until they are completely burned, except for the losses sent for vitrification with the FP.

A key reprocessing parameter is the proportion $\alpha$ of the salt reprocessed each equivalent full power day (efpd), giving the time to reprocess the whole core: $T = 1/\alpha$. In the MSBR project, loaded with thorium fuel, this time was very short (less than 10 efpd) in order to let most of the $^{233}Pa$ decay outside the core. This major constraint was imposed by the search for a conversion factor which was as high as possible. In our studies, which did not include this goal, this constraint was removed and we chose a full core reprocessing time of 300 efpd, allowing reprocessing of a far smaller amount of salt every day. Moreover, it is certainly possible to increase this value in an optimised operating process, even if a longer time means more FP poisoning.

**Figure 2. Schematic diagram of AMSTER's fuel cycle**

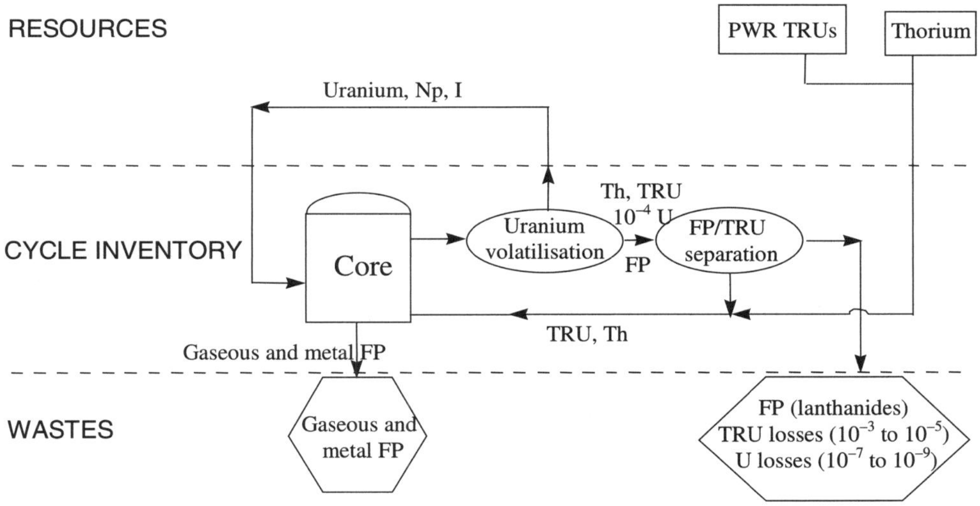

**A molten salt reactor using the thorium cycle – why?**

In this section we describe in detail the logic showing that a thorium-based molten-salt reactor is a candidate for the development of sustainable nuclear power.

**Figure 3. Reasons for choosing the characteristics of the AMSTER concept**

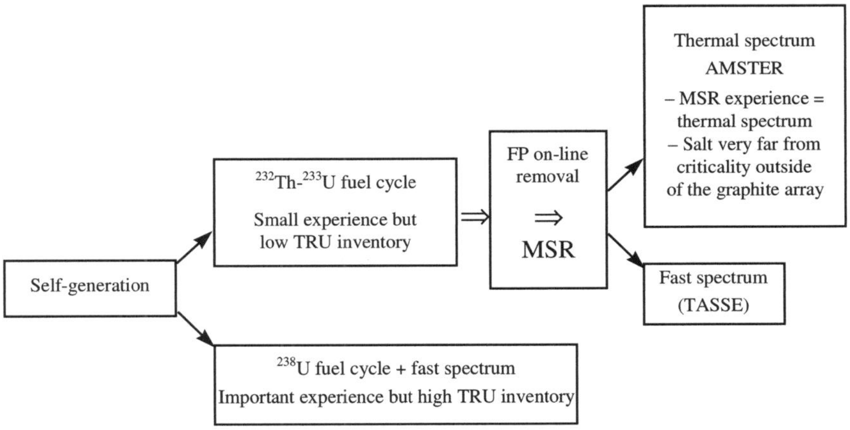

*The thorium cycle*

We began our studies on the thorium cycle for two main reasons:

1) *For incineration of the minor actinides.* The transuranium elements' inventory in the core is smaller than the cycle using $^{238}U$ as the fertile core. The radiotoxicity of the cycle is thus lower than the U/Pu cycle.

2) *The thorium cycle can ensure breeding of fertile fuel.* This breeding is possible in both the thermal and fast spectrum, whereas the $^{238}U$ cycle can only be a breeder in the fast spectrum.

In addition, the U/Pu cycle has been closely studied, while use of the thorium cycle has yet to be completely studied or optimised. We thus considered that it could be useful to examine the thorium cycle.

### The key advantages of the MSR fuel cycle

As the investment needed to develop and industrialise a nuclear reactor technology is extremely high, it is worth proving that any new technology is necessary before taking it to the industrial stage. Current technologies do not allow fast breeding for the thorium cycle as neutron economy is not as good as in the $^{238}$U cycle. It thus becomes essential to remove the fission products from the core at relatively short intervals. The core is only reasonably accessible in a liquid fuel reactor with on-line reprocessing.

This constraint more than any other should lead to the consideration of molten-salt reactor technology.

### Changes with respect to the MSBR concept

The fuel cycle chosen for the AMSTER concept is derived from that of the MSBR, but has been simplified and resized to take account of the second constraint: the reduction in long-lived waste. To reduce the production of transuranium element waste, the salt must be reprocessed as little as possible (as transuranium element losses are proportional to the quantities of salt reprocessed). We are therefore looking for a minimum core reprocessing rate, provided that we regenerate as much $^{233}$U as we burn thorium.

The MSBR involved extremely fast fuel reprocessing (in about 10 days). Reprocessing as fast as this was chosen in order to obtain maximum conversion of the thorium into $^{233}$U (which is not our priority, as simple regeneration of the fuel is enough to avoid having to enrich it). Our calculations show that it is sufficient to reprocess all the salt in about 300 days, which considerably reduces the size and the throughput of the reprocessing installation when compared with the MSBR concept.

### Additional advantages linked to the fuel form and reprocessing method

Molten salt reactors have other potential advantages over solid fuel reactors. They are examined here:

- *Reprocessing in situ.* The fact that reprocessing takes place *in situ* does away with the transportation of highly radioactive materials between the power plants and the reprocessing plant. This advantage becomes even more marked when developing self-regenerating technologies (producing as much fissile material as they consume thorium). In the same way as in breeder reactors running with the U/Pu cycle, AMSTER uses an external sub-moderated zone, regenerating more $^{233}$U than the rest of the core. This sub-moderation is obtained simply by increasing the radius of the salt hole in this zone. In a solid fuel reactor, the fuel in the fertile zone must be reprocessed in order to recover the fissile nuclei created there, to manufacture new fuel. In the molten-salt reactor, this recovery is effortless, as the fuels in the fissile and fertile zones are permanently mixed due to salt circulation.

- *"Continuous" reactivity adjustment.* As the fuel is reprocessed on-line, the reactivity can be adjusted either by modulating the input of fissile nuclei, or by altering the reprocessing rate. This results in neutron economy, as there is no need for an in-core reactivity margin. Furthermore, the fact of not having an in-core reactivity margin reduces the potential causes of accidents.

- A *"passive" safety device: the emergency drain.* In a thermal spectrum reactor such as AMSTER, the salt is easily subcritical outside the graphite array ($k_\infty \approx 0.4$). The core thus needs simply to be drained to stop the chain reaction. As the fuel is liquid, it can be extracted from the core at any time. To do this, we have adopted a concept proposed by EDF and the CEA which consists of placing a drain tank under the core, permanently connected to it. The salt is contained in the core by helium back-pressure (Figure 4). Consequently, one need simply interrupt the electricity supply to the He compressor to drain the core by gravity. This feature, allied with the considerable thermal inertia of the reactor and the problem with rapid insertion of reactivity, should lead to excellent reactor safety.

**Figure 4. Core drainage principle**

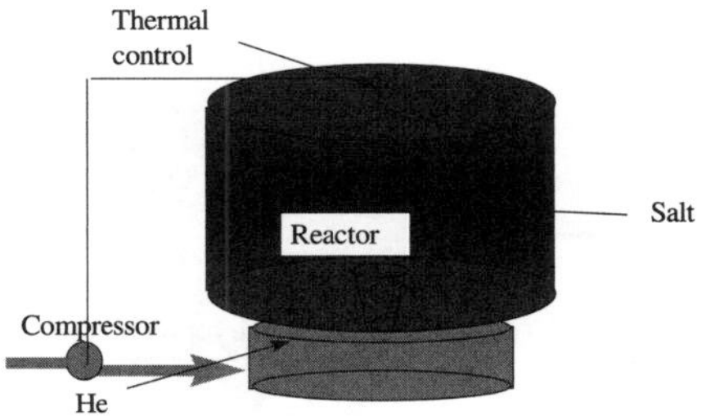

This type of arrangement is only of use in a thermal neutron spectrum. No problems with salt criticality in the heat exchangers or in the reprocessing installation are to be feared, unlike with a fast neutron reactor. These considerations, along with the fact that experience with molten-salt reactors has been with a thermal spectrum, naturally led to the choice of a thermal spectrum for the AMSTER project.

**Numerical simulation principle**

**Figure 5. The evolution of the fuel isotopic composition is simulated with the numerical method schematically shown below**

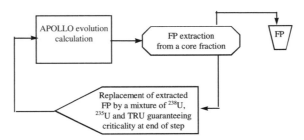

The fuel evolution under irradiation is computed by APOLLO-2 during a given time step. After this time step, a part of the FP is removed (reprocessing) and replaced by heavy nuclei, the enrichment of which is calculated to keep the reactor critical. This numerical simulation of the fuel evolution transient leads in many cases to equilibrium of the fuel composition.

Given that a core calculation is relatively lengthy, this method demands a large amount of computing time. Our aim is to quickly determine whether breeding may or may not be achieved in a reactor with a given geometry. In this way, reactor geometry optimising breeding can be determined. This optimisation requires large amounts of equilibrium calculations and we therefore use another method in order to speed up the definition of the equilibrium state.

Assuming that the reactor is exactly self-generating in fissile uranium (breeding factor = 1), it is possible to determine the equilibrium composition by an iterative matrix-based method. The $k_{eff}$ of the core loaded with this composition is then obtained by an APOLLO-2 core calculation; if the $k_{eff}$ is higher than 1, the reactor is a breeder (the higher the $k_{eff}$, the greater the breeding capacity of the reactor).

The isotopic composition $N_{eq}$ at equilibrium is computed using a matrix-based method, described in Figure 6 (the effective cross-sections are determined using cell calculations with the APOLLO-2 transport code).

**Figure 6. Direct computation of the equilibrium by a matrix iterative method**

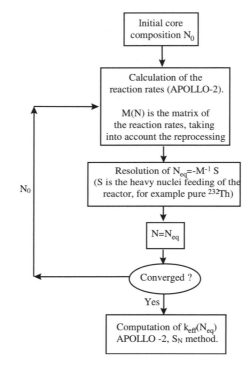

## AMSTER configuration leading to incineration of transuranium element waste and then long-term energy production

A thorium support has two major advantages over a uranium support: it produces far less TRU (the thorium chain begins with isotope 232) and consumes far less $^{235}U$ since the thorium chain can breed $^{233}U$ with an epithermal neutron spectrum.

### *The thorium support TRU incinerating configuration*

The reactor uses a fuel made of thorium and transuranium elements taken from PWR spent fuel. The proportion of transuranium elements introduced is determined so that the reactor is kept just critical. At equilibrium, the TRU inventory stabilises itself at around 2 t. It then consumes 22 kg/TWhe of TRU and 76 kg/TWhe of $^{232}$Th. Figure 7 gives the materials balance of this type of reactor at equilibrium. There is a high proportion of curium.

**Figure 7. Materials balance of an incinerating reactor at equilibrium**

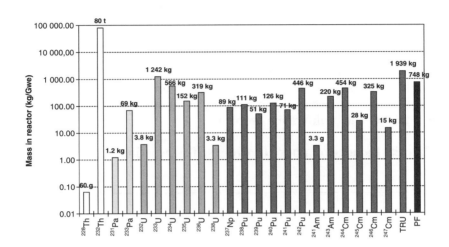

This configuration has the following advantages:

- A TRU consumption of 22 kg/TWhe requires a TRU inventory of only 2 t/GWe.

- The decay period (half-life decay period of the existing mass of TRU in the event of scheduled closure of the programme) of this reactor is short (6 years).

- No fuel enrichment is needed.

- The mass handled to burn 1 kg of TRU is small (12 kg).

### *The TRU self-consuming and uranium self-generating configuration on a thorium-uranium support*

*Feasibility of an AMSTER supplied only with thorium*

In the case of a thorium support, the consumption of fissile uranium is low enough for it to be possible to produce it in the form of $^{233}$U in an additional region of the core. This fertile region, located on the core periphery, is under-moderated by increasing the diameter of the salt hole. In this concept, the size of the fertile region would be adapted so that the reactor is only just self-generating in $^{233}$U (the production of the fertile region would exactly compensate the consumption by the fissile region).

We examined the influence on $^{233}$U production of the core height, noted H, the relative salt volume of the fertile blanket (or $V_1$ the salt volume of the fissile region, $V_2$ the salt volume of the fertile region, the ratio being $V_2/V_1$, the total volume of salt in the core being assumed to be constant and equal to $V_1 + V_2 = 30$ m$^3$), the radius of the salt hole in the fissile region, noted $r_1$ (which determines the moderating ratio of the fissile region), the radius of the salt hole in the fertile region, noted $r_2$ (which determines the moderating ratio of the fertile region).

The following domain was systematically explored: H = 400 cm to 600 cm, $V_2/V_1$ = 0.25 to 2.5, $r_1$ = 4 to 6cm, $r_2$ = 8 to 9.5 cm. In all, equilibrium was calculated for more than 700 geometries, which was only possible thanks to the matrix method, which proved to be a powerful exploratory tool. A practical optimum was determined in this range, for H = 550 cm (the core height is limited), $V_2/V_1$ = 1 (the salt volume in the blanket is limited), $r_1$ = 5 cm, $r_2$ = 9.5 cm. The results of the parametric study around this point showed that the significant parameters are the height of the core and the diameter of the hole in the fertile region. The ratio between the volume of the fertile region and the volume of the fissile region and the diameter of the fissile region hole have less influence. Figure 8 gives the material balance for a two-region core on a thorium support only, with the core at equilibrium. One should note the small quantity of transuranium elements and curium at equilibrium. This core has the advantage of consuming the theoretical minimum of $^{232}$Th and of producing very little TRU, but the uranium extracted in reprocessing is highly enriched in fissile isotopes (63%), which could pose criticality and proliferation problems.

**Figure 8. Material balance of a pure thorium self-generating AMSTER at equilibrium**

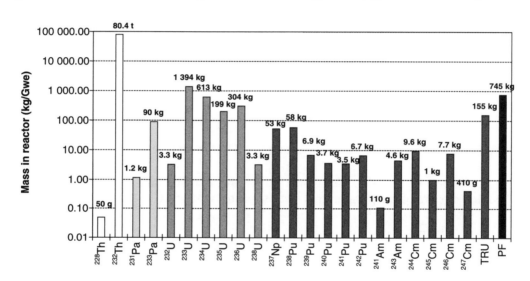

*Case of mixed thorium-uranium fuel*

It is possible to add uranium to the support, in the form of $^{nat}$U, depleted uranium, or reprocessed uranium (URT). This increases the quantity of TRU at equilibrium, while remaining within a reasonable range (400 kg/GWe). This strategy enables $^{238}$U (depleted U in stock) to be burned, leading to savings in $^{232}$Th. In addition, the fissile nuclei enrichment of the uranium in the core is reduced, which helps make the reactor non-proliferating. On the other hand, the production of $^{233}$U will have to be increased.

For this, we have taken the characteristics of one of the most regenerative reactors with thorium alone: height of 550 cm, $V_1 = V_2 = 15.2$ m$^3$, $r_1$ = 5 cm and $r_2$ = 9.5 cm, and we used a support consisting of 10% $^{238}$U and 90% $^{232}$Th. The reactor is still self-generating (it should be noted that the entire optimisation process needs to be repeated, as the optimum changes with respect to a 100% $^{232}$Th fuel).

The material balance of the fuel at equilibrium is given in Figure 9.

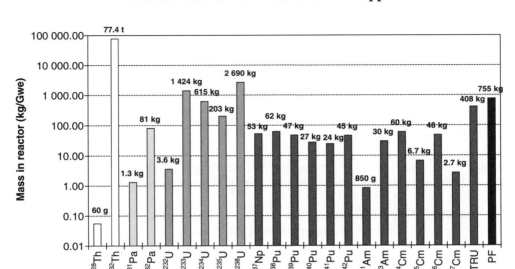

### Figure 9. Isotopic composition of the self-generating AMSTER with a 90% $^{232}$Th-10% $^{238}$U support

## Conclusion

The AMSTER configurations with a thorium support:

- produce very little TRU;

- require no $^{235}$U enrichment (savings in natural uranium, no risk of proliferation);

- can both:

  - incinerate the TRU highly efficiently;

  - produce energy with minimal consumption of heavy nuclei (100 kg /TWhe).

The efficiency of the AMSTER concept is to a large extent due to its extremely particular fuel cycle, the performance of which on an industrial scale needs to be checked by experimentation. Finally, we should point out that only the neutron aspect was considered and considerable R&D work is still required to establish the technological feasibility and safety of this concept.

# REFERENCES

[1]    "Conceptual Design Study of a Single Fluid Molten-salt Breeder Reactor", Oak Ridge National Laboratory Report, ORNL-4541, June 1971.

[2]    J. Vergnes, D. Lecarpentier, P. Barbrault, Ph. Tetart, H. Mouney, "The AMSTER (Actinides Molten-salt TransmutER) Concept", Physor'2000, Pittsburgh, May 2000.

[3]    D. Lecarpentier, "Contributions aux travaux sur la transmutation des déchets nucléaires, voie des réacteurs à sels fondus : le concept AMSTER, aspects physiques et sûreté", PhD Thesis of the Conservatoire National des Arts et Métiers, Paris, June 2001.

# THE SPHINX PROJECT: EXPERIMENTAL VERIFICATION OF DESIGN INPUTS FOR A TRANSMUTER WITH LIQUID FUEL BASED ON MOLTEN FLUORIDES

**Miloslav Hron, Jan Uhlir**
Nuclear Research Institute Řež plc
CZ-250 68 Řež, Czech Republic

**Jiri Vanicek**
Czech Power Company
Jungmannova 29, CZ-111 48 Praha 1, Czech Republic

## Abstract

The current proposals for high-active long-lived (more then $10^4$ years) waste from spent nuclear fuel disposal calls forth an increasing societal mistrust towards nuclear power. These problems are highly topical in the Czech Republic, a country which is operating nuclear power and accumulating spent fuel from PWRs and is further located on an inland and heavily populous Central European region.

The proposed project, known under the acronym SPHINX (*SP*ent *H*ot fuel *I*ncineration by *N*eutron flu*X*) deals with a solution to some of the principle problems through a very promising means of radioactive waste treatment. In particular, high-level wastes from spent nuclear fuel could be treated using this method, which is based on the transmutation of radionuclides through the use of a nuclear reactor with liquid fuel based on molten fluorides (Molten Salt Transmutation Reactor – MSTR) which might be a subcritical system driven by a suitable neutron source. Its superiority also lies in the fact that it makes possible to utilise actinides contained, by others, in spent nuclear fuel and so to reach a positive energy effect. After the first three-year stage of R&D which has been focused mostly on computer analyses of neutronics and corresponding physical characteristics, the next three-year stage of this programme will be devoted to experimental verification of inputs for the design of a demonstration transmuter using molten fluoride fuel.

The R&D part of the SPHINX project in the area of fuel cycle of the MSTR is focused in the first place on the development of suitable technology for the preparation of an introductory liquid fluoride fuel for MSTR and subsequently on the development of suitable fluoride pyrometallurgical technology for the separation of the transmuted elements from the non-transmuted ones.

The idea of the introductory fuel preparation is based on the reprocessing of PWR spent fuel using the Fluoride Volatility Method, which may result in a product the form and composition of which might be applicable as a starting material for the production of liquid fluoride fuel for MSTR. Consequently, the objective is a separation of a maximum fraction of uranium components from Pu, minor actinides and fission products. Final processing of the fuel for MSTR is proposed through the use of electro-separation methods in the fluoride melt medium, and the partitioning technologies used after passing the fuel through the MSTR should be on the basis of electrowinning.

The adjusting of a solution development and obtained results with world-wide trends will certainly be a project benefit. The project encourages a solution for the spent nuclear fuel issue as well as a limitation of their potentially negative influence upon the environment. Simultaneously, it contributes to the introduction of a prospective energy source endowed with a nearly waste-less technology.

## Introduction

After a few decades experience in the development and operation of nuclear power on an industrial scale and namely after several attempts to close the nuclear fuel cycle, there has been a renaissance of the so-called "partitioning and transmutation" (P&T) technology recognised at the end of the first nuclear era. The reason for the reappearance of this idea of a final solution of the key issue of nuclear power – spent fuel – is (to certain degree) a barrier which the so far proposed final disposal has been stuck in for the past couple of decades. The main problem is the accumulated inventory of minor actinides in spent fuel from nuclear power with uranium-plutonium fuel cycle. The broad and deep analyses that have been made in the recent stage of the first nuclear era are demonstrate with a high level of creditability that the roots of this near crisis are found in the solid fuel concept that has been thus far exclusively employed in the industrial nuclear power era. The results of the analyses mentioned above, either qualitative or quantitative, have been presented at a number of workshops and meetings devoted to innovative forms of nuclear fuel cycle and rector concepts. There is no need to repeat the whole series of sometimes very elementary ideas and conclusions namely of qualitative character. The main result and recommendation worth repeating and emphasising is the conclusion regarding the necessity to make a principle change in the nuclear reactor fuel concept – most notably, the shift from a solid fuel to a fluid fuel concept. There have been several different concepts of fluid nuclear fuel employed in various projects of innovative nuclear reactor (namely P&T) technology. The preliminary results, however sufficiently representative, of world-wide experience accumulated up to now have focused the main attention towards liquid fuel based on molten fluorides.

The problems associated with spent fuel from nuclear power plants are highly topical in the Czech Republic for several reasons. As a nuclear-power-operating country that is accumulating spent fuel from PWRs, the Czech Republic has the further considerations of being located in an inland and heavily populous Central European region. Thus, the great interest in adopting an available technology providing a possible and efficient solution to the spent fuel problem is not surprising.

An intensive study of a world-wide knowledge and experience during the first half of the nineties culminated in a close collaboration with the group of Charles Bowman, which in 1995 developed the ADTT system employing liquid fuel based on molten fluorides [1]. This concept was elaborated as a basis for a national programme of new technology development for spent nuclear fuel incineration.

Toward the end of 1996, a consortium of four of the national leading bodies in nuclear research [Nuclear Research Institute (NRI) Řež plc, Nuclear Physics Institute (NPI) of the Academy of Sciences in Řež, SKODA Nuclear Machinery (NM) Ltd. in Pilsen and the Faculty of Nuclear Science and Engineering (FNSPE) of the Czech Technical University in Prague] was established. The staffs of these organisations (grouped in the "Consortium Transmutation") elaborated a detailed plan for a national R&D programme and proposed it into a public competition launched by the Ministry of Industry and Trade of the Czech Republic in 1997. The proposal was accepted and financial support was provided for the first three-year stage of the proposed R&D programme. This first stage (1998-2000) focused on verification of the viability of the concept and the main physical and chemical characteristics at the zero power level (the project was labeled LA-0). On the basis of the results obtained, the second stage of the project, labelled LA-10, was approved in the year 2000 for an experimental verification of design inputs for a demonstration transmuter with a power output of the order of 10 MWs over the period 2001-2003. Simultaneously, it was recommended to propose an incorporation of the programme into a suitable form of multinational co-operation in the field of question. The first publication of intention was presented in mid-1996 at the 2$^{nd}$ International Conference ADTT&A'96 in Kalmar, Sweden and the results of the first half of the first stage in mid-1999 at the 3$^{rd}$ International

Conference ADTT&A'99 [2]. The complex programme is known under the acronym SPHINX (*SP*ent *H*ot fuel *I*ncineration by *N*eutron flu*X*), and was proposed as part of the European Commission's 5th Framework Programme in the framework of the MOST project in the year 2001.


## Principle features of the SPHINX project

A schematic drawing of the SPHINX project is shown in Figure 1. The main feature of the project is the front end of the fuel cycle – the dry technology of spent-fuel reprocessing from commercial reactors based on a fluoride volatility process will be employed. Thus the preparation process of the transmuter fuel in the form of molten fluorides for the transmuter fuel cycle supposing a continuous circulating of liquid fuel and its at least quasi-continuous cleaning up during the transmuter operation are undertaken simultaneously. It should be emphasised that the majority (about 95%) of the spent-fuel inventory from the uranium-plutonium fuel cycle of conventional power reactors (all uranium inventory as well as all short-lived fission products) is supposed to be taken out from the rest of the transmuter fuel cycle where all transuranium as well as long-lived fission products should remain in the mentioned form of fluorides. The second main feature of the SPHINX project is the developed concept of the transmuter, and in particular its blanket (Figure 2).

This concept is based on the idea of utilisation of an efficient burning of transuranium nuclei in the large fuel channel in the epithermal energy range where majority of their isotopes have resonances in their neutron cross-sections. The system can be either critical or subcritical. If a critical system is undertaken, the reactivity as well as power is controlled by a specific for the flowing liquid fuel control systems (based upon a control of liquid fuel composition or its flow characteristics). If the system is to be subcritical (with a subcriticality margin of the order of several betas), reactivity (keeping the prescribed level of subcriticality) is controlled in the same manner as in the above-described case of a critical system and the system is kept in a steady state and its power is controlled (driven) by an external neutron source. The subcritical fuel channels –arranged in either a critical or subcritical system – driven by an external multi-source represent in any case an intensive source of energy as well as of epithermal neutron flux which might be slowed down in graphite blocks surrounding the fuel channel and being equipped by coaxial tubes in which molten fluorides of long-lived radionuclides may flow allowing an efficient incineration by an intensive flux of well-thermalised neutrons.

The dual purpose of the proposed blanket system makes it possible to be used as an efficient nuclear incinerator and actinide burner and, simultaneously, as an efficient energy source. Let us note that once such a system is developed and proved, it might serve as a new, clean source of energy just by switching from a uranium-plutonium to a uranium-thorium fuel cycle, thus excluding a generation of actinides. The main features of the SPHINX concept and the proof of individual processes as well as the corresponding technological and operational units of that nuclear "jeep" should be verified by an operation of a demonstration complex with a power output of the order of 10 MWs. Before we are able to design such a device a broad experimental verification of design inputs for a credible design of a demonstration transmuter must be performed. This process of experimental verifications has started in the frame of the LA-10 project just recently on the basis of a preliminary preparation of an ideology of the experimental programme which was formulated earlier (in the framework of the LA-0 project, 1999-2000) and some of the principle features were also proved during that period. The experimental programs and projects developed and started up to now will be described in the following section.


## Main experimental programmes started in the frame of the SPHINX project

The two specific features of the SPHINX project mentioned above, the chemical technology of the front end of fuel cycle process and the neutronics (both static and kinetic) of the multi-purpose blanket with flowing liquid fuel, represent sufficiently new concepts to be calculated by modified computer codes that should be experimentally verified. In the case of the fluoride volatility process we base upon experience accumulated in the 80s when, in a close collaboration with the Russian Kurchatov Institute (RKI) in Moscow, such a technology was developed and a pilot line was built and prepared for testing by processing hot spent fuel from the BOR60 fast reactor at the Atomic Reactor Research Institute (ARRI) in Dimitrovgrad (Figure 3). During the 90s this line has been innovated and installed in NRI Řež and the first stage of testing for the purpose of the SPHINX project has begun.

The new blanket concept was developed and proved by broad computer analyses in the second half of the 90s. Simultaneously, experimental programmes for the verification of main neutronic characteristics, including time behaviour, have been proposed. To this purpose, several large experimental devices are operated in the framework of the Consortium Transmutation. First of all, the experimental reactor LR-0 in NRI Řež and VR-1 at the FNSPE are at our disposal, as is the NPI cyclotron accelerator that has been equipped with a new target serving as an intensive neutron source. The VR-1 experimental reactor has been equipped with an oscillator for the purpose of SPHINX-type blanket testing (namely time behaviour and safety important characteristics). The first stage of experimental verification of some basic neutronic characteristic as well as some reactor equipment, materials technology and measurement techniques has already started by experiments with inserted zones containing materials typical for our blanket concept and simulating its neutronic features in a simplified model (Figure 4).

The most important experimental programme, based on experimental irradiation of the blanket samples in the high neutron flux of research reactors, has begun just recently. This programme is being carried out in close contracted collaboration with RKI and contains a development and validation of a modified reactor computer code based on a stochastic Monte Carlo model of neutron field behaviour in the liquid-fuelled system. The experimental irradiations are simultaneously performed at NRI and RKI in the first stage. The second stage, representing irradiation of samples containing fluorides of actinides (Np, Pu, Am and Cm), has been agreed to be performed and finalised jointly at RKI over the period 2002-2003.

The technology of the new non-traditional materials in the SPHINX blanket are the subject of a broad R&D programme which includes experimental testing. For this purpose a series of technological loops called ADETE were erected by SKODA NM, starting with laboratory equipment and going on to a semi-pilot scale (Figure 5). The same is valid for the development of a technology for continuous cleaning-up of circulating liquid fuel in the transmuter internal fuel cycle based on electro-separation methods (Figure 6).

In 2000, the Technical University in Brno (TUB) began an association with the Consortium Transmutation for studying and testing the secondary cooling circuits of the LA-10 transmuter. The first step in an experimental programme in that field, a measuring bench has been erected that will serve for both measurement of basic technological characteristics of the fluoride compositions and verification of the design input for an auxiliary circuit of a molten fluoride media pantry in the large loop testing complexes being developed jointly by SKODA NM and Energovyzkum (EVM – a daughter company of TUB) as well as the LA-10 demonstration transmuter.

## Summary and conclusions on the status of the SPHINX project

The complex project of a demonstration transmuter with a power output level of the order of 10 MWs initially dedicated to nuclear incineration of spent fuel from PWRs (with a future perspective for a clean generation of nuclear energy employing the thorium-uranium cycle) based on a specific concept of liquid fuel (molten fluorides), blanket arrangement and front end as well as back end of its fuel cycle begain in the framework of a national consortium in the Czech Republic in the second half of the 90s. The first (three-year) stage (known as LA-0) having a basic R&D programme character was performed for the chosen system at the zero power level between 1998-2000. On the basis of the results obtained mainly by computer analyses performed during that period, the second stage having mainly a character of an experimental verification of design inputs for the design of a demonstration transmuter with a power output of the order 10 MWs (called LA-10) was begun in 2001 and is will continue up to 2003.

The relatively broad experimental programme having been proposed and developed during the first stage LA-0 has already started preliminary verification of basic neutronic as well as technological characteristics of individual blocks of the complex system called SPHINX (*SP*ent *H*ot fuel *I*ncineration by *N*eutron flu*X*). Namely, the results of the neutronic experiments will serve for modified computer code validation. Nevertheless, the results of chemical, material and technological testing will be used for the verification of a set of design inputs and a very important safety analyses of the LA-10 demonstration transmuter project. The whole complex SPHINX project will be reviewed in the framework of European collaboration (project MOST of the 5th Framework Programme of EC approved for two-year period commencing end 2001) in which some US and Japanese labs are also taking part. The project is also partly based on contracted collaboration with Russian partners and is of course open for other suitable forms of multinational co-operation.

## REFERENCES

[1]    M. Hron, "A Preliminary Design Concept of the Experimental Assembly LA-0 for Accelerator-driven Transmuter Reactor/Blanket Core Neutronics and Connected Technology Testing", Report LA-UR 95-3761, Los Alamos National Laboratory (1995).

[2]    M. Hron, "National R&D Program of Nuclear Incineration of PWR Spent Fuel in a Transmuter with Liquid Fuel as being Developed on the Czech Republic", ADTT&A'99, Prague-Pruhonice, 7-11 June 1999.

# Figure 1. Principle scheme of a transmuter for the SHINX project

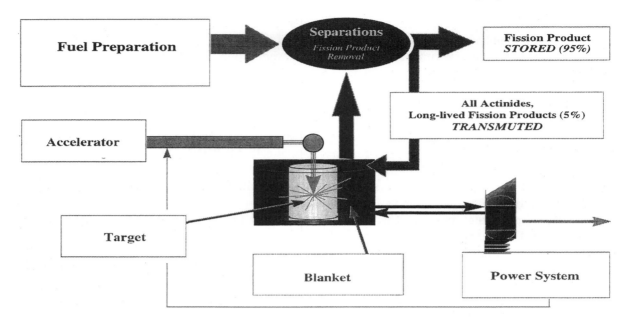

# Figure 2. The elementary module of the LA-10 transmuter for the SPHINX project

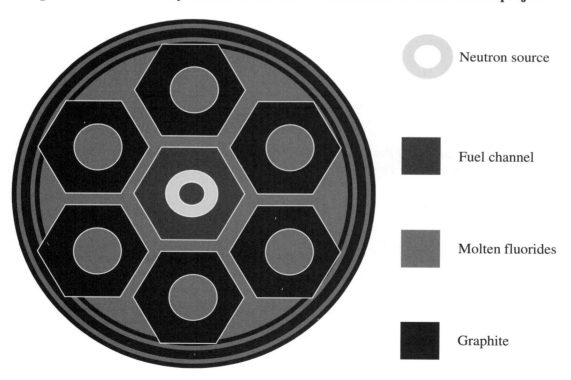

**Figure 3. NRI Řež pilot line for spent hot fuel dry reprocessing by fluorination while tested in hot cells of ARRI in Dimitrovgrad by BOR 60 spent fuel**

**Figure 4. Transmuter LA-10 blanket model for testing in experimental reactor VR-1**

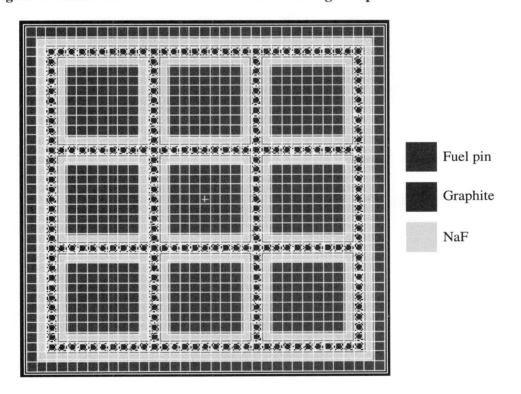

**Figure 5. The ADETE-0 loop from the series of
SKODA NM Ltd. technological loops with molten fluoride salts**

**Figure 6. Pilot separation electrolyser SKODA NM Ltd.**

# PROPOSAL OF A MOLTEN-SALT SYSTEM FOR LONG-TERM ENERGY PRODUCTION

**V. Berthou\*, I. Slessarev, M. Salvatores**
Commisariat a l'Énergie Atomique
CEN-Cadarache
F-13108 St. Paul-lez-Durance Cedex, France

## Abstract

Within the framework of nuclear waste management studies, the "one-component" concept is considered to be an attractive option for the long-term perspective. This paper proposes a new system called TASSE ("*T*horium-based *A*ccelerator-driven *S*ystem with *S*implified fuel cycle for long-term *E*nergy production") destined for the current French park renewal. The main idea of the TASSE concept is to simplify both the front end and the back end of the fuel cycle. Its major goal is to provide electricity with low waste production and economical competitiveness.

---

\* Current address: Delft University of Technology, Interfaculty Reactor Institute, Mekelweg 15, 2629 JB Delft, The Netherlands, E-mail: v.berthou@iri.tudelft.nl

# Introduction

The long-term perspective of nuclear power is an important subject of discussion nowadays. In France, within the framework of the Bataille law, numerous investigations are being conducted to find a range of solutions for the effective management of nuclear waste.

In the present paper, we propose a new concept called TASSE ("*Th*orium-based *A*ccelerator-driven *S*ystem with *S*implified fuel cycle for long-term *E*nergy production") that has been designed as a part of a long-term strategy for the renewal of the current reactor park.

The TASSE design goals have been chosen to fulfil long-term energy requirements such as: to reduce nuclear waste production, to eliminate already-accumulated waste, to improve economical competitiveness, to use natural resources in an optimal way and to enhance non-proliferation [1,2]. These features, as well as the principle of park renewal, are presented in the first part. In the second part, the equilibrium state of TASSE is presented followed by some optimisations of the concept. Then we will show some outstanding results such as the reduction of nuclear waste toxicity. Finally, the results of the transition phase are provided, in terms of PWR waste incineration performances.

# The TASSE concept

In France, numerous investigations are conducted within the framework of nuclear waste management. One approach consists of replacing the PWR reactor itself by an innovative system, which is able to produce electricity with a very low transuranium (TRU) production. This radical long-term strategy involving the renewal of the reactor park is called the "one-component" approach. The TASSE concept, which is presented in this paper, takes part in these studies [1,2].

The main idea of the TASSE concept is to simplify both the front end and the back end of the fuel cycle (e.g. neither fuel enrichment nor fuel reprocessing are foreseen, or only fission-product separation). This aspect can improve economic competitiveness as well as public acceptance.

With the purpose of a radical reduction of the waste toxicity level (at least up to 10 000 years after storage), the thorium cycle has been chosen for TASSE. Indeed, the thorium cycle has the advantage of a low production of actinides, which are the major contributors to waste toxicity. Moreover, the thorium cycle is endowed with important natural resources, which could allow a longer utilisation of nuclear energy [3].

Operating with high burn-up is crucial for the economic and environmental aspects. To reach a high burn-up level, molten-salt fuel has been considered. In addition, since the liquid fuel stays longer in the core, one can reach a higher burn-up and extract more energy from the fuel. This characteristic is crucial from the economic point of view.

The fast neutron spectrum is chosen because of its potentially superior neutron economy.

Finally, TASSE operates in subcritical mode. The reason for the subcriticality is rather particular. Since the use of a natural thorium cycle is associated with a tight neutron economy, the use of an external source is mandatory. Thus, in this concept, the utilisation of the subcritical mode is original, as it is associated with a strategic choice instead of, for example, safety enhancement reasons.

For the transition from the present reactor park to the TASSE system, the idea is to use the discharge of PWRs added to natural thorium as the initial inventory for a new TASSE reactor. In this case, two goals can be reached: an increase of the low reactivity of TASSE at the beginning (due to the natural fertile thorium fuel) and the transmutation of the "waste" already accumulated by the present reactor park.

When the park consists only of TASSE reactors, we can consider two fuel cycles: a once-through (no reprocessing of the irradiated fuel) version and a so-called "closed cycle" version (on-line separation of the fission products). In both cases, no fuel enrichment is foreseen. In the once-through cycle option, the discharge is either going to the disposal or it is used as an initial inventory to start a new TASSE reactor if the goal is to further develop the reactor park.

## Equilibrium phase

For historical and chemical reason [4], the studied molten salts are essentially fluoride ones. The main characteristics of the different molten salts studied for TASSE (in order to find an optimal one) are gathered in Table 1 [5].

**Table 1. TASSE molten salts characteristics**

| Composition (molar %) | ThF$_4$ 30%<br>NaF 24.5%<br>PbF$_2$ 45.5% | ThF$_4$ 32%<br>NaF 13.5%<br>LiF 54.5% | ThF$_4$ 12%<br>BeF$_2$ 16%<br>LiF 72% | ThF$_4$ 22%<br>NaF 11%<br>KF 67% |
|---|---|---|---|---|
| Density (g/cm$^3$) | 7 | 3.31 | 3.35 | 2.64 |
| Melting point (°C) | 600 | 525 | 500 | 535 |

The equilibrium condition is determined by two important parameters: the neutron spectrum and the choice of the burn-up level. In this case, one can find the optimum (which means as high as possible) neutron multiplication for the set of equilibrium states.

### Spectrum

Figure 1 shows the spectrum of TASSE for one molten salt (TASSEs for all the different molten salts have the same spectrum) compared to the SPX spectrum (reference for the fast spectrum). The TASSE neutron spectrum is fast, but with a significant epithermal component. This shift towards the epithermal energy range is due to the fluorine. This element is present in all salts in large quantities and has a significant inelastic cross-section (around 4 barns) in the fast energy range (Figure 2).

### The burn-up choice

For molten-salt fuel, the burn-up (BU) is defined as the percentage of fission products (FP) present in the salt. Because of the "on-line" charge and discharge of the fuel, there is a specific equilibrium state for each burn-up level. An essential feature of the TASSE concept is the choice of the appropriate burn-up. The main idea is to choose a burn-up (for the once-through cycle) corresponding to an "optimal" value of reactivity, which is approximately kept constant throughout the cycle. Figure 3 shows the relation between $k_\infty$ and the burn-up (% h.a.) level both for the case of the once-through cycle and for the "closed" cycle.

**Figure 1. TASSE spectrum compared to that of SPX**

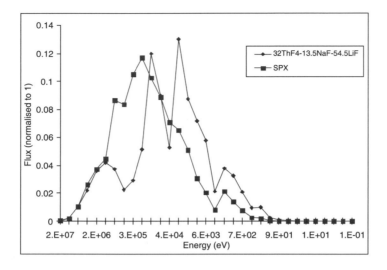

**Figure 2. Inelastic cross-section of some light isotopes contained in molten salt**

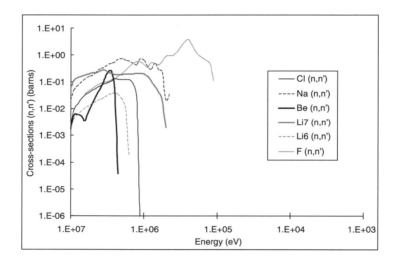

**Figure 3. Reactivity at equilibrium as a function of burn-up**

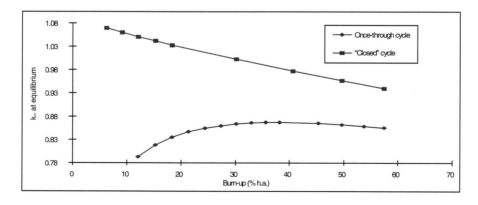

As the discharge rate is inversely proportional to the burn-up, when the BU is too low, the discharge rate is high, which makes it difficult to accumulate $^{233}$U (in the once-through cycle). Then, one can observe that the reactivity level at equilibrium increases with burn-up. However, with high BU, the discharge rate is low, and the core becomes poisoned by FPs. The neutronic balance at equilibrium is once again unfavourable. Hence, the reactivity at equilibrium has an optimal value corresponding to a certain burn-up level. In a "closed" cycle, all the heavy nuclides return to the core after FP separation. Thus, the highest reactivity level is reached for the lowest BU. Hence, the choice of the burn-up level is a compromise between the reactivity level and the discharge rate.

### Choice of the optimal salt

Among the four molten salts studied, the best neutronic results are obtained with the following two: 32ThF$_4$-13.5NaF-54.5LiF and 30ThF$_4$-24.5NaF-45.5PbF$_2$. One obtains an infinite multiplication factor k$_\infty$ in the vicinity of 0.9 for the once-through cycle with a maximum burn-up of 38% (% h.a.), and a k$_\infty$ of 1.05 for the "closed" cycle with a burn-up of 12%.

From the chemical point of view, 32ThF$_4$-13.5NaF-54.5LiF gives the best results for the solubility aspect [5]. Its major drawback consists of the tritium production by the reaction ($^6$Li + n $\rightarrow$ $^4$He + $^3$H). Nevertheless the capture cross-section of 6Li in the epithermal and fast energy range is quite low ($10^{-4}$ barns).

From the neutronic point of view, 30ThF$_4$-24.5NaF-45.5PbF$_2$ is also quite interesting, but the Pb is very corrosive.

The results concerning the reactivity level are worse with 12ThF$_4$-16BeF$_2$-72LiF, which does not contain enough heavy nuclides than the other salt.

The lowest reactivity level is obtained with 22ThF$_4$-11NaF-67KF, due to the large absorption by K.

## Optimisations

A low reactivity level is obtained with the once-through cycle, due to the non-enrichment of the fuel and the loss of fissile material during discharge. In order to improve the results for the once-through cycle some optimisations have been carried out.

### Optimisation of the flux level

Because of the "protactinium effect" the low flux gives the best reactivity level. Indeed the $^{233}$Pa can produce either $^{233}$U by decay or $^{234}$Pa by neutron capture as is shown below:

$$^{233}\text{Pa} \quad \begin{cases} \xrightarrow{\beta-(27\,j)} {}^{233}\text{U} \\ \\ \xrightarrow{(n,\gamma)} {}^{234}\text{Pa} \xrightarrow{\beta-(6.7\,h)} {}^{234}\text{U} \end{cases}$$

When the flux level increases, the production of [234]Pa by capture also increases. [234]Pa produces [234]U by decay with a half-life of 6.7 hours. This means that the concentration of an absorbing nuclide ([234]U) increases instead of the concentration of a fissile one ([233]U).

Table 2 shows the optimum of reactivity, corresponding burn-up and different flux levels for $32ThF_4$-$13.5NaF$-$54.5LiF$.

**Table 2. Optimum of the reactivity level at equilibrium for different flux level**

| Flux (n/cm$^2$) at equilibrium | $9.2 \cdot 10^{14}$ | $1.9 \cdot 10^{15}$ | $3.8 \cdot 10^{15}$ | $9.9 \cdot 10^{15}$ |
|---|---|---|---|---|
| $k_\infty$ at equilibrium | 0.88 | 0.87 | 0.85 | 0.81 |
| Burn-up | 39% | 38% | 37% | 41% |

*Optimisation of the salt component proportions*

This study aims to show the impact of the salt component proportion variations on the spectrum and on the reactivity level. In Table 3 are gathered the characteristics of one molten salt existing with three different proportions of the components [5].

**Table 3. Characteristics of $ThF_4$-$BeF_2$-$LiF$ with different proportions of the components**

| Composition | Density | Melting point |
|---|---|---|
| $ThF_4$ 8%-$BeF_2$ 22.5%-$LiF$ 69.5% | 3.1 g/cm$^3$ | 427 |
| $ThF_4$ 12%-$BeF_2$ 16%-$LiF$ 72% | 3.35 g/cm$^3$ | 500 |
| $ThF_4$ 20%-$BeF_2$ 9%-$LiF$ 71% | 2.97 g/cm$^3$ | 540 |

When the fraction of heavy nuclides in the salt increases, the spectrum shifts slightly towards higher energies. A better reactivity level is obtained with a high quantity of heavy nuclides in the salt. Nevertheless the reactivity gap between the case with 8% of $ThF_4$ and that with 20% of $ThF_4$ is only 6 000 pcm.

Molten salts with a significant proportion of heavy atoms are preferred, but real improvement in terms of neutronics are related to the chemistry possibilities. Indeed, too many heavy atoms are already present in the best molten salt ($32ThF_4$-$13.5NaF$-$54.5LiF$), and this value of 30% is now considered as a limit [5].

*Optimisation of the spectrum: chloride salt*

In order to optimise the spectrum, a chloride salt has been investigated. Since the chloride inelastic cross-section is not as large as the fluorine one, a fast spectrum can be expected, as can a better neutronic balance.

A study of $63NaCl$-$36ThCl_3$ (density 3.8 g/cm$^3$, melting point 453 degrees) was undertaken, and it was compared to the $32ThF_4$-$13.5NaF$-$54.5LiF$ salt.

Figure 4 shows the spectrum of TASSE for the two different molten salts and for the fast spectrum reference SPX. The chloride salt has a spectrum as hard as that of the SPX.

**Figure 4. Spectra of TASSE with a chloride and a fluoride molten salt, and SPX spectrum**

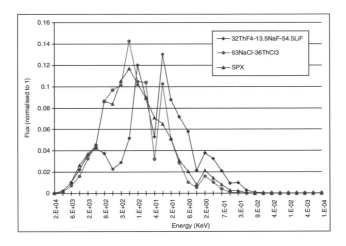

The fast spectrum increases the neutron production due to better fission cross-sections. Despite this effect, the optimum of reactivity for the chloride salt is 0.87 for a burn-up level of 44%. No improvement can be reached compared to the fluoride salt. This is due to the fact that natural chloride salt fuel suffers from intensive neutron capture by the chlorine itself.

The calculations have been done with natural chloride containing 75% of $^{35}$Cl and 25% of $^{37}$Cl. The $^{35}$Cl has a large absorption cross-section, and it produces $^{36}$Cl, which is radiotoxic with a half-life of 300 000 years. A solution could be the isotopic separation of the chlorine in order to get rid of the $^{35}$Cl. In this view we could reach a better neutronic balance for the TASSE with chloride salt using the once-through cycle.

## Outstanding results

Figure 5 shows the mass proportion of the most important heavy nuclides. Two cases are taken into account: TASSE molten salt (32ThF$_4$-13.5NaF-54.5LiF) in a "closed" cycle, and in a once-through cycle. The major isotopes are, as expected, the $^{232}$Th (around 80-90%), and the $^{233}$U (8-9%). The very low production of transuranium (less than 1%) can be noted.

**Figure 5. Mass proportions for the major heavy nuclides in TASSE molten salt at equilibrium**

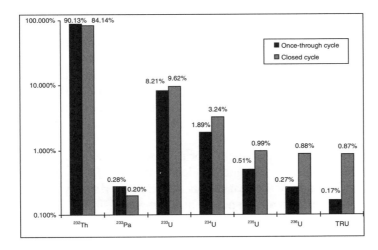

The low quantity of transuranium in the fuel at equilibrium has an important impact on the waste toxicity. For the TASSE concept, in both cycles, a significant gain in terms of toxicity is obtained (in comparison with PWR discharge). In the once-through cycle, a factor 20 to 1 000 (depending on the time scale) is foreseen, and the toxicity reduction for the "closed" cycle is expected to be $10^4$ (Figure 6).

**Figure 6. Toxicity (Sv/GWe.years) as a function of the time for TASSE in once-through and closed cycles compared to a PWR discharge**

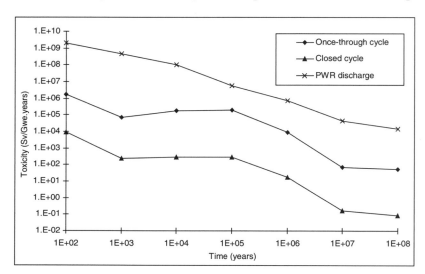

Finally, it should once again be emphasised that operating at a subcritical level is mandatory. The strategic choice of the fuel cycle simplification leads to an original utilisation of the ADS concept. In this respect, we can consider the use of the accelerator as a way to eliminate a complicated fuel cycle.


**Transition phase**

The transition phase is the period in which the PWR park could be replaced by the TASSE park.

Natural thorium added to the transuranium (TRU) discharged from PWRs is taken as the initial inventory for the first TASSE reactor. The aim is to increase the initial reactivity level (which is too low with only natural thorium) and to incinerate the PWR waste. Only natural Th composes the feed.

This study has been performed with the $32ThF_4$-$13.5NaF$-$54.5LiF$ salt using a once-through cycle. The maximal quantity of TRU in the initial inventory is determined by the chemistry. Indeed, for solubility reasons, the value of 6% (molar) of $TRUF_4$ in the fuel is considered a limit [5]. This leads to an initial reactivity of 0.9, which is kept constant during the path to equilibrium. In terms of mass this corresponds to an insertion of 10 tonnes of TRU in the fuel. At equilibrium, only 0.4 tonnes of TRU are present.

Figure 7 shows the time evolution of the major isotopes' concentration. One can see the evolution towards equilibrium for the $^{233}U$ and $^{234}U$, and the decrease of the TRU concentration. Finally, 95% of the TRU are burnt in approximately 45 years. This last characteristic is also important regarding the necessity to reduce the TRU stock in a time period as short as possible.

**Figure 7. Time evolution of major isotope concentrations**

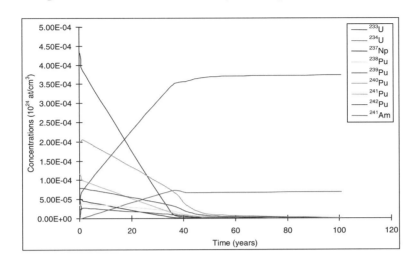

## Conclusion

This paper presents new developments concerning a new reactor concept known as TASSE ("*T*horium-based *A*ccelerator-driven *S*ystem with *S*implified fuel cycle for long-term *E*nergy production"). TASSE is an electricity producer, and, in a long-term perspective, could be an option for the renewal of the current French park. TASSE is capable, in an extreme case, of operating without either fuel enrichment or fuel reprocessing. More realistically, only the fission product separation is currently foreseen.

An important result should be pointed out: the strategic choice of the fuel cycle simplification leads to an original utilisation of the ADS concept. In this respect, we can consider the use of the accelerator as a way to simplify the fuel cycle.

In once-through cycle the reactivity level at equilibrium is quite low and TASSE would likely need a large proton current. None of the optimisations could lead to an efficiency improvement. Only two possibilities are foreseen: either an isotopic separation of the natural chlorine in order to eliminate the $^{35}$Cl, or the use of fluoride salt with a higher actinide concentration in order to obtain a harder neutron spectrum.

In terms of performances, TASSE has demonstrated its burner capabilities. During the transition phase towards a renewal of the park, TASSE incinerates transuranium elements already accumulated by the current PWR park. In a 1 GWe TASSE, about 10 tonnes of TRU can be loaded. They are incinerated within about 50 years. This feature is important regarding the necessity to reduce the TRU stock in a time period as short as possible.

Due to the low production of transuranium elements, waste toxicity is drastically reduced. TASSE, with the once-through cycle, offers a gain of 20 to 1 000 (depending on the time scale) in comparison with PWR discharged fuel in a reprocessing scenario. For the "closed" cycle, the toxicity reduction factor is expected to be $10^4$.

Finally, TASSE could have non-proliferation features since its fuel cycle is free of fuel enrichment processes, while reprocessing (if any) does not imply isotopic or element separation.

# REFERENCES

[1]    M. Salvatores, I. Slessarev, V. Berthou, "Role and Key Research & Development Issues for Accelerator-driven Systems", *Progress of Nuclear Energy*, Vol. 38, Issues 1-2, pp. 167-178 (2001).

[2]    V. Berthou, I. Slessarev, M. Salvatores, "The TASSE Concept (*T*horium-based *A*ccelerator-driven *S*ystem with *S*implified Fuel Cycle for Long-term *E*nergy Production)", GLOBAL'01, Paris, Palais des Congres, 9-13 September 2001.

[3]    H. Gruppelaar, J-P. Schapira, "Thorium as a Waste Management Option" Petten, 8 Sept. 1999, 21125/99.27350/C.

[4]    A. Weinberg, M. Rosenthal, R. Briggs, P. Kasten, *et al*., "Molten-salt Reactors", *Nuclear Applications and Technology*, Vol. 8, February 1970.

[5]    P. Faugeras, private communications.

# ACCELERATOR-DRIVEN SYSTEMS

## *Chair: Ch. DeRaedt*

# EXPERIMENTAL INVESTIGATIONS OF THE ACCELERATOR-DRIVEN TRANSMUTATION TECHNOLOGIES AT THE SUBCRITICAL FACILITY "YALINA"

**Sergei E. Chigrinov, Hanna I. Kiyavitskaya, Ivan G. Serafimovich, Igor L. Rakhno*,**
**Christina K. Rutkovskaia, Yurij Fokov, Anatolij M. Khilmanovich, Boris A. Marstinkevich,**
**Victor V. Bournos, Sergei V. Korneev, Sergei E. Mazanik, Alla V. Kulikovskaya,**
**Tamara P. Korbut, Natali K. Voropaj, Igor V. Zhouk, Mikhail K. Kievec**
Radiation Physics & Chemistry Problems Institute, National Academy of Sciences
Minsk-Sosny, 220109, Republic of Belarus
E-mail: anna@sosny.bas-net.by

## Abstract

The investigations on accelerator-driven transmutation technologies (ADTT) focus on the reduction of the amount of long-lived wastes and the physics of a subcritical system driven with an external neutron source. This paper presents the experimental facility "Yalina" which was designed and created at the Radiation Physics and Chemistry Problems Institute of the National Academy of Sciences of Belarus in the framework of the ISTC project #B-070 to study the peculiarities of ADTT in thermal spectrum. A detailed description of the assembly, neutron generator and a preliminary analysis of some calculated and experimental data (multiplication factor, neutron flux density distribution in the assembly, transmutation rates of some long-lived fission products and minor actinides) are presented.

---

* Present address: Fermilab, MS 220, P.O. Box 500 Batavia, IL 60510, USA.

## Introduction

Many theoretical papers have been published in which the basic aspects of accelerator-driven systems (ADS) have been addressed: production of energy, transmutation of radioactive waste (ADTT), tritium production and incineration of weapons-grade plutonium [1-4]. The experimental research in this field is rather scarce because the experiments on available high-energy accelerators are difficult and expensive, and in some case are simply not feasible. In this regard experimental research concerning various aspects of ADS and ADTT on the basis of low-energy ion accelerators is of great importance. A subcritical assembly with thermal neutron spectrum driven by the neutron generator NG-12-1 was designed and constructed at the Radiation Physics and Chemistry Problem Institute of the National Academy of Sciences of Belarus in the framework of the ISTC project #B-070 to investigate the peculiarities of transmutation technologies and the neutronics of ADS [5-6].

## Experimental facility "Yalina"

The experimental facility "Yalina" consists of a subcritical uranium-polyethylene assembly, a high-intensity neutron generator and vital support systems (Figure 1).

### Figure 1. The subcritical facility "Yalina"

*1 – Deuteron accelerator, 2 – Ti-$^3$H target system, 3 – The subcritical assembly*

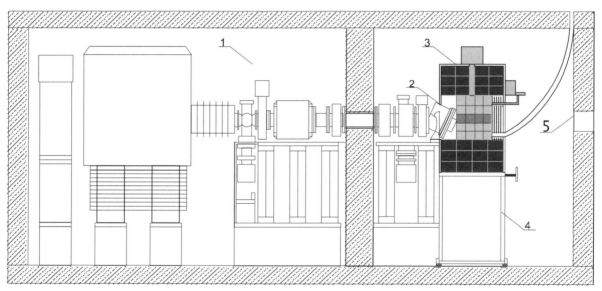

The neutron generator NG-12-1 [I ~ $10^{10}$ n/s in (d,d) and ~$10^{12}$ n/s in (d,t) modes] operates under two regimes: continuous and pulsed. The neutron pulse duration can be varied from 5 to 100 microseconds and the pulse repetition can be changed from 5 to 1 000 Hz.

The subcritical uranium-polyethylene assembly is a zero power subcritical assembly driven with a high-intensity neutron generator. The core is a rectangular parallelepiped 40 cm width, 40 cm length and 57.0 cm height. It is assembled from polyethylene blocks with the channels to place the fuel pins. The core has a square lattice with 2.0 cm pitch. The central part of the subcritical assembly is a neutron-producing lead target (8 cm × 8 cm × 60 cm).

There are four channels of 50 mm diameter each for location of detectors of the neutron flux monitoring system at the boundaries of the core and three experimental channels with diameters of 25 mm for radii 5, 10 and 16 cm for placing different types of samples or $^{252}$Cf source inside the core.

The core is surrounded by a 40-cm thick high-purity graphite reflector and a thin (1.5 mm) Cd layer. There is one radial experimental and two axial channels with diameters of 25 mm in the graphite reflector. $UO_2$ dispersed in Mg matrix fuel (enrichment is equal 10%) was loaded in the core of the subcritical assembly.

The number of fuel pins providing a $k_{max}$ of less than 0.98 of the subcritical assembly with the chosen material composition and geometry and a lead target introduced into the central part of the assembly is 280. Figure 2 shows the calculated neutron spectra in the experimental channels of the assembly with 2.5 MeV neutrons as an external source (I ~ $10^{10}$ n/s). It should be noted that the neutron spectrum in experimental channels of the uranium-polyethylene subcritical assembly is unique and differs from the neutron spectra in any other reactor.

**Figure 2. Calculated neutron spectra in the experimental channels of the subcritical assembly**

*It is supposed that of neutron source intensity equals $10^{12}$ n/s*

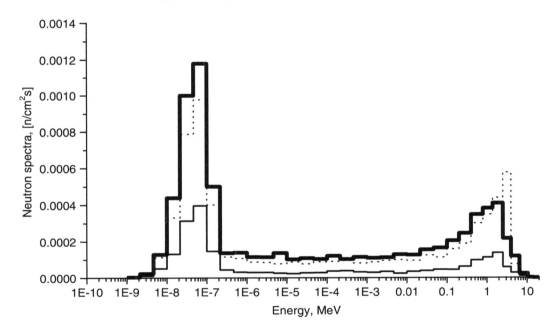

**Experimental measurements**

Up to now the experiments at the subcritical facility "Yalina" were carried out in (d,d) mode for neutron generator and with a $^{252}$Cf source (I ~$10^7$ n/s). Measurement of $K_s$ with a $^{252}$Cf source at the loading of 280 fuel pins has been performed using a source multiplication method and is equal to 0.9675 with an accuracy of within 2%. It should be noted that measured and calculated (MCNP-4B code [7]) $K_s$ are in good agreement.

The axial distributions of $^{232}$Th relative fission rates were calculated [7] and measured using solid track nuclear detectors (Figure 3). The high value of the $^{232}$Th fission rate in the first channel (R = 5 cm) shows the influence of external neutrons (2.5 MeV) on the spectrum in this channel.

**Figure 3. $^{232}$Th fission rates axial distributions in the subcritical assembly irradiated with neutrons from (d,d) reaction**

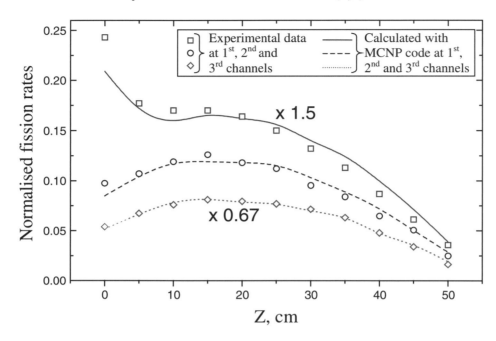

The measurements of neutron flux density dependence versus the type and position of an external neutron source and in the core have been performed at the level of subcriticality $K_s = 0.9675$. Figures 4 and 5 show axial distribution neutron flux density at irradiation with 2.5 MeV neutrons [NG-12-1 was working in (d,d) mode] and with a $^{252}$Cf source placed in the core at Z = 25 cm.

**Figure 4. Axial distribution of neutron flux density in the experimental channels of the assembly driven with neutron generator**

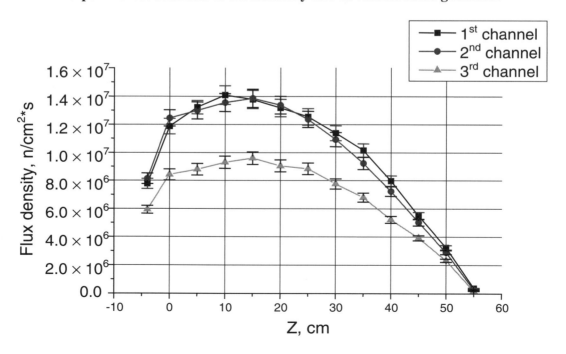

**Figure 5. Axial distribution of neutron flux density in the experimental channels of the assembly at the irradiation with $^{252}$Cf source (Z = 25 cm)**

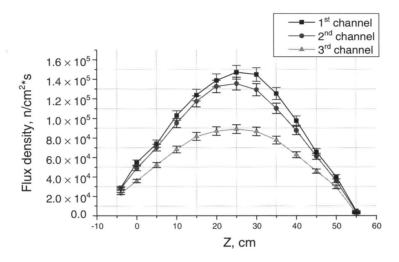

The first measurements of transmutation rates for the samples $^{129}$I, $^{237}$Np and $^{243}$Am were performed at the level of subcriticality $K_s$ = 0.9675. Table 1 shows the characteristics of the samples being used in the experiments. The irradiation was carried out over two hours with 2.5 MeV neutrons and measurements were performed during a one-hour period using an HPGe detector. The samples were irradiated in the experimental channels (R = 5, 10 and 16 cm) and at Z = 10, 25 and 40 cm. The gamma spectra of the iodine sample after irradiation are shown in Figure 6. The comparison of calculated and measured distributions of axial reaction rates for $^{237}$Np, $^{129}$I and $^{243}$Am in the experimental channels of the subcritical assembly "Yalina" are presented in Figures 7-9 (preliminary results). One can see satisfactory agreement between calculated and measured data.

**Table 1. Characteristics of irradiated samples**

| Sample | Activity, $10^8$ Bq | Mass, mg | Admixture |
|--------|--------|--------|--------|
| NaI | 0.044 | 1 | < 17% $^{127}$I |
| NpO$_2$ | 0.113 | 366 | < 0.2% $^{239}$Pu |
| AmO$_2$ | 1.100 | 14.8 | < 0.2% $^{239}$Pu |

**Figure 6. Gamma spectrum of the$^{129}$I samples irradiated (during 2 hours) in the subcritical assembly driven with neutron generator working with deuterium target**

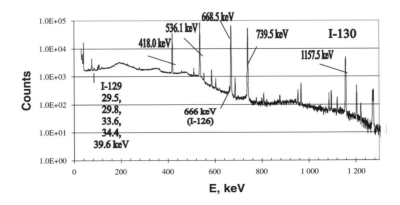

**Figure 7. Axial distribution of $^{237}$Np(n,γ)$^{238}$Np reaction rates in 2$^{nd}$ experimental channel (R = 10 cm) of the subcritical assembly driven with neutron generator**

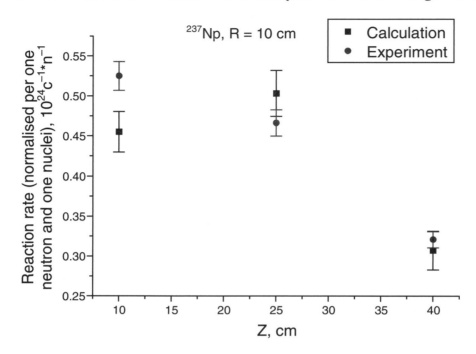

**Figure 8. Axial distribution of $^{129}$I(n,γ)$^{130}$I reaction rates in 1$^{st}$ experimental channel (R = 5.0 cm) of the subcritical assembly driven with neutron generator**

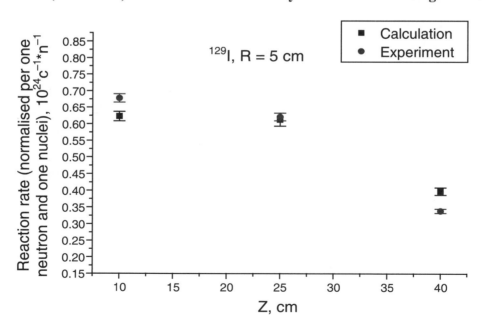

**Figure 9.** Axial distribution of $^{243}$Am(n,γ)$^{244}$Am reaction rates in $1^{st}$ experimental channel (R = 5.0 cm) of the subcritical assembly driven with neutron generator

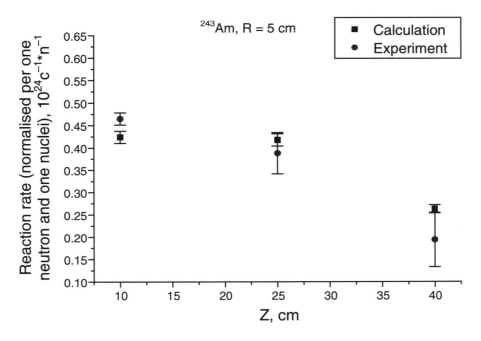

## Conclusion

The subcritical assembly with thermal neutron spectrum driven by the neutron generator NG-12-1 was designed and constructed at the IRPCP NAS B in the framework of the ISTC project #B-070 to investigate the peculiarities of transmutation technologies LLRW and neutronics of ADS with thermal spectra. It was shown that calculated and measured characteristics are in satisfactory agreement. The preliminary experimental data on the transmutation rates for $^{129}$I and $^{237}$Np in the subcritical facility as a prototype of feature ADS were presented. The pulse mode of operation of neutron generators allows to study dynamic characteristics of ADS using the subcritical facility "Yalina". These investigations will be continued at the facility driven with a 14-MeV neutron source in the near feature.

*Acknowledgements*

The authors are grateful for EU funding of the ISTC (Moscow, Russia) project No. B-070-98.

# REFERENCES

[1]    F. Venneri, C.D. Bowman, S. Wender, "The Physics Design of Accelerator-driven Transmutation Systems", GLOBAL'95, Versailles, France, Vol. 1, pp. 474-481, 11-14 September (1995).

[2]    M. Salvatores, I. Slesarev, M. Uematsu, "Global Physics Approach to Transmutation of Radioactive Nuclei", *Nucl. Sci. and Technol.*, 116, 1 (1994).

[3]    W. Gudowski, "Accelerator-based Technologies – Swedish, European and other Activities", Proc. 2nd Int. Symposium, Liblice, 24-25 September (1995).

[4]    C. Rubbia, *et al.*, "Experimental Study of the Phenomology of Spallation Neutrons in a Large Lead Block", European Organisation for Nuclear Research, CERN/SPSLC 95-17, SPSC/P291.

[5]    S. Chigrinov, A. Kievitskaia, I. Serafimovich, *et al.*, "Experimental and Theoretical Research on Transmutation of Long-lived Fission Products and Minor Actinides in a Subcritical Assembly Driven by a Neutron Generator", ADTTA'99, Prague, 7-11 June 1999, Mo-O-B14 (CD-room edition).

[6]    S. Chigrinov, A. Kievitskaia, I. Serafimovich, *et al.*, "A Small-scale Set-up for Research of some Aspects of Accelerator-driven Transmutation Technologies", Mo-O-C5, ibid.

[7]    J.F. Briesmeister, *et al.*, "MCNP – A General Monte Carlo Code for Neutron and Photon Transport", Report LA-7396-M, Los Alamos (1989).

# MYRRHA, A MULTI-PURPOSE ADS FOR R&D: PRE-DESIGN PHASE COMPLETION

**H. Aït Abderrahim, P. Kupschus, Ph. Benoit, E. Malambu, K. Van Tichelen,**
**B. Arien, F. Vermeersch, Th. Aoust, Ch. De Raedt, S. Bodart, P. D'hondt**
SCK•CEN, Boeretang 200, B-2400 Mol, Belgium

**Y. Jongen, S. Ternier, D. Vandeplassche**
IBA, Louvain-la-Neuve, Belgium

## Abstract

SCK•CEN, the Belgian Nuclear Research Centre, in partnership with IBA s.a., Ion Beam Applications, is designing an ADS prototype, MYRRHA, and is conducting an associated R&D programme. The project focuses primarily on research on structural materials, nuclear fuel, liquid metals and associated aspects, on subcritical reactor physics and subsequently on applications such as nuclear waste transmutation, radioisotope production and safety research on subcritical systems.

The MYRRHA system is intended to be a multi-purpose R&D facility and is expected to become a new major research infrastructure for the European partners presently involved in the ADS demo development.

## Introduction

One of the SCK•CEN core competencies is and has at all times been the conception, design and realisation of large nuclear research facilities such as BR1, BR2, BR3, VENUS reactors, LHMA hot cells, or HADES underground research laboratory (URL) for waste disposal. SCK•CEN has then proceeded to successfully operate these facilities thanks to the high degree of qualification and competency of its personnel and by inserting these facilities in European and international research networks, contributing hence to the development of crucial aspects of nuclear energy. One of the main SCK•CEN research facilities, BR2, is now arriving at an age of 40 years just like the major materials testing reactors (MTR) in the world and in Europe [HFR (EU-Petten), OSIRIS (F-Saclay), R2 (Sw-Studsvik), etc.]. MYRRHA has been conceived as a potential facility for replacing BR2 and to be a fast spectrum facility complementary to the thermal spectrum Réacteur Jules Horowitz (RJH) facility in France. This situation would give Europe a full research capability in terms of nuclear R&D.

Furthermore, the disposal of radioactive wastes resulting from industrial nuclear energy production has still to find a fully satisfactory solution, especially in terms of environmental and social acceptability. As a consequence, most countries with significant nuclear power generating capacity are currently investigating various options for the disposal of their nuclear waste. Scientists are looking for ways to drastically reduce (by a factor of 100 or more) both the volume and the radiotoxicity of the high-level waste (HLW) to be stored in deep geological repositories, and to reduce the time needed to reach the radioactivity level of the fuel originally used to produce energy.

This can be achieved via the development of the partitioning and transmutation and burning MAs and, to a lesser extent, LLFPs in ADS. The MYRRHA project contribution will be particular in helping to demonstrate the ADS concept at reasonable power level and the demonstration of the technological feasibility of MA and LLFP transmutation in real conditions.

## Principle choices of MYRRHA facility

### Introduction

Taking into account what is said above, the MYRRHA team has developed the MYRRHA project based on the coupling of an upgraded commercial proton cyclotron with a liquid Pb-Bi windowless spallation target, surrounded by a Pb-Bi-cooled subcritical neutron-multiplying medium in a pool-type configuration (Figure 1). The spallation target circuit is fully separated from the core coolant as a result of the windowless design presently favoured in order to utilise low-energy protons without drastically reducing the core performances.

The core pool contains a fast spectrum core, cooled with liquid Pb-Bi, and several islands housing thermal spectrum regions located in in-pile sections (IPS) at the periphery of the fast core or in the fast core. The fast core is fuelled with typical fast reactor fuel pins with an active length of 600 mm arranged in hexagonal assemblies. The central hexagon position (or the three central hexagons, depending on the adopted fuel assembly design) is (are) left free for housing the spallation module. The core is made of fuel assemblies made of MOX typical fast reactor fuel with Pu contents of 30% and 20%.

Taking into account the above-summarised MYRRHA tasks profile, the MYRRHA project team is developing the MYRRHA project as a multi-purpose neutron source for R&D applications on the basis of an accelerator-driven system (ADS). This project is intended to fit into the European strategy towards an ADS demo facility for nuclear waste transmutation. It is also intended to be a European, fast neutron spectrum, irradiation facility allowing various applications, such as:

- *ADS concept demonstration.* Coupling of the three components at rather reasonable power level (20 to 30 MWth) to allow operation feedback and reactivity effects mitigation.

- *MAs transmutation studies.* High fast flux level needed ($\Phi_{>0.75 \text{ MeV}} = 10^{15}$ n/cm$^2$·s).

- *LLFPs transmutation studies.* High thermal flux level needed ($\Phi_{\text{th}} = 1$ to $2.10^{15}$ n/cm$^2$·s).

- *Radioisotopes for medical applications.* High thermal flux level needed ($\Phi_{\text{th}} = \sim 2 \cdot 10^{15}$ n/cm$^2$·s).

- *Material research.* Need for large irradiation volumes with high constant fast flux level ($\Phi_{>1 \text{ MeV}} = 1 \sim 5.10^{14}$ n/cm$^2$·s).

- *Fuel research.* Need for irradiation rigs with adaptable flux spectrum and level ($\Phi_{\text{tot}} = 10^{14}$ to $10^{15}$ n/cm$^2$·s).

- *Safety studies for ADS.* To allow beam trip mitigation, subcriticality monitoring and control, optimisation of restart procedures after short or long stops, feedback to various reactivity injections.

- Initiation of *medical* and new *technological* applications such as proton therapy and proton material irradiation studies.

**Figure 1. MYRRHA subcritical reactor with proton beam injection from the top**

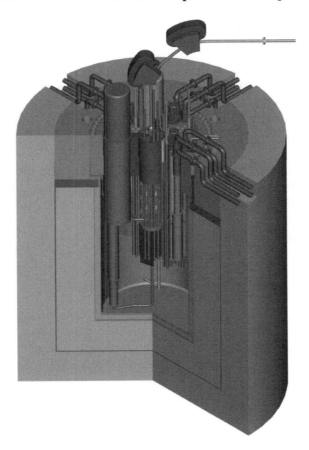

The flexibility and the versatility of the applications it would allow determine the present MYRRHA concept. Some choices are also conditioned by the timing of the project. Indeed as we intended to achieve the operability of MYRRHA before 2010, the project team has favoured the mature technologies or the less demanding in terms of development, for example concerning fuel and accelerator. Nevertheless, all the components of MYRRHA are not existing off the shelf. Therefore, a thorough R&D support programme for the risky points of the project has been started and is summarised in this report.

## MYRRHA: Critical reactor versus ADS?

Having the listed applications above that are considered in MYRRHA except those related to ADS demonstration, one can ask why not in a critical reactor? Indeed, nowadays material and fuel research are conducted in critical MTRs, radioisotopes are produced in these machines, transmutations studies could be also conducted in critical reactors *but*:

- Using critical thermal reactor technology, even when going to very high dense cores and very high enriched fuel (93% w/o $^{235}U$), the highest total flux level one can achieve is $10^{15}$ n/cm$^2$·s, mainly dominated by thermal flux (about 80%). In principle, these flux levels can be used for LLFPs transmutation experiments but they will already require very long irradiation times (1 to 2 years) for obtaining burn-up levels of the 1~2% that are needed for performing relevant radiochemistry analysis with uncertainties limited to a few %. These are the minimum transmutation levels needed for integral validation experiments. Since in an ADS the absolute flux level is determined by the spallation neutron source intensity, one can overcome these intrinsic limitation of the critical reactor.

- Due to the same limitation for the thermal flux level, radioisotope production will be limited to the classical isotopes produced via single capture or fission reactions used mainly for diagnostic purposes in nuclear medicine departments. But if one is considering the curative radioisotopes like α-emitters that are produced by double capture or long-lived generators such as $^{188}W$ for $^{188}Re$ that is also produced by double capture, one needs to go to thermal flux levels above $10^{15}$ n/cm$^2$·s. Therefore the ADS option would also be an asset here.

- MA transmutation demands fast spectrum irradiation, as one should favour fission reactions as compared to capture reactions. In fast spectrum critical reactors once again the total flux level is about $10^{15}$ n/cm$^2$·s, whereas the most effective part of the neutron spectrum for favouring the fission of the MAs lies above 0.75 MeV. Therefore, as in the case of the transmutation of LLFPs, the ADS whose flux level is driven by the spallation source intensity will permit higher flux levels than the critical fast reactor. This is certainly an advantage, especially for an experimental facility where one would be conducting development experiments and thus would desire to run accelerated experiments as compared to the real MA burner machine.

- For structural material research under irradiation or fuel research and development, these research topics can be conducted evenly in MTRs as well as in an experimental ADS. Nevertheless, one can mention here the need for a fast spectrum irradiation facility that would be complementing a thermal spectrum one that would be available in 2010. Indeed, hardening the irradiation spectrum of a thermal MTR such as RJH by spectrum tailoring techniques would be very demanding and will never reach the flux levels that can be obtained in a fast spectrum ADS.

- Besides these applications that can be completed in a MTR but with better performances in an ADS, one should mention that all the R&D activity related to the ADS development can be triggered only if one is developing and building such a facility. The challenging aspect of such a project should be stressed here as an incentive for maintaining the know-how and the expertise in the nuclear field. Nuclear energy is contributing to a large extent to the electricity provision of Europe and will still do so for the next two decades at least, even if the present phase-out strategy is maintained as the future policy in Europe. The development of such an innovative project will be an asset for attracting a new generation of scientists and engineers towards the nuclear sector.

For all these reasons and particularly the complementarity with a future European MTR, such as the RJH project, it seems that choosing the ADS orientation is the most relevant option for developing a new fast spectrum R&D facility.

### Required design parameters of MYRRHA

The performances of an ADS in terms of flux and power levels are dictated by the spallation source strength that is proportional to the proton beam current at a particular energy, and the subcriticality level of the core. Thus having in mind the targeted performances required by the different applications considered in MYRRHA system, as summarised below:

| | |
|---|---|
| $\Phi_{>0.75\,MeV}$ | $= 1.0 \times 10^{15}$ n/cm$^2$·s at the locations for MAs transmutation |
| $\Phi_{>1\,MeV}$ | $= 1.0 \times 10^{13}$ to $1.0 \times 10^{14}$ n/ cm$^2$·s at the locations for structural material and fuel irradiation |
| $\Phi_{th}$ | $= \sim 2.0 \times 10^{15}$ n/ cm$^2$·s at locations LLFPs transmutation or radioisotope production |

Taking into account that the MYRRHA facility is intended to be put into operation before 2010, it is obvious that one should go towards not-too-revolutionary options for the accelerator that is to deliver the needed proton beam as well as for the fuel option.

Considering the above-mentioned constraints, we had to fix the subcriticality level of the subcritical assembly in order to define the needed beam power intensity to achieve the above performances. A subcriticality level of 0.95 has been considered an appropriate level for a first of kind medium-scale ADS. Indeed, this is the criticality level accepted by the safety authorities for fuel storage. Besides this aspect, we considered various incidental situations that can lead to reactivity variation such as: Doppler effect, realistic water intrusion, temperature effect, voiding effect of the spallation module, voiding effect of the coolant in the core and core compaction. We found that the majority of those effects would bring a negative reactivity injection or a limited positive reactivity injection not leading to criticality in any case when starting at a $K_s$ of 0.95.

To design a subcritical core having a $K_s$ of 0.95 and a fast spectrum it was obvious to go towards existing fast reactor MOX fuel technology for keeping the design time and building time within the time frame of the project. Thus the upper limit of the fuel enrichment we put to ourselves was 30% w/o total Pu with the Pu vector of reprocessed fuel from PWRs.

## Nuclear gain and spallation source intensity

Having fixed the subcriticality level – determining the nuclear gain – as well as the desired neutron flux in the position of the irradiation location for MA transmutation, this will determine the required strength of the neutron spallation source. Nevertheless, one still has a degree of freedom in achieving the needed performances in the core via the geometrical design of the core, especially the central hole in the core that will be housing the spallation target module. In order to achieve the above-mentioned performances in terms of fast flux levels for relevant MAs transmutation experiment at the lowest possible total power of the SC, we had to limit the central hole diameter to a maximum of 120 mm, thus putting the above irradiation location for the MAs at roughly 5 cm radius. As a consequence of this constraint and on the other hand having the need of a minimum lateral Pb-Bi volume for allowing an effective spallation process, intra-nuclear cascade as well as evaporation processes, the proton beam external diameter is limited to ~70 mm. The required strength of the spallation source to produce the desired neutron flux at this location is approaching $2 \cdot 10^{17}$ n/s. At the energy choice discussed in the next section, this requires 5 mA of proton beam intensity and this in turn would lead to a proton current density on an eventual beam window of the order of 150 µA/cm$^2$. This is by at least a factor of 3 exceeding the current density of other attempted window design for spallation sources which are already stressed to the limit and have high uncertainties with regard to material properties suffering from swelling and radiation embrittlement. As a result of these constraints and the comparison with other designs, we favoured the windowless spallation target design in the MYRRHA project.

## Required accelerator

Having fixed the subcriticality level, we started the design work with a 250 MeV × 2 mA cyclotron as advised by the AMSC and that would have been a slightly upgraded machine from the cyclotron developed by IBA for proton therapy application. However, this beam power level does not allow to meet the neutronic performances demanded from the core. Therefore, we first had to increase the proton energy to 350 MeV, as the gain on the neutron intensity due to the energy increase is more than linear. Indeed, the neutron multiplicity at 250 MeV is ~ 2.5 n/p whereas it is ~5-6 n/p at 350 MeV. Despite this energy increase of the incident protons we also had to increase the proton current to 5 mA to arrive at the required source strength. The final proton beam characteristics of 350 MeV × 5 mA then permitted to reach a fast neutron flux of $1.10^{15}$ n/s (E > 0.75 MeV) at an acceptable MA irradiation position under the geometrical and spatial restrictions of subcritical core and spallation source. This upgrade is also regarded as being within the extrapolable performance of IBA cyclotron technology. Also compared to the largest continuous wave (CW) neutron source – SINQ at PSI with its cyclotron generated proton beam of 590 MeV and 1.8 mA – it is a modest extrapolation and well within the conceptual extrapolation of the SINQ to the "PSI Dream Machine" with the proton parameters of 1 GeV × 10 mA.

The MYRRHA cyclotron would consist of four magnet segments of about 45° (cf. Figure 2) with two acceleration cavities at ca 20 MHz RF frequency. The diameter of the active field would be of the order of 10 m and the diameter of the physical magnets of order 16 m with a total weight exceeding 5 000 t. The handling of only part of the components requires lifting capabilities of at least 125 t, and according provisions in the building need to be made.

## Subcritical core configuration

As already mentioned above, due to the objective of conceiving a fast spectrum core, and due to the fact that no revolutionary options were considered, we naturally started the neutronic design of the

## Figure 2. The MYRRHA cyclotron

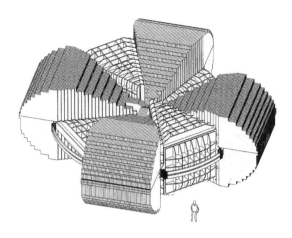

subcritical core based on MOX fast reactor fuel technology. As we wanted to limit the technological development to the choice of cladding material being compatible with Pb-Bi, the initial request was to limit the Pu enrichment to maximum 30% in weight and the maximum linear power to 500 W/cm. Due to the low proton energy chosen (350 MeV), leading to a spallation neutron source length of about 13 cm (penetration depth of protons), it was also decided to limit the core height to 60 cm. This height is compatible with the purpose of MYRRHA to be an irradiation facility for technological developments. The fuel assembly design had to be adapted to the coolant characteristics especially for its higher density as compared to the Na. A first core configuration with a typical SuperPhenix hexagonal fuel assembly (122 mm plate-to-plate with 127 fuel pins per assembly) with a modified cell pitch answering the requested performances has been conceived. Nevertheless, this configuration is subject to the large radial burn-up and mechanical deformation stress gradients that will render fuel assembly reshuffling difficult or even impossible. That is why we also conceived in parallel a smaller fuel assembly, 85 mm plate-to-plate, with 61 fuel pins per assembly allowing a larger flexibility in the core configuration design. The active core height is kept at 60 cm and the maximum core radius is 100 cm with 99 hexagonal positions. All the positions are not filled with fuel assemblies but could contain moderating material (ZrH$_2$ pins, 6 hexagons) around a hexagonal position that becomes a thermal neutron flux trap ($\Phi_{th}$ = ~2.10$^{15}$ n/cm$^2$·s). There are 19 such positions able to house thermal flux traps. These positions are accessible through the reactor cover. They could house fast neutron spectrum experimental rigs equipped with their own operating conditions control on top of it via the reactor cover. All the other position can be housing either fuel assemblies or non-instrumented experimental rigs. About half of the positions should be filled with fuel assemblies to achieve a K$_s$ of 0.95.

It is worth mentioning here one particularity of the Pb-Bi as a coolant. Indeed due to its high density (10.737), the fuel assemblies will be floating in this coolant. Therefore, we decided not to plug the fuel assemblies in a supporting plate but to implant them from beneath in the top core plate that is fixed to the diaphragm separating the cold zone from the hot zone of the primary circuit of the reactor. The fuel assemblies as well as any non-instrumented experimental rig or moderating assembly will then be manipulated from beneath for its positioning in the core. By doing so one is easing the access to the experimental position from the top of the reactor.

Two interim storages for the used fuel are foreseen inside the vessel on the side of the core. They are dimensioned for housing the equivalent of two full core loadings, ensuring in this way that no time-consuming operations must take place in the out-of-vessel transfer of fuel assemblies or waiting for the (about) 100 days of cool-down.

## Operation fuel cycle

The MYRRHA operation cycle will be determined by the $K_s$ drop as a function of the irradiation time or core burn-up. Taking into account the power density distribution in the core, we ran evolution calculations for the core and we observed the following: an initial $K_s$ sharp drop of about 1 800 pcm ($K_s$: $0.95 \rightarrow 0.932$) after five days of irradiation time due to fission-product build-up. After reaching a sort of equilibrium we then observe a smooth decrease of $K_s$ with a slope of 5 pcm/day.

Thus the first thing to do would be to overcome the initial $K_s$ drop, either by an active compensation (higher initial $K_s$ that can be compensated by partial coolant voiding as the voiding coefficient is negative) or by a passive one (conditioning the fuel assemblies thanks to a pre-irradiation outside the core for a longer period than five days). Both approaches are presently under consideration.

The targeted operating regime is three months of operations and one month for core reshuffling and loading. This will lead to a $K_s$ drop per cycle of 450 pcm at maximum, as this value does not take into account the coupled effect of the linear power drop during operation. This will correspond to a multiplication factor drop from 20 to 18.3, thus about ~9%. Core reshuffling bringing the less-burned peripheral fuel assemblies towards the core centre would allow to partially compensate for this loss of $K_s$.

The intermediate cooling time between two irradiation cycles does bring an extra accumulation of absorbing materials via delayed radioactive decay. The objective is to maintain the $K_s$ drop within the 10% range by using core reshuffling and partial reloading of fresh fuel with a total residence time of the fuel in the core of three years, i.e. 810 EFPD (equivalent full power days). This objective looks to be achievable but requires more investigation.


## MYRRHA subcritical reactor configuration

Due to the main objective of the MYRRHA facility of obtaining very high fast flux levels, it was obvious that we should go towards a design of a fast reactor core. As we wanted to realise our objectives within a limited time development and due to the high linear power to be achieved, it is obvious that a gas fast reactor option would be very difficult to realise. Indeed, at normal operation conditions, the thermal-hydraulic problems related to use of helium (or carbon dioxide) coolant in the MYRRHA subcritical core could be resolved only by using high pressures (100-150 bar) and then by optimising the operation parameters and the fuel rod bundle design. However, even at such high pressure, the power of circulation in the gas loop is very high (~2 to 4 MW for $CO_2$ or He as compared to 0.2 MW for Pb-Bi). This power level is comparable with that needed for the proton beam. Further, a gas-cooled ADS is less robust under accidental conditions than an ADS cooled by liquid metal. A depressurisation accident is the major safety concern. Very special means must be anticipated in order to cool down the core at the reduced pressures of gas. MYRRHA being intended as a flexible experimental facility made the use of gas as impractical as incompatible with this goal. Therefore we discarded the gas option in our design.

When considering the liquid-metal option two designs were possible: the loop and the pool options. The loop option has been discarded due to the very high vessel exposure and thus the radiation damage it would undergo, the risk of LOC and LOF accidents, the difficulty of the inter-linking of the spallation target loop with the primary reactor cooling loop due to the above-mentioned optimisation process. Finally, one should mention the desired flexibility in loading and unloading experimental devices.

The pool design has been favoured because it avoids the penetration from beneath the spallation target circuit into the main vessel and thus enhances the safety of the design. It also allows having an internal interim storage, easing the fuel handling. The natural circulation (free convection) for the extraction of the residual heat removal in case of loss-of-heat-sinks (LOHS) is certainly easier to achieve, particularly with the large thermal inertia that is also an argument in favour of this design.

## *Safety considerations*

Even if for ADS one of the main characteristics that is desired is to achieve an inherent safety of the system, one should not underestimate the safety considerations for preparing the licensing of such an innovative system. The following reactivity perturbation initiating conditions have been studied in the MYRRHA system: power increase leading to average temperature increase, Doppler effect, spallation source level positioning (leading to voiding or filling of the central channel in the core with Pb-Bi), partial core voiding due to fuel elements blockage. All these situations lead to negative reactivity effects. Whereas the following situation: pitch compaction, loading faults (30% enriched fuel assemblies loaded instead of 20%), water ingress from in the core, could lead to slight or heavy reactivity increase and thus have been integrated in the design to avoid their occurrence.

From the safety point of view, the aim is to reduce both the probability of the events and their associated off-site consequences in order to avoid the need for extensive countermeasures and to offer the authorities the possibility of simplifying or declaring unnecessary off-site emergency planning. This is the well-known "in-depth defence safety approach" that is adhered to in the MYRRHA design.

One of the main accidents to be considered is a loss of flow accident resulting from the failure of the circulation pumps. In such a case, natural convection will take place and the following question arises immediately: is the natural circulation sufficient to remove the decay heat released by the core after reactor shutdown? "Sufficient", in this context, signifying a total lack of fuel damage.

When the emergency cooling was studied in a first approach, the design of the MYRRHA subcritical reactor was not very far advanced and it was therefore impossible to simulate the free circulation accurately, so only very rough models could be used. The purpose of the study was to provide very general indications on the possibility of cooling the reactor by natural convection.

Due to the lack of detailed information on the whole reactor design, a parametric approach has been followed. Three main free parameters were used: the total pressure drop in the primary circuit, the cross-sectional area of the pipes simulating the primary circuit and the difference of height between the core and the heat exchanger.

The main conclusions of the study are as follows:

- Even in the worst cases, the coolant temperature remains much lower than the Pb-Bi boiling point, so no loss of heat transfer caused by vapour formation at the clad-coolant interface has to be feared.

- The fuel behaviour is fully safe, because the power drop in the reactor very rapidly reduces the fuel temperature, averting any risk of melting.

- Concerning the clad behaviour, the situation is less comfortable: a temperature peak is observed at the beginning of the transient, proportional to the flow deceleration, and a maximum temperature nearing 700°C is reached in the present design configuration.

- Lowering of temperatures in the fuel rods, in particular in the cladding, can be obtained:

    – by minimising the pressure drops in the circuit, e.g. by reducing as much as possible the local pressure losses.

    – by increasing the difference of elevation between the heat exchanger and the core, but this is limited by design constraints, especially with the pool reactor concept.

For the future, those results will have to be confirmed:

- First by refining the data for which some uncertainties subsist, such as the law of time power decrease.

- Secondly, when the circuit design is better defined, by using more sophisticated tools, like RELAP5 adapted for lead-bismuth for instance, which could allow to model and simulate the system much more accurately.

### Examples of the performances of the proposed design

Both present designs of MYRRHA (large fuel assembly and small fuel assembly) are delivering the expected performances in terms of fast and thermal fluxes, linear power in the core and total power. Table 1 summarises the main parameters of both configurations of MYRRHA.

**Table 1. Summary of the main parameters of both configurations of MYRRHA**

| | Neutronic parameters | Units | Values | |
| | | | Large assembly configuration | Small assembly configuration |
|---|---|---|---|---|
| **Spallation source** | Ep | MeV | 350 | |
| | Ip | mA | 5 | |
| | n/p yield | | 6.0 | |
| | Intensity | $10^{17}$ n/s | 1.9 | |
| **Subcritical core** | $k_{eff}$ | | 0.948 | |
| | $K_s$ | | 0.959 | 0.965 |
| | Importance factor | | 1.29 | |
| | $MF = 1/(1 - K_s)$ | | 24.51 | 28.64 |
| | Thermal power | MW | 32.2 | 35.5 |
| | Average power density | $W/cm^3$ | 231.5 | |
| | Peak linear power | W/cm | 475.4 | 582 |
| | Max flux > 1 MeV<br>  Close to the target<br>  First fuel ring | $10^{15}$ n/cm$^2$s | 0.94<br>0.83 | <br>0.85 |
| | Max flux > 0.75 MeV<br>  Close to the target<br>  First fuel ring | $10^{15}$ n/cm$^2$s | 1.30<br>1.17 | <br>1.16 |
| | Number of fuel pins | | 2 286 | 2 745 |

## MYRRHA project time schedule

The main milestones of the project time schedule are given below:

- Mid-2002, at which time a decision is needed for the start of the detailed engineering phase and a budget increase would be needed not only for the funding of the team to be devoted to this detailed design but also for testing the required large-scale elements such as the heat exchanger or fuel assemblies mock-up in Pb-Bi.

- End 2003, by which the detailed engineering design as well as the business planning should be sufficiently advanced to allow a decision as to when to begin the building phase of the subassemblies. In parallel the R&D support programme for corrosion and spallation module design should have delivered their results to allow the start of the construction.

- End 2006, termination of the building of the sub-components (accelerator, spallation, module and subcritical core elements) and the start-up of their commissioning individually.

- Mid-2008 will be the integration of the sub-components and commissioning of the full ADS.

- Beginning 2010, will be the start-up of MYRRHA operation at full power for routine use of the facility.


## Conclusion

ADS can become an essential and viable solution to the major remaining problems of nuclear energy production, largely related to the disposal of radioactive waste, especially in terms of environmental and social acceptability.

MYRRHA can provide the indispensable first ADS step without freezing all options for the larger scale ADS: liquid Pb-Bi versus gas, subcriticality level, mitigation tools for reactivity effects, mitigation tools for proton beam issues, licensing issues, etc.

MYRRHA is an innovative project that will trigger research in the fields of accelerator reliability, transmutation, development of new materials and fuels, structural material corrosion and embrittlement and safety of nuclear installations.

# OPTIMISATION OF CONCEPTUAL DESIGN OF
# CASCADE SUBCRITICAL MOLTEN-SALT REACTOR

**A. Vasiliev, P. Alekseev, A. Dudnikov, K. Mikityuk, S. Subbotin**
Russian Research Centre "Kurchatov Institute", Kurchatov sq., 123182, Moscow, Russia
Phone: +7-095-196-70-16, Fax: +7-095-196-37-08, E-mail: avas@dhtp.kiae.ru

## Abstract

Some of the results of a study aimed at the selection of main parameters and optimisation of a Cascade subcritical molten-salt reactor concept are presented in the paper. This accelerator-driven system is envisaged for integration into the nuclear power structure for minor-actinide burning and transmutation of long-lived fission products. A feature of this project is the increase in the number of external neutrons due to their multiplication in the reactor central zone with a high $k_\infty$ value. Proposals regarding the composition of fluoride salts/fuel carriers, equilibrium vector of nuclides to be burned, reactor design, etc., are considered in the paper. The main limitations are formulated and problems to be solved at subsequent stages are stated. The study is undertaken in the framework of ISTC#1486 Project.

# Introduction

A concept of Cascade subcritical molten-salt reactor (CSMSR) [1] is currently being considered at the RRC "Kurchatov Institute". The primary circuit of the CSMSR option with circulated fuel molten-salt compositions, proposed at a given stage of the study, consists of:

- a graphite vessel with a cylindrical cavity of the main core, designed for minor-actinide (MA) burning and transmutation of long-lived fission products;

- a cylindrical zone, inserted in the reactor centre and designed for multiplication of external neutrons, with its own fuel molten-salt composition and circulation circuit;

- a proton beam pipe and solid target, inserted in the centre of the central zone;

- intermediate heat exchangers (IHX) built in the graphite vessel;

- bypass technological circuit, etc.

The CSMSR concept can be considered as an accelerator-driven system (ADS) option, optimised for both MA burning efficiency and a target design. It should be taken into account that the main CSMSR task is not energy production, which is a by-product of the CSMSR operation, but effective burning of MA for minimisation of equilibrium amounts of the most dangerous elements in the nuclear power structure.

External neutrons in CSMSR are generated in the central zone as a result of protons interacting with the target nuclei. The CSMSR central zone is designed to increase the number of these external neutrons through their multiplication in the molten salt of the central zone with a high content of dissolved fissile material. To obtain the highest multiplication of external neutrons and the highest neutron leakage from the central zone to the MA burning zone possible while not exceeding the selected reactor subcriticality level, the central zone should have a fast neutron spectrum and contain as much fissile material as possible for heat removal, as low a concentration of neutron moderator as possible and an absorber of thermal neutrons dissolved in the molten salt [1].

The peripheral zone is designed for minor-actinide burning and transmutation of long-lived fission products. The main requirement for the peripheral zone is provision of neutron flux, which is optimal for MA burning. Preliminary estimates show that the optimal neutron flux should be higher than $2 \cdot 10^{15}$ n/cm$^2$s [2].

Fluoride salts are considered the most appropriate carriers for CSMSR fuel compositions for both zones due to their good thermal-physical properties and their natural combination with non-aqueous reprocessing technologies for spent fuel of solid-fuel power reactors.

Proposals regarding compositions of fluoride salts/fuel carriers, on the vector of nuclides to be burned, on reactor design, etc., are considered in the paper. The main limitations are formulated and problems to be solved at subsequent stages are addressed. The study is undertaken in the framework of ISTC#1486 Project.

# Molten-salt compositions and structural materials

Following requirements for fuel molten-salt compositions of the CSMSR central and MA burning zones formulated in [1], the following fuel salts were preliminarily chosen:

- 57 mole % LiF + 38 mole % NaF + 5 mole %($XF_3$ + $SmF_3$) for the central zone, where X signifies the fuel nuclides of the central zone;

- 23 mole % LiF + 22 mole %$BeF_2$ + 54 mole % NaF + 1 mole % $YF_3$ for MA burning zone, where Y is the vector of heavy nuclides to be burned.

A salt carrier of fuel composition for the central zone was selected, mainly in order to meet the requirement of a high concentration of fuel nuclides (high $k_\infty$). The chosen salt composition allows for a solution of up to 10 mole % of trivalent elements, while its main drawback is a high melting point (about 600°C). To exclude freezing of the salt in IHXs of the central zone circuit, the temperature of the fuel salt in this circuit is supposed to be above 650°C. The plutonium isotopes $^{239}$Pu and $^{240}$Pu are considered as fuel nuclides of the central zone in the given study. At further stages of the study some isotopes of minor actinides can be considered for utilisation in the central zone.

Dissolution of a thermal neutron absorber in the central zone salt allows to diminish the power density peaking factor on the boundary between the central core salt and graphite wall. Besides, the thermal neutron absorption in the central zone could provide the necessary reactor subcriticality and increase neutron yield per fission. Samarium is considered in this study as the thermal neutron absorber. A review of other possible thermal neutron absorbers and the use of absorber films on the wall surface in combination with dissolution of absorber in the salt are planned for the next stages of the study.

The solubility of trifluorides of heavy nuclides in the salt composition of the MA burning zone is limited by 2 mole %, and its melting point is estimated to be 350-400°C. For both zones, the content of trifluorides of heavy nuclides in the salts is chosen to be two times lower than the peak solubility. This margin is assumed to be enough for total dissolution of fission products, generated in irradiation, taking into account uncertainties in burn-up calculations. This assumption requires additional verification and improvement.

The intermediate salt composition is chosen to be 92 mole % $NaBF_4$ + 8 mole % NaF. The melting point of this salt is 385°C.

Depending on the temperature and concentration of trifluorides of heavy nuclides, the density of the considered fluoride salts can vary over the range 2.0-2.7 g/cm$^3$. The thermal conductivity of the considered compositions is about 1 W/mK and requires further experimental verification. High specific heat of fluoride salts (1.5-2 kJ/kgK) is an important advantage of the chosen compositions from the viewpoint of the core cooling.

Following from the high temperature level in the CSMSR core, graphite, which is suitably compatible with fluoride salts, was chosen as one of the main structural materials of the primary circuit. Based on the positive experience of MSRE operation, Hastelloy-N is supposed to be used as a main material of heat exchangers and pipelines. Fuel salt temperature should not exceed 850°C to provide reliable operation of Hastelloy-N [3].

Thus, the range of permissible temperature variation in the central zone is 650-850°C. The salt velocity in the central zone is assumed to be below 5 m/s to diminish the erosion effect.

## Equilibrium concentrations of heavy nuclides and average neutron flux in MSR burner

Optimisation calculations of equilibrium state of the closed fuel cycle were performed for the nuclear power (NP) structure, consisting of power thermal reactors (TR), power fast reactors (FP) and molten-salt reactor burners (MSR). The optimisation was aimed at minimisation of equilibrium amounts of minor actinides and long-lived fission products throughout the NP system [2].

A calculational model of this fuel cycle consists of five homogeneous calculational areas, in which isotopic compositions change and nuclides flow between them. The five calculational areas simulate the cores of TR, FR, MSR burner, as well as cooling and reprocessing of fuel for TR and FR. Discrete fuel transition from process to process is simulated by continuous nuclide flows with specified rates. Average neutron fluxes in the TR, FR and MSR burner cores were assumed in the calculations to be $3\cdot10^{14}$ n/cm$^2$s, $2\cdot10^{15}$ n/cm$^2$s and $5\cdot10^{15}$ n/cm$^2$s, respectively.

The high value required for the average neutron flux in the MSR burner is explained as follows. Neutron absorption by heavy nuclides causes an increase in concentration of curium isotopes (the number of neutrons per fission in fast and resonance ranges for curium isotopes is $\nu \approx 3.4$-$3.5$, fission cross-section is relatively high), while decay of heavy nuclides leads to an increase in concentration of neptunium and uranium isotopes ($\nu \approx 2.4$-$2.8$, fission cross-section is relatively low). Thus, the neutron absorption-to-radiation decay rate ratio for MA nuclides should be increased to improve the neutron balance in the MA burning zone. A preliminary estimation of the optimal neutron flux was in the range of $2\cdot10^{15}$-$5\cdot10^{15}$ n/cm$^2$s [2]. According to the conducted estimates, this range provides a minimal amount of MA and long-lived fission products in the three-component NP system (TR+FR+MSR), when there is fuel breeding in the system, compared to other variants of the NP structure (FR, FR+TR, FR+MSR, etc.). A vector of concentrations of heavy nuclides in MSR, corresponding to equilibrium state of the MSR fuel cycle, was obtained among other results [2] (Figure 1).

### Figure 1. Fractions of equilibrium concentrations of heavy nuclides in MSR in the three-component NP system

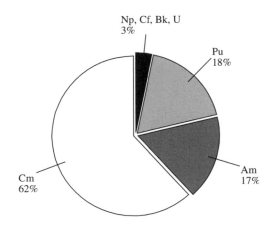

One of the most important problems to be solved for the continuation of the CSMSR concept development is the estimation and improvement of neutron data for minor actinides, which are the basis of fuel in this reactor. As an illustration of CSMSR physical features, fractions of fission rates for different isotopes in the CSMSR MA burning zone are given on Figure 2. Calculational neutron spectra in infinite medium with isotopic composition of the CSMSR central zone and the MA burning zone are shown in Figure 3.

## Figure 2. Fractions of fission rates for different isotopes in the CSMSR MA burning zone

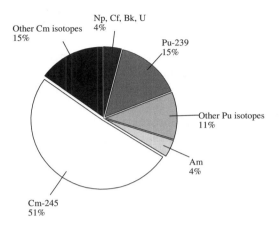

## Figure 3. Neutron spectra in infinite medium with isotopic composition of the CSMSR zones

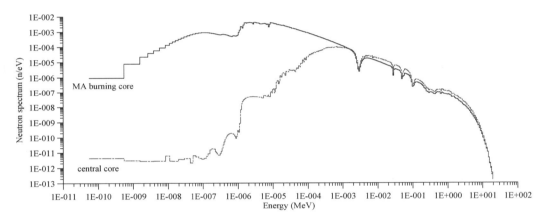

The significant influence of neutron data uncertainties on calculations of reactor systems with such a vector of heavy nuclides is demonstrated, in particular, by the fact that divergence of $k_\infty$ can reach 3-4% when using the evaluated nuclear data files for [245]Cm from the JENDL-3.2, JEF-2.2 and ENDF/B-VI libraries. According to Figure 2, fissions of this isotope will compose more than half of all fissions in CSMSR.

## Safety and economics

As one of the main assumptions, the CSMSR subcriticality of 2% ($k_{eff} = 0.98$) was assumed to be sufficient so as to provide high reactor safety. The required CSMSR subcriticality can be corrected once new test data on uncertainties in neutron data of minor actinides are obtained and on the basis of calculational estimates of reactivity change with burn-up during reactor operation.

A final conclusion concerning the validity of the accepted assumptions can be made on the basis of accident analysis, planned for the next stages of the study. However, some qualitative remarks can already be made. In particular, a feature of reactors with circulating fuel to be taken into account in accident analysis is a reduced number of delayed neutrons in the core, provided by the removal of a part of precursors of delayed neutrons from the core with circulating salt and their decay out of the core.

A loss-of-flow accident in the central zone with operating external neutron source seems one of the most dangerous potential accidents in CSMSR. This situation can be caused by a trip of the central zone circuit pumps and/or failure of the wall, dividing downcomer and chimney of the central zone. Loss of flow would be accompanied by a sudden temperature increase because of high power density in the central zone. A passive system should be provided in the reactor bottom to mitigate the consequences of this kind of accident. When temperature increases, this system should actuate fuel salt drainage from the circuit to tanks, cooled by natural circulation of environmental air. This system could be based, for instance, on the use of a fusible membrane. In addition, another safety passive system can be used on the basis of neutron absorbers, kept above the central zone by the rising fuel salt flow. When flow drops, the absorbers under gravitational force passively go down, entering the central core and reducing the reactor power. Further, a passive trip of accelerator power should be provided, in case of deviations from CSMSR operational regimes.

A failure of the CSMSR central zone wall and mixture of the central zone salt, which has a high content of fissile material, with the peripheral zone salt should also be considered. In this case, positive reactivity to be inserted should be compensated by absorber of thermal neutrons, which is a part of the central zone salt composition.

As a first approximation, the economic effect from the use of the two-zone scheme for subcritical MSR can be estimated by calculating the efficiency of cascade neutron multiplication [1]. In more detailed consideration, this parameter can be determined as a ratio of external neutron source intensities in CSMSR and in a homogeneous subcritical MSR, which provide the same average neutron flux in the MA burning zone at the same initial subcriticality and reactor dimensions. At the next stages of the study the cost of introducing the additional circulation circuit for the CSMSR central zone should be compared to the profit provided by the decrease in the accelerator power due to the neutron cascade multiplication.

**Scheme of fuel salt circulation**

The main requirements for the scheme of the CSMSR fuel salt circulation include:

- The possibility should be provided to periodically extract the central zone for replacement of structural materials due to high neutron and proton fluences.

- Subcriticality of the fuel salt volume in the IHX cavity should be provided after extraction of the IHX from the reactor.

Following these requirements, the fuel salt of the CSMSR central zone was chosen to circulate in the coaxial tube, installed in the centre of the reactor. Additionally, the central zone salt with high plutonium concentration was accepted to circulate inside the IHX tubes. The proposed option of the CSMSR circulation scheme is presented in Figure 4.

**Optimisation of CSMSR core dimensions**

The core height was designated as 400 cm and requires further optimisation, based on analysis of spatial distribution of external neutron source. The thickness of the graphite wall, dividing chimney and downcomer of the central zone was determined at 1 cm, while the thickness of the graphite wall between two zones was set at 5 cm. These values require further optimisation, based on analysis of thermal-mechanical behaviour of graphite walls under conditions of high temperatures, neutron and proton fluence. The CSMSR scheme used for neutronics calculations is shown in Figure 5.

**Figure 4. The proposed option of the CSMSR circulation scheme**

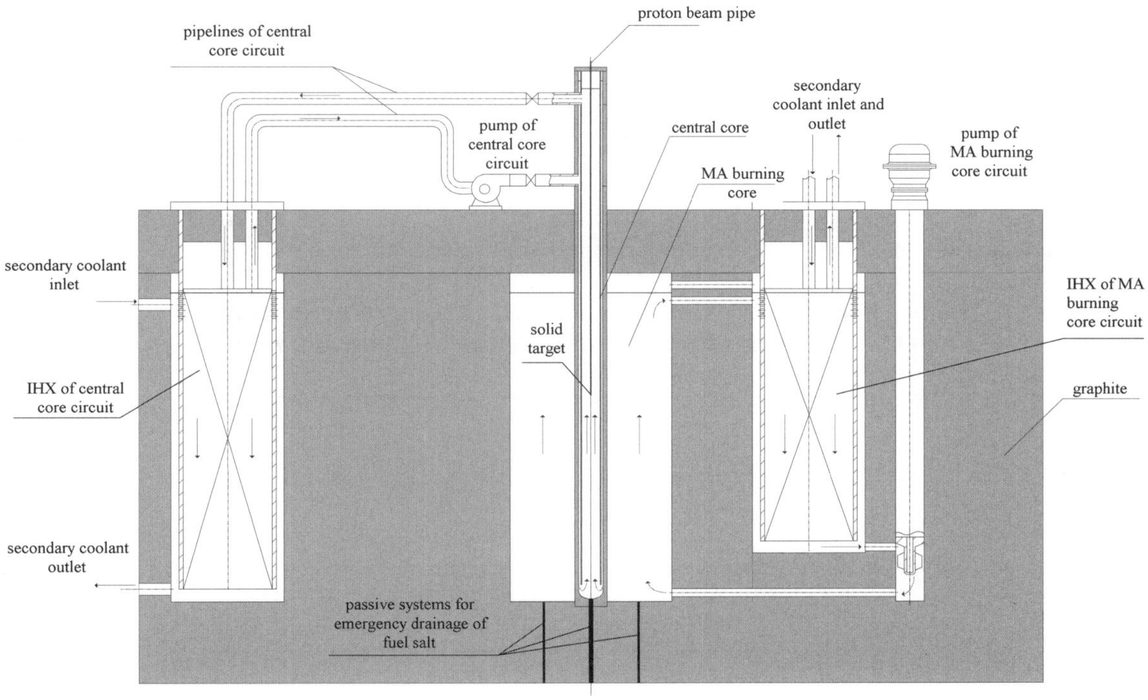

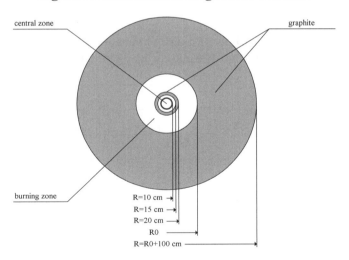

pipelines of central core circuit

proton beam pipe

pump of central core circuit

secondary coolant inlet and outlet

central core

MA burning core

pump of MA burning core circuit

secondary coolant inlet

solid target

IHX of MA burning core circuit

IHX of central core circuit

graphite

secondary coolant outlet

passive systems for emergency drainage of fuel salt

**Figure 5. Calculational diagram of CSMSR**

central zone

graphite

burning zone

R=10 cm
R=15 cm
R=20 cm
R0
R=R0+100 cm

Optimisation analysis showed that theoretically the highest effect of decreasing the external source intensity due to cascade neutron multiplication is reached by increasing the plutonium concentration in the central zone salt. In this case, the chosen reactor subcriticality is provided by reducing the central zone radius and/or increasing the thermal neutron absorber concentration. The search for an optimal correlation between these parameters is a subject of further studies. Preliminary calculations showed that there is an optimal ratio between these parameters, which provides the highest efficiency of the cascade neutron multiplication.

Note that decrease of the central zone radius is limited by the heat removal capability at the chosen temperature range (650-850°C) and fuel salt velocity limit (< 5m/s). The central zone radius was selected to be 15 cm in calculations. Thus, for the chosen salt composition and geometry the central zone power is limited by 160 MWt.

An average neutron flux in the MA burning zone grows when the radius of this zone decreases. The radial distributions of axially averaged neutron flux and power density peaking factor in CSMSR with different radii of the MA burning zone are given in Figures 6 and 7, respectively. All calculations were made with an external neutron source intensity of $6.5 \cdot 10^{17}$ n/s. The required subcriticality for each reactor radius was obtained by correcting the samarium concentration in the central zone salt.

**Figure 6. Radial distributions of axially averaged neutron flux
in CSMSR with different radius of the MA burning zone**

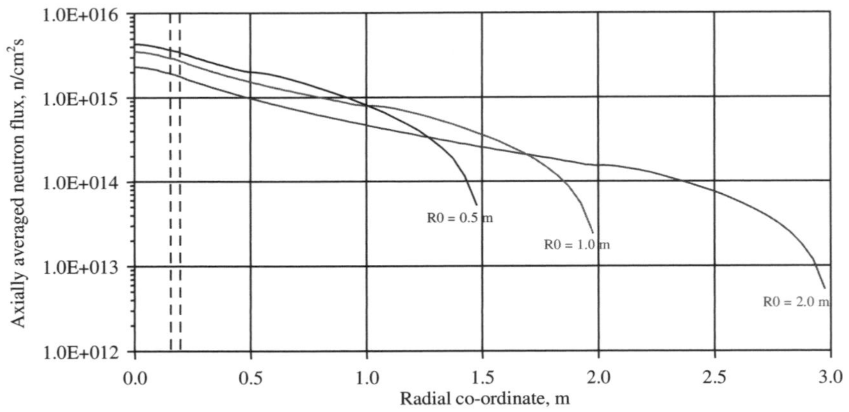

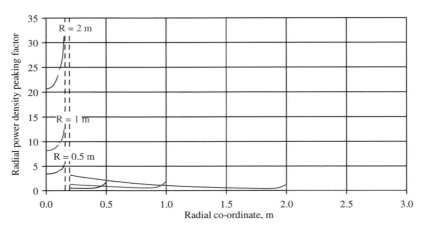

**Figure 7. Radial distributions of axially averaged power density peaking
factor in CSMSR with different radius of the MA burning zone**

Average neutron fluxes and power fractions of separate zones in CSMSR for different radii of the MA burning zone are given in Table 1. As seen from the table, a decrease in the MA burning zone radius leads to an increase of the ratio of the average neutron flux in this zone to the central zone power. Note that for the MA burning zone radii of 1 m and 2 m the selected limitations on the central zone power (below 160 MWt) were not reached. Thus, an increase in the external neutron source intensity, until the central zone power is 160 MWt, allows to obtain in the MA burning zone $1.7 \cdot 10^{15}$ n/cm²s and $0.8 \cdot 10^{15}$ n/cm²s for the MA burning zone radii of 1 m and 2 m, respectively.

**Table 1. Average neutron fluxes and power fractions of separate zones in CSMSR for different radii of the MA burning zone**

| MA burning zone | | Central zone | | Whole reactor |
|---|---|---|---|---|
| Radius, m | Average flux, n/cm²s | Power, MW | Average flux, n/cm²s | Power, MW |
| 2.0 | $0.4 \cdot 10^{15}$ | 82 | $2.1 \cdot 10^{15}$ | 655 |
| 1.0 | $1.3 \cdot 10^{15}$ | 126 | $3.3 \cdot 10^{15}$ | 625 |
| 0.5 | $2.5 \cdot 10^{15}$ | 156 | $4.0 \cdot 10^{15}$ | 428 |

Thus, only the CSMSR option with the MA burning zone radius of 0.5 m meets the stated criteria on the central core power (below 160 MWt) and the average neutron flux in the MA burning zone (above $2 \cdot 10^{15}$ n/cm²s). As a first approximation, this option was accepted as a reference one – a subject for further optimisation. The main parameters of this CSMSR option are given in Table 2.

**Table 2. Main parameters of the chosen CSMSR option**

| Parameter | Value |
|---|---|
| Reactor thermal power, MWt | 428 |
| $k_{eff}$ | 0.980 |
| Source intensity, n/s | $6.5 \cdot 10^{17}$ |
| Core height, cm | 400 |
| **Central zone** | |
| Composition, mole % | 57LiF 38NaF 5(PuF$_3$+SmF$_3$) |
| Thermal power, MWt | 156 |
| Inner tube radius, cm | 10 |
| Outer radius, cm | 15 |
| Average power density, MW/m³ | 552 |
| Inlet temperature, °C | 650 |
| Outlet temperature, °C | 850 |
| Flow rate, kg/s | 488 |
| Average velocity, m/s | 5.3 |
| Average neutron flux, n/cm²s | $4.0 \cdot 10^{15}$ |
| **MA burning zone** | |
| Composition, mole % | 23LiF 22BeF$_2$ 54NaF 1XF$_3$ |
| Thermal power, MWt | 272 |
| Inner radius, cm | 20 |
| Outer radius, cm | 50 |
| Average power density, MW/m³ | 103 |
| Inlet temperature, °C | 650 |
| Outlet temperature, °C | 850 |
| Flow rate, kg/s | 618 |
| Average velocity, m/s | 0.4 |
| Average neutron flux, n/cm²s | $2.5 \cdot 10^{15}$ |

## Conclusions

Some of the results obtained at the current stage of the study of the concept of Cascade subcritical molten-salt reactor (CSMSR), performed in the framework of ISTC#1486 Project at the RRC "Kurchatov Institute", are summarised in this paper:

- The chosen fuel compositions, including fluoride salts and heavy nuclides, are presented.

- The option of the CSMSR core and circulation circuit was developed on the basis of criteria obtained in the framework of optimisation studies.

- Some directions for further activities are formulated.

The following tasks require further consideration:

- estimates of uncertainties and improvement of neutronics data for minor actinides, which are the basis of fuel in CSMSR reactor;

- correction of the vector of heavy nuclides on the basis of calculations of equilibrium state of the closed fuel cycle for three-component NP structure with the chosen CSMSR option as a MA burner;

- calculational evaluation of efficiency of MA burning in the CSMSR fuel salt;

- coupled neutronics/thermal-hydraulics calculations with account for feedback between spatial fields of neutron fluxes, power densities, velocities, temperatures and densities of the fuel salts;

- further optimisation of dimensions and compositions of the core for decrease of fuel salt velocity in the central zone, increase in the neutron flux in the MA burning zone;

- estimate of behaviour of structural materials in the core with account for high levels of thermal stresses, neutron and proton fluences;

- cost estimate of introducing the additional circulation circuit for the CSMSR central zone and the profit provided by the decrease in the accelerator power;

- calculational analysis of accidents.

In addition to these tasks, possibility and expediency should be studied to modify the proposed CSMSR design, aimed at increasing the radwaste burning rate. In particular, the prospect of introducing the channels in the reactor graphite vessel around the MA burning zone should be considered. In this case, some fraction of fuel salt of the MA burning zone passes these channels with rather high neutron flux and thermal neutron spectrum and mixes at core outlet with the salt, which passed the main cavity of the core. As is well known, there is a unique optimal neutron spectrum for each heavy nuclide, in which this nuclide could be burned or transmuted in the most effective way. The introduction of the additional channels for fuel salt circulation could significantly expand the neutron spectral range in CSMSR, increase MA loading in the reactor, decrease neutron leakage from the reactor, decrease the power fraction generated in the central zone, etc.

# REFERENCE

[1]    A.V. Vasiliev, *et al.*, "Features of Cascade Subcritical Molten-salt Reactor – Burner of Long-lived Radioactive Wastes", Proc. of GLOBAL'2001.

[2]    A. Dudnikov, *et al.*, "Status of Benchmark Calculations of the Neutron Characteristics of the Cascade Molten-salt ADS for the Nuclear Waste Incineration", Proc. of GLOBAL'2001.

[3]    "Hastelloy N Alloy Principal Features", Haynes International, see the following web address: http://www.haynesintl.com/H2052AN/HastelloyNPF.htm.

# TRANSMUTATION AND INCINERATION OF MAs IN LWRs, MTRs AND ADSs

**Ch. De Raedt, B. Verboomen, Th. Aoust, H. Aït Abderrahim, E. Malambu, L.H. Baetslé**
SCK•CEN, Boeretang 200, B-2400 Mol, Belgium
charles.de.raedt@sckcen.be

## Abstract

The paper examines the neutron-induced evolution of MAs irradiated in LWRs, MTRs and ADSs, viz. in a typical 1 000 MWe MOX-fuelled PWR, in the high flux materials testing reactor BR2, and in the multi-purpose R&D ADS MYRRHA. The transformation of the actinides by $(n,\gamma)$ reactions into higher isotopes ("transmutation") and their disappearance out of the actinide family by $(n,f)$ reactions (actinide fission or "incineration") are discussed. While fast spectrum systems such as the proposed ADS immediately incinerate the MAs, but at relatively low rates because of the small cross-sections, thermal spectrum systems, with large $(n,\gamma)$ cross-sections, first transmute the MAs into higher isotopes, most of which ultimately are also fissile in the thermal energy domain. In the case of MTRs, in which high (thermal) fluxes prevail, large fractions of MAs are thus transmuted and ultimately incinerated in relatively short times, but at the cost of several neutron captures before fission occurs and therefore with bad neutron economy.

# Introduction

In the framework of partitioning and transmutation (P&T), transmutation and incineration is the only technology which is capable of accelerating the natural decay sequence and hence of reducing the radiotoxic inventory of some actinides for which the natural decay reactions take hundred thousands of years to reach the initial uranium ore toxicity level. If partitioning of minor actinides (MAs) is successful, the way to transmutation/incineration is open. Transmutation/incineration in thermal or fast neutron spectra has been thoroughly studied and both have their merits. Thermal neutrons are very effective for fissile transuranium (TRU) nuclides ($^{239}$Pu, $^{241}$Pu, $^{242}$Am, $^{245}$Cm) whereas fast neutron spectra (in FRs and ADS) are indispensable for fissioning the fertile ($^{237}$Np, $^{241}$Am, $^{243}$Am) and even mass-number nuclides ($^{238}$Pu, $^{240}$Pu, $^{242}$Pu, $^{244}$Cm, etc.).

The present paper continues the study [1] that made an analysis of the possibilities of thermal and fast spectrum approaches in relation with technological demonstration experiments. These could be performed in BR2, and later on in the planned MYRRHA ADS facility, to investigate the issues related to target optimisation, cladding selection and structural material behaviour under intense irradiation. In addition, the present paper also examines irradiations in a 25% MOX-fuelled PWR.

# Irradiation devices considered

## 25% MOX-fuelled PWR

The PWR irradiations considered in this paper are similar to the concept of "heterogeneous recycling – target pins in corner rod positions of MOX assemblies" studied in [2,3,4] by BELGONUCLÉAIRE and EDF. In this concept, the target rods occupy the 12 corner rod positions of each MOX assembly in a 1 000 MWe PWR reactor loaded for 25% with MOX assemblies.

Figure 1 shows the model used in the calculations described further: one-fourth of the MOX fuel assembly occupies the left upper quarter of the "supercell" while the three surrounding fourths with $UO_2$ rod assemblies occupy the other quarters of the "supercell". The "supercell" is assumed to have "reflective" boundary conditions and hence represents a 1 MOX-3 $UO_2$ assembly system. The dark circles represent the MA targets, the dark grey circles MOX-H rods [with 8.7% Pu/(Pu+U)], the light grey circles MOX-M rods [with 6.0% Pu/(Pu+U)] and the very light grey circles the $UO_2$ rods (with 4.0% $^{235}$U/U) in the assemblies surrounding the MOX assemblies. The large circles correspond to guide tube positions and – in the four corners of the "supercell" – to the central holes of the assemblies.

## The high flux materials testing reactor BR2

BR2 is a heterogeneous thermal high flux materials testing reactor [5-8]. Figure 2 (left-hand side) shows a horizontal cross-section of the reactor core at the reactor midplane with a typical loading. Each BR2 fuel element has a 762 mm active fuel length. The presently used 6n-G fuel elements (Figure 2, right-hand side) contain, when fresh, 400 g $^{235}$U in the form of $UAl_x$ (1.3 g U/cm$^3$) + 3.8 g boron ($B_4C$) + 1.4 g samarium ($Sm_2O_3$). The reactor core is loaded with 10 to 13 kg $^{235}$U (30 to 40 fuel elements, not all fresh). The concentration at discharge of the fuel elements is about 50% of the initial fissile content value. The present nominal heat flux at the hot spot is 470 W/cm$^2$ and the maximum value allowed for nominal cooling conditions (probable onset of nucleate boiling) is 600 W/cm$^2$. Typical neutron fluxes (in the reactor hot spot plane) are shown in the table below.

| Thermal conventional neutron flux | Fast neutron flux |
|---|---|
| $v_0 n = v_0 \int_0^{0.5\ \text{eV}} n(E)dE$ : 2 to $4 \cdot 10^{14}$ n/cm$^2$s in the reactor core, 2 to $9 \cdot 10^{14}$ n/cm$^2$s in the reflector and core flux trap (H1) | $\Phi_{>0.1\ \text{MeV}} = \int_{0.1\ \text{MeV}}^{\infty} \varphi(E)dE$ : 4 to $7 \cdot 10^{14}$ n/cm$^2$s in the reactor core |

**Figure 1. Calculational model ("supercell") for the MA irradiations in corner positions of a MOX assembly in a 25% MOX-fuelled PWR**

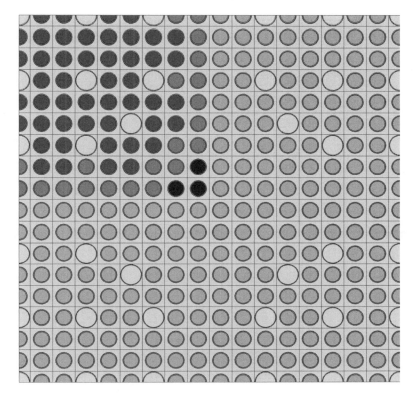

**Figure 2. Horizontal cross-section (left) of the BR2 reactor at the reactor midplane with a typical loading and (right) of a type 6n-G fuel element with central aluminium plug**

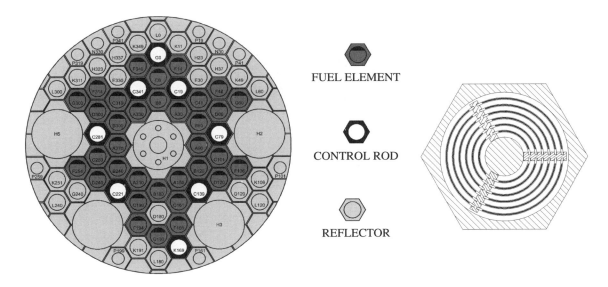

## MYRRHA, a Multi-purpose hYbrid Research Reactor for High-tech Applications

MYRRHA, in its present development stage, is described in another paper of this workshop [9]. It is based on the coupling of a proton cyclotron with a liquid Pb-Bi windowless spallation target, surrounded by a subcritical neutron multiplying medium in a pool-type configuration [10,11]. Ion Beam Applications (IBA) is in charge of the design of the accelerator. The accelerator parameters presently considered are a 5 mA continuous proton beam at 350 MeV energy. The proton beam will impinge on the spallation target from the top. To meet the goals of material studies, fuel behaviour studies, radioisotope production, transmutation of MAs and LLFPs, the MYRRHA facility should include two spectral zones: a fast neutron spectrum zone and a thermal spectrum one.

The core pool contains the fast spectrum core zone, cooled with liquid Pb-Bi, and islands housing thermal spectrum regions located in in-pile sections (IPSs) at one side of the fast core. In its present design phase [12], the fast core is fuelled with typical fast reactor fuel pins (triangular pitch: 10.0 mm) with an active length of 600 mm arranged in hexagonal assemblies with 87 mm pitch. Each hexagonal assembly contains 61 fuel pins. The central three hexagon positions are left free for housing the spallation source in its pressure tube. The fast subcritical MYRRHA "reference core", without thermal island, (see Figure 3 for horizontal cross-section) is made of 45 MOX fuel assemblies, 30 of which have a Pu content of 30% and 15 a Pu content of 20%. Table 1 illustrates the neutronic performances calculated for the "reference core" [12].

**Figure 3. Horizontal cross-section of the fast subcritical core of MYRRHA ("reference core") [12]**

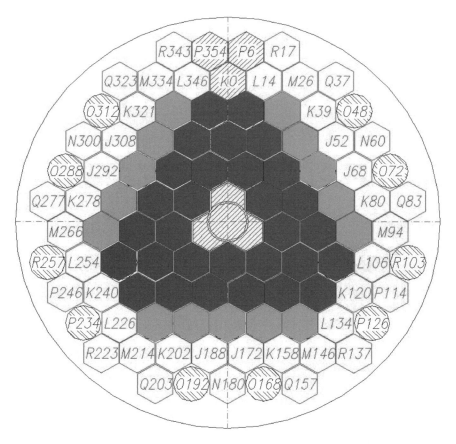

## Table 1. Neutronic design parameters of the MYRRHA facility ("reference core") [12] – irradiation targets and irradiation positions

| Zone | Parameter | Value |
|---|---|---|
| Spallation source | $E_p$ | 350 MeV |
| | $I_p$ | 5 mA |
| Fast subcritical core | $k_s$ | 0.956 |
| | $MF = 1/(1 - k_s)$ | 22.52 |
| | Thermal power | 36.5 MW |
| | Peak linear power | 431 W/cm |
| | Max. $\Phi_{>1\,MeV}$: nearest to the spallation target | $0.65\ 10^{15}$ n/cm$^2$s |
| | Max. $\Phi_{>0.75\,MeV}$: nearest to the spallation target | $0.90\ 10^{15}$ n/cm$^2$s |

The targets considered for the irradiations in the 25% MOX-fuelled PWR, in BR2 and in MYRRHA were assumed to have an outer diameter of 8.36 mm and to be clad with HT-9 steel with an outer diameter of 9.5 mm. The target compositions are indicated in Table 2. The targets were given quite a long active length, but only the 200 mm positioned symmetrically with respect to the axial maximum flux level of the three irradiation systems were considered in the results reported.

## Table 2. Composition of the various targets

| MA | Chemical form of the target | Density (%TD) |
|---|---|---|
| $^{237}$Np | 20 vol.% $NpO_2$ + 40 vol.% $MgAl_2O_4$ + 40 vol.% Al | 90 |
| 77.9 wt.% $^{241}$Am + 22.1 wt.% $^{243}$Am | 20 vol.% $Am_2O_3$ + 40 vol.% $MgAl_2O_4$ + 40 vol.% Al | 90 |

The targets irradiated in the 25% MOX-fuelled PWR occupy, as mentioned above, the twelve corner rod positions of each MOX assembly (25% of the total core loading: see Figure 1, representing the calculational model ("supercell") with one-fourth MOX fuel assembly and the three remaining fourths with UO$_2$ fuel assemblies).

The BR2 irradiation targets were assumed to be introduced into a loop consisting of concentric aluminium tubes with in-between cooling water circulation. The outer diameter of the outer tube was 25.4 mm, allowing the loop to be substituted for the central aluminium plug (also with diameter 25.4 mm) of a standard BR2 fuel element such as shown in Figure 2 (right-hand side). As irradiation position in BR2 the high-flux channel B180 (see Figure 2, left-hand side) was selected.

For the irradiations in MYRRHA, the targets with their cladding were assumed to be introduced into one of the three irradiation spaces between the tube surrounding the spallation source and the first ring of nine hexagonal assemblies of the fast subcritical core (see Figure 3), at a radial distance of 54.1 mm from the MYRRHA main axis.

## Calculated transmutation yields

The transmutation rates of the MAs in the 25% MOX-fuelled PWR and in BR2 were calculated with the aid of the Monte Carlo code MCNP-4C [13]. For the MYRRHA calculations, the Monte Carlo code MCNPX-2.1.5 [14], combining MCNP-4B and LAHET, was used. The continuous energy neutron cross-section library used was ENDF/B-VI, except for the nuclides not present in ENDF/B-VI; for these, ENDF/B-V data were taken. The model indicated in Figure 1 was adopted for the irradiations in the 25% MOX-fuelled PWR: it presents a typical situation in a large (assumption made here: infinitely large) reactor core. To each of the four sides of the elementary "supercell" shown in Figure 1, "reflective" boundary conditions were imposed. The way BR2 was modelled is explained in [15,16] (e.g. the meat and the cladding of each of the 6 × 3 fuel plates of each of the fuel elements of the BR2 loading were considered as separate zones). Also for MYRRHA the calculations were performed in great detail: each of the 2 745 fuel pins (with for each: fuel, gap and cladding described separately) was modelled.

The fission and "disappearance" reaction rates are indicated in Table 3 for the various minor actinides considered in the present study. The values are averaged over the 200 mm high target volumes. By "disappearance" is meant the sum of all nuclear reactions leading to the removal of the MA considered from its (A,Z) position in the table of isotopes. The "disappearance" reaction is hence, practically, the sum of processes 16 (n,2n), 17 (n,3n), 18 (total fission) and 101 [neutron disappearance, viz. mainly (n,$\gamma$), (n,p), (n,d), (n,t), (n,$^3$He) and (n,$\alpha$)], where the numbers refer to the MT numbers in the ENDF/B format. It should be noted that in the present MYRRHA calculations neither the very high-energy tail of the spallation neutron spectrum (above 20 MeV) nor the proton-induced transmutations were taken into account. These contributions are expected to be negligible outside the spallation target.

**Table 3. Target-volume-averaged direct fission and disappearance reaction rates in the various targets irradiated in the 25% MOX-fuelled PWR, in BR2 and in MYRRHA**

| Target | | $NpO_2$-$MgAl_2O_4$-Al | $Am_2O_3$-$MgAl_2O_4$-Al | |
|---|---|---|---|---|
| Nuclide | | $^{237}$Np | $^{241}$Am | $^{243}$Am |
| Reaction rates in PWR | Direct fiss. (s$^{-1}$) | 3.63·10$^{-10}$ | 4.62·10$^{-10}$ | 3.07·10$^{-10}$ |
| | Disapp. (s$^{-1}$) | 1.29·10$^{-8}$ | 2.05·10$^{-8}$ | 1.27·10$^{-8}$ |
| | Direct fiss./disapp. | 0.028 | 0.023 | 0.024 |
| Reaction rates in BR2 | Direct fiss. (s$^{-1}$) | 5.17·10$^{-10}$ | 9.13·10$^{-10}$ | 4.43·10$^{-10}$ |
| | Disapp. (s$^{-1}$) | 4.37·10$^{-8}$ | 7.87·10$^{-8}$ | 2.91·10$^{-8}$ |
| | Direct fiss./disapp. | 0.012 | 0.012 | 0.015 |
| Reaction rates in MYRRHA | Direct fiss. (s$^{-1}$) | 1.62·10$^{-9}$ | 1.40·10$^{-9}$ | 1.10·10$^{-9}$ |
| | Disapp. (s$^{-1}$) | 5.05·10$^{-9}$ | 5.23·10$^{-9}$ | 4.39·10$^{-9}$ |
| | Direct fiss./disapp. | 0.32 | 0.27 | 0.25 |

## Comparison of the transmutation/incineration performances of the 25% MOX-fuelled PWR, of BR2 and of MYRRHA

From Table 3 one can easily deduce that for $^{237}$Np the disappearance reaction rate in BR2 is 3.4 times larger than that in the 25% MOX-fuelled PWR and 8.7 times larger than that in MYRRHA. For Am (77.9 wt.% $^{241}$Am + 22.1 wt.% $^{243}$Am) the disappearance reaction rate in BR2 is 3.6 times larger than that in the 25% MOX-fuelled PWR and 13.4 times larger than that in MYRRHA. The larger disappearance rates of the MAs in thermal spectrum (25% MOX-fuelled PWR and BR2) irradiations than in MYRRHA result mainly from the fact that the "disappearance" cross-sections are much larger in the thermal and epithermal energy regions than in the fast energy region. As is known (see Table 4), the (n,$\gamma$) reactions in the MAs considered in the present study lead to the formation of other actinides, mainly (for $^{237}$Np and for Am) to $^{238}$Pu and the further Pu family, and (for Am) to $^{244}$Cm and the

**Table 4. Main transmutation reactions occurring in the targets considered**

$^{237}$Np  (n,γ)  $^{238}$Np  $\xrightarrow{\text{2.1 d}}$  $^{238}$Pu  (n,γ)  $^{239}$Pu  (n,γ)  $^{240}$Pu  (n,γ)  $^{241}$Pu
                                                              (n,f)  FP  (n,f)  FP

                                    (n,f)  FP

(n,f)  FP

$^{241}$Am  (n,γ)  $^{242m}$Am  (n,γ)  $^{243}$Am
       ≈10%  ↓141 y

       (n,γ)  $^{242g}$Am  $\xrightarrow{\text{16 h}}$  $^{242}$Cm  $\xrightarrow{\text{163 d}}$  $^{238}$Pu  (n,γ)  $^{239}$Pu  (n,γ)  $^{240}$Pu
       ≈90%        82.7%                       (n,f)  FP

                     $\xrightarrow{\text{16 h}}$  $^{242}$Pu
                     17.3%

                     (n,f)  FP

(n,f)  FP

$^{243}$Am  (n,γ)  $^{244}$Am  $\xrightarrow{\text{16 m...10 h}}$  $^{244}$Cm  (n,γ)  $^{245}$Cm  (n,γ)  $^{246}$Cm  (n,γ)  $^{247}$Cm...
                                               (n,f)  FP

(n,f)  FP

further Cm family. Only the fission process allows complete removal of the MAs out of the actinide family. For this process, ADS-type reactors such as MYRRHA, with a fast neutron spectrum, seem, at first sight, the most attractive. Table 3 indicates that at the start of the irradiations about 30% of the disappearance of MAs in MYRRHA are due to "direct" fissions, while for BR2 the figure is less than 2% and in the 25% MOX-fuelled PWR less than 3%. The fissions indicated here are the "direct" fissions of $^{237}$Np, $^{241}$Am and $^{243}$Am and do not include the "secondary" fissions, viz. those occurring in the fissile actinides formed by (n,γ) reactions during the irradiation, possibly followed by natural decay. In the case of irradiations in BR2 (and in PWRs for long irradiation periods), these "secondary" fissions become very important as the irradiations proceed.

This phenomenon is illustrated in Table 5, where the concentrations (in atom %) of the main nuclides present in the $^{237}$Np and the $^{241,243}$Am targets are indicated, as calculated for irradiations of 200, 400 and 800 EFPD in the 25% MOX-fuelled PWR, in BR2 and in MYRRHA, followed by a cooling period of five years. "Fissium" represents the nuclides that have disappeared from the actinide family and is hence equal to half the total number of FP atoms.

451

**Table 5. Atom per cent concentration of the various nuclides in MA targets irradiated in the 25% MOX-fuelled PWR, in BR2 and in MYRRHA during 200, 400 and 800 EFPD, followed by five years cooling**

| Target | | 0 EFPD | PWR | | | BR2 | | | MYRRHA | | |
|---|---|---|---|---|---|---|---|---|---|---|---|
| | | | 200 EFPD | 400 EFPD | 800 EFPD | 200 EFPD | 400 EFPD | 800 EFPD | 200 EFPD | 400 EFPD | 800 EFPD |
| $^{237}$Np | $^{234}$U | 0 | 0.7 | 1.1 | 1.5 | 1.0 | 0.7 | 0.2 | 0.2 | 0.4 | 0.8 |
| | **$^{237}$Np** | **100.0** | **80.0** | **64.0** | **41.0** | **47.0** | **22.1** | **4.9** | **91.6** | **84.0** | **70.5** |
| | $^{238}$Pu | 0 | 16.0 | 24.9 | 30.4 | 22.9 | 16.0 | 4.4 | 5.2 | 9.4 | 15.5 |
| | $^{239}$Pu | 0 | 1.4 | 3.3 | 5.2 | 6.2 | 5.0 | 1.5 | 0.1 | 0.3 | 0.9 |
| | $^{240}$Pu | 0 | 0.2 | 0.5 | 1.0 | 1.7 | 1.6 | 0.8 | ~0 | ~0 | ~0 |
| | $^{241}$Pu | 0 | 0.1 | 0.4 | 1.3 | 1.2 | 1.5 | 0.5 | ~0 | ~0 | ~0 |
| | $^{242}$Pu | 0 | ~0 | 0.1 | 0.6 | 0.6 | 2.0 | 1.8 | ~0 | ~0 | ~0 |
| | $^{241}$Am | 0 | ~0 | 0.1 | 0.4 | 0.3 | 0.4 | 0.1 | ~0 | ~0 | ~0 |
| | $^{243}$Am | 0 | ~0 | ~0 | 0.2 | 0.1 | 0.8 | 1.5 | ~0 | ~0 | ~0 |
| | $^{244}$Cm | 0 | ~0 | ~0 | 0.1 | ~0 | 0.3 | 1.5 | ~0 | ~0 | ~0 |
| | **Fissium** | **0** | **1.6** | **5.3** | **17.9** | **19.0** | **49.4** | **81.4** | **2.9** | **5.9** | **12.2** |
| | *Sum* | *100.0* | *100.0* | *99.7* | *99.6* | *100.0* | *99.8* | *98.6* | *100.0* | *100.0* | *99.9* |
| Am | $^{234}$U | 0 | 0.6 | 1.0 | 1.4 | 1.3 | 1.2 | 0.5 | 0.1 | 0.3 | 0.5 |
| | $^{237}$Np | 0 | 0.5 | 0.4 | 0.2 | 0.2 | 0.1 | ~0 | 0.6 | 0.6 | 0.6 |
| | $^{238}$Pu | 0 | 15.3 | 24.4 | 30.5 | 34.2 | 30.6 | 11.4 | 3.5 | 6.4 | 10.7 |
| | $^{239}$Pu | 0 | 0.2 | 1.2 | 3.8 | 2.1 | 4.7 | 3.3 | ~0 | 0.1 | 0.4 |
| | $^{240}$Pu | 0 | 0.7 | 1.3 | 2.1 | 1.7 | 2.9 | 2.9 | 0.2 | 0.4 | 0.8 |
| | $^{241}$Pu | 0 | ~0 | 0.1 | 0.5 | 0.2 | 0.8 | 0.9 | ~0 | ~0 | ~0 |
| | $^{242}$Pu | 0 | 3.0 | 4.3 | 4.7 | 6.2 | 5.4 | 3.1 | 0.8 | 1.4 | 2.5 |
| | **$^{241}$Am** | **78.08** | **54.4** | **38.2** | **19.0** | **19.9** | **5.3** | **0.6** | **70.7** | **64.5** | **53.8** |
| | $^{242m}$Am | 0 | 0.6 | 0.5 | 0.2 | 0.2 | ~0 | ~0 | 0.4 | 0.8 | 1.1 |
| | **$^{243}$Am** | **21.92** | **18.3** | **15.9** | **12.7** | **16.0** | **12.6** | **7.5** | **20.3** | **18.9** | **16.3** |
| | $^{243}$Cm | 0 | 0.2 | 0.4 | 0.5 | 0.7 | 0.7 | 0.2 | ~0 | ~0 | 0.1 |
| | $^{244}$Cm | 0 | 3.2 | 5.4 | 7.8 | 6.4 | 9.6 | 10.6 | 1.0 | 1.7 | 2.9 |
| | $^{245}$Cm | 0 | 0.3 | 0.8 | 1.8 | 0.5 | 0.9 | 1.1 | ~0 | 0.1 | 0.2 |
| | $^{246}$Cm | 0 | ~0 | 0.1 | 0.3 | 0.1 | 0.4 | 1.2 | ~0 | ~0 | ~0 |
| | **Fissium** | **0** | **2.6** | **5.9** | **14.3** | **10.1** | **24.5** | **55.7** | **2.3** | **4.8** | **10.0** |
| | *Sum* | *100.0* | *99.9* | *99.9* | *99.8* | *99.8* | *99.7* | *99.0* | *99.9* | *100.0* | *99.9* |

The absolute total flux levels (total flux $= \int_0^\infty \varphi(E)dE$) in the two types of targets and three irradiation environments, as calculated, amount to:

- 25% MOX-fuelled PWR: $6.91 \; 10^{14}$ n/cm²s in the $^{237}$Np target, $6.54 \; 10^{14}$ n/cm²s in the Am target;

- BR2: $1.20 \; 10^{15}$ n/cm²s in the $^{237}$Np target, $1.05 \; 10^{15}$ n/cm²s in the Am target;

- MYRRHA: $3.28 \; 10^{15}$ n/cm²s in the $^{237}$Np target, $3.37 \; 10^{15}$ n/cm²s in the Am target.

In Figures 4 and 5, the atom percentages of the various nuclides formed in the $^{237}$Np and the Am targets are shown for burn-up values of the MAs ranging up to 95%. The percentage fissions is represented by the "horn-shaped" area with vertical hatching just above the $^{237}$Np or Am linear decrease line. One observes in Table 5 and in Figures 4 and 5 the important percentage of fissions in BR2,

**Figure 4. Atom per cent concentration of the various nuclides in $^{237}$Np
targets irradiated in the 25% MOX-fuelled PWR, in BR2 and in MYRRHA,
as a function of the $^{237}$Np burn-up (each time followed by five years cooling)**

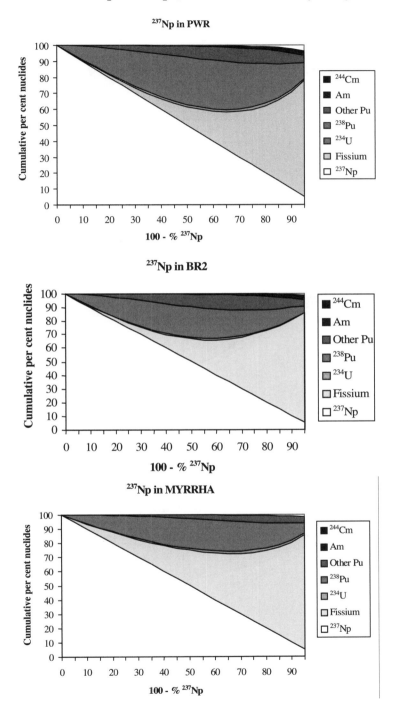

# Figure 5. Atom per cent concentration of the various nuclides in Am targets irradiated in the 25% MOX-fuelled PWR, in BR2 and in MYRRHA, as a function of the Am burn-up (each time followed by five years cooling)

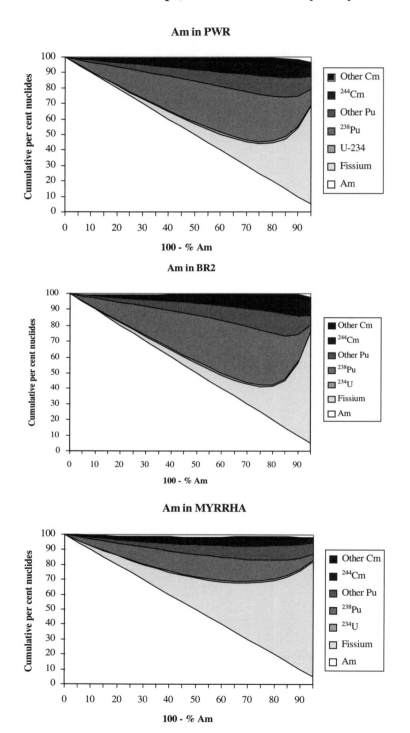

"direct" + (mainly) "indirect". This contribution increases strongly with irradiation time, due to the quadratic (or higher) order of the build-up curve of nuclides in long formation chains (the values given in Table 5 and in Figures 4 and 5 are only approximate as both the neutron flux levels and the microscopic cross-sections were assumed to remain constant during the irradiations and as the build-up of fission products was not taken into account. Control calculations nevertheless show that the trends indicated remain valid, in particular that the percentage fissions remains important in BR2). The same trends as in BR2 can be observed in the 25% MOX-fuelled PWR, but with 3 to 4 times longer irradiation periods because of the lower flux levels.

It appears clearly in Figures 4 and 5 that, at the early stage of the irradiation (up to, say, 60% $^{237}$Np disappearance and 80% Am disappearance), the disappearance of the MAs due to fissions is much more important in MYRRHA than in BR2 or in the 25% MOX-fuelled PWR, due to the direct fission contribution (in the fast spectrum domain) mentioned above. Above about 60% $^{237}$Np and 80% Am disappearance, fission plays an increasingly important role in BR2 and in the 25% MOX-fuelled PWR, as also mentioned above.

One observes that in a fast spectrum the generation of toxic long-lived actinides is practically inexistent for the irradiation of $^{237}$Np.

According to the present calculations, the following EFPD irradiation periods are needed to achieve 80% MA disappearance (these figures are only approximate as both the neutron flux levels and the microscopic cross-sections were assumed to remain constant during the irradiations – see above):

- In the $^{237}$Np targets: about 1 450 days in the 25% MOX-fuelled PWR, 425 days in BR2 and 3 700 days in MYRRHA. It should be noted (see Figure 4) that this 80% disappearance corresponds to 42% fissions in the 25% MOX-fuelled PWR, to 52% fissions in BR2 and to 54% fissions in MYRRHA.

- In the Am targets: about 1 200 days in the 25% MOX-fuelled PWR, 360 days in BR2 and 3 800 days in MYRRHA. It should be noted (see Figure 5) that this 80% disappearance corresponds to only 25% fissions in the 25% MOX-fuelled PWR, to only 22% fissions in BR2 but to 50% fissions in MYRRHA.

A more interesting comparison is that of the EFPD irradiation periods needed to achieve 80% MA incineration (here also the figures are only approximate, and only the irradiations in BR2 and in MYRRHA were considered):

- In the $^{237}$Np targets: about 770 days in BR2 and 6 800 days in MYRRHA.

- In the Am targets: about 1 470 days in BR2, and 7 900 days in MYRRHA.

For the disappearance times needed the ratio (EFPD in MYRRHA/EFPD in BR2) is hence about 8.7 for $^{237}$Np and 10.5 for Am, while for the incineration times needed this ratio remains about 8.8 for $^{237}$Np but is reduced to about 5.4 for Am. One of the reasons for the latter reduction is the larger formation of $^{243}$Am from $^{241}$Am in thermal spectrum systems.

In any case, in fast neutron systems, and in particular ADS devices such as MYRRHA, the irradiation period needed to fully deplete a MA target is longer than in thermal neutron systems for the neutron flux levels considered in this paper. Higher flux levels, e.g. in ADS systems, would obviously shorten the irradiation times needed to achieve transmutation, probably with a lower fissile material inventory. When comparing the performances of MYRRHA and BR2, one should also take into account the operation regime. Currently, BR2 only operates about 105 days per year while MYRRHA is to operate about nine months per year: the MYRRHA utilisation factor would hence be 2.6 times that of BR2.

**Comparison of the 25% MOX-fuelled PWR, of BR2 and of MYRRHA, from the neutron economy point of view**

From the neutron economy point of view, which is an essential topic when applying transmutation on an industrial scale, ADSs (and FRs) have the undeniable advantage over thermal reactors of needing shorter chains (i.e. less neutrons) to achieve fission. Larger quantities of fissile material will be needed in thermal critical systems to compensate the reactivity. In order to analyse in a quantitative way the neutron economy aspects of transmutation/incineration, the number of neutrons needed to obtain fission in each chain leading to fission was calculated and the average value over all chains leading to fission considered in this study was derived. In Table 6 these values are indicated by <nf>. The average number of neutrons consumed in all transmutation chains not leading to fission was also calculated, and is indicated by <ng>. The sum of both, <n> = <nf> + <ng>, is also shown in Table 6 and is in fact the only pertinent figure to be considered. As the irradiations progress in time, the average number <nf> tends to reach the average number <n>, as practically all transmutations ultimately end in fission reactions.

**Table 6. Average number of neutrons needed to obtain one fission in the transmutation/fission chains leading to fission <nf>, and in all chains <n>**

| MA | PWR | | | BR2 | | | MYRRHA | | |
|---|---|---|---|---|---|---|---|---|---|
| | % fiss | <nf> | <n> | % fiss | <nf> | <n> | % fiss | <nf> | <n> |
| $^{237}$Np | 1.61 | 1.98 | 14.6 | 5.83 | 2.53 | 7.99 | 1.43 | 1.04 | 3.05 |
| | 5.26 | 2.47 | 9.5 | 19.1 | 2.91 | 5.55 | 12.2 | 1.24 | 2.72 |
| | 17.9 | 2.91 | 6.08 | 49.5 | 31.8 | 4.31 | 31.4 | 1.45 | 2.41 |
| | 33.2 | 3.10 | 4.99 | 81.5 | 3.31 | 3.83 | 58.1 | 1.64 | 2.15 |
| | 69.9 | 3.28 | 3.96 | 87.1 | 3.33 | 3.73 | 85.4 | 1.80 | 1.97 |
| Am | 2.57 | 1.81 | 10.6 | 4.85 | 1.97 | 9.09 | 1.16 | 1.03 | 3.37 |
| | 5.91 | 1.99 | 8.62 | 10.1 | 2.22 | 7.55 | 10.01 | 1.24 | 2.99 |
| | 14.3 | 2.33 | 6.48 | 24.5 | 2.67 | 5.4 | 27.04 | 1.48 | 2.57 |
| | 33.2 | 2.73 | 4.73 | 55.7 | 3.06 | 3.98 | 52.24 | 1.69 | 2.26 |
| | 66.9 | 3.08 | 3.69 | 66.3 | 3.14 | 3.76 | 80.40 | 1.88 | 2.07 |

In Figure 6, <nf> and <n> have been drawn as a function of the percentage of fissions for each type of irradiation. One clearly sees that:

- For small percentages of fission (say up to 25%) <n>, the number of neutrons needed in the 25% MOX-fuelled PWR and in BR2 to achieve one fission is very large (20 ... 5) while in MYRRHA it is 3.5 ... 2.5. This is due to the much longer length, in thermal systems, of the transmutation chains need to achieve fission, as mentioned above.

- For high MA burn-up both in the $^{237}$Np and in the Am targets <n>, the number of neutrons needed to achieve one fission is about 3.5 when the irradiation is carried out in a 25% MOX-fuelled PWR and in BR2, while <n> only amounts to about 2.0 in MYRRHA.

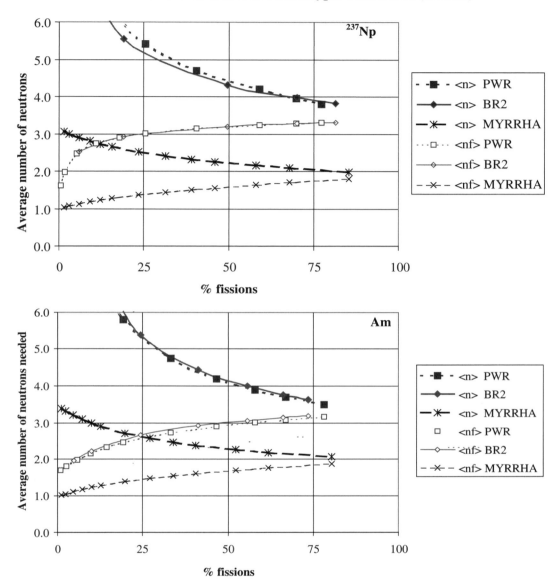

The smaller <n> values are a substantial advantage of ADS systems over thermal spectrum systems. Indeed, to obtain higher numbers of neutrons one obviously has to fission more primary fissile material. This is a drawback for thermal systems, which will produce more FPs and hence also more long-lived fission products (LLFPs), assuming the same scaling factor as in Table 6 and Figure 6 discussed above.

The inverse quantities of <nf> and <n>, viz <ef> and <e>, could be called "neutron economy factors". They may be considered as figures of merit. Ideally, they should be equal to 1.00. In the 25% MOX-fuelled PWR and in BR2 <e> tends to the value 0.28 for high MA burn-up while in MYRRHA <e> tends to 0.50.

## Comparison of the performances of accelerator-driven systems and fast reactors

In [17], the performances of FRs and ADS systems were compared, the conclusion being that the neutron-flux-averaged cross-sections governing the transmutation of MAs and LLFPs do not lead to very important differences in the performances of ADS devices compared to FRs. The absolute neutron flux levels, on the other hand, which are proper to each individual device, strongly influence the transmutation capacity. In the case of fast reactors the neutron flux levels can only vary within certain limits, while in the case of ADS the neutron flux levels are directly proportional to the proton beam current delivered by the accelerator and to the multiplication factor of the subcritical system, and also depend on the energy of the proton beam (the higher the energy, the higher the neutron/proton ratio in the spallation reaction). From the core reactivity control point of view, the ADS devices present undeniable advantages in the case of variable core loadings with large amounts of MAs.

## Conclusions

While the disappearance of the MAs in thermal systems (BR2 and the 25% MOX-fuelled PWR considered in the present study) in the early irradiation stages mainly consists of transmutation into higher MA isotopes, it gradually evolves towards incineration as the role of the "secondary" fissions increases. In MYRRHA the part taken by incineration in the disappearance of the MAs is already important in the early stages of the irradiation (higher "direct" fission-over-total-disappearance rate in the fast neutron spectrum). From the neutron economy point of view, MA irradiations in fast systems (MYRRHA) are therefore more advantageous than in thermal systems (BR2 and PWRs). As a result, smaller quantities of fissile material are needed to carry out the same MA incineration.

Concerning the irradiation times needed to achieve a given value of MA ($^{237}$Np and Am) disappearance, say 80%, irradiation times about three times shorter are needed when the transmutation/ incineration occurs in a MTR such as BR2 than when it occurs in a PWR (mainly because of the higher flux level prevailing in a MTR than in a PWR ) and about 9-10 times shorter irradiation times than when it occurs in an ADS such as MYRRHA (because of the much larger cross-sections in thermal spectrum systems than in fast spectrum systems). Nevertheless, the MA disappearance in BR2 and in PWRs contains a higher transmutation fraction and consequently a lower incineration fraction than in MYRRHA, at low MA disappearance, but also at high (say 80%) disappearance in the case of Am irradiation. In the case of Am irradiation, the MYRRHA irradiation time needed to achieve 80% Am incineration is only five times longer than the BR2 irradiation time, according to the present schematic calculations.

Of the two aspects considered, neutron economy and irradiation time, neutron economy is the more important one and hence irradiations in fast neutron spectra appear to be most interesting.

Other important aspects, such as the amount of MA that can be loaded in a thermal or fast system and the control possibilities of the various systems have not been examined here.

The transmutation capacity of BR2 can be used for investigating, at the technological scale, the formation of transmutation products ($^{238}$Pu, $^{239}$Pu, FPs, etc.) in a thermal neutron spectrum with a large contribution of epithermal and fast neutrons as well as the metallurgical behaviour of the targets. In particular, the calculated high fission-over-total-disappearance rate in the $^{237}$Np and Am targets could be checked if the irradiations are carried out during a long period. It is indeed essential to take into account the total length of the transmutation chains when performing calculations for thermal systems and in particular for high flux reactors.

MYRRHA, on the other hand, as a multi-purpose ADS for R&D, is an interesting tool for the investigation of transmutation/incineration of MAs in a fast neutron environment, having the advantage, with respect to FRs, of a larger versatility and an improved core reactivity control. In addition, MYRRHA is expected to operate with a high utilisation factor.

### *Acknowledgements*

Messrs B. Lance, Th. Maldague and S. Pilate of Belgonucléaire are thanked for providing us with most of the data concerning geometry, composition, temperature and linear power used as the starting point for the present 25% MOX-fuelled PWR calculations.

# REFERENCES

[1] Ch. De Raedt, B. Verboomen, Th. Aoust, A. Beeckmans de West-Meerbeeck, H. Aït Abderrahim, E. Malambu, Ph. Benoit, L. H. Baetslé, "MA and LLFP Transmutation in MTRs and ADSs. The Typical SCK•CEN Case of Transmutation in BR2 and MYRRHA. Position with Respect to Global Needs", 6th Information Exchange Meeting on Actinide and Fission-product Separation and Transmutation, Madrid, 11-13 Dec. 2000.

[2] Th. Maldague, S. Pilate, A. Renard, A. Harislur, H. Mouney, M. Rome, "Core Physics Aspects and Possible Loadings for Actinide Recycling in Light Water Reactors", GLOBAL'1995, International Conference on Evaluation of Emerging Nuclear Fuel Cycle Systems, Versailles, 11-14 Sept. 1995.

[3] Th. Maldague, S. Pilate, A. Renard, A. Harislur, H. Mouney, M. Rome, "Homogeneous or Heterogeneous Recycling of Americium in PWR/MOX Cores?" International Conference on the Physics of Nuclear Science and Technology, Long Island (New York), 5-8 Oct. 1998.

[4] Th. Maldague, S. Pilate, A. Renard, A. Harislur, H. Mouney, M. Rome, "Recycling Schemes of Americium Targets in PWR/MOX Cores", 5th International Information Exchange Meeting on Actinide and Fission Product Partitioning and Transmutation, Mol, 25-27 Nov. 1998.

[5] Brochure "BR2, Multipurpose Materials Testing Reactor. Reactor Performance and Irradiation Experience", SCK•CEN, Nov. 1992.

[6] J.M. Baugnet, Ch. De Raedt, P. Gubel, E. Koonen, "The BR2 Materials Testing Reactor. Past, Ongoing and Under-study Upgradings", 1st Meeting of the International Group on Research Reactors, Knoxville, Tennessee, 28 Feb.-2 March 1990.

[7] Ch. De Raedt, H. Aït Abderrahim, A. Beeckmans de West-Meerbeeck, A. Fabry, E. Koonen, L. Sannen, P. Vanmechelen, S. Van Winckel, M. Verwerft, "Neutron Dosimetry of the BR2 Aluminium Vessel", 9th International Symposium on Reactor Dosimetry, Prague, 2-6 Sept. 1996.

[8]  E. Malambu, Ch. De Raedt, M. Wéber, "Assessment of the Linear Power Level in Fuel Rods Irradiated in the CALLISTO Loop in the High Flux Materials Testing Reactor BR2", 3rd International Topical Meeting "Research Reactor Fuel Management (RRFM)", Bruges, 28-30 March 1999.

[9]  H. Aït Abderrahim, P. Kupschus, Ph. Benoit, E. Malambu, K. Van Tichelen, B. Arien, F. Vermeersch, Th. Aoust, Ch. De Raedt, S. Bodart, P. D'hondt, Y. Jongen, S. Ternier, D. Vandeplassche, "MYRRHA, a Multi-purpose ADS for R&D: Pre-design Phase Completion", these proceedings.

[10] H. Aït Abderrahim, P. Kupschus, E. Malambu, Ph. Benoit, K. Van Tichelen, B. Arien, F. Vermeersch, P. D'hondt, Y. Jongen, S. Ternier, D. Vandeplassche, "MYRRHA: A Multi-purpose Accelerator-driven System for Research and Development", *Nuclear Instruments & Methods Physics Research*, A 463, pp.487-494 (2001).

[11] P. Govaerts, P.D'hondt, H. Aït Abderrahim, P. Kupschus, "MYRRHA, a Multi-purpose Accelerator-driven System (ADS) for Research and Development, SCK•CEN R-3523, May 2001.

[12] E. Malambu, Th. Aoust, N. Messaoudi, G. Van den Eynde, Ch. De Raedt, "Status of Neutronics Analysis of the MYRRHA ADS. Progress Report September 2001", Draft Version, Internal Report SCK•CEN, Sept. 2001.

[13] "MCNP, a General Monte Carlo N-particle Transport Code, Version 4C", J.F. Briesmeister, ed., LA-13709-M, April 2000.

[14] "MCNPX User's Manual, Version 2.1.5", Revision 0, L.S. Waters, ed., 14 Nov. 1999.

[15] Ch. De Raedt, E. Malambu, B. Verboomen, "Increasing Complexity in the Modelling of BR2 Irradiations", PHYSOR 2000 International Topical Meeting, Advances in Reactor Physics and Mathematics and Computation into the Next Millennium, Pittsburgh, PA, 7-11 May 2000.

[16] B. Verboomen, A. Beeckmans de West-Meerbeeck, Th. Aoust, Ch. De Raedt, "Monte Carlo Modelling of the Belgian Materials Testing Reactor BR2: Present Status", Monte Carlo 2000, International Conference on Advanced Monte Carlo for Radiation Physics, Particle Transport Simulation and Applications, Lisbon, 23-26 Oct. 2000.

[17] Ch. De Raedt, L.H. Baetslé, E. Malambu, H. Aït Abderrahim, "Comparative Calculation of FR-MOX and ADS-MOX Irradiations", International Conference on Future Nuclear Systems, GLOBAL'99, Jackson Hole (WY), 30 Aug.-2 Sept. 1999.

# DETERMINATION OF THE $^{233}$Pa(n,f) REACTION CROSS-SECTION FOR THORIUM-FUELLED REACTORS

**Fredrik Tovesson,[1,2] Franz-Josef Hambsch,[1] Andreas Oberstedt,[2]**
**Birger Fogelberg,[3] Elisabet Ramström,[3] Stephan Oberstedt[1]**
[1]EC-JRC-Institute of Reference Materials and Measurements (IRMM)
Retieseweg, B-2440 Geel, Belgium
[2]Örebro University, Department of Natural Sciences, SE-70182 Örebro, Sweden
[3]Uppsala University, Department of Radiation Sciences, SE-61182 Nyköping, Sweden

## Abstract

A direct measurement of the energy-dependent neutron-induced fission cross-section of $^{233}$Pa has been performed for the first time. The $^{233}$Pa isotope plays a key role in the thorium fuel cycle, serving as an intermediate isotope in the formation of the uranium fuel material. Since fission is one of the reactions determining the balance of nuclei at a given time, the cross-section is of vital importance for any calculation of a thorium-fuel-based nuclear-power device. In a first measurement series, four energies between 1.0 and 3.0 MeV were measured. The resulting average above-threshold cross-section found is lower than all literature values.

## Introduction

The concept of using thorium-based fuel as an alternative to the conventional uranium-plutonium cycle in nuclear reactors has been under discussion for quite some time. The advantages of employing the thorium cycle are many, i.e. less long-lived radiotoxic waste, higher fuel burn-up, etc. Thorium-based fuel has been considered for a number of different systems ranging from traditional thermal reactors to fast spectrum reactors (e.g. the accelerator-driven reactor design known as the energy amplifier suggested by Rubbia, *et al.* [1]). A crucial point in the design and development of new advanced reactor concepts is the access to accurate neutron reaction data. Even though most of the required cross-section data for the thorium cycle exists in the different evaluated data libraries such as ENDF/B-6 and JENDL-3, the neutron-induced reactions of some of the important isotopes are not seriously studied for different reasons. One such example is $^{233}$Pa, for which no reliable data exist for the capture and fission cross-sections.

$^{233}$Pa plays an important role in the thorium cycle as an intermediate isotope between the thorium source material and the uranium fuel material (see Figure 1). In a thorium-fuelled reactor, $^{233}$Th is formed by neutron capture in $^{232}$Th and decays by β-emission with a 22 minute half-life to $^{233}$Pa. The $^{233}$U fuel isotope is then produced as a result of the β-decay of this protactinium isotope. The $^{233}$Pa decays with a half-life of 27 days, which is a rather long time in this context. In the meantime, there are other reactions competing with the natural decay to $^{233}$U. If a $^{233}$Pa nucleus undergoes fission or neutron capture it means that a $^{233}$U nucleus is lost as well. The $^{233}$Pa thus plays a role for the balance of fissile nuclei in a thorium fuel-based system.

**Figure 1. The most important reactions in the thorium fuel cycle**

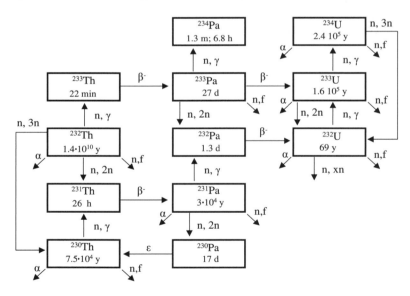

The "slow" decay of $^{233}$Pa to the fissionable $^{233}$U also has an important impact on safety aspects of a thorium-fuelled reactor. In case of a reactor stop (intentional or non-intentional) the fraction of $^{233}$Pa present in the fuel is continuously decaying to the fissile $^{233}$U. This means that the reactivity is increasing with time, possibly causing criticality a long time after the initial shutdown takes place. This is known as the protactinium effect [2]. It is interesting to note that no corresponding effect exists for the uranium-plutonium cycle.

The competition between fission and neutron capture on one hand and natural decay of [233]Pa on the other, is dependent on the neutron energy spectrum. In a thermal flux capture totally dominates with respect to fission, meaning that the fission cross-section in this energy region is of minor importance. In a fast system the situation is different. At higher neutron energies the capture cross-section decreases and the fission threshold is reached. Hence, for these types of systems accurate fission cross-section data are needed for reliable modelling and calculations. The required accuracy of the [233]Pa(n,f) cross-section is given in a recent IAEA report [3] to be 20% for fast systems.

The only direct experimental data on the [233]Pa(n,f) that has existed up to now was a measurement from 1967 by Van Gunten, *et al.* [4], where a reactor spectrum was used as neutron source. An average above-threshold fission cross-section of 775±190 mb was found, but the effective neutron spectrum was not very well known. The reason that so little experimental data exist on neutron-induced reactions in [233]Pa is due to some experimentally challenging properties, i.e. its short half-life, high β-activity and in-growth of [233]U.

Another technique of indirectly measuring the [233]Pa(n,f) was used in a more recent work [5]. In this case the fission probability of [234]Pa was determined from the substitution reaction [232]Th([3]He,pf)[234]Pa. Model calculations were then used to estimate the compound nuclear formation cross-section. The drawback of this method is of course that it is relying on an accurate model description. This measurement indicated a much lower cross-section compared to the measurement of Van Gunten, *et al.* with an average above-threshold value of about 450 mb.

When looking to the theoretical evaluations for the [233]Pa(n,f) cross-section available from neutron data libraries such as ENDF/B-VI and JENDL-3 large differences are exhibited. In the two mentioned evaluations, for example, the average above-fission threshold cross-section differs by a factor of two and there is also a significant difference in the threshold energy (1.5 compared to 0.8 MeV for ENDF/B-VI [6] and JENDL-3 [7], respectively).

In order to solve the existing discrepancies, an experiment was set up to directly measure the [233]Pa(n,f) reaction cross-section with quasi-monoenergetic neutrons. In the first measurement campaign that is reported here, four neutron energy points were measured between 1.0 and 3.0 MeV.

**Experiment**

The sample preparation was carried out at the Studsvik Neutron Research Laboratory in Nyköping, Sweden. A batch of thorium nitrate was irradiated in a reactor neutron flux in order to produce [233]Pa by neutron capture followed by β-decay. Repeated chemical washing procedures were then performed, resulting in a highly purified organic [233]Pa solution. The solution was then deposited onto a tantalum backing, giving a sample of 0.564±0.025 μg of [233]Pa.

The cross-section measurements took place at the Van De Graaff accelerator facility of the Institute of Reference Materials and Measurements (IRMM) in Geel, Belgium. Accelerated protons impinging on a tritium target produced the quasi-monoenergetic neutron beam in the desired energy range. The fission detector was a twin Frisch gridded ionisation chamber [8]. A schematic view of the chamber and the associated electronics is given in Figure 2.

The cross-section measurement was performed relative to the well-known fission cross-section of [237]Np. Hence, a [237]Np sample was placed in a back-to-back geometry with the [237]Pa sample on the common cathode. The anode and sum signals were stored by the acquisition system for each fission event. In this way both the fission fragment energy and a signal proportional to its range and emission

**Figure 2. Schematic view of the chamber and associated electronics**

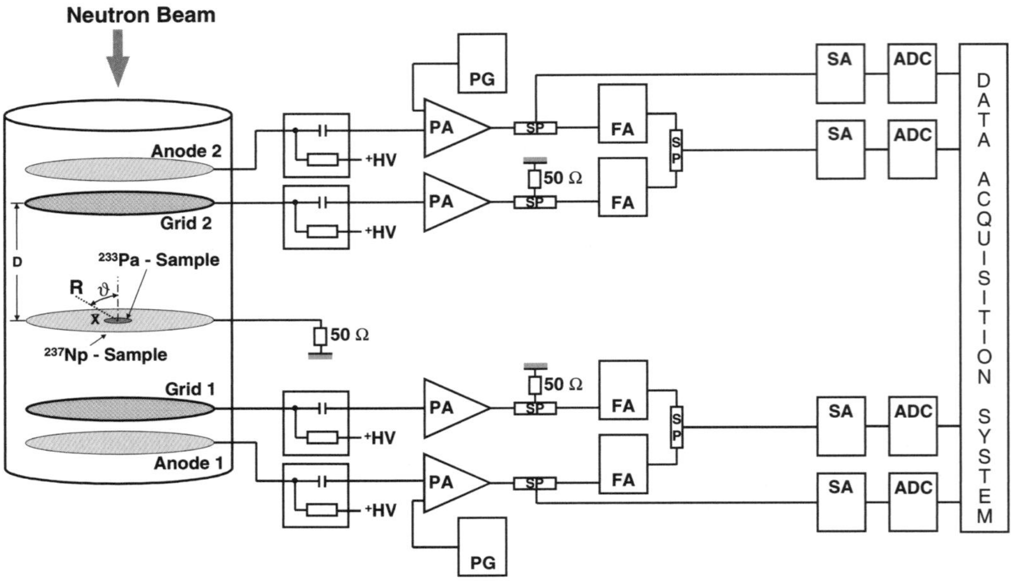

angle was available. By using this information, a very efficient background reduction could be achieved. This is certainly an important feature of the detector system in order to provide accurate counting of the number of fission events from $^{233}$Pa, since its $\beta$-activity is so high that without special experimental measures the piled-up $\beta$-events would overlap with the fission fragment spectrum.

The measurements were carried out at four incident neutron energies: 1.0, 1.6, 2.0 and 3.0 MeV. For each energy point irradiation time was about 100 hours to achieve the required accuracy.

## Analysis

The first step in the data analysis was to carry out the background reduction. This was achieved by plotting the energy signal vs. the signal proportional to the range and angular orientation [Figure 3(a)] in a two-dimensional representation. In this plot the area containing the piled-up $\beta$-events could be clearly identified and by setting a region of interest around, these events are cut out from the fission fragment spectrum. The spectra from the two chamber sides after background reduction are shown in Figure 3(b).

When the fission fragments had been clearly identified, the number of fission events coming from $^{233}$U which is increasingly present in the $^{233}$Pa sample over the duration of the measurement had to be corrected for. The amount of $^{233}$U at any given time can be easily calculated from the usual exponential decay law, since at production time the $^{233}$U present in the sample was separated out. The number of fission events attributed to $^{233}$U for a specific measurement run can then be calculated as:

$$C_U = \frac{N_U \cdot \sigma_U}{N_{Np} \cdot \sigma_{Np}} \cdot C_{Np}$$

where $C$ is the number of fission fragments, $N$ is the number of nuclei in the respective samples, $\sigma$ is the fission cross-section and the subscripts $U$ and $Np$ stand for $^{233}$U and $^{237}$Np. The number of fission events from the $^{233}$Pa side of the fission chamber can then be corrected for this in-growth of $^{233}$U.

# Figure 3

*(a) The plotting representation used for the background separation. The energy is plotted versus the signal proportional to the particle range and angular orientation.*

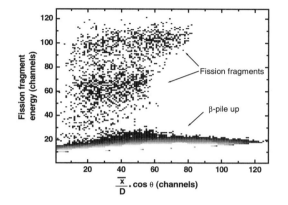

*(b) Fission fragment energy spectra from $^{233}$Pa (above) and $^{237}$Np (below) after background reduction has been performed*

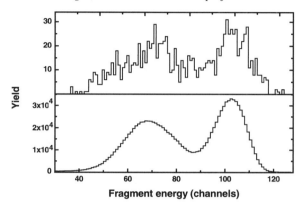

The fission cross-section of $^{233}$Pa is deduced according to the following equation:

$$\sigma_{Pa} = \frac{N_{Np} \cdot C_{Pa}}{N_{Pa} \cdot C_{Np}} \cdot \sigma_{Np}$$

It should be noted, however, that this formula only holds under the assumption that the efficiencies of the two chamber sides are the same. Since the two chamber halves are actually identical, the two samples have the same deposition area and the mounting of the samples is done in exactly the same way, the assumption holds.

The resulting cross-section points are shown in Figure 4. The data from the present measurement are compared to the evaluations from ENDF/B-VI [6] and JENDL-3 [7] as well as the indirect measurement of Barreau, *et al.* [5]. As can be seen in the figure, our results show a cross-section that is lower by as much as 50% at 2.0 MeV compared to the data of Barreau, *et al.* This is quite a lot, considering the accuracy requirement of 20% [3] for the fission cross-section. When comparing with the evaluations, the measured average above-threshold cross-section agrees most closely with the evaluation of JENDL-3. For the fission threshold energy, the best agreement is with the evaluation of ENDF/B-VI.

## Conclusions

The neutron-induced cross-section of $^{233}$Pa has been measured at four incident neutron energies ranging from 1.0 to 3.0 MeV. The results point to a fission threshold energy of about 1.3 MeV and an average above-threshold cross-section value of about 365±30 mb. This cross-section is lower than the theoretical evaluations as well as the indirect measurement of Barreau, *et al.* [5] and the integral measurement of von Gunten, *et al.* [4].

This investigation is to be followed by at least one more measurement series in the near future. The present priorities would be to include an investigation of the second chance fission region, as well as additional studies of the first chance fission threshold to map this out in more detail.

**Figure 4. The measured $^{233}$Pa(n,f) cross-section. Also shown are the evaluations from ENDF/B-VI [6] and JENDL-3 [7] as well as the indirect measurement of Barreau, *et al.* [5].**

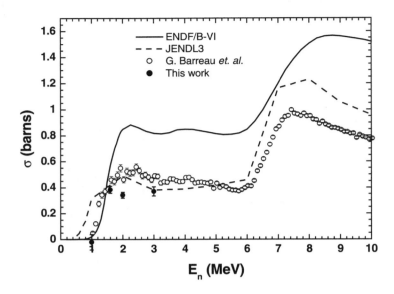

**REFERENCES**

[1]   F. Carminati, *et al.*, "An Energy Amplifier for Cleaner and Inexhaustible Nuclear Energy Production Driven by a Particle Beam Accelerator", CERN/AT/93-47(ET), 1 Nov. (1993).

[2]   C. Rubbia, *et al.*, "Conceptual Design of a Fast-neutron-operated High-power Energy Amplifier", CERN/AT/95-44(ET), 29 Sept. (1995).

[3]   V.G. Pronyaev, "Summary Report of the Consultants' Meeting on Assessment of Nuclear Data Needs for Thorium and other Advanced Cycles", IAEA report INDC(NDS)-408 (1999).

[4]   H.R. von Gunten, R.F. Buchanan, A. Wyttenbach and K. Behringer, *Nucl. Sci. and Eng.*, 27, 85-94 (1967).

[5]   G. Barreau, *et al.*, private communication.

[6]   Mathews, R. Kinsey, P.C. Young, BNL-325 (Ed. 3) (1978), Brookhaven National Laboratory, USA, ENDF/B-VI evaluation, MAT #9137.

[7]   T. Ohsawa, M. Inoue, in Proceedings of the International Conference Nuclear Data for Science and Technology, Mito, Japan, 30 May-3 June (1988), S. Igarasi, ed., Saikon Publishing Co. Ltd., JENDL-3.2 evaluation, MAT #9137.

[8]   C. Budtz-Jørgensen, *et al.*, *Nucl. Inst. and Meth.*, A256 (1987).

# MISCELLANEOUS THEMES

*Chair: K. Hesketh*

# THE KEY ROLE OF CRITICAL MOCK-UP FACILITIES FOR NEUTRONIC PHYSICS ASSESSMENT OF ADVANCED REACTORS: AN OVERVIEW OF CEA CADARACHE TOOLS

**G. Bignan, D. Rippert, P. Fougeras**
Commissariat à l'Energie Atomique, CE Cadarache
Service de Physique Expérimentale
F-13108 St. Paul-lez-Durance Cedex, France
E-mail: bignan@cea.fr

## Abstract

The Experimental Physics section of CEA Cadarache operates three critical facilities devoted to neutronic studies of advanced reactors (EOLE, MINERVE and MASURCA) covering a large scope of interests. These include 100% MOX core in ABWR qualification, knowledge improvement of basic nuclear data for heavy nuclides for new options of the fuel cycle – especially the multi-recycling of plutonium – and accelerator-driven systems neutronic behaviour for transmutation studies. The paper describes these facilities, the scientific programmes associated and the progressive improvement of experimental techniques, the aim being to significantly reduce the uncertainties regarding the evaluation of the physical parameters.

# Introduction

At the present time, research and development concerning nuclear energy in France is motivated by two major objectives:

- *Competitivity of the nuclear fuel cycle.* This field of research leads to various studies concerning innovative fuels and reactors, multi-recycling of plutonium, use of burn-up credit approach, innovative neutron absorbers, etc.

- *Back-end of fuel cycle management.* This field of research leads to various studies concerning waste treatment and monitoring, actinide transmutation [especially in accelerator-driven systems (ADS)], etc.

From a neutronic physic point of view, all these studies require experimental qualification of calculation tools, improvement of basic nuclear data, and understanding of the neutronic behaviour of specific cores. To achieve these goals, the Experimental Physics section of CEA Cadarache operates three critical mock-up facilities which have many advantages in terms of fuel availability, flexibility and instrumentation. After a description of these reactors and of the scientific programmes associated, we present new developments which are under way for the improvement of experimental techniques.

## The critical mock-up facilities

### *The EOLE reactor* [1]

This reactor is dedicated to the neutronic study of moderated lattices (LWR fuels).

It is an easily adaptable facility set up in a building maintained in depression (air leakage controlled building). The reactor (maximum power 100 W) is composed of:

- a structure in concrete offering biological shielding for a flux level up to $10^9$ neutrons/cm$^2$/sec in the core;

- a cylindrical vessel in AG3 (diameter = 2.3 m, height = 3 m) with an overstructure in stainless steel, able to contain various types of core and related structures;

- control rods (four safety rods, one pilot rod) linked to the overstructure, the position and composition which can be adapted as required to the studied cores;

- water circuits in order to fill up and to empty with a moderator to introduce soluble boron and to control the moderator temperature between 5 and 85°C;

- an up-to-date command-control with a numerical treatment of the signals, programmable automatic control and operator assistance through visualisation of the reactor status on a screen;

- modern and computerised equipment for experimental data treatment.

At the present time, the BASALA programme is carried out in the EOLE reactor in the framework of an international co-operation between NUPEC (Japan), CEA and COGEMA. The purpose of this programme is to provide experimental results for the qualification of the methods used for design calculations of BWR cores fully loaded with MOX fuels. These 100% MOX cores are being considered for future Japanese BWR advanced type reactors (ABWR). In addition to the fuel composition, they present physical properties slightly different from current BWRs.

In particular, reactivity control management requires an increase in the water proportion in the fuel assembly (moderation ratio) or even the use of new absorbers.

The BASALA experimental programme is based on two different experimental cores:

- the BASALA H core, representative of a 100% MOX ABWR in hot condition;

- the BASALA C core, representative of a 100% MOX ABWR in cold condition.

Each core will comprise various configurations enabling the study of neutronic parameters representative of such BWR 100% MOX assemblies:

- in the reference condition;

- in the presence of burnable poison absorbers;

- in higher void conditions;

- in the presence of control blades ($B_4C$ and or Hf);

- fuel substitution.

Experimental measurements will concern reactivity worth, radial and axial power distribution, temperature coefficient and boron worth as burnable poisons or control blades.

Subsequent to the BASALA programme, the EOLE reactor will be used in support of a LWR new fuel concept for plutonium stockpile optimisation (CORAIL assembly, APA assembly, etc.).

### *The MINERVE reactor* [1]

This reactor is devoted to the neutronic study of lattices of different reactor types. It is located in the same building as EOLE.

The reactor is a pool-type reactor operating at a maximum power of 100 W. The core, submerged in 3 meters of water, is used as a driver zone for the different experiments located in a central square cavity (chimney). The coupled lattices in its experimental zone are built in such a way that they can reproduce various neutronic spectra: PWR UOx spectrum, PWR 100% MOX spectrum, epithermal spectrum, fast spectrum.

The core is built in a parallelepiped pool of stainless steel of about 140 $m^3$. The moderator is de-mineralised water and the cooling is performed only by natural convection. The reactor is controlled by control rods made of hafnium.

One of the main advantages of MINERVE is its oscillating device which allows to perform very precise reactivity measurements of small samples and so to improve knowledge of basic nuclear data for heavy nuclides, structural material and long-life fission products.

The oscillation technique consists of oscillating samples that contain the studied actinide in the centre of the experimental lattice in order to measure the associated reactivity variation with an accuracy better than 1% (at 1 σ). Each sample is placed in an oscillation rod and moved periodically and vertically between two positions located inside and outside the experimental core zone by an oscillator, as shown in Figure 1.

**Figure 1. Movement of the oscillation sample inside the MINERVE facility**

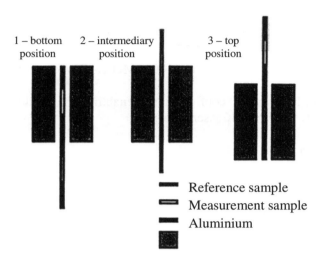

The study sample is compared to a reference sample that differs just by the lack of actinide and that is placed in the bottom of the oscillation rod. Each sample is measured at least four times with the aim of significantly decreasing systematic errors. A measurement corresponds to 20 oscillations of 60 s each.

The reactivity variations due to the oscillation are compensated by a rotary automatic pilot rod using cadmium sectors. The latter is calibrated by $^{235}$U and $^{10}$B samples.

The final experimental accuracy of the reactivity value, taking into account the uncertainties on the experiment, on the material balance and on the calibration of the pilot rod, is around 3% (at 1 σ).

MINERVE is currently the subject of a renovation programme which is planned to continue until the end of 2002 (notably with a brand-new command-control device). Between 2003 and 2009 the OSMOSE programme (oscillation in MINERVE of isotopes in "Eupraxic" spectra) will be carried out. This programme aims at obtaining, in different experimental lattices, a single and accurate experimental database for separated heavy nuclides. It will study a large majority of actinides appearing in the fuel cycle (front-end, reactor, back-end). Its main goals [2] are (see Table 1):

- improvement of knowledge of the integral absorption cross-sections of actinides (from $^{232}$Th to $^{245}$Cm) in thermal, epithermal and fast spectra (related to JEFF-3 project);

- achievement of experimental data to extend the validity domain of the criticality calculation tools (extension of the burn-up credit approach to actinides);

- knowledge of integral absorption cross-sections for transmutation and incineration of minor actinide studies.

**Table 1. OSMOSE programme – isotopes of interest**

| | JEFF-3 validation | Criticality burn-up credit | Pu recycling | Transmutation and incineration | Decay heat power | Sub-surface and long time storage | Reactivity loss per cycle | Thorium cycle |
|---|---|---|---|---|---|---|---|---|
| $^{232}$Th | ⊗ | | | | | | | ⊗ |
| $^{233}$U | ⊗ | | | | | | | ⊗ |
| $^{234}$U | ⊗ | ⊗ | | | | | ⊗ | |
| $^{235}$U | ⊗ | ⊗ | | | | ⊗ | ⊗ | |
| $^{236}$U | ⊗ | ⊗ | | | | | ⊗ | |
| $^{238}$U | ⊗ | ⊗ | | | | | ⊗ | |
| $^{237}$Np | ⊗ | ⊗ | | ⊗ | | ⊗ | ⊗ | |
| $^{238}$Pu | ⊗ | ⊗ | ⊗ | ⊗ | ⊗ | ⊗ | ⊗ | |
| $^{239}$Pu | ⊗ | ⊗ | ⊗ | ⊗ | ⊗ | ⊗ | ⊗ | |
| $^{240}$Pu | ⊗ | ⊗ | ⊗ | ⊗ | ⊗ | ⊗ | ⊗ | |
| $^{241}$Pu | ⊗ | ⊗ | ⊗ | ⊗ | ⊗ | ⊗ | ⊗ | |
| $^{242}$Pu | ⊗ | ⊗ | ⊗ | ⊗ | ⊗ | ⊗ | ⊗ | |
| $^{241}$Am | ⊗ | ⊗ | ⊗ | ⊗ | ⊗ | ⊗ | ⊗ | |
| $^{243}$Am | ⊗ | ⊗ | ⊗ | ⊗ | ⊗ | ⊗ | ⊗ | |
| $^{244}$Cm | ⊗ | ⊗ | ⊗ | ⊗ | ⊗ | ⊗ | ⊗ | |
| $^{245}$Cm | ⊗ | ⊗ | | ⊗ | ⊗ | ⊗ | ⊗ | |

## *The MASURCA reactor* [1]

This reactor is dedicated to the neutronic study of fast reactor lattices. Its main characteristics are the large volume of the core (6 000 litres), the availability of various plutonium fissile material [$UO_2$-$PuO_2$ (25%); Pu metal (100%), etc.], the fast neutron spectrum with a possibility of local moderation and the flexibility of the core building. The materials of the core are contained in cylinder rodlets, along with square platelets. These rodlets or platelets are put into wrapper tubes having a square section (4 inches) and which are about 3 meters in height. These tubes are hanged vertically from a horizontal plate supported by a structure of concrete. To build such cores the tubes are introduced from the bottom in order to avoid that the fall of a tube corresponds to a positive step in reactivity (Figure 2).

The reactivity control is assumed by absorber rods in varying number depending on core type and size. The control rods are composed of fuel material in their lower part, so that the homogeneity of the core is kept when the rods are withdrawn. The core is cooled by air and is surrounded by a biological shielding in heavy concrete allowing operation up to a flux level of $10^9$ n/cm$^2$/sec. The core and biological shielding are inside a reduced-pressure vessel, relative to the outside environment.

MASURCA currently plays a key role in the support of ADS studies for transmutation systems (French parliament law of 30 December 1991 regarding waste management). Such ADSs could be helpful for the reduction of harmfulness and toxicity of long-life radioactive waste. For that purpose, CEA, in a large national and international co-operative effort (CNRS, EDF, FRAMATOME, European Community, SCK Mol, FZK, FZJ, CIEMAT, NRG, BNFL, ENEA, Stockholm University, Chalmers University, PSI, DOE, JAERI), is leading the MUSE (*MU*ltiplication par *S*ource *Ex*terne) programme [3]. It consists of experimentally studying the physical problems due to subcritical multiplying media using an external source, in this case, a deuteron accelerator called GENEPI supplied by the *Institut des Sciences Nucléaires* – CNRS Grenoble.

**Figure 2. The MASURCA reactor**

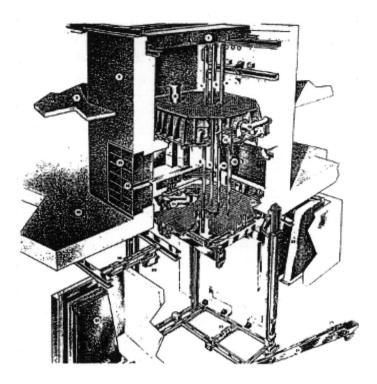

**Table 2. GENEPI deuteron accelerator characteristics**

| | | |
|---|---|---|
| Beam energy | 140 to 240 keV | |
| Peak current | 50 mA | |
| Frequency | 10 to 5 000 Hz | |
| Minimal pulse duration | 700 ns | |
| Focusing point dimension | $\simeq 20$ mm | |
| Pulse reproductibility | $\simeq 1\%$ | |
| Neutronic sources | | |
|    Deuterium deposit | $E_n = 2.7$ MeV | Number of n/s: $2 \cdot 10^8$ |
|    Tritium deposit | $E_n = 14$ MeV | Number of n/s: $2.5 \cdot 10^{10}$ |

The main objectives of the MUSE programme are:

- the understanding of neutron behaviour of multiplying subcritical media;

- the development of reference neutronic calculation tools using a C/E benchmark under the auspices of the OECD/NEA;

- the development of specific experimental techniques for dynamic measurements (coupling of the MASURCA core and the accelerator).

The main advantage of MUSE is a very good knowledge of the source in terms of spectra, intensity and location which will allow precise measurements.

Two phases are planned in the MUSE programme:

| Phase | Fuel | Cooling | Reactivity configurations | | | |
|---|---|---|---|---|---|---|
| MUSE 4 | UO₂-PUO₂ | Na | Critical | Subcritical 1 0.994 | SC2 0.97 | SC3 0.95 |
| MUSE 5 | UO₂-PUO₂ | "Gas" | Clean | Critical | SC1 | SC2 |

The main physical parameters which will be measured and the target uncertainty (1 σ) are listed in Table 3.

Table 3. MUSE physical parameters

| Parameter | Target uncertainty (1 σ) |
|---|---|
| Absolute reactivity ($\rho$) | ±0.2% |
| Effective delayed neutron fraction ($\beta_{eff}$) | ±3% |
| Neutron lifetime $\Lambda$ | $< \pm10\%$ |
| Axial and radial flux distribution | ±2% |
| Minor actinide fission rate | ±2% |
| Spectrum indices | ±2% |

The MUSE 5 configuration represents a step towards high-efficiency gas-cooled reactor studies which is, for the CEA, a promising candidate for GENERATION 4 forum (combination of high thermodynamic conversion yield – HTR advantage – and high nuclear burner yield using fast spectrum – advantage in waste incineration and Pu stockpile control). The development of such a gas-cooled fast reactor will require:

- good knowledge of basic nuclear data;

- validation of new fuel conception;

- validation of neutronic account (production, leaks, etc.);

- shield dimensioning.

These requirements will be necessary for qualification of neutronic calculation tools for which MASURCA will play a key role in the future.

## New developments in experimental techniques

### *New generation of fission chambers*

Miniature fission chambers are the most widely used in-core neutron detectors in research reactors due to their robustness, prompt answer and large availability of fissile material deposit (from ²³²Th to ²⁴⁶Cm). A great effort has been made these last few years for the design and the fabrication of such detectors [4,5] in order to better master their fabrication and thus decrease sources of uncertainties. For any new requirements of measurements, the chamber geometry, the fissile deposit and the gas pressure are optimised using a computer code to foresee the detector current-voltage characteristics and to calculate the evolution of the isotopic composition of the fissile material during irradiation.

Concerning the manufacturing, the main steps are realised under a quality assurance programme:

- fissile material elaboration: the fissile deposit (from $^{232}$Th to $^{246}$Cm) is electrolysed in a special glove box, and good monitoring of the deposit mass and its associated uncertainty (about 1%) are ensured using chemical analysis – mass spectrometry – before and after electrolysis;

- the chamber assembly;

- the filling-up with gas (argon, helium, ...) in a specific glove box;

- the good behaviour tests: electrical insulation resistance ($> 1 \times 10^{12}$ ohm), helium-tightness.

Three kinds of geometries are currently used in the critical mock-ups described earlier: 4 mm cylindrical detector which operates in pulse mode and with a deposit mass between 10 and 300 µg, 8 mm cylindrical diameter with a deposit mass between 1 and 100 mg which also operates in pulse mode and 1.5 mm cylindrical detector intended for use in current mode (fissile mass of about 200 µg, 30 meters cable, see Figure 3).

**Figure 3. Schematic views of the φ1.5 mm cylindrical fission chamber**

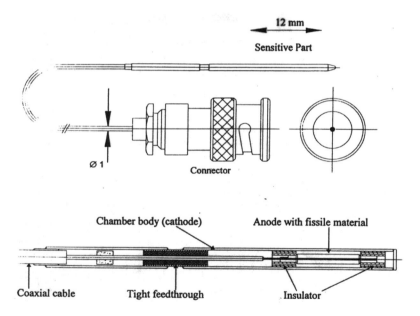

The preliminary tests of this new generation of fission chambers indicate a significant reduction in the uncertainties on physical parameters obtained through neutron detection.

### Calibration cell

In order to calibrate *in situ* and to recalibrate the fission chambers periodically, a non-destructive assay using the differential die-away technique (DDT) is under development [6]. This device will be located in the EOLE MINERVE building. It will use a 14 MeV neutron generator. The neutrons produced in over short time periods (~ 15 µs) slow down inside the cavity which surrounds the fission chamber to be calibrated and also induce fission in the deposit.

[3]He detectors, surrounded by a low-energy neutron-absorbing sheet, are located on the sides of the cell. They are mainly used for neutron generator flux monitoring.

A schematic view of the calibration cell is presented in Figure 4.

**Figure 4. Schematic view of the fission chamber calibration cell**

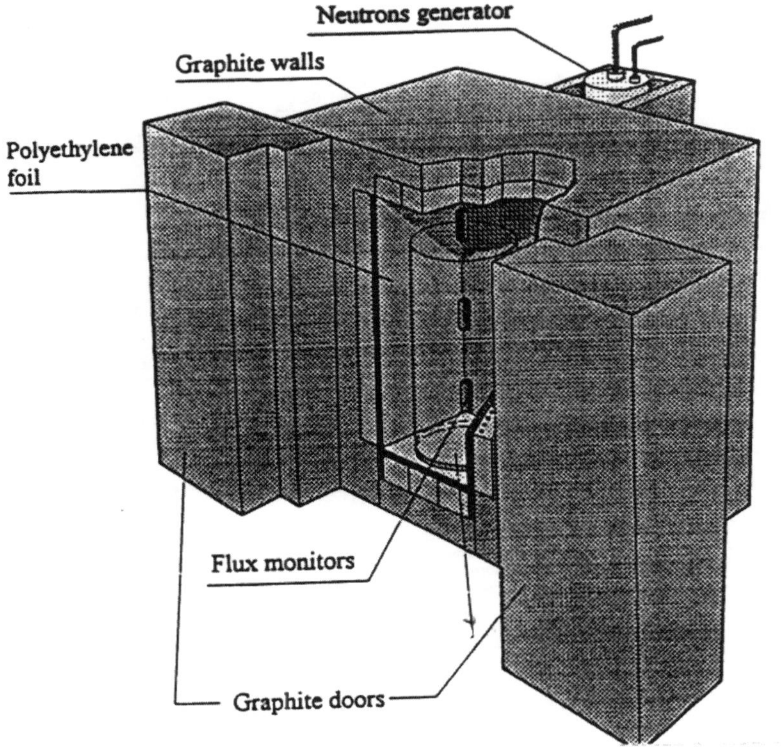

This device will allow a calibration of the fission chambers in terms of mass and of sensitivity (count $s^{-1}/ncm^{-2}s^{-1}$) at the reception of new detectors and periodically before starting a new programme.

*Neutronic pulse dating system*

The MUSE programme described above requires the investigation of dynamic measurement methods due to the coupling between the accelerator and the core (pile noise method, reactor transfer function method, frequency modulation method, pulse source method). To achieve this investigation a new acquisition system called DITER (*D*atation neutron*I*que *TE*mps *R*eel) in under development. This system will allow neutronic pulse dating (temporal information of neutron pulses); that is to say, for an experimental run, several analyses will be possible. The DITER system will be able to monitor 12 pulse-mode-operated fission chambers, two silicon detectors, the GENEPI accelerator signal (deuteron burst emission) and the deuteron current into the target (see Figure 5).

The main goal of DITER is to time-mark TTL pulses outgoing from the neutron detection and GENEPI related electronics. For one acquisition run, the data flow to the storage device can be huge (60 Mbytes/s). To reach this goal, no "trigger-type" electronics are used, but "time to digital" converter with a 50 MHz clock and a real-time operating system. This DITER project allows to obtain an on-line measurement on the physical parameters and to "replay" the experiment as often as required [7].

477

## Figure 5. Viewgraph of DITER

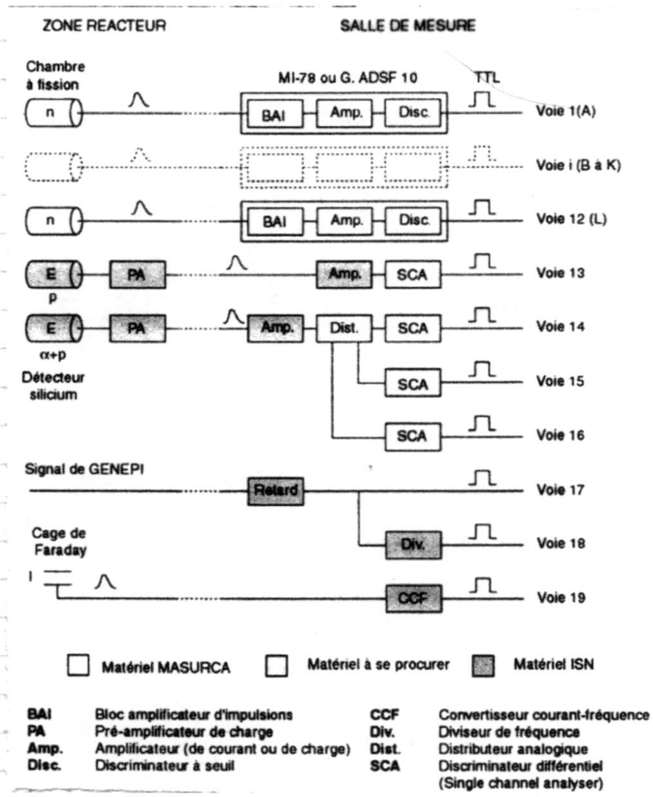

**Conclusion**

In this paper, we have outlined the ambitious programmes either currently under way or planned in the near future in the three critical mock-ups of CEA Cadarache. From a nuclear safety point of view, up-to-date reviews are permanently performed in order to satisfy the request of the French safety authority. At the present time, CEA considers that these three research reactors will continue to operate until 2015.

# REFERENCES

[1]  P. Fougeras, *et al.*, "The Use of the CEA Critical Facilities for the Assessment of the Physics", ENS Class 1 Topical Meeting on Research Facilities for the Future of Nuclear Energy, Belgium, June 1996.

[2]  J.P. Hudelot, *et al.*, "The OSMOSE Experimental Program in MINERVE for the Qualification of the Integral Cross-sections of Actinides", Nuclear Data 2001, JAERI Conference, Tsuskuba, Japan.

[3]  R. Soule, *et al.*, "Validation of Neutronic Methods Applied to the Analysis of Fast Subcritical Systems: The MUSE Experiments", GLOBAL'1997, Japan.

[4]  G. Bignan, *et al.*, "Direct Experimental Test and Comparison between Sub-miniature Fission Chambers and SPND for Fixed In-core Instrumentation of LWR", OECD/NEA Specialists Meeting on In-core Instrumentation, IN-CORE'96, Mito, Japan (1996).

[5]  C. Blandin, *et al.*, "Development and Modelling of Neutron Detectors for In-core Measurement Requirements in Nuclear Reactors", 10th International Symposium on Reactor Dosimetry, ASTM STP 1398, Osaka, Japan (1999).

[6]  J.P. Hudelot, "Dévelopement, amélioration et calibration des mesures de taux de réactions neutroniques : élaboration d'une base de techniques standards", Ph. D., Grenoble 1 University, 19 June 1998.

[7]  C. Jammes, "DITER: An On-line Neutron Pulse Dating for ADS Studies", 5th FWP EEC Contract N° FIKW-CT2000-0063 – Working group meeting, Rome, March 2001.

# ADVANCED CONCEPTS FOR WASTE MANAGEMENT AND NUCLEAR ENERGY PRODUCTION IN THE EURATOM 5<sup>TH</sup> FRAMEWORK PROGRAMME*

**M. Hugon, V.P. Bhatnagar, J. Martín Bermejo**
European Commission
Rue de la Loi, 200
B-1049 Brussels
E-mail: Michel.Hugon@cec.eu.int

## Abstract

This paper summarises the objectives of the research projects on partitioning and transmutation (P&T) of long-lived radionuclides in nuclear waste and advanced systems for nuclear energy production in the key action on nuclear fission of the EURATOM 5th Framework Programme (FP5) (1998-2002). As these FP5 projects cover the main aspects of P&T, they should provide a basis for evaluating the practicability, on an industrial scale, of P&T for reducing the amount of long-lived radionuclides to be disposed of. Concerning advanced concepts, a cluster of projects is addressing the key technical issues to be solved before implementing high-temperature reactors (HTRs) commercially for energy production. Finally, the European Commission's proposal for a New Framework Programme (2002-2006) is briefly outlined.

---

* This paper was presented at GLOBAL'2001, International Conference on the Back-end of the Fuel Cycle: From Research to Solutions, Paris, France, 9-13 September 2001.

# Introduction

The 5th Framework Programme (FP5) (1998-2002) of the European Atomic Energy Community (EURATOM) has two specific programmes on nuclear energy, one for indirect research and training actions managed by the Research Directorate General (DG) and the other for direct actions performed by the Joint Research Centre of the European Commission (EC). The strategic goal of the former programme, "Research and Training Programme in the Field of Nuclear Energy", is to help exploit the full potential of nuclear energy in a sustainable manner, by making current technologies even safer and more economical and by exploring promising new concepts [1]. This programme includes a key action on controlled thermonuclear fusion, a key action on nuclear fission, research and technological development (RTD) activities of a generic nature on radiological sciences, support for research infrastructure, training and accompanying measures. The key action on nuclear fission and the RTD activities of a generic nature are being implemented through indirect actions, i.e. research co-sponsored and managed by DG Research of the EC, but carried out by external public and private organisations as multi-partner projects. The total budget available for these indirect actions during the 5th Framework Programme is 191 millions €.

The key action on nuclear fission comprises four areas:

(i)  operational safety of existing installations;

(ii) safety of the fuel cycle;

(iii) safety and efficiency of future systems;

(iv) radiation protection.

In the safety of the fuel cycle, waste and spent fuel management and disposal and partitioning and transmutation are the two larger activities, as compared to the decommissioning of nuclear installations. The safety and efficiency of future systems covers two sub-areas:

(i)  innovative or revisited reactor concepts;

(ii) innovative fuels and fuel cycles.

The implementation of the key action on nuclear fission has been made through targeted calls for proposals with fixed deadlines. Two calls have been made, one in March 1999 and another in October 2000. A final call will be made in October 2001. Following the calls for proposals made in 1999, about 140 proposals covering all areas of the key action and of the generic research have been accepted for a total funding of around 110 million €. Most of the projects were started in 2000. The October 2000 call resulted in the selection of about 75 proposals for another 50 million €. The new research contracts are being negotiated and the projects are expected to start in the autumn of 2001. All information concerning the nuclear fission programme is available on the CORDIS web site (www.cordis.lu/fp5-euratom).

This paper provides a general overview of the projects selected for funding by the EC in the field of advanced concepts for waste management and nuclear energy production in the EURATOM FP5. First, the research projects related to partitioning and transmutation (P&T) of long-lived radionuclides in nuclear waste are summarised after a brief outline of the goals of P&T. Then, a presentation is made of the projects on future systems based on nuclear fission for energy production (new and revisited reactor concepts) and other applications (e.g. desalination). Finally, the main goals and the content of the proposal made by the EC for a New Framework Programme (2002-2006) in February 2001 are briefly summarised.

## Partitioning and transmutation (P&T)

### *Goals of P&T and the organisation of R&D*

Spent fuel and high-level waste contain a large number of radionuclides ranging from short-lived to long-lived, requiring geological disposal for very long time periods. The long-lived radionuclides are mainly the actinides and some fission products. Partitioning and transmutation aims at reducing the inventories of long-lived radionuclides in radioactive waste by transmuting them into radionuclides with a shorter lifetimes [2].

If successfully achieved, P&T will produce waste with a shorter lifetimes. However, as the efficiency of P&T is not 100%, some long-lived radionuclides will remain in the waste, which will have to be disposed of in a deep geological repository. P&T is still at the research and development (R&D) stage. Nevertheless, it is generally accepted that the techniques used to implement P&T could alleviate the problems linked to waste disposal.

The interest for P&T in the EU is reflected in the increased funding for this area over the EURATOM Framework Programmes, 4.8, 5.8 and almost 28 million € for the 3rd, 4th and 5th Framework Programmes respectively.

In the EURATOM 4th Framework Programme (1994-1998), progress was made in the development of aqueous processes for partitioning (chemical separation) of minor actinides from liquid high-level waste. In addition, P&T strategy studies concluded that the feasibility of subcritical reactors coupled to accelerators, the so-called accelerator-driven system (ADS), should be more thoroughly investigated for transmutation of nuclear waste [3].

The objective of the research work carried out under FP5 is to provide a basis for evaluating the practicability, on an industrial scale, of partitioning and transmutation for reducing the amount of long-lived radionuclides to be disposed of. The two major aims of this work are:

(i) to develop efficient processes for the chemical separation of long-lived radionuclides from liquid high-level waste;

(ii) to gather all scientific and technical data necessary to carry out a detailed engineering design of an ADS demonstrator in the next Framework Programme.

The organisation of the FP5 research projects in the field of P&T is shown in Figure 1. The projects are subdivided into four groups (or clusters):

(i) partitioning;

(ii) basic studies on transmutation;

(iii) technological support for transmutation;

(iv) fuel for transmutation.

# Figure 1. FP5 projects on advanced options for partitioning and transmutation

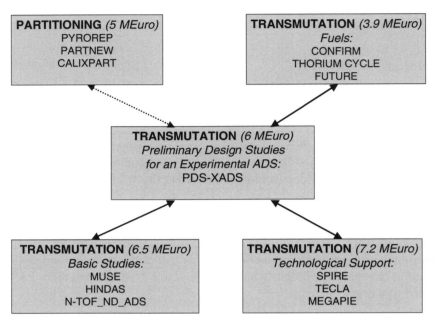

The preliminary design studies for an ADS demonstrator are mainly linked to the three clusters on transmutation. Finally, all FP5 P&T projects are part of a network. The R&D projects on P&T are briefly presented in the following sections.

## Cluster on partitioning (PARTITION)

The cluster on partitioning includes three projects, PYROREP, PARTNEW and CALIXPART. PYROREP aims at assessing flow sheets for pyrometallurgical processing of spent fuels and targets. Two methods, salt/metal extraction and electrorefining, will investigate the possibility of separating actinides from lanthanides. Materials compatible with corrosive media at high temperatures will be selected and tested. It is worth noting that one of the partners of this project is CRIEPI, the research organisation of the Japanese utilities.

The two other projects deal with the development of solvent extraction processes of minor actinides (americium and curium) from acidic high-level liquid waste (HLLW) issuing from the reprocessing of spent nuclear fuel. In PARTNEW, the minor actinides are extracted in two steps. They are first co-extracted with the lanthanides from HLLW (DIAMEX processes), then separated from the lanthanides (SANEX processes). Basic studies will be performed for both steps, in particular synthesis of new selective ligands and experimental investigation and modelling of their extraction properties. The radiolytic and hydrolytic degradation of the solvents will also be studied and the processes will be tested with genuine HLLW.

The CALIXPART project deals with the synthesis of more innovative extractants. Functionalised organic compounds, such as calixarenes, will be synthesised with the aim of achieving the direct extraction of minor actinides from HLLW. The extracting capabilities of the new selective compounds will be studied together with their stability under irradiation. The structures of the extracted species will be investigated by nuclear magnetic resonance (NMR) spectroscopy and X-ray diffraction to provide an input to the molecular modelling studies carried out to explain the complexation data.

## *Cluster on basic studies on transmutation (BASTRA)*

Three projects are grouped in the cluster on basic studies on transmutation: MUSE, HINDAS and n-TOF-ND-ADS. The MUSE project aims to provide validated analytical tools for subcritical neutronics including recommended methods, data and a reference calculation tool for ADS study. The experiments will be carried out by coupling a pulsed neutron generator to the MASURCA facility loaded with different fast neutron multiplying subcritical configurations. The configurations will have MOX fuel with various coolants (sodium, lead and gas). The cross-comparison of codes and data is foreseen. Experimental reactivity control techniques, related to subcritical operation, will be developed. Argonne National Laboratory (ANL) is also participating in the MUSE project.

Two other projects deal with nuclear data, one at medium and high energy (HINDAS), and the other encompassing the lower energy in resonance regions (n-TOF-ND-ADS).

The objective of the HINDAS project is to collect most of the nuclear data necessary for ADS application. This will be achieved by basic cross-section measurements at different European facilities, nuclear model simulations and data evaluations in the 20-200 MeV energy region and beyond. Iron, lead and uranium have been chosen to have a representative coverage of the periodic table, of the different reaction mechanisms and, in the case of iron and lead, of the various materials used for ADS.

The n-TOF-ND-ADS project aims at the production, evaluation and dissemination of neutron cross-sections for most of the radioisotopes (actinides and long-lived fission products) considered for transmutation in the energy range from 1 eV up to 250 MeV. The project is starting with the design and development of high-performance detectors and fast data acquisition systems. Measurements will be carried out at the TOF facility at CERN, at the GELINA facility in Geel and using other neutron sources located at different EU laboratories. Finally, an integrated software environment will be developed at CERN for the storage, retrieval and processing of nuclear data in various formats.

## *Cluster on technological support for transmutation (TESTRA)*

The cluster on technological support for transmutation has three projects: SPIRE, TECLA and MEGAPIE. The SPIRE project addresses the irradiation effects on an ADS spallation target. The effects of spallation products on the mechanical properties and microstructure of selected structural steels (e.g. martensitic steels) will be investigated by ion beam irradiation and neutron irradiation in reactors (HFR in Petten, BR2 in Mol and BOR 60 in Dimitrovgrad). Finally, data representative of mixed proton/neutron irradiation will be obtained from the analysis of the SINQ spallation target at the Paul Scherrer Institute (PSI) in Villigen.

The objective of TECLA is to assess the use of lead alloys both as a spallation target and as a coolant for an ADS. Three main topics are addressed: corrosion of structural materials by lead alloys, protection of structural materials and physico-chemistry and technology of liquid lead alloys. A preliminary assessment of the combined effects of proton/neutron irradiation and liquid metal corrosion will be done. Thermal-hydraulic experiments will be carried out together with numerical computational tool development.

The MEGAPIE project has the aim of developing and validating expertise for the design and operation of a heavy liquid-metal (Pb-Bi) spallation target producing a high neutron flux. It is planned to be coupled to the proton beam of the cyclotron accelerator ($\approx$1 MW power) in PSI in Villigen in 2004. The project will provide a comprehensive database from several experiments testing a single component of the target, such as the beam window, the heat exchanger, the corrosion control system and from a full-scale thermal-hydraulic simulation experiment. The safety and reliability aspects will be assessed for the whole system. An outlook on the extrapolation and applicability of the results of the MEGAPIE project for an ADS spallation target will be given.

### Cluster on fuel for transmutation (FUETRA)

Fuel issues for ADS are addressed in three projects: CONFIRM, THORIUM CYCLE and FUTURE. In the CONFIRM project, computer simulation of uranium-free nitride fuel irradiation up to about 20% burn-up will be made to optimise pin and pellet designs. Other computations will be performed especially concerning the safety evaluation of nitride fuel. Plutonium zirconium nitride [(Pu,Zr)N] and americium zirconium nitride pellets will be fabricated and their thermal conductivity and stability at high temperature will be measured. (Pu,Zr)N pins of optimised design will be fabricated and irradiated in the Studsvik reactor at high linear power ($\approx$70 kW/m) with a target burn-up of about 10%.

The objective of the project THORIUM CYCLE is to investigate the irradiation behaviour of thorium/plutonium (Th/Pu) fuel at high burn-up and to perform full core calculations for thorium-based fuel with a view to supplying key data related to plutonium and minor actinide burning. Two irradiation experiments will be carried out:

(i)  Four targets of oxide fuel (Th/Pu, uranium/plutonium, uranium and thorium) will be fabricated, irradiated in HFR in Petten and characterised after irradiation.

(ii) One Th/Pu oxide target will also be irradiated in KWO Obrigheim.

The FUTURE project aims at studying the feasibility of irradiation of innovative actinide-based oxide fuels for transmutation. These fuels contain compounds of the type $(Pu,Am)O_2$, $(Th,Pu,Am)O_2$ and $(Pu,Am,Zr)O_2$ homogeneously. These compounds will be synthesised and characterised (their thermal and chemical properties will be investigated at different temperatures). Fabrication processes will be tested. Modelling codes will be developed to assess the fuel performance, bearing in mind the large helium release and the degraded thermal properties. The safety performance of the fuel forms under normal, transient and accident conditions will be modelled with existing codes in the case of ADS.

### Preliminary design studies

Preliminary design studies of a European experimental ADS, PDS-XADS, are aiming toward the selection of the most promising technical concepts, at addressing the critical points of the whole system (i.e. accelerator, spallation target unit, reactor housing the subcritical core), at identifying research and development in support, at defining the safety and licensing issues, at making a preliminary assessment of the cost of the installation and finally at consolidating the road mapping for its development. Two types of accelerator will be investigated: cyclotron and linac. For the spallation target unit, two main options are considered depending whether or not the target liquid heavy metal is separated from the accelerator by a physical barrier (window). Three concepts for the subcritical core will be studied: a small core of about 20-40 MW cooled by lead-bismuth eutectic (LBE), a larger core of approximately 80 MW cooled by LBE and a gas-cooled core.

### Networking

A thematic network on *AD*vanced *O*ptions for *P*artitioning and *T*ransmutation (ADOPT) is intended to co-ordinate the FP5 R&D activities on P&T. The partners of ADOPT are European research organisations and industries, which are either co-ordinating or playing a significant role in the FP5 projects described above. The objectives of the ADOPT network are to suggest actions suited to promote consistency between FP5 projects and national programmes, to review the overall results, to identify the gaps, to give rise to future research proposals and to maintain relations with international organisations and countries outside the EU involved in P&T and ADS development.

### *ADS-related research activities in the framework of the International Science and Technology Centre (ISTC)*

A Contact Expert Group (CEG) on ADS-related ISTC projects was created in January 1998. Its main objectives are to review proposals in this field and to give recommendations for their funding to the ISTC Governing Board, to monitor the funded projects and to promote the possibilities of future or joint research projects through the ISTC. Five topics have been identified for the ADS-related projects:

(i)   accelerator technology;

(ii)  basic nuclear and material data and neutronics of ADS;

(iii) targets and materials;

(iv)  fuels related to ADS;

(v)   aqueous separation chemistry.

Because the funding parties primarily respond to local scientific/political interests and pressure, it was decided in January 2000 to reorganise the CEG into "local" CEGs (EU, Japan, Korea and USA) with some inter-co-ordination. This inter-co-ordination should foster information exchanges between ISTC projects in the same field, even if they are supported by different funding parties.

The EU CEG is helping in the development of co-operation between ISTC and FP5 EU funded projects. In fact, collaboration has already started between EU scientists and CIS research teams both in the preparation of ISTC proposals and in the follow-up of projects in some specific areas. Links between ISTC and FP5 EU funded projects are being established in the areas of basic studies on transmutation and aqueous processes for partitioning.

### Safety and efficiency of future systems

The objective of the area "safety and efficiency of future systems" is to investigate and evaluate new or revisited concepts (both reactors and alternative fuel cycles) for nuclear energy that offers potential longer-term benefits in terms of cost, safety, waste management, use of fissile material, less risk of diversion and sustainability. In the medium-term increasing competitiveness is the main priority, whereas in the long-term the main priority is to develop a sustainable energy system. Two sub-areas are covered, namely "innovative or revisited reactor concepts" and "innovative fuels and fuel cycles".

The majority of the FP5 research projects selected for funding deal with reactor concepts, with one exception, which is related to the sub-area of fuels and fuel cycles. This project, THORIUM CYCLE, has been grouped in the FUETRA cluster for convenience (see above). Concerning the new or revisited reactors, the nine research projects related to high-temperature reactors (HTR) are grouped in a cluster. They address the main technical issues to be solved before using the HTRs on the industrial scale to produce energy: fuel technology, fuel cycles, waste, reactor physics, materials, components, systems, safety approach and licensing issues. Four other projects are assessing the sate-of-the-art and R&D needs of other reactor concepts and of other applications of nuclear energy, such as high-performance light water reactors, gas-cooled fast reactors, molten-salt reactors and sea water desalination. Finally, a thematic network is addressing the competitiveness and sustainability of nuclear energy in the European Union. These research projects are outlined below.

### *The HTR cluster*

A number of HTRs were developed throughout the 1960s and 1970s (i.e. Peach Bottom and Fort St. Vrain in the US, AVR and THTR in Germany, and Dragon in the UK), but then abandoned. However, this reactor concept has been kept alive due its inherent safety features, its potential for use in high-temperature industrial processes and the possibility of using direct cycle gas turbines. It is today the subject of a renewal of interest in Europe as well as in other parts of the world (Japan, China, South Africa, USA, Russia).

The conclusions and recommendations of a project of the EURATOM 4th Framework Programme, which assessed the HTR key technologies, provided input for the R&D proposals submitted for funding in FP5.

The nine projects of the HTR cluster have a total EC funding of about 8.3 million € [4]. They address the following technical issues:

- *Fuel technology* investigated in the HTR-F/F1 projects. The objective is to restore (and improve) the capability of fabricating coated fuel particles in Europe. Existing irradiation data will be analysed. Fuel particles, and in particular German pebbles, will be irradiated in HFR in Petten at high burn-up and examined after irradiation. Heat-up tests under normal and accident conditions will be performed in the new KÜFA facility at JRC-ITU in Karlsruhe. A code modelling the thermo-mechanical behaviour of coated fuel particles will be developed and validated. First batches of U- and Pu-bearing kernels and coated particles will be fabricated. Finally, innovative fuels and alternative coating materials will be studied.

- *Reactor physics, waste and fuel cycles* in the HTR-N/N1 projects. The main aims are to provide numerical nuclear physics tools for the analysis and design of innovative HTR cores, to investigate different fuel cycles that can minimise the generation of long-lived actinides and optimise the Pu-burning capabilities, and to analyse the HTR-specific waste and the disposal behaviour of spent fuel. Existing core physics codes will be validated against tests carried out in HTTR in Japan and in HTR-10 in China. Basic HTR fuel designs (hexagonal block type and pebble-bed) will be studied. HTR specific operational and decommissioning waste streams will be analysed. Corrosion and leaching tests will be performed to model the geo-chemical behaviour of spent fuel dissolution under different conditions.

- *Materials*. The main objective of the HTR-M/M1 projects is to set up a database for the material properties of the HTR key components, such as the reactor pressure vessel (RPV), high-temperature areas (internal structures and turbine) and graphite structures. Activities include: compilation and review of existing data about materials for the above-mentioned components; thermo-mechanical tests on RPV welded joints, irradiated specimens, control rod claddings, and turbine disk and blade materials; oxidation tests at high temperatures on a fuel matrix graphite and on advanced carbon-based materials; and long-term irradiation tests for the graphite components. As the graphites used previously are no longer available, the models describing the graphite behaviour under irradiation will be verified as well as the screening tests of recent graphite properties.

- *Components and systems* investigated in the HTR-E project. These are the helium turbine, the recuperator heat exchanger, the active and permanent magnetic bearings, the leak-tightness rotating seal, the sliding parts (tribology) and the helium purification system. The programme contains design studies and also experiments (e.g. magnetic bearing tests at the University of Zittau, thermal-hydraulics test on recuperator heat exchanger at CEA or tribological investigations at FRAMATOME).

- *Safety approach and licensing.* The HTR-L project proposes a safety approach for a licensing framework specific to modular high-temperature reactors and a classification for the design basis operating conditions and associated acceptance criteria. Special attention is placed on the confinement requirements and the rules for system, structure and component classification as well as a component qualification level being compatible with economical targets.

- *Co-ordination.* The co-ordination, integration and the quality of the work to be performed in the eight HTR-related projects are assured by the HTR-C project. HTR-C will also organise a world-wide technological watch and develop international co-operation, first with China and Japan, which have at present the only research HTRs in operation in the world.

A European Network on "High Temperature Reactor Technology", HTR-TN, has been set up by a multi-partner collaboration agreement between 18 EU organisations in 2000. The agreement does not involve cash flow between the members and all contributions are made in kind. The EC-sponsored projects described above are the initial core from which HTR-TN will depart. The general objective of this network is to co-ordinate and manage the expertise and resources of the participant organisations in developing advanced technologies for modern HTRs, in order to support the design of these reactors.

A Contact Expert Group (CEG) on ISTC projects related to the gas-turbine modular-helium reactor (GT-MHR) has been set up with the same objectives as the CEG on ADS (see above). It is hoped that the exchange of information between ISTC and FP5 EU funded projects on HTRs will be greatly improved in the near future.

### Other reactor concepts and other applications of nuclear energy

The total EC funding for the five projects, which are summarised below, is 2.5 million €.

The overall objective of the HPLWR project is to assess the merit and economic feasibility of a high-performance LWR operating in thermodynamically supercritical regime (i.e. at a temperature and pressure above the water critical point). This project should also provide a thorough state of the art of this reactor concept, an identification of the main difficulties for its future development, and, if the concept is found to be feasible, recommendations for future R&D programmes. The University of Tokyo is participating actively in this project.

The main objective of the GCFR project is to produce a state-of-the-art summary report on gas-cooled fast neutron reactors including:

(i) review of earlier work and existing applicable technology;

(ii) evaluation of the safety feasibility of the GCFR based on EUR and EFR safety requirements;

(iii) integration of GCFR into the overall fuel cycle.

In May 2001, the Japan Nuclear Cycle Development Institute (JNC) joined the project as an observer because of its strong interest in gas-cooled fast neutron reactors.

Molten-salt reactors (MSR) present a number of advantages in terms of waste management (high burn-up and on-site reprocessing), system efficiency, use of fissile materials and non-proliferation issues. The objectives of the MOST concerted action are to evaluate the knowledge accumulated on MSRs between the 60s and the 80s in the USA, Europe and the former Soviet Union. The best options will be chosen (e.g. salt composition, structural materials), the weak points will be identified and a future R&D programme will be proposed. Oak Ridge National Laboratory (ORNL) is a partner of this project. Tight links are established with a similar ISTC project.

The EURODESAL concerted action aims at assessing the technical and economical feasibility of the production of potable or irrigating water through sea water desalination with innovative nuclear reactors (with emphasis on HTRs). The expected outcome is a thorough strategic study built on the available experience. The study should identify the main safety, technological and economic issues related to the coupling of nuclear and non-nuclear systems. A preliminary economic evaluation should also permit a comparison with fossil and renewable energy sources.

The Michelangelo Network, MICANET, has the objective of proposing a R&D strategy to keep the option of nuclear fission energy open in the 21$^{st}$ century in Europe. Its partners are the main European industrial companies and research organisations involved in nuclear energy. The Network will identify R&D needs and establish roadmaps for developing innovative reactors and fuel cycles. Connections will be established with the FP5 projects on future systems and the possibility of hydrogen production by nuclear energy will be assessed. MICANET will also establish a European partnership with the American Generation IV (Gen-IV) initiative by enabling the participation of experts from other member states than France and the UK, which are already members of Gen-IV, in its working groups and by actively co-operating in its future development.

## New Framework Programmes for research and innovation in Europe (2002-2006)

The Commission is aiming at the adoption of the new Framework Programmes (2002-2006) and their specific programmes by both the Council and the European Parliament by June 2002.

In January 2000, the Commission proposed a "European Research Area" [5]. The intention was to contribute to the creation of better overall working conditions for research in Europe. The starting point was that the situation concerning research in Europe is worrying, given the importance of research and development for future prosperity and competitiveness.

In October 2000, the Commission adopted a communication for the future of research in Europe, which sets out guidelines for implementing the "European Research Area" initiative, and more particularly the Research Framework Programme [6]. It is proposed to change the approach for the next Framework Programme, based on the following principles:

- concentrating on a selected number of priority research areas in which EU action can add the greatest possible value;

- defining the various activities in such a way as to enable them to exert a more structuring effect on research conducted in Europe through a stronger link with national, regional and other European initiatives;

- simplifying and streamlining the implementation arrangements, on the basis of intervention methods and decentralised management procedures.

In February 2001, the Commission made proposals for two new EC and EURATOM research Framework Programmes (2002-2006) aimed at contributing towards the creation of the European Research Area for total financial amounts of 16.27 billion € and 1.23 billion € respectively [7].

The EURATOM FP includes four areas: waste treatment and storage (150 million €), controlled thermonuclear fusion (700 million €), other EURATOM activities (50 million €) and JRC's EURATOM activities (330 million €).

Waste treatment and storage will cover R&D work on:

(i)  long-term disposal in deep geological repositories;

(ii) reduction of the impact of waste by developing new concepts of reactors producing less waste and partitioning and transmutation techniques.

The other EURATOM activities are:

(i)  research on radiation protection, more particularly on low levels of exposure;

(ii) studies of new and safer processes for the exploitation of nuclear energy;

(iii) education and training in nuclear safety and radiation protection.


## Conclusion

The research projects on advanced concepts for waste management and nuclear energy production in the EURATOM 5th Framework Programme have already begun. European research organisations and the nuclear industry are actively participating in these activities. In addition, international co-operation with Japan, Russia and USA has begun through various means: direct involvement of some Japanese and American R&D organisations in FP5 research projects, establishment of a European partnership with the American Generation IV initiative and links between FP5 and ISTC EU funded projects.

All the important aspects of partitioning and transmutation are covered by the FP5 research projects presented in this paper: chemical separation of long-lived radionuclides, basic nuclear data and subcritical neutronics, irradiation and corrosion effects, neutron spallation target, fuel issues and preliminary design studies of an ADS demonstrator. These projects should significantly contribute to fulfilling the objective of the programme, which is to provide a basis for evaluating the practicability, on an industrial scale, of partitioning and transmutation so as to reduce the amount of long-lived radionuclides to be disposed of. This challenging field also has the merit of attracting young researchers, which is essential for the preservation of the expertise in nuclear fission energy.

Concerning future systems, the high-temperature reactor (HTR) is raising at present a lot of interest world-wide because of its excellent safety features and its potential for efficient nuclear energy production. This interest is reflected in the cluster of HTR projects in FP5, which should provide a preliminary answer to the problems to be solved and requiring further R&D before deploying these reactors on the industrial scale.

The Commission made proposals for two new EC and EURATOM research Framework Programmes (2002-2006) in February 2001 and is expecting their adoption by June 2002.

# REFERENCES

[1]     "Council Decision of 25 January 1999 Adopting a Research and Training Programme (EURATOM) in the Field of Nuclear Energy (1998 to 2002)", *Official Journal of the European Communities*, L 64, 12 March 12 1999, p. 142, Office for Official Publications of the European Communities, L-2985 Luxembourg.

[2]     "Actinide and Fission-product Partitioning and Transmutation – Status and Assessment Report", OECD Nuclear Energy Agency, Paris, France (1999).

[3]     "Overview of the EU Research Projects on Partitioning and Transmutation of Long-lived Radionuclides" Report EUR 19614 EN, Office for Official Publications of the European Communities, L-2985 Luxembourg (2000).

[4]     J. Martin Bermejo, M. Hugon and G. Van Goethem, "Research Activities on High-temperature Gas-cooled Reactors (HTRs) in the 5th EURATOM RTD Framework Programme", Proceedings of the 16th International Conference on Structural Mechanics in Reactor Technology, SMiRT 16, 12-17 August 2001, Washington, DC, USA.

[5]     "Towards a European Research Area", Communication from the Commission, COM (2000) 6, 18 January 2000, http://europa.eu.int/comm/research/area.html.

[6]     "Making a Reality of the European Research Area: Guidelines for EU Research Activities (2002-2006)", Communication from the Commission, COM (2000) 612, 4 October 2000, http://www.cordis.lu/rtd2002/fp-debate/cec.htm.

[7]     "Proposals for Decision of the European Parliament and of the Council Concerning the Multi-annual EC and EURATOM Framework Programmes 2002-2006 Aimed at Contributing Towards the Creation of the European Research Area", COM (2001) 94 final, 21 February 2001, http://www.cordis.lu/rtd2002/fp-debate/cec.htm.

## LIST OF PARTICIPANTS*

## BELARUS

KIYAVITSKAYA, Hanna
Project Manager
Institute of Radiation Physics
and Chemistry Problems
Acad. Krasin str. 99
220109 Minsk-Sosny

Tel:  +375 172 46 74 58
Fax: +375 172 46 73 17
Eml: anna@sosny.bas-net.by

## BELGIUM

DE RAEDT, Charles
Project Leader
SCK•CEN Nuclear Research Centre
Boeretang, 200
B-2400 Mol

Tel:  +32 14 33 22 71
Fax: +32 14 32 15 29
Eml: cdraedt@sckcen.be

D'HONDT, Pierre Joseph
SCK•CEN
200 Boeretang
B-2400 Mol

Tel:  +32 14 33 22 00
Fax: +32 14 32 15 29
Eml: pdhondt@sckcen.be

## CZECH REPUBLIC

HRON, Miloslav
Nuclear Research Institute Rez, plc
CZ-25068 Řež

Tel:  +420 (0)2 66172370
Fax: +420 (0)2 20940156
Eml: hron@nri.cz

UHLIR, Jan
Head, Fluorine Chem. Dept.
Nuclear Research Institute Řež, plc
CZ-25068 Řež

Tel:  +420(0)2 6617 3548
Fax: +420(0)2 209 405 52
Eml: uhl@ujv.cz

---

\* Fourteen (14) countries, 3 international organisations, 40 establishments or institutes, 84 registered participants
(65 actually attended, 10 had their paper presented, and 9 others could not attend because of travel restrictions
at that time).

* VESPALEC, Rudolf
Head of Department
Reactor Physics & Fuel Management
Czech Power Works
Nuclear Power Plant Dukovany
CZ-67550 Dukovany

Tel: +420 618 81 4608
Fax: +420 618 86 6111
Eml: vespar1.edu@mail.cez.cz

ZEZULA, Lubor
Research Worker
Nuclear Research Institute Řež, plc
CZ-250568 Řež

Tel: +420 2 6617 2083
Fax: +420 2 2094 0954
Eml: zez@ujv.cz

**FRANCE**

* BERNARDIN, Bruno
Engineer
DER/SERSI, Bât. 212
CEA Cadarache
F-13113 Saint-Paul-lez-Durance

Tel: +33 4 42 25 63 66
Fax: +33 4 42 25 40 46
Eml: bernardin@drncad.cea.fr

BIGNAN, Gilles
CEN Cadarache
DER/SSAR/LDMN
Bât. 238
F-13108 Saint-Paul-lez-Durance

Tel: +33 4 42 25 48 16
Fax: +33 4 42 25 70 25
Eml: bignan@cea.fr

* BOUILLOUX, Yves
CEN Cadarache
DRN/DEC
Bât. 155
F-13108 Saint-Paul-lez-Durance Cedex

Tel: +33 4 4225 4816
Fax:
Eml: ybouilloux@cea.fr

CHANTOIN, Pierre
Projet RJH
Bât.780
Centre de Cadarache
F-13108 Saint-Paul-lez Durance Cedex

Tel: +33 4 42 25 76 86
Fax: +33 4 42 25 62 99
Eml: pierre.chantoin@cea.fr

CHAUVIN, Nathalie
CEA Cadarache
DEN/DTAP/SPI
Bât. 780
F-13108 Saint-Paul-lez Durance Cedex

Tel: +33 4 42 25 48 10
Fax: +33 4 42 25 74 60
Eml: nchauvin@cea.fr

* DELPECH, Marc
CE Cadarache
DER/SERI/LCSI
Bât. 212
F-13108 Saint-Paul-lez-Durance Cedex

Tel: +33 (0) 4 42 25 28 82
Fax: +33 (0) 4 42 25 71 87
Eml: marc.delpech@cea.fr

---

* Regrets not being able to attend.

DUMAZ, Patrick
CEA
CEA Cadarache
F-13108 Saint-Paul-lez-Durance Cedex

Tel:  +33 (0)4 42 25 40 98
Fax:
Eml: Patrick.Dumaz@cea.fr


* GAUTIER, G.M.
CEA
F-13108 Saint-Paul-lez-Durance

Tel:  +33 4 4225 4098
Fax:
Eml: gmgautier@cea.fr


GRENECHE, Dominique
COGEMA
DSP/DIST
1, rue des Herons Montigny Le Bretonnneux
F-78182 St-Quentin-en-Yvelines Cedex

Tel:  +33 1 39 48 51 88
Fax: +33 1 39 48 51 64
Eml: dgreneche@cogema.fr


LECARPENTIER, David
EDF R&D Departement PhR
1, Av. du Général de Gaulle
F-92141 Clamart Cedex

Tel:  +33 1 4765 4195
Fax: +33 1 4765 4399
Eml: david-n.lecarpentier@edf.fr


LORIETTE, Philippe
Project Manager
EDF
1, Av. Charles de Gaulle
F-92141 Clamart Cedex

Tel:  +33 1 4765 5792
Fax: +33 1 4765 3499
Eml: philippe.loriette@edf.fr


MOUNEY, Henri
EDF-POLE INDUSTRIE
Division Ingénierie et Services
CAP AMPERE
1, place Pleyel
F-93282 Saint-Denis Cedex

Tel:  +33 1 4369 0452/4369 0463
Fax: +33 1 43 69 04 55
Eml: henri.mouney@edf.fr


* PORTA, Jacques
CEA/DRN/DER/SERSI
CEN Cadarache
Bât. 211
F-13108 Saint-Paul-lez-Durance Cedex

Tel:  +33 (4) 42 25 36 31
Fax: +33 (4) 42 25 40 46
Eml: jporta@cea.fr


ROUAULT, Jacques
Département d'Études des Réacteurs
Service d'Études des Réacteurs Innovants
Bât. 212 CE Cadarache
F-13108 Saint-Paul-lez-Durance

Tel:  +33 4 42 25 7265
Fax: +33 4 42 25 48 58
Eml: jrouault@drncad.cea.fr

_____

* Regrets not being able to attend.

## GERMANY

PORSCH, Dieter
FRAMATOME ANP
Dept. NBTI
Postfach 3220
D-91050 Erlangen

Tel: +49 9131 18 95542
Fax: +49 9131 18 92453
Eml: Dieter.Porsch@framatome-anp.de

ZWERMANN, Winfried
Gesellschaft für Anlagen- und
Reaktorsicherheit (GRS) mbH
Forschungsgelaende
D-85748 Garching

Tel: +49 (89) 320 04 425
Fax: +49 (89) 320 04 599
Eml: zww@grs.de

## JAPAN

* AIZAWA, Otokiko
Musashi Institute of Technology
Tamazutsumi 1-28-1
Setagaya-ku
Tokyo 158-8557

Tel: +81 03 3703 3111 (ext. 3805)
Fax: +81 03 5707 1172
Eml: oaizawa@eng.musashi-tech.ac.jp

AKIE, Hiroshi
Advanced Reactor Systems
Dept. Nuclear Energy System
JAERI
Tokai-Mura, Naka-gan
Ibaraki-ken 319-1195

Tel: +81 (29) 2 82 6939
Fax: +81 (29) 2 82 6805
Eml: akie@mike.tokai.jaeri.go.jp

* HIRAIWA, Kouji
System Design and Engineering Dept.
Toshiba Corporation Yokohama Facility
Isogo Nuclear Engineering Centre
8 Shinsugita-cho, Isogo-ku
Yokohama 235-8523

Tel: +81 45 770 2047
Fax: +81 45 770 2179
Eml: kouji.hiraiwa@toshiba.co.jp

* KITAGAWA, Kenichi
Representative Director
Nuclear Fuel Industries Ltd.
Europe Office
10, rue de Louvois
F-75002 Paris

Tel: +33 1 44 86 05 75
Fax: +33 1 44 86 05 80
Eml: k-kitagw@nfi.co.jp

KUSAGAYA, Kazuyuki
Fuel Safety Research Laboratory
Dept. of Reactor Safety Research
JAERI
2-4 Shirakata-shirane, Tokai-mura
Naka-gun, Ibaraki-ken 319-1195

Tel: +81 29 282 5279
Fax: +81 29 282 5429
Eml:

---

* Regrets not being able to attend.

* MATSUMOTO, Takaaki
Hokkaido University
Dept. of Nuclear Engineering
Kita-13, Nishi-8
Sapporo-shi 060-0813

Tel: +81 11 706 6682
Fax: +81 11 706 7888
Eml: mtmt@qe.eng.hokudai.ac.jp

MORO, Satoshi
The Institute of Applied Energy
Shinbashi SY Bldg.
14-2 Nishishinbashi 1-chome
Minato-ku
Tokyo, 105-0003

Tel: +81 3 3508 8894
Fax: +81 3 3501 1735
Eml: moro@iae.or.jp

SEKIMOTO, Hiroshi
Tokyo Institute of Technology
Research Lab. for Nuclear Reactors
2-12-1 O-okayama, Meguroku
Tokyo 152-8550

Tel: +81 3 5734 3066
Fax: +81 3 5734 2959
Eml: hsekimot@nr.titech.ac.jp

SHIMAZU, Yoichiro
Division of Quantum Energy
Graduate School of Engineering
Hokkaido University
Kita 13, Nishi 8, Kita-ku
Sapporo 060-8628

Tel: +81 11 706 6676
Fax: +81 11 707 7888
Eml: shimazu@qe.eng.hokudai.ac.jp

YAMASHITA, Toshiyuki
Dept. of Nuclear Energy System
Research Group for Advanced Fuel
JAERI
Tokai-mura, Ibaraki-ken 319-1195

Tel: +81 29 282 6951
Fax: +81 29 282 5935
Eml: yamasita@analchem.tokai.jaeri.go.jp

## KOREA (REPUBLIC OF)

LEE, Young-Woo
Principal Research Scientist
Korea Atomic Energy Research Institute
P.O. Box 105
Yuseong, Taejon 305-600

Tel: +82 42 868 2129
Fax: +82 42 868 8868
Eml: ywlee@kaeri.re.kr

## THE NETHERLANDS

BAKKER, Klaas
NRG
Nuclear Chemistry Dept.
Postbus 25
NL-1755 ZG Petten

Tel: +31 (224) 56 43 86
Fax: +31 (224) 56 36 08
Eml: k.bakker@nrg-nl.com

---

* Regrets not being able to attend.

BERTHOU, Veronique
Interfaculty Reactor Institute
Technology University of Delft
Mekelweg 15
NL-2629 JB Delft

Tel: +31 15 278 21 63
Fax:
Eml: v.berthou@iri.tudelft.nl

KUIJPER, Jim C.
Fuels, Actinides and Isotopes
NRG
Postbus 25
NL-1755 ZG Petten

Tel: +31 (224) 56 4506
Fax: +31 (224) 56 3490
Eml: kuijper@nrg-nl.com

## RUSSIAN FEDERATION

ALEKSEEV, Pavel N.
Executive Director
Nuclear Power Development Institute
RRC Kurchatov Institute
Kurchatov Square, 1
123182 Moscow

Tel: +7 (095) 196 7621
Fax: +7 (095) 196 3708
Eml: apn@dhpt.kiae.ru

KALUGIN, Mikhail A.
RRC Kurchatov Institute
Dept. Physical and Technical Research
of Advanced Reactors
Kurchatov Square, 1
123182 Moscow

Tel: +7 (095) 1969279
Fax: +7 (095) 1969944
Eml: kalugin@adis.vver.kiae.ru

KOUZNETSOV, Vladimir
A.A. Bochvar, All-Russia Research
Institute of Organic Material VNIINM
Rogova St. 5
P.O.Box 369
123060 Moscow

Tel: +7 095 190 8135
Fax: +7 095 882 5720 or 190 5116
Eml: kvi@bochvar.ru

MIKITYUK, Konstantin
RRC Kurchatov Institute
Kurchatov Square, 1
123182 Moscow

Tel: +7 095 196 7016
Fax: +7 095 196 3708
Eml: kon@dhtp.kiae.ru

VASILIEV, Alexander
Senior Researcher
RRC Kurchatov Institute
Kurchatov Square, 1
123282 Moscow

Tel: +7 (095) 1967016
Fax: +7 (095) 1963708
Eml: avas@dhtp.kiae.ru

## SWEDEN

HELMERSSON, Sture
Technology and Development Mgr
Westinghouse Atom
SE-72163 Västerås

Tel: +46 (21) 107169
Fax: +46 (21) 182737
Eml: sture.helmersson@se.westinghouse.com

## SWITZERLAND

DEGUELDRE, Claude
Paul Scherrer Institute
OHLA/131
CH-5232 Villigen PSI

Tel:  +41 (0)56 3104176
Fax: +41 (0)56 3102205
Eml: claude.degueldre@psi.ch

KASEMEYER, Uwe
Paul Scherrer Institute
CH-5232 Villigen PSI

Tel:  +41 56 310 20 46
Fax: +41 56 310 23 27
Eml: uwe.kasemeyer@psi.ch

STREIT, Marco
Paul Scherrer Institute
Laboratory for Materials Behaviour
OHLB/410
CH-5232 Villigen PSI

Tel:  +41 (0)56 310 4385
Fax: +41 (0)56 310 4438
Eml: Marco.Streit@psi.ch

## TURKEY

*   COLAK, Uner
    Nuclear Engineering Dept.
    Hacettepe University
    Onder Caddesi 3/7 Mebusevleri
    Beytepe 06532 Ankara

Tel:  +90 (312) 297 7300
Fax: +90 (312) 299 2078
Eml: uc@nuke.hun.edu.tr

*   KADIROGLU, Osman Kemal
    Nuclear Engineering Dept.
    Hacettepe University
    Onder Caddesi 3/7 Mebusevleri
    Beytepe 06532 Ankara

Tel:  +90 312 2220774
Fax: +90 312 2992078
Eml: okk@alum.mit.edu

## UNITED KINGDOM

*   ABRAM, Timothy J.
    Research and Technology
    BNFL Springfields
    Salwick, Preston PR4 OXJ

Tel:  +44 (0)1772 764696
Fax: +44 (0)1772 762470
Eml: tim.j.abram@bnfl.com

ACKROYD, Ron T.
Computational Physics and Geophysics
Imperial College of Science,
Technology and Medicine
Prince Consort Road
London SW7 2BP

Tel:  +44 20 7594 9319
Fax: +44 20 7594 9341
Eml: r.transport@ic.ac.uk

BARKER, Fred
Nuclear Policy Analyst
4 St. John's Close
Hebden Bridge HX7 8DP

Tel:  +44 (0) 1 422 847 189
Fax:
Eml: fbarker@gn.apc.org

---

* Regrets not being able to attend.

BEAUMONT, Heather
NNC Ltd.
Booths Hall
Chelford Road, Knutsford
Cheshire WA16 8QZ

Tel: +44 15656 843232
Fax: +44 15656 843878
Eml: heather.beaumont@nnc.co.uk

BLUE, Roger
Group Manager, International Affairs
BNFL
Risley, Warrington
Cheshire WA3 6AS

Tel: +44 1925 832 668
Fax: +44 1925 833 704
Eml: roger.p.blue@bnfl.com

BUCKTHORPE, Derek
NNC Ltd.
Booths Hall
Chelford Road, Knutsford
Cheshire WA16 8QZ

Tel:
Fax:
Eml: Derek.Buckthorpe@nnc.co.uk

*  CLEGG, Richard
Manager, Waste Disposal Res.
British Nuclear Fuels plc
Building B229
Sellafield, Seascale
Cumbria CA20 1PG

Tel: +44(19467) 75925
Fax: +44(19467) 76984
Eml: richard.clegg@bnfl.com

CROSSLEY, Steven J.
British Nuclear Fuels plc
Springfields, Preston
Lancashire PR4 0XJ

Tel: +44 1772 762216
Fax: +44 (1772) 76 38 88
Eml: steven.j.crossley@bnfl.com

EATON, Matthew
Imperial College
T.H. Huxley School
Prince Consort Road
London SW7 2BP

Tel: +44 171 594 9323
Fax: +44 171 594 9341
Eml: m.eaton@ic.ac.uk

EVERY, Denis P.
British Nuclear Fuels plc
Springfields, Preston
Lancashire PR4 0XJ

Tel: +44 1772 762842
Fax: +44 1772 763888
Eml: denis.p.every@bnfl.com

FRANKLIN, Simon
Director of Reactor Operations & Safety
Imperial College Reactor Centre
Silwood Park, Buckhurst Road
Ascot, Berkshire SL5 7TE

Tel: +44 (0) 207 5942291
Fax: +44 (0) 1344624931
Eml: s.franklin@ic.ac.uk

---

* Regrets not being able to attend.

GODDARD, Anthony
Imperial College
T.H. Huxley School
Royal School of Mines Building
Prince Consort Road
London SW7 2BP

Tel: +44 171 594 9320
Fax: +44 171 594 9341
Eml: a.goddard@ic.ac.uk


GOMES, Jefferson
Imperial College
T.H. Huxley School
Royal School of Mines Building
Prince Consort Road
London SW7 2BP

Tel: +44 171 594 9322
Fax:
Eml: j.gomes@ic.ac.uk


HESKETH, Kevin (Chairman)
British Nuclear Fuels plc
Research and Technology
B709 Springfields
Salwick Preston
Lancashire PR4 0XJ

Tel: +44 1772 76 23 47
Fax: +44 1772 76 24 70
Eml: Kevin.W.Hesketh@bnfl.com


LENNOX, Tom
NNC Ltd.
Booths Hall
Chelford Road, Knutsford
Cheshire WA16 8QZ

Tel: +44 15656 843470
Fax: +44 15656 843878
Eml: tom.lennox@nnc.co.uk


MIGNANELLI, Michael
AEA Technology plc
Nuclear Science
220 Harwell, Didcot
Oxfordshire, OX11 0RA

Tel: +44 1235 43 4300
Fax: +44 1235 43 6314
Eml: mike.mignanelli@aeat.co.uk


MURGATROYD, Julian
NNC Ltd.
Booths Hall
Chelford Road, Knutsford
Cheshire WA16 8QZ

Tel: +44 1565 843696
Fax: julian.murgatroyd@nnc.co.uk
Eml:


NEWTON, Tim
Reactor Analysis Team Leader
Serco Assurance
Winfrith Technology Centre 303/A32
Dorchester

Tel: +44 (0) 1 305 20 2061
Fax: +44 (0) 1 305 20 2194
Eml: tim.newton@sercoassurance.com


PAIN, Christopher
Imperial College
T.H. Huxley School
Royal School of Mines Building
Prince Consort Road
London SW7 2BP

Tel: +44 171 594 9322
Fax: +44 171 594 9341
Eml: c.pain@ic.ac.uk

SMITH, Peter
Serco Assurance
378/A32
Risley, Warrington
Cheshire WA3 6AT

Tel: +44 (0) 1 305 20 3706
Fax: +44 (0) 1 305 20 2194
Eml: peter.j.smith@sercoassurance.com

SUNDERLAND, Richard
Reactor Systems and Advanced Technology
NNC Ltd.
Chelford Road, Knutsford
Cheshire WA16 8QZ

Tel: +44 (0) 1565 843061
Fax: +44 (0) 1565 843878
Eml: richard.sunderland@nnc.co.uk

THETFORD, Roger
Serco Assurance
424 Harwell
Didcot, Oxfordshire OX11 0QJ

Tel: +44 1235 433975
Fax: +44 1235 436798
Eml: roger.thetford@sercoassurance.com

WILKINSON, William L.
Department of Chemical Engineering
and Chemical Technology
Imperial College
Prince Consort Road
London SW7 2BP

Tel:
Fax:
Eml:

WORRALL, Andy (Local Organisation)
British Nuclear Fuels plc
MOX Design and Licensing
B709, Research & Technology Dept.
Springfields, Salwick
Preston, Lancashire, PR4 0XJ

Tel: +44 1772 764474
Fax: +44 1772 762470
Eml: andrew.worrall@bnfl.com

## UNITED STATES OF AMERICA

* CARELLI, Mario D.
  Manager, Energy Systems
  Westinghouse Electric Company
  1344 Beulah Road
  Pittsburgh, PA 15235-5083

Tel: +1 412 256 1042
Fax: +1 412 256 2444
Eml: carellmd@westinghouse.com

DOWNAR, Thomas J.
School of Nuclear Engineering
Purdue University
1290 Nuclear Engineering Bldg.
W. Lafayette, IN 47907-1290

Tel: +1 (765) 494 5752
Fax: +1 (765) 494 9570
Eml: downar@ecn.purdue.edu

---

* Regrets not being able to attend.

GEHIN, Jess C.
Oak Ridge National Laboratory
Building 6025, MS-6363
Bethel Valley Road
P.O.Box 2008
Oak Ridge, TN 37831-6363

Tel: +1 865 576 5093
Fax: +1 865 574 9619
Eml: gehinjc@ornl.gov

* HAYES, Steven L.
Manager, Reactor Fuels & Materials Section
Argonne National Laboratory-West
P.O. Box 2528
Idaho Falls, ID 83403-2528

Tel: +1 208 533 7255
Fax: +1 208 533 7863
Eml: steven.hayes@anl.gov

INGERSOLL, Daniel T.
Oak Ridge National Laboratory
P.O. Box 2008, Bldg. 6025
Bethel Valley Road
Oak Ridge, TN 37831-6363

Tel: +1 865 574 6102
Fax: +1 865 574 9619
Eml: ingersolldt@ornl.gov

PETROVIC, Bojan
Senior Scientist
Westinghouse Science and Technology
1344 Beulah Road
Pittsburgh, PA 15235

Tel: +1 412 256 1295
Fax: +1 412 256 2444
Eml: PetrovB@westinghouse.com

* TODOSOW, Michael
Brookhaven National Laboratory
Associated Universities, Inc.
Bldg. 475B, PO Box 5000
Upton, Long Island NY 11973-5000

Tel: +1 631 344 2445
Fax: +1 631 344 7650
Eml: todosowm@bnl.gov

* WALTERS, Leon
Director, Engineering Div.
ANL-West
P.O. Box 2528
Idaho Falls, ID 83403-2528

Tel: +1 (208) 533 7384
Fax: +1 (208) 533 7340
Eml: leon.walters@anl.gov

## INTERNATIONAL ORGANISATIONS

ANGERS, Laetitia
Nuclear Engineer
IAEA
Wagramerstrasse 5
P.O. 100
A-1400 Wien

Tel: +43 1 2600 22770
Fax:
Eml: l.g.angers@iaea

TOVESSON, Fredrik
EC-JRC-IRMM
Retieseweg
B-2440 Geel

Tel: +32 14 571350
Fax: +32 14 571376
Eml: fredrik.tovesson@irmm.jrc.be

---

* Regrets not being able to attend.

SARTORI, Enrico (Secretary)
OECD/NEA Data Bank
Le Seine-Saint Germain
12, boulevard des Iles
F-92130 Issy-les-Moulineaux

Tel: +33 1 45 24 10 72 / 78
Fax: +33 1 45 24 11 10 / 28
Eml: sartori@nea.fr

SOMERS, Joseph
Institute for Transuranium Elements
Postfach 2340
D-76125 Karlsruhe

Tel: +49 (07247)951 359
Fax: +49 (07247)951 566
Eml: somers@itu.fzk.de

*Annex 2*

# TECHNICAL PROGRAMME

*Monday, 22 October (Plenary Sessions)*

## Opening Session

*Chairman: Kevin Hesketh*

9:00    Opening Address
*A. Worrall, K. Hesketh*

9:30    Barriers and Incentives to Introducing New Reactors in the Deregulated Market
*W.L. Wilkinson*

10:00    The Need to Preserve Nuclear Fuels and Materials Knowledge
*L.C. Walters* (ANL-W, Idaho, USA), *J. Graham* (ETCetera Assessments LLP, USA, presented by K. Hesketh)

## High Temperature Gas Reactors

*Chairman: Derek Buckthorpe*

11:00    European Collaboration on Research into High-temperature Reactor Technology
*T.J. Abram* (BNFL, UK), *D. Hittner* (FRAMATOME ANP, France), *W. von Lensa* (FZ, Jülich, Germany), *A. Languille* (CEA, France), *D. Buckthorpe* (NNC, UK), *J. Guidez* (EC-JRC), *J. Martín-Bermejo* (EC DG-Research)

11:30    Design of $B_4C$ Burnable Particles Mixed in LEU Fuel for HTRs
*V. Berthou, J.L. Kloosterman, H. Van Dam, T.H.J.J. Van der Hagen* (IRI, Delft, The Netherlands)

## Design and Performance of Innovative Fuels I

*Chairman: Roger Thetford*

13:30    Extensive Characterisation of a Material for Understanding its Behaviour as a Nuclear Fuel: The Case of a Zirconia/Plutonia Inert-matrix Fuel
*C. Degueldre, F. Ingold, C. Hellwig, P. Heimgartner* (PSI, Villigen, Switzerland), *S. Conradson* (LANL, USA), *M. Döbeli* (PSI-ETH-Zurich, Switzerland), *Y.W. Lee* (KAERI, Republic of Korea)

14:00    Annular Plutonium Zirconium Nitride Fuel Pellets
*M. Streit, F. Ingold* (PSI, Villigen, Switzerland), *L.J. Gauckler* (ETH, Zürich, Switzerland), *J-P. Ottaviani* (CEA Cadarache, France)

14:30　Rock-like Oxide Fuels for Burning Excess Plutonium in LWRs
*T. Yamashita, K. Kuramoto, H. Akie, Y. Nakano, N. Nitani, T. Nakamura, K. Kusagaya* (JAERI, Tokai-mura, Japan), *T. Ohmichi* (RIST, Tokai-mura, Japan)

## Evolutionary and Modular Water Reactors I

*Chairman: Dieter Porsch*

15:30　Innovative Features and Fuel Design Approach in the IRIS Reactor
*B. Petrovic, M. Carelli* (WEC-STD, USA), *E. Greenspan, H. Matsumoto* (UC Berkeley, USA), *E. Padovani, F. Ganda* (Politecnico, Milan, Italy)

16:00　Core Concepts for Long Operating Cycle Simplified BWR (LSBWR)
*K. Hiraiwa, N. Yoshida, M. Nakamaru, H. Heki* (Toshiba Corp., Japan), *M. Aritomi* (TIT, Japan) *(not presented orally)*

16:30　Core Design Study on Reduced-moderation Water Reactors (RMWRs)
*H. Akie, Y. Nakano, T. Shirakawa, T. Okubo and T. Iwamura* (JAERI, Japan)

## *Tuesday, 23 October (Parallel Sessions)*

## Design and Performance of Innovative Fuels II

*Chairman: Young-Woo Lee*

9:00　Application of Ceramic Nuclear Fuel Materials for Innovative Fuels and Fuel Cycles
*Y-W. Lee* (KAERI, Republic of Korea)

9:30　Innovative MOX Fuel for Fast Reactor Applications
*K. Bakker, H. Thesingh* (NRG-Petten, Netherlands), *T. Ozawa, Y. Shigetome, S. Kono, H. Endo* (JNC, Japan), *Ch. Hellwig, P. Heimgartner, F. Ingold, H. Wallin* (PSI, Villigen, Switzerland)

10:00　Behaviour of Rock-like Oxide Fuels Under Reactivity Initiated Accident Conditions
*K. Kusagaya, T. Nakamura, M. Yoshinaga, H. Akie, T. Yamashita and H. Uetsuka* (JAERI, Japan)

## Design and Performance of Innovative Fuels III

*Chairman: Klas Bakker*

11:00　Theoretical Requirements to Tolerances to be Imposed on Fuel Rod Design Parameters for RBEC Lead-bismuth Fast Reactor
*A. Vasiliev, P. Alekseev, K. Mikityuk, P. Fomitchenko, A. Shestopalov* (RRC-KI, Russian Federation)

11:30　Advanced Plutonium Assembly (APA): Evolution of the Concept, Neutron and Thermal-mechanic Constraints
*J. Porta, B. Gastaldi, C. Krakowiak-Aillaud, L. Buffe* (CEA Cadarache,France)

## Evolutionary and Modular Water Reactors II

*Chairman: Pierre D'hondt*

9:00    The Utilisation of Thorium Fuel in a Generation IV Light Water Reactor Design
       T.J. Downar, Y. Xu (Purdue University, W. Lafayette, USA)

9:30    *Thorium Fuel in LWRs – An Option for an Effective Reduction of Plutonium Stockpiles*
       D. Porsch (FRAMATOME ANP, Erlangen, Germany), D. Sommer (KKW, Obrigheim, Germany)

10:00   PWRs Using HTGR Fuel Concept with Cladding for Ultimate Safety
       *Y. Shimazu* (Hokkaido University, Japan), *H. Tochihara* (EDC, Japan),
       *Y. Akiyama* (MHI, Japan), *K. Itoh* (NDC, Japan)

## Fast Spectrum Reactors I

*Chairman: Pavel Alekseev*

11:00   A Simplified LMFBR Concept (SFR)
       *D.V. Sherwood, T.A. Lennox* (NNCL, Knutsford, UK)

11:30   The Design and Flexibility of the Enhanced Gas-cooled Reactor (EGCR)
       *H.M. Beaumont, A. Cheyne, J. Gilroy, G. Hulme, T.A. Lennox, J.T. Murgatroyd,*
       *R.E. Sunderland, E.K. Whyman* (NNC), *S.J. Crossley, D.P. Every* (BNFL)

## Molten-salt Reactors/HTGR II

*Chairman: Winfried Zwermann*

13:30   Molten-salt Reactor for Burning of Transuranium Nuclides Forming in Closed Nuclear Fuel Cycle
       *P.N. Alekseev, A.A. Dudnikov, V.V. Ignatiev, N.N. Ponomarev-Stepnoy, V.N. Prusakov,*
       *S.A. Subbotin, A.V. Vasiliev, R.Ya. Zakirov* (RRC-KI, Moscow, Russian Federation)

14:00   AMSTER: A Molten-salt Reactor Concept Generating its Own $^{233}$U and Incinerating Transuranium Elements
       *D. Lecarpentier, C. Garzenne, J. Vergnes, H. Mouney* (EDF, France), *M. Delpech* (CEA Cadarache, France)

14:30   Thorium and Plutonium Utilisation in Pebble-bed Modular Reactor
       *U.E. Sikik, H. Dikmen, Y. Çeçen, Ü. Çolak, O.K. Kadiroglu* (Hacettepe University, Turkey, presented by K. Hesketh)

15:00   A New Approach for the Systems Dedicated to the Transmutation: The Reactor with Compensated Beta
       *B. Bernardin* (CEA Cadarache, France, presented by P. Dumaz)

## Evolutionary and Modular Water Reactors III

*Chairman: Thomas Downar*

16:00   Feasibility of Partial LWR Core Loadings with Inert-matrix Fuel
       *U. Kasemeyer, C. Hellwig, R. Chawla* (PSI, Villigen, Switzerland), *D.W. Dean* (Studsvik Scandpower, USA), *G. Meier* (KKW-Gösgen-Däniken, Switzerland),
       *T. Williams* (EG, Laufenburg, Switzerland)

16:30    CEA Studies About Innovative Water-cooled Reactor Concepts
*P. Dumaz, A. Bergeron, G.M. Gautier, J.F. Pignatel, G. Rimpault, G. Youinou*
(CEA Cadarache, France)

## Accelerator-driven Systems

*Chairman: Charles De Raedt*

13:30    Experimental Investigations of the Accelerator-driven Transmutation Technologies at
the Subcritical Facility "YALINA"
*S.E. Chigrinov, H.I. Kiyavitskaya, I.G. Serafimovich, C.K. Rutkovskaia, Y. Fokov,
A.M. Khilmanovich, B.A. Marstinkevich, V.V. Bournos, S.V. Korneev, S.E. Mazanik,
A.V. Kulikovskaya, T.P. Korbut, N.K. Voropaj, I.V. Zhouk, M.K. Kievec, I.L. Rakhno*
(RPh&ChPI, Minsk, Belarus)

14:00    MYRRHA, A Multi-purpose ADS for R&D – Pre-design Phase Completion
*H. Aït Abderrahim, P. Kupschus, Ph. Benoit, E. Malambu, K. Van Tichelen, B. Arien,
F. Vermeersch, Th. Aoust, Ch. De Raedt, S. Bodart, P. D'hondt* (SCK•CEN, Belgium)

14:30    Optimisation of Conceptual Design of Cascade Subcritical Molten-salt Reactor
*A. Vasiliev, P. Alekseev, A. Dudnikov, K. Mikityuk, S. Subbotin*
(RRC-KI, Russian Federation)

## Fast Spectrum Reactors II

*Chairman: Hiroshi Sekimoto*

15:30    RBEC Lead-bismuth-cooled Fast Reactor: Review of Conceptual Decisions
*P. Alekseev, P. Fomichenko, K. Mikityuk, V. Nevinitsa, T. Shchepetina, S. Subbotin,
A. Vasiliev* (RRC-KI, Moscow, Russian Federation)

16:00    Design and Performance Studies for Minor Actinide Target Fuels
*T.D. Newton, P.J. Smith* (AEAT, Winfrith, UK)

16:30    Applications of "CANDLE" Burn-up Strategy to Several Reactors
*H. Sekimoto* (TIT, Japan)

## Poster Session

Some Views on the Design and Fabrication of Targets or Fuels Containing Curium
*J. Somers, A. Fernandez, R.J.M. Konings* (JRC-ITU, Karlsruhe, Germany),
*G. Ledergerber* (KKW, Leibstadt, Switzerland)

Thermophysical and Chemical Properties of Minor-actinide Fuels
*M.A. Mignanelli, R. Thetford* (Serco Assurance, UK)

Determination of the $^{233}$Pa(n,f) Reaction Cross-section for Thorium-fuelled Reactors
*F. Tovesson, F-J. Hambsch, S. Oberstedt* (JRC-IRMM, Geel, Belgium), *A. Oberstedt*
(Örebro University, Sweden), *B. Fogelberg, E. Ramström* (Studsvik, Nyköping, Sweden)

Advanced Fuel Cycle for Long-lived Core of Small-size Light Water Reactor
of ABV Type
*A. Polismakov, V. Tsibulsky, A. Chibinyaev, P. Alekseev* (RRC-KI, Russian Federation)

Proposal of a Molten-salt System for Long-term Energy Production
*V. Berthou* (IRI, Delft, The Netherlands), *I. Slessarev, M. Salvatores* (CEA Cadarache,
France)

A Conceptual Fluidised Particle-bed Reactor – Application of Space-dependent Kinetics
*C.C. Pain, J.L.M.A. Gomes, C.R.E. de Oliveira, M.D. Eaton, A.J.H. Goddard* (Imperial College, UK), *H. van Dam, T.H.J.J. van der Hagen and D. Lathouwers* (IRI, Delft, The Netherlands)

Transmutation and Incineration of MAs in LWRs, MTRs and ADSs
*Ch. De Raedt, B. Verboomen, Th. Aoust, H. Aït Abderrahim, E. Malambu, L.H. Baetslé* (SCK•CEN, Mol, Belgium)

BARS: BWR with Advanced Recycle System
*K. Hiraiwa, Y. Yamamoto, K. Yoshioka, M. Yamaoka* (Toshiba Corp., Japan), *A. Inoue, J. Mimatu* (Gifu University, Japan)


## *Wednesday, 24 October (Plenary Sessions)*

### Miscellaneous Themes

*Chairman: Kevin Hesketh*

9:00    The Key Role of Critical Mock-up Facilities for Neutronic Physics Assessment of Advanced Reactors: An Overview of CEA Cadarache Tools
        *G. Bignan, D. Rippert, P. Fougeras* (CEA Cadarache, France)

9:30    The SPHINX Project (Experimental Verification of Design Inputs for a Transmuter with Liquid Fuel Based on Molten Fluorides)
        *M. Hron, J. Uhlir* (NRI Řež, Czech Republic), *J. Vanicek* (CPC, Czech Republic)

10:00   Advanced Concepts for Waste Management and Nuclear Energy Production in the EURATOM 5th Framework Programme
        *M. Hugon, V.P. Bhatnagar, J. Martín Bermejo* (EC, Brussels, presented by J. Somers)

### Panel Discussion

*Chairman: Kevin Hesketh*

*Panelists: P. D'hondt, H. Mouney, H. Sekimoto, J. Somers, R. Sunderland, E. Sartori*

1.  Is there a gap between vendors' and utilities' fuel research programmes designed to support operation and the advanced concept research such as that presented at this conference? If so, what can the research community do to narrow the gap? In other words, does the field need to be made more relevant to the utilities?

2.  The benefits of advanced concepts are usually in areas such as safety, proliferation resistance, environmental impact/radiotoxic burden, strategic and so on. A major weakness is that these are "soft" issues for which there is no agreed measure of the benefit. Are there any actions the research community could take to promote agreed metrics in these areas?

3.  What should be our strategy for partitioning and transmutation given the intractability of processing and destroying curium? Should there be a policy of encapsulating curium for eventual disposal?

4.  In the context of the objectives of initiatives such as Generation IV (particularly sustainability), how would once-though fuel cycles such as those HTGRs fit it? What role would once-through fuel cycles fit in?

# ORDER FORM

**OECD Nuclear Energy Agency, 12 boulevard des Iles, F-92130 Issy-les-Moulineaux, France**
Tel. 33 (0)1 45 24 10 15, Fax 33 (0)1 45 24 11 10, E-mail: nea@nea.fr, Internet: www.nea.fr

| Qty | Title | ISBN | Price | Amount |
|-----|-------|------|-------|--------|
|  |  |  |  |  |
|  |  |  |  |  |
|  |  |  |  |  |
|  |  |  |  |  |
|  |  |  |  |  |
|  |  |  |  |  |
|  |  |  |  |  |
|  |  |  |  |  |
|  |  |  | **Total\*** |  |

\* Prices include postage fees.

❑ Payment enclosed (cheque payable to OECD Publications).

Charge my credit card    ❑ VISA    ❑ Mastercard    ❑ Eurocard    ❑ American Express

| Card No. | Expiration date | Signature |
|----------|-----------------|-----------|
| Name |  |  |
| Address | Country |  |
| Telephone | Fax |  |
| E-mail |  |  |

OECD PUBLICATION, 2, rue André-Pascal, 75775 PARIS CEDEX 16
PRINTED IN FRANCE
(66 2002 13 1 P) – No. 52681  2002